Contents

THE GIANT BOOK OF ELECTRONICS PROJECTS

BY THE EDITORS OF 73 MAGAZINE

TAB
TAB BOOKS
Blue Ridge Summit, PA

FIRST EDITION
THIRTEENTH PRINTING

Library of Congress Cataloging-in-Publication Data

Main entry under title:

The Giant book of electronics projects.

Includes index.
1. Electronics—Amateurs' manuals. I. 73 magazine for radio amateurs.
TK9965.G498 621.381 81-18243
ISBN 0-8306-0078-7 AACR2
ISBN 0-8306-1367-6 (pbk.)

TAB Books offers software for sale. For information and a catalog, please contact TAB Software Department, Blue Ridge Summit, PA 17294-0850.

Introduction

This book of electronic projects represents some of the very best projects to appear in *73 Magazine*. Here is one volume you will find projects for electronics experimenters, amateur radio operators, and just anyone who has an interest in the exciting world of electronics. These projects cover the whole gamut of electronics. From power supplies to test equipment and from receivers to transmitters you'll find it all in these pages. You can build your own antenna or your own amplifier, and there are even some simple one-evening projects to help you get started the easy way! We've tried our best to give you a real smorgasbord of tried and tested electronics projects that everyone from the beginner to the old pro will find interesting and useful. So dig out your junk boxes, heat up your soldering iron, and enjoy!

Chapter 1
Power Supplies and Regulators

If you're going to build electronic projects you'll need some way to power them. Of course, batteries can be used for some projects, and others can be powered directly from the ac line, but for many projects you'll need a permanent power supply that provides just the right voltage. In this chapter you'll find a variety of power supplies and regulators that can be constructed exactly as designed or modified to fit your exact needs.

The 13.8-volt ham radio transceiver has really come of age. Many of these units are great for mobile operation, but when it comes to fixed-station use, the transceivers can really come up short—primarily because of the comparatively high current values they draw on peaks.

The two-meter FM and police-scanner industries have given us a variety of 13.8-volt, low-current power supplies which in many cases can be bought more cheaply than built. However, the seemingly rarified 13.8-volt, fifteen-amp (or higher) power supply is not that easy to come by, which really limits the possibilities with transceivers drawing anywhere from sixteen to 20 amperes.

The Circuit

The power-supply circuit uses 2N3055 transistors. You can tailor your current capability by the number of pass transistors you use. I wanted a 25-Amp supply, so I used five 2N3055s. You can figure roughly one transistor for every 5 Amps you'll be drawing. In a 10-Amp supply, only two transistors would be used, and so on. See Fig. 1-1 and Table 1-1.

The 2N3055 is an NPN power device built into a TO-3 case. The 3055 is one of the more easily come by transistors and is very cheap. Because of the power these little devils are going to be dissipating, heat sinks should be employed. I used a heat sink with approximately 27 square inches of surface area with four half-inch fins, which cools nicely. Extreme heat can quickly mess up the transistor junction (not to mention a nice paint job). Before securing the transistors to the heat sink, apply some silicone thermal compound between the 3055s and the surface of the heat sink to provide a good positive heat transfer.

I have always used the rule of thumb that if you can't touch it, you can blow it. If you don't care to go heat-sink shopping, use a cooling fan. If you use a fan in addition to the heat sink, be sure the air circulates in line with the fins. Blowing air perpendicularly to the fins sets up standing waves—the aerodynamic kind—and turbulence and the cooling effect is minimal.

Transistor-mounting hardware is nice, but I didn't feel that it was necessary. I attached the transistors directly to the heat sink and then mounted the whole heat-sink assembly on a sheet of Plexiglas™ attached to four standoffs. Since the transistor case is common to the collector, I tapped a screw into one of the heat-sink fins and this became my common collector tiepoint. It is important to keep all lead lengths constant. After drilling matching holes for the base and emitter pins in the heat sink, heavy-gauge wire was soldered (carefully) to each emitter pin through a 0.25-ohm resistor, and then a second piece of wire was attached to each base pin. I then had only to connect the rest of the circuit to either the heat sink or one of the two bus wires.

You may or may not have difficulty locating a suitable transformer capable of taking 120 V ac and squeezing it down to 17 to 24 V ac. I was lucky enough to locate an old, beat-up, ex-battery-charger transformer at a hamfest which gave me 120/17 V ac. I think you will find old battery chargers to be a good source for the transformer you will need. Remember, the transformer must be capable of carrying the current you are going to draw from your power supply. I paid $3.00 for my transformer and felt robbed; I have seen them for a dollar. Yes, you do take a risk, but remember, even if the transformer is no good, it is an excellent source of #14 AWG antenna wire (or larger)!

In the rectifier circuit there are two avenues to follow. You can buy four diodes and make your bridge or you can do as I did and use one of the nifty one-inch-square epoxy bridge rectifiers. The little one-inch jobs are convenient because you don't have to mess around figuring which end is the anode and which is the cathode. Ordinarily, the expoxied bridges are simply marked ac, ac, +, and −.

As always, no matter what you do for rectification, be sure your rectifier is rated for the current you will be needing. Most of

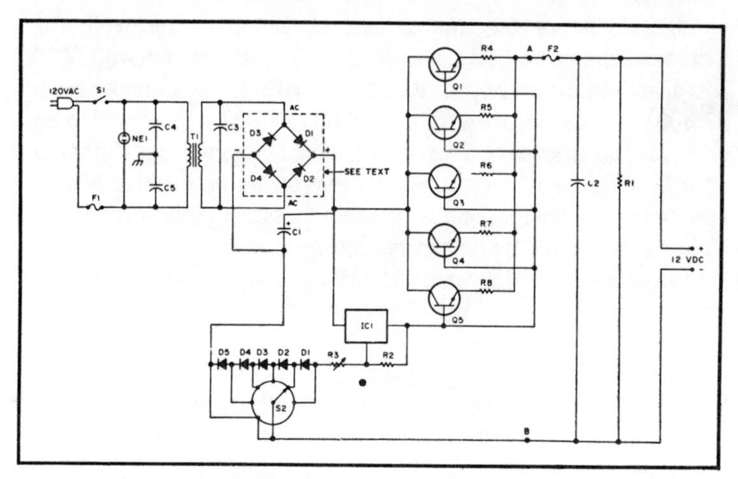

Fig. 1-1. Power supply schematic diagram.

Table 1-1. Parts List For 25 Amp Supply.

C1—13,000-uF, 25-V electrolytic capacitor
C2—10-uF, 25-V electrolytic capacitor
C3—0.22-uF, 100-V tubular capacitor
C4, C5—0.01-uF, 500 V ceramic capacitor
D1-D4—25-A diodes or epoxy bridge rectifier (see text)
D1-D5—1N4004 diodes
F1—Fuse, 5 Amp
F2—Fuse, 30 Amp
Q1, Q2, Q3, Q4, Q5—2N3055 transistors
R1—120-Ohm 4-W resistor
R2—3000-Ohm, ½-W resistor
R3—500-Ohm, 1-W potentiometer
R4, R5, R6, R7, R8—0.25-Ohm, 1-W resistor
IC1—7812 voltage regulator
S1—SPST switch
S2—6-position wafer switch
T1—120/17-24-V ac power transformer (see text)
Miscellaneous: NE1 neon bulb, binding posts, line cord, 0-25-V dc voltmeter, 0-30-A ammeter, heat sinks, chassis, blower, fuseholders, and bulb socket.

the little square bridges are rated between 20 and 35 amps. I am using a Semtech-Alpac 7905 only because I happened to have one on hand. Motorola, International Rectifier, VARO, and EDI make excellent equivalents.

Voltage regulation depends on adequate filtering and an IC known as a 7812. After much experimentation. I found that my voltage regulation (as well as hum attenuation) improved as I increased the value of filter capacitor C2. Starting out with 2000 μF, I worked my way upward to 13,000 μF. Though I now have a 37,000-μF filter capacitor in the circuit, 13,000 μF seemed to be enough. The amount of filtering achieved by going from 13,000 to 37,000 μF is very, very slight and detectable only with a scope. Obviously one can't ignore the thought that if 13,000 μF is good, a higher value would be better, but let me caution you enthusiastic high-capacity freaks against installing 150,000-μF capacitors without limiting inrush current. I haven't experimented beyond 37,000 μF.

The 7812 voltage regulator is an IC device capable of maintaining excellent regulation as long as the input voltage falls between 14.6 and 19 volts nominally. A number of companies are producing the 7812 and it generally has some sort of prefix or suffix, but the digits remain the same.

In this circuit, the 7812 is above ground through a 200- to 500-ohm resistor. I don't put an exact value on this because it is not that critical.

As was the case with the pass transistors, I mounted the 7812 on a heat sink affixed to a small piece of Plexiglas on standoffs (to simplify its isolation from the chassis). The heat sink (see Fig. 1-2) is made of four strips of one-inch-wide aluminum cut at varying lengths and bent up a half-inch at each end. I then placed each one "inside" a larger one. To keep the strips aligned, a hole was drilled which also served to attach the 7812.

While it isn't necessary, you can build in a selectable voltage feature by connecting any number of 1N4004 diodes on a wafer-type switch. This switch goes between pin 3 of the 7812 and ground. (If this seems like a lot of hooey to you, you may disregard the above and connect pin 3 of the 7812 to ground through R3. You will see a voltage change of approximately 0.7 V with each position on the switch. With my supply, I have the capability of as much as 15 V or so, and the switch permits me to "switch down to" the proper voltage I desire (13.8 V dc).

The value of bleeder resistor R1 across the output is not critical either but have something there for your protection.

By varying the resistance of R3, your output voltage will vary considerably. I believe a potentiometer instead of a fixed-value

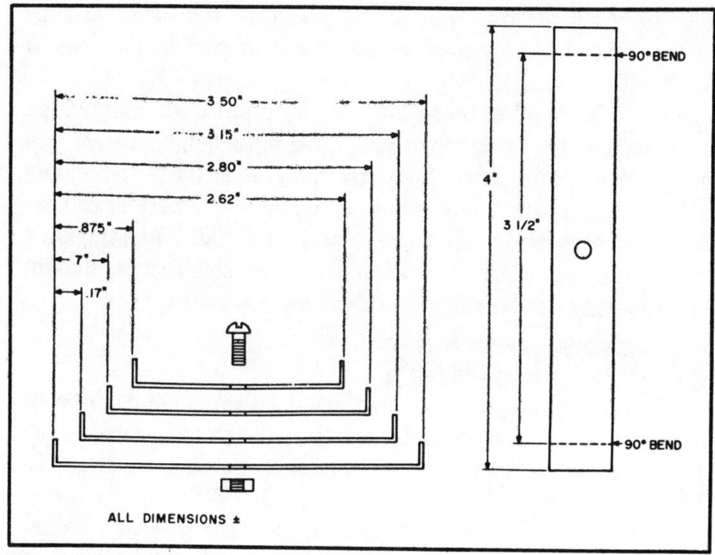

Fig. 1-2. Heat sink construction details.

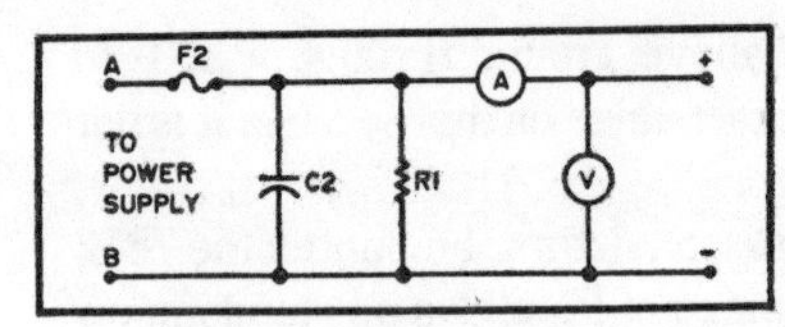

Fig. 1-3. Power supply metering arrangement.

resistor is a better route so that more flexibility is available for future voltage needs which now might not be considered. As in my case, if you are receiving 16.8 V from your transformer, 250 ohms is sufficient to yield the 13.8 V dc you want.

Should you be supplying your rectifier with 16 to 18 volts and not be getting a stable 13 volts or so, check to be sure that you are not losing (dropping) all of your voltage in your rectifier diodes or epoxy bridge. Some of the epoxy bridge rectifiers are poor in the area of voltage consistency. Try a different one, even of the same manufacturer. Another place to watch for voltage losses is in your wiring. The more current you draw, the higher your voltage drops may become in your transformer, rectifier, filter capacitor, or wiring. Wire which is too small may cause substantial voltage drops. I would suggest using #14 AWG wire at least.

Hum Problems

My first test of the power supply was disastrous. Not only was the regulation terrible, but the audio was 80% hum, 20% ham. Two things lead to the elimination of hum: First (and already covered), I placed my voltage regulator above ground on the Plexiglas support; second, I connected all of my chassis ground connections to one point.

As with my other home-brew endeavors, I first mounted the power supply on an open chassis. Bread-boarding can save you much agony when it comes time to actually fitting the power supply in a permanent box. I was able to come up with a perfect cabinet (which formerly was a microvolt meter) for $1.00. When shopping for an enclosure, don't overlook old, non-working test equipment etc. Metering can be added easily as shown in Fig. 1-3.

A 12 VOLT POWER SUPPLY

This design will allow you to adjust it to fit your needs. Most of the parts used are available from your junk box or local parts store. The hardest item to locate at a reasonable cost is a transformer. The voltage requirements for a good 13.5-V dc supply are a minimum of 36 V ac, center-tapped, or a single 18-V ac winding (see Fig. 1-4). For a 13.5-V dc regulator to perform, we have an upper voltage-sag limitation of 18 to 18.5 V dc under a maximum load. What we are asking the regulator to do is to maintain regulation at 13.5 V dc with a difference voltage of 4 to 5 V dc (13.5 V dc regulated + 5 V dc = 18.5 V dc unregulated). Keep in mind that this is the minimum voltage needed to maintain regulation. If you choose a transformer that yields a higher voltage difference regulated to unregulated), the product of this difference voltage and load current, in power (heat), must be dissipated in the regulator, which we will discuss later.

If you have a solid-state rig you wish to power check the manufacturer's specifications for current consumption. Choose a transformer which will handle that load current with a voltage level sufficient enough to maintain the unregulated supply requirements.

Select a diode assembly which will handle the I_{dc} output. My requirement was a maximum of 20 amps to power a repeater and amplifier. My diode assembly will therefore have to handle 20 amps. Each diode used will have a voltage drop of approximately 1 volt across it, and, at 20 amps, $P = 1 \times E$, or 20 amps × 1 volt = 20 watts. Heat-sink these devices well to dissipate this energy.

The filter capacitor may be gauged by a simple rule of thumb. For every amp I_{dc} delivered, a minimum of 3000 μF of capacitance is required. You can have ripple in your supply and never notice it at the output of the regulator as long as the maximum ripple component never drops below the minimum unregulated voltage of 18 to 18.5 V dc. This capacitor value is arrived at mathematically, but, for simplicity, let's stick to the rule of thumb.

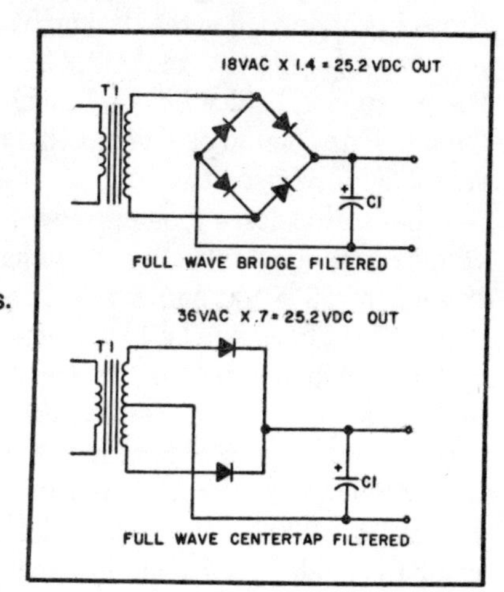

Fig. 1-4. Power supply schematics.

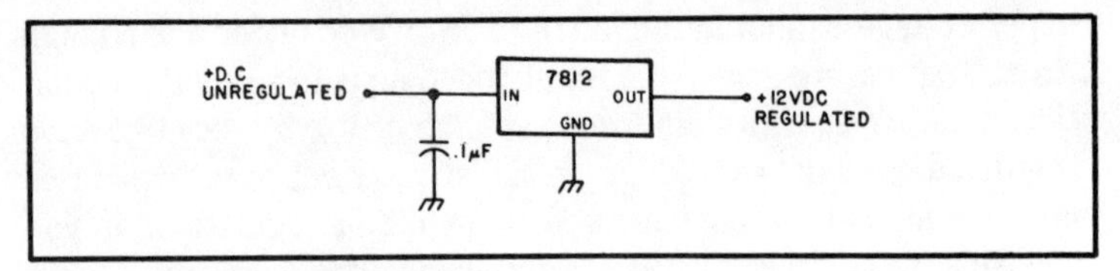

Fig. 1-5. MC7812 regulator circuit.

Now let's get to work on the heart of the supply. The key part of our regulated supply is a simple 3-lead positive regulator, an MC7812 that you can get at Radio Shack. This device will handle a maximum of 1 Amp alone, and has designed-in current limiting and short-circuit protection. See Fig. 1-5.

There are many manufacturers of this device who use prefixes other than MC, but 7812 is the device number. 78 is the design series and 12 is the regulated output voltage. You are about to ask a question! If I want 13.5 V dc, what am I doing with a 12-V dc fixed regulator? It is very simple. To increase the voltage of the regulator, we add one diode in series with the ground lead for every .6-V dc increase desired. See Fig. 1-6. These regulators vary slightly in regard to their actual regulated output voltage, but the additional diodes will allow us to select the actual voltage needed.

In Figs. 1-5 and 1-6, I have used a .1-μF capacitor. This capacitor is needed to stabilize the regulator from ground loops. Attach this capacitor as close to the regulator chip as possible.

As I mentioned earlier, this 3-lead regulator is capable of 1-Amp maximum output current. To achieve a higher current capability, we add a pass transistor. This device will give the current gain needed in the design. The pass transistor, or transistors, must handle the total output current of the supply. For this 20-amp supply, I selected two 15-amp PNP power transistors to do the job. One 20-amp device would do it, but for a heat dissipation safety factor, I used two.

Let's stop for a moment now and talk about the difference voltage I mentioned earlier. If we have an unregulated dc supply voltage of 25 V dc and a regulated output of 13.5 V dc, the difference voltage will be 11.5 volts. The product of the difference voltage and the load current will be the power dissipated, in watts, by the pass transistor. For example, 11.5 volts × 2 amps (load current) = 23 watts of heat in the transistor. See Fig. 1-7.

My supply circuit requirement was 20 amps. Now, 20 amps × 11.5 volts is 230 watts! That is a lot of power! The transformer is going to help in this dissipation, though. Fortunately, the unregu-

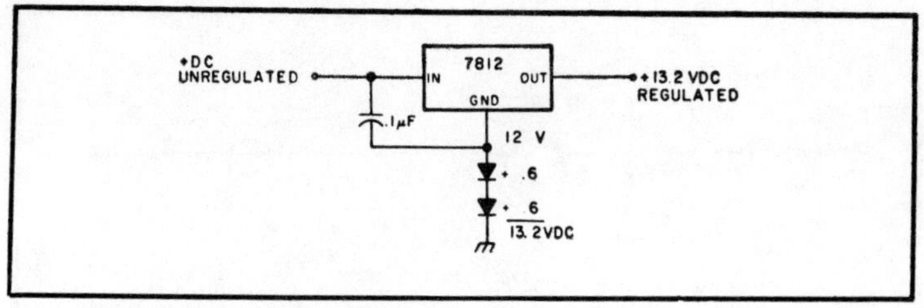

Fig. 1-6. Diodes added to the regulator circuit.

lated voltage will sag, and we have selected a transformer that will only deliver 18.5 volts unregulated at 20 amps. The difference between 13.5 volts regulated and the unregulated 18.5 volts is 5 volts. So, a 5-volt difference × 20 amps = 100 watts, which is a big difference! Using two pass transistors, we can dissipate 50 watts in each device. With 50 watts of heat to get rid of, you must use a good heat sink to pass this power into the air effectively. We must keep the junction temperature of the transistor below its maximum rating to keep from destroying the device. In this 20-amp design, I used a 120-CFM muffin fan and two heat sinks that would handle 80 watts using natural convection. By using forced air instead of natural convection, I could mount the heat sinks in any position. For natural convection, position the heat sink fins in a vertical direction.

Heat-sink considerations should be given to the 7812 regulator, also. At 20 amps, and a maximum beta per transistor of approximately 50, we will have to handle a combined base current of 400 mA. The difference voltage of 5 volts will require a power dissipation of .4 amps × 5 volts = 2 watts.

Fig. 1-8 illustrates the complete regulator, including the 6.8-ohm, ½-watt resistor which is used to establish the bias of the regulator and pass transistors. Using collector feedback with this

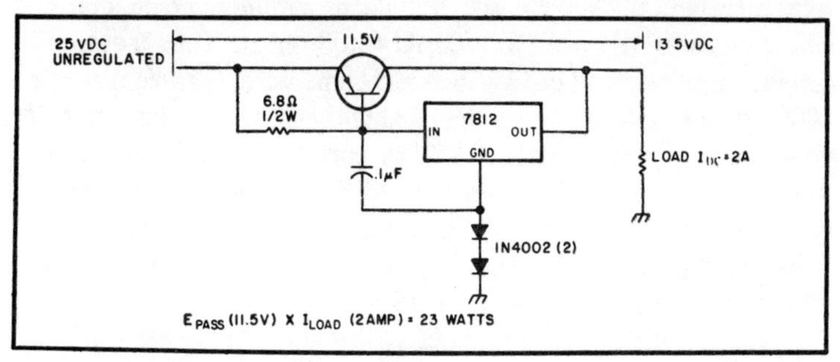

Fig. 1-7. Pass transistor added to circuit.

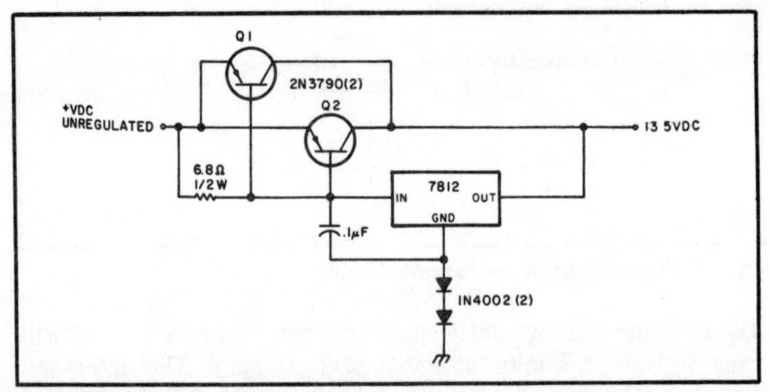

Fig. 1-8. Complete regulator circuit.

regulator chip proves to be very effective and stable. The supply will even be free of rf instability unless the regulator chip is involved in a concentrated rf field.

In the area of wiring, the only point of note is to use the proper size wire to handle the current in the high-current areas. Also, keep the two wire leads to the emitters of the pass transistors equal in length to allow balanced emitter current.

We have discussed the regulator; now let's examine a current limiter. See Fig. 1-9. This current limiter operates instantaneously and it will only reset after the power is turned off and the supply bleeds down, or a reset push-button, normally closed, is added at point X. C2 should be sized to allow a delayed dropout, so relay CR1 does not buzz when it reaches the current limit. Resistor "IR" is selected so that the current-limit relay will trip at 20 amps and higher, depending on the setting of the 100-ohm pot. To operate, a 6-V dc drop is needed between the base and emitter of Q3. This voltage provides bias current which permits collector current to flow which, in turn, energizes CR1. CR1 must handle the total current of the supply even if it means paralleling contacts. To select resistor IR, use .6 V/20 A = .03 ohms. This "resistor" turned out to be a coil of 14-gauge nichrome wire. By adjusting the 100-ohm pot, we may now select a slightly higher current limit. If your supply is smaller, use .6 V/your current = IR. For example, .6 V/5 A = .12 ohms. This resistor will drop some voltage, so rate the power carefully: P = .6 × 5 = 3 watts, and remember the voltage drop. This .6 volts may not seem like much, but the minimum unregulated voltage is 18 V dc.

The last control circuit to design is called a "crowbar" circuit. Its intent is to protect your gear from overvoltage due to a regulator failure. See Fig. 1-10. In this circuit, we will select an overvoltage of approximately 15 volts. This is done by adjusting control R1 until the lamp just lights, and then backing the control off slightly until the lamp does not light. Remember, to turn off the lamp you must reset the SCR by momentarily disconnecting the lamp or turning off the supply. After this calibration, remove the lamp and connect the anode of the SCR directly to the output of the supply. If, during use, an overvoltage condition occurs, the SCR will conduct, the output will be short-circuited, and the current limit circuit will turn off the supply.

Figure 1-11 shows the complete power supply schematic. Figures 1-12 and 1-13 show what the completed power supply should look like.

A 20-AMP, ADJUSTABLE, REGULATED DC SUPPLY

This supply is versatile in that one has a reasonable amount of control over the heat dissipation of the pass transistors despite the amplitude of voltage to be controlled.

Two paralleled power transistors are used as pass transistors; the 2N3772 and 2N3773 might be described as extra-heavy-duty 2N3055s. I use this as a comparison because the 2N3055 is the well-known workhorse. The 3772 and 3773 can dissipate 150 watts, as compared to the 3055's 115-watt capacity. These transistors can effectively control voltage amplitudes as low as 1.4 volts. Each 3772 can very safely handle currents to 15 amperes. The 3773 handles 10 amperes, again as compared to the 3055's 5-ampere safe capacity. These currents must, of course, be kept within the power

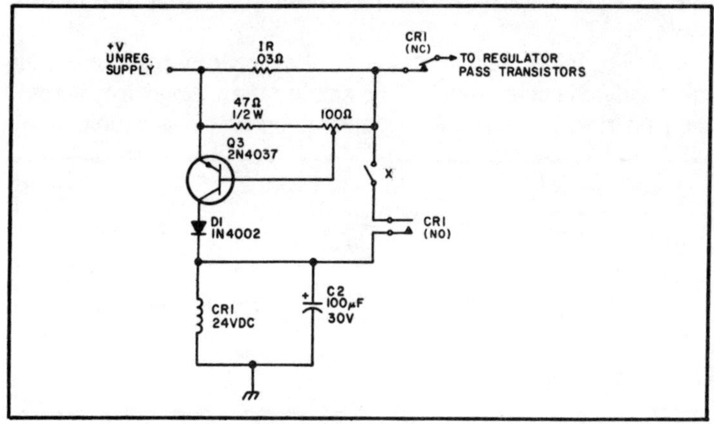

Fig. 1-9. Current limiter.

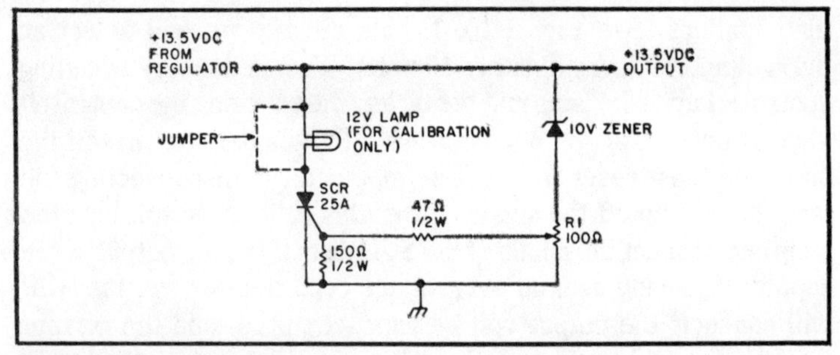

Fig. 1-10. Crowbar circuit.

dissipation capability of the unit. One would not, for example, attempt to drop 30 volts across a 3772 while pulling 15 amperes, as this is a 450-watt dissipation and the transistor has a rating of 150 watts. Using two of these in parallel, the dissipation capability is increased to 300 watts. (Even 300 watts takes a lot of heat sinking to dissipate the heat.)

Refer to the schematic in Fig. 1-14. Let's examine some practical extremes. Say we wish to regulate 13.0 volts at 20 amperes, and the input voltage (at the choke output) into the regulators, using the full secondary, is 35 volts; this would result in a drop of 22 volts across the regulators. 22 volts × 20 amperes = 440 watts. Obviously, things would start to melt after a very short operating period. Therefore, switches S1 and S2 have been incorporated so that one can obtain the required voltage and current and hold the dissipation within reasonable levels. Thus, setting S2 in the low position, we are only using ¾ of the secondary, or 22.5 V ac, possibly delivering 25 volts (loaded) to the pass transistors. 25 less 13 = 12 volts across the pass transistors, or a total of 240 watts.

Following this same reasoning, it is apparent that the higher the regulated output voltage, the smaller the voltage drop across the pass transistors and the smaller the power dissipation.

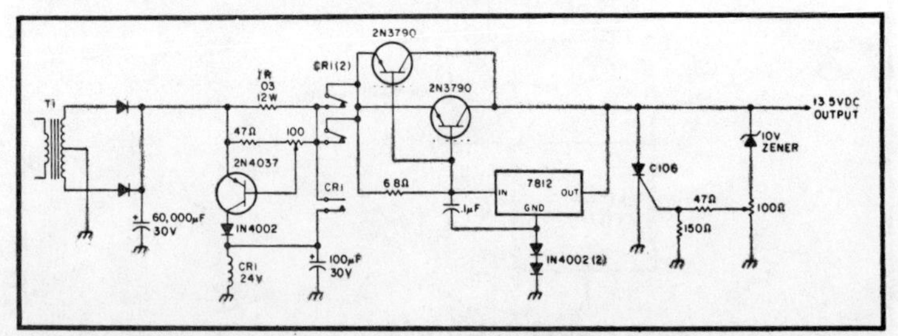

Fig. 1-11. Complete power supply schematic diagram.

Fig. 1-12. Top view of completed supply.

Note that, when using multiple secondary transformers, one has the capability of choosing the appropriate voltages. The switching arrangement shown is the one I use. It may be convenient for some applications to tap the secondaries at other points, i.e., tapping at the 15-volt ac position will result in approximately a 60-watt dissipation for the above stated example.

The transformers I used were purchased from army surplus (WW II). They are hermetically sealed units and have an operating

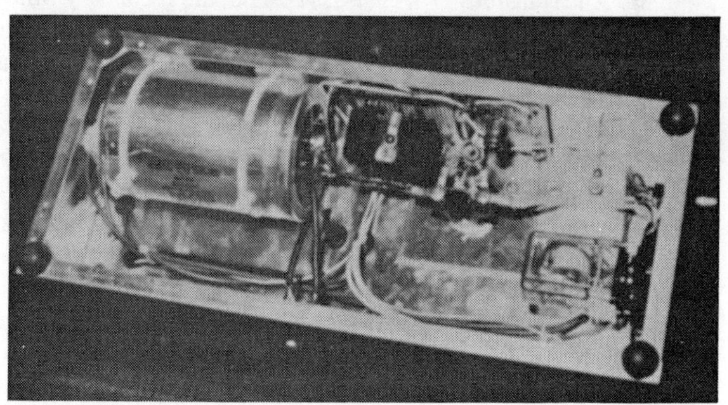

Fig. 1-13. Bottom view of completed supply.

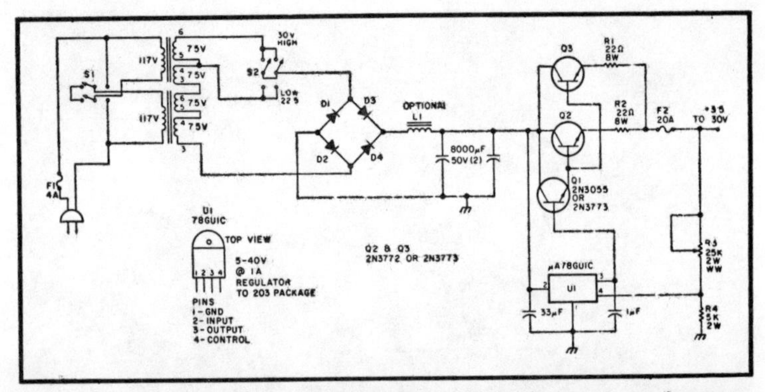

Fig. 1-14. 17 Ampere super-regulated low-voltage universal power supply—3.5-30.0 volts. S1 and S2—10 Amp contacts @ 120 V; D1-D4—50 V, 30 Amp or equivalent; Q2 and Q3—2N3772 or 2N3773; Q1—2N3055 or 2N3773; U1—μA78GUIC; R1 and R2—paralleled .44 Ω, 4 watt or equivalent; R3—2.0 wtt, 20k to 25k wire-wound pot; R4—5k, 2 W.

ceiling of 60,000 ft. The continuous operating current is called out as 10 amperes. I have used the supply for over an hour of continuous service at 17 amperes and the cases stayed at room temperature. The transformers each have two 7.5-volt secondaries. As can be seen in Fig. 1-15, I use the windings in series and parallel configurations. The choke possibly can be eliminated with somewhat higher peak voltages being present. Any possible increase in ripple voltage will be smoothed out by the gain in the regulator chip. The choke was used in a previous unregulated power supply built on this same chassis; therefore, it was retained rather than removed. This choke was built from a Triad filament transformer I had on hand. Originally, the transformer had four 6.3 V ac 4-amp windings. The transformer was disassembled, the secondary winding removed and about 3 or 4 layers of no. 10 or 12 wire wound onto the core. The laminations were reinserted with all of the "E" s in one direction and the "I"s placed at the end.

The regulator chip, a Fairchild uA78GUIC, is the whole key to the success of regulation. It drives a 2N3772; a 2N3773 or even a 2N3055 will work equally well. The rectifiers are 1N3209s, 100-volt, 15-amp units. They happened to be something I had in my junk box—suitable substitutions can be used, incidentally, they ran cool at 17 amps; the heat sinks were cut from heavy aluminum heat sink rails. Two heat sinks were used side by side and isolated from each other, one section for the rectifiers, the other for the pass transistors. For better dissipation, the radiator fins should be vertical. However I have had no problems mounting them horizontally.

The divider composed of R3 and R4 only requires about 1.0 mA of current as the control current to the regulator is only 5 to 8 μA under worst-case conditions. The control voltage is 5.0 volts. The fixed values for a given voltage output at pin 3 can be calculated from the formula $V_{out} = [(R3 + R4)/R4]\,V$, where V control on the 78GUIC = 5 V.

Figures 1-16, 1-17 and 1-18 show different views of the finished power supply.

DUAL VOLTAGE POWER SUPPLY

The power supply uses two type 309K three terminal regulator integrated circuits. These consist of an in, out and common pin connection and thus this IC is the simplest possible device to work with. The K suffix designates the type TO-3 package. This regulator is also available in the type TO-5 package, but we will use the higher power package to obtain output currents in excess of one ampere. This is a particularly fine power regulator specifically built for five volt output for TTL use but a simple connection allows the regulator to furnish higher output voltages with equally excellent regulation.

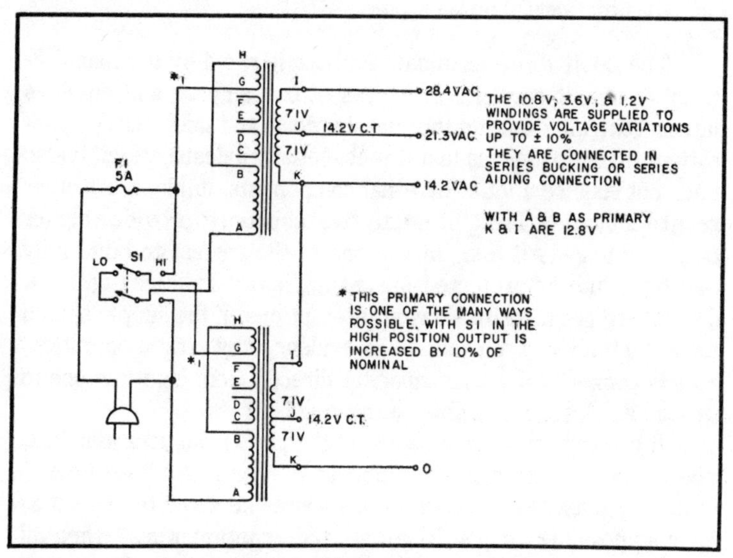

Fig. 1-15. ITC transformer, part #BP-6242, size 4.5″×5.0″ × 7.0″, hermetically sealed, conservatively rated at 20.0 amperes continuous operation. A pair of these can be substituted for the transformers in the text and offer even more versatility. They are available for $19.50 each from Hiway Company, 305 W. Wisconsin Ave., Oceanside, CA 92054.

Fig. 1-16. Front view of power supply.

The 309K three terminal regulator is rated by the manufacturer at output currents in excess of one ampere and employs internal current limiting, thermal shutdown and safe-area compensation. All of this means that it is essentially indestructible. It also does not require a lot of external components, unlike most other regulator circuits. It requires only two resistors to provide a higher output voltage. All told, this is one of the easiest to build fully regulated and self-protected power supplies around. See Fig. 1-19.

There are no critical layout precautions. If the supply should exhibit a tendency to oscillate as evidenced by erratic operation, simply connect a 0.22 μF capacitor directly from input pin one to ground as close as possible to the regulator.

If you can scrounge up some of the parts from your junk box, you are already ahead of the game, but later on you will see how all of the parts may be obtained from a source near you to make it as easy as possible for you to get started experimenting. When all parts are on hand, you can assemble and package them in any way you like.

The power supply schematic may look a little strange to you, but do not be alarmed. This is a solid state version of the so-called "economy" power supply, more popular several years ago than it is today. It is a combination of the well known bridge and full wave rectifier circuit configurations. One regulator is connected to the transformer secondary winding center tap to yield five volts at the output for TTL circuits, and makes use of two of the diodes in the bridge rectifier to work as a standard full wave rectifier circuit. The other regulator is connected to the output of the bridge rectifier to yield about 14 volts at the output for CMOS and linear circuits. This regulator is biased up from ground by the two resistor divider network across the output. The regulated output voltage is adjustable by virtue of the variable resistor. The regulator requires a minimum voltage differential of about two volts between the regulated output and the dc voltage at input pin 1 from the output of the rectifier diodes and filter capacitor.

Most CMOS integrated circuits are rated by the manufacturer at 15 volts maximum, and it is wise to limit the output voltage of the power supply to not over, say, 14 volts. A voltage between 12 and 14 is good in order to take full advantage of the exceptionally high

Fig. 1-17. Left side view of power supply.

Fig. 1-18. Right side view of power supply.

noise immunity of CMOS integrated circuits, and is also a good operating voltage for linear IC projects. The variable resistor may be replaced with a fixed resistor of equal in-circuit value, if it is not desired to vary the output voltage. Temporarily connect a variable resistor and adjust it to set the regulator output voltage as read by a voltmeter. Shut off the power supply, remove and measure the value of the resistor and solder in a fixed ½ watt resistor of the nearest standard value.

Notice that the common terminal of the 309K is also the case of the regulator, so it is necessary to insulate the higher voltage regulator from the heat sink. For this purpose use the insulator found in the power transistor mounting hardware kit. Spread a very thin coat of heat sink compound (silicone grease) on each side of the insulator before mounting the regulator to the heat sink. Use the power transistor sockets to save soldering directly onto the pins of the regulator. An important reason for using the sockets is that they have self-aligning insulated hubs that center the pins as they pass through the holes in the heat sink. The socket also insulates the mounting screws from the heat sink and prevent shorts from this cause.

ANOTHER DUAL-VOLTAGE SUPPLY

There are two dc voltages that I generally need supplies to provide: +5 volts for TTL and +13.6 volts as a car battery eliminator. A 12.6 V ac filament transformer will not work well for these dc voltages. Those of you who have tried were frustrated, I'm sure, by the attempt.

Take the +5-volt supply, for instance. Full-wave rectification with the c-t grounded will yield a peak dc on the filter capacitor of $(6.3 \times 1.414) - 0.5 = 8.4$ volts. Now, an LM309K needs +7 volts to regulate (a 2-volt regulator margin). This means that the maximum ripple is 1.4 volts. Let's say you need 1 amp from this supply and you chose a 1- or 2-amp transformer. Now, to get the ripple to less than 1.4 volts, you start to pile on the capacitance and the full-load ripple comes down. Then as you add more capacitance, the ripple goes up! What is going on here? As you add more capacitance, the phase angle over which you draw current decreases. This means that you are not continuously drawing 1 amp from the transformer, but, instead, you are drawing many amps over a short time to yield a continuous load current of 1 amp. Transformer core saturation and winding resistance drops are causing the problem. The truth is that this transformer will not work except for small currents. The +13.6-volt supply using a bridge rectifier will yield the same picture—not enough margin for

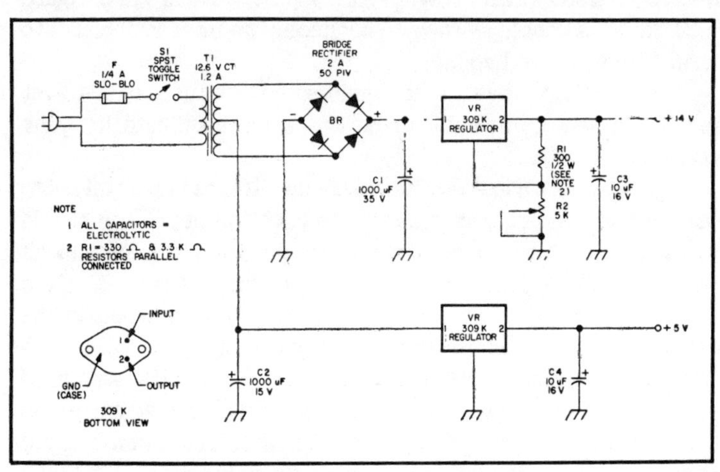

Fig. 1-19. Fully regulated power supply.

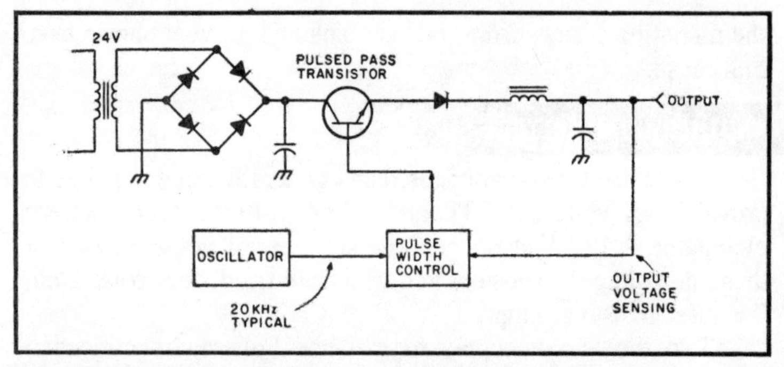

Fig. 1-20. Switching regulator showing basic functional blocks.

the regulator to operate. The next highest filament transformer is 24 volts—too much voltage and too much power to be thrown away via the heat sink.

A switching regulator, Fig. 1-20, has some advantages inasmuch as the utilization of transformers is concerned. The pulse width is varied to the pass transistor to control the output voltage as the load current changes. The efficiency can be very high with this configuration. The disadvantages are formidable. Switching transients clutter up the spectrum so that much shielding and filtering is necessary. Transient load changes may generate output voltage variations amounting to ±20 percent or more. As you might gather, I don't recommend this circuit for the ham shack.

What we *really* need in a transformer is 18 V c-t. This voltage is perfect for both the +5-volt and +13.6-volt supplies. Few distributors stock them so you really have to look around. Figure 1-21 shows a couple of ways to connect filament transformers to furnish the required voltage.

Figure 1-22 shows a really great supply that uses easy-to-get parts. I have included the Radio Shack part numbers for your convenience.

Now, a word about the filter capacitor. It is as important not to have too much capacitance as it is to have too little. You should select a capacitor so that at maximum current you have enough margin for the regulator to work. In the circuit of Fig. 1-22, C1 is selected so that the ripple voltage is 5 or 6 volts. This allows the transformer to furnish the 3 amperes of output current over a much wider phase angle. I have used the formula $E = 1/120C$ with good results. E is the peak-to-peak ripple voltage, I, the dc current in amperes, and C, the filter capacitor in farads. This formula is not exact but should be quite close for most applications.

Resistors R1 and R2 in the circuit of Fig. 1-22 cause the current to divide between the regulator and the pass transistor. For three amperes at the load, two will flow through the pass transistor and one through the LM309K. If you use a germanium pass transistor, you can omit D2, as D2 compensates for the base-to-emitter junction voltage drop of the pass transistor. Zener diodes D4 and D5 form an overvoltage protection that will blow the output fuse in case of a regulator failure. If you don't put in this protection, you'll be sorry! An insulating washer is needed for U1 as the case is above ground by the drop across D3. The case of U2 can be bolted to ground. Use a common heat sink having a thermal resistance to ambient of less than one degree per watt. The addition of panel meters to monitor the current would be a nice improvement. The fuses would not be expected to blow as the power supply is protected against short-circuiting and overload. Of course, you can't load both regulators to their full capacity at the same time, as this would overload the transformer.

HEAVY-DUTY POWER SUPPLY

This power supply disconnects itself from an over-current load or short circuit, protecting its components without blowing any fuses. It also disconnects it, for any reason, it should attempt to provide over-voltage output, protecting expensive equipment

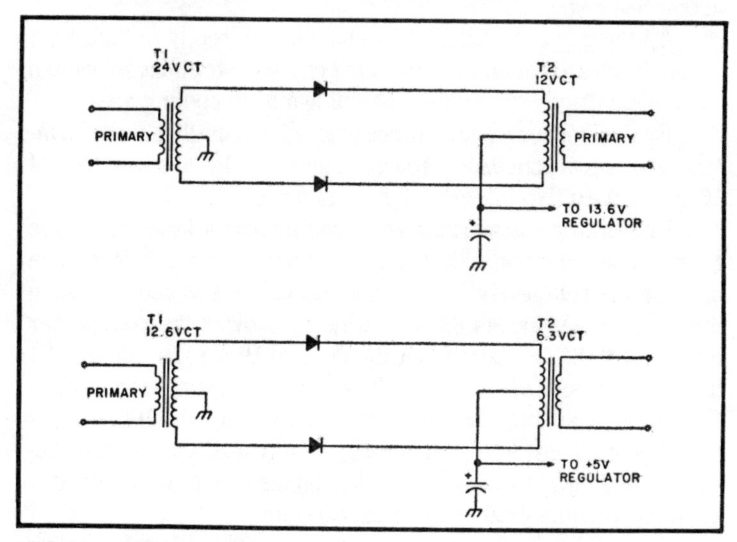

Fig. 1-21. Filament transformers connected to yield the required input voltages to the regulator.

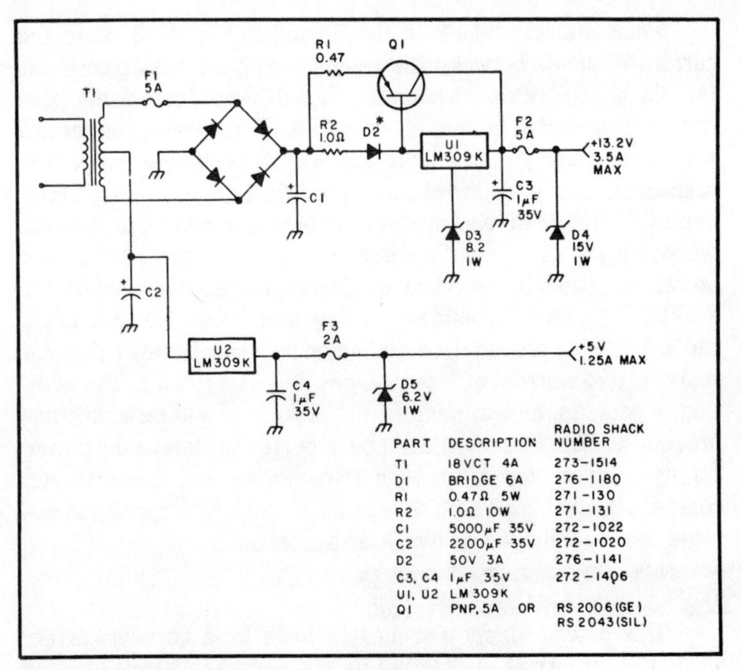

PART	DESCRIPTION	RADIO SHACK NUMBER
T1	18VCT 4A	273-1514
D1	BRIDGE 6A	276-1180
R1	0.47Ω 5W	271-130
R2	1.0Ω 10W	271-131
C1	5000μF 35V	272-1022
C2	2200μF 35V	272-1020
D2	50V 3A	276-1141
C3, C4	1μF 35V	272-1406
U1, U2	LM 309K	
Q1	PNP, 5A OR	RS 2006 (GE) RS 2043 (SIL)

Fig. 1-22. Power supply using easy-to-find parts.

which it may be powering. Again, no fuses blow; it just disconnects. See Fig. 1-23.

All the power transistors can be bolted directly to their heat sinks, which are bolted directly to the chassis, providing maximum thermal conductivity and best use of available cooling area.

Equivalent source impedance is less than 5 milliohms, equivalent to a zero impedance source connected by about a foot of 16-gauge wire. No integrated circuits are used.

The power transformer was rebuilt from a large tube-type color TV set and had a core area of about 3 square inches. The primary was retained, all other windings being removed. Counting the burns of a previous 6.3 volt winding showed the voltage per turn to be 0.62, indicating that the desired 19.4 volts required 31 turns on each side of center-tap. There was just room to snugly use #12 in the available window. A pair of 100-ampere, 100-volt piv diodes were used. An experimental dummy load showed the secondary holding up to about 18.5 volts and the trough of the ripple at 17 volts with the 43,000 μF filter capacitor.

About 26 volts dc was present at no load, comfortably below the 30-volt rating of the capacitor. A surplus computer power supply heat sink with three NPN TO-3 transistors having about 120 square inches of cooling fins was used for the pass transistor output stage. Mounted on the sink were the three 0.18-ohm emitter-balancing resistors which assure an extremely good current sharing between the three NPN transistors. However, this accounts for a whole volt of drop at 20 amperes. Probably, 0.1-ohm, 5-watt resistors would be just fine and would conserve about half a volt of drop. It was determined by experiment that 10 milliamperes into the base of the Darlington-connected driver power transistor would maintain a 20-ampere load, indicating a beta product of about 2000, which is somewhat higher at more moderate loads.

About a volt of high-frequency oscillation was observed at output under moderate to heavy load. Adding the 24-ohm resistor and 0.1 μF capacitor to ground spoiled the high-frequency gain enough to clear this up. Output ripple is less than 5 millivolts rms.

Experience had shown that a "crowbar" circuit was worthwhile. This is merely an SCR connected across the output which is triggered into conduction by a voltage higher than normal. Most

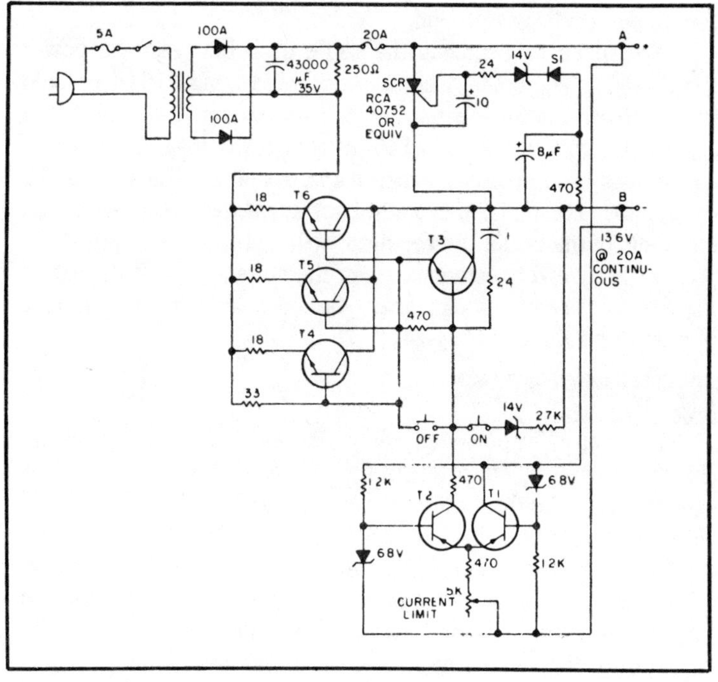

Fig. 1-23. Heavy-duty regulated power supply. T1-2—2N2906; T3—2N3715 or equivalent; T6-8—RCA 40325, 2N5302, or an equivalent 15-Amp NPN.

circuits of this type connect the SCR across the filter capacitor and blow a fuse. I had intended to do this, but the negative terminal of the capacitor is at a variable voltage depending upon load, so this was impractical. As it happens now, if the SCR fires in addition to shorting the output, it disables the differential amplifier and removes base current from the pass transistors, shutting down the supply so rapidly that no fuse blows. The SCR I had available was a real monster, but, under this condition of operation, the peak current rating need be no more than 10 or 15 amperes.

Experiment a bit with this to make sure that the SCR conducts with the voltage output somewhere between 14 and 15 volts. I had to add one diode drop to the voltage of my nominal 14 V zener to achieve this. A 470-ohm, 1-watt resistor serves as a minimum load on the supply so that it doesn't accidentally turn on without having depressed the start button. A small filter capacitor across it keeps noise from the load from getting back into the supply. A little rf bypassing on the way into the differential amplifier would be a good idea so that rectified rf from that 2 meter amplifier doesn't bias the power supply silly. Bypassing is simple with the negative side of the output at chassis ground.

Not to be forgotten is the cable drop between the power supply and its equipment load. A 3-foot cable using #16 wire would yield .024 Ohms for the pair, which seems small enough but it represents half a volt drop and 10 watts of power loss! It's better to use #12 or heavier. If the sense leads, A and B, which can be 22 gauge, are extended and brought back separately from the equipment end of the cable, the regulator will increase its output at the power supply end to compensate for the voltage drop in the cable. But try to use heavy enough wire to hold the cable drop to a quarter of a volt or so. See Fig. 1-24.

FM POWER SUPPLY

I built my FM power supply inside a 4″ × 8″ × 2″ aluminum chassis. Getting the transformer into the 2″ dimension was a tight squeeze; 2½″ would have been better, and 5″ × 10″ × 3″ would have been spacious. The cord, switch, LED pilot light, output socket, fuseholder and the T0-3 power transistor went on one 2″ × 4″ end, leaving five faces blank. Inside, after finding room for two surplus electrolytic capacitors and the transformer, I wired the circuit on a couple of Cinch-Jones 2012 terminal strips.

Using a bigger transformer or more (thousands of) microfarads could only improve things. The power transistor was bolted directly to ground and the aluminum box provided adequate cooling. If you use a steel chassis, however, you may need some sort of heat sink, but remember that a silicon transistor will work all right even when its case is uncomfortably hot.

Figure 1-25 shows the schematic of the 12 V supply. The transformer is nominally 35 V at 1.5 amps, center-tapped. It weighs about one kilogram. An 18 V, 2.5 amp unit of about the same weight and size, used with a bridge rectifier, would probably be satisfactory. I wouldn't try to use a 12 V transformer, as there isn't enough reserve for low line voltage.

With the F-54X at a line voltage of only 100 V, the supply put out 3 amps before losing regulation; at 105 line volts, 4 amps; at 110 V and up, 5 amps. For continuous service, two amps is probably the limit, but since FM transceivers spend most of the time on receive, the low current works OK. The above figures are for 12.6 V out. For higher outputs, the transmit power will increase, but the current capability at any particular primary voltage will be less.

Each diode specified is rated for 300 amps surge, 3 amps average and will stand a short circuit long enough for the primary fuse to blow. No heat sinking is needed.

The filter capacitors may see more than 30 V, so you need capacitors rated at either 35 working volts or 25 working volts/40 V surge rated. They should have at least 2000 μF per amp of rated load (more is better).

The pass transistor I used was a 2N3715 (similar to a 2N3055). The plastic MJE 3055 (or any one of the many T0-3 size silicon NPN power transistors) should be satisfactory.

Fig. 1-24. Front view of heavy-duty power supply.

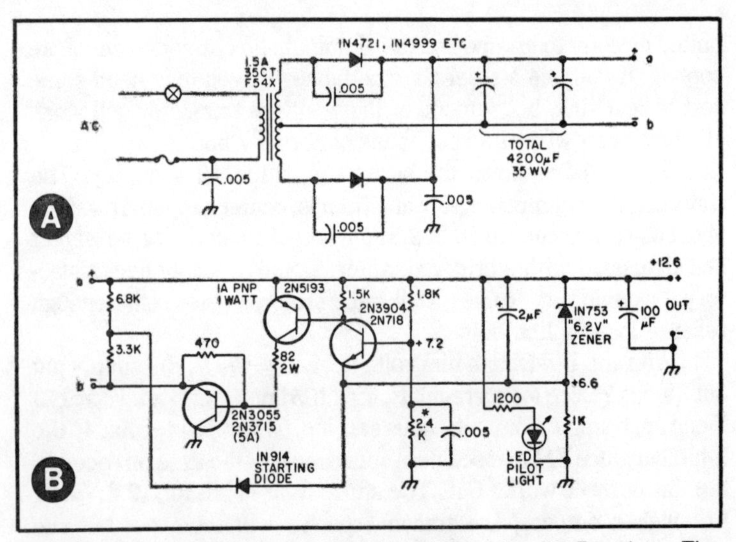

Fig. 1-25. 3 A regulated supply (12.6 V). (a) Rectifier. (b) Regulator. The capacitor on the output is a tantalum slug type, of any value over 10 μF. The 3 A unit will carry a typical 10 or 15 watt 2m transceiver. The 82 Ohm resistor constitutes the short circuit protection. *Adjust to set output voltage.

Although the regulator circuit is simple, it continues working down to less than a volt between input and output—this is unusual. To keep the voltage drop down, the pass transistor is driven from the regulated output voltage, but this means that a separate starting circuit (two resistors and a diode of no particular type) is needed to turn things on initially. Once the starting diode has disconnected (during operation, its cathode is more positive than its anode), the regulator is unaffected by the magnitude of the unregulated input voltage—which is why it wouldn't start without the extra circuit.

Multistage regulators may oscillate. Stabilize this one by putting 10 μF or more (tantalum type preferred) on the output. I used a CS13AE 101K (100 μF, 20 V), because I had a lot of them.

The PNP driver could probably be a 2N2905, but the 1 amp plastic power transistor (2N4918) I used is harder to blow out and has good gain down to 1 mA. I mounted it inside the chassis with a 4-40 metal screw using the mica washer provided. The 82 ohm, 2 watt resistor protects both the transistor and the power supply in the event of a short on the output. If you use higher resistances, you'll get a lower maximum base drive to the pass transistor and less short circuit current.

Depending on the actual voltage of your reference (zener) diode, you may have to adjust the voltage divider in the base of the NPN amplifier. I prefer setting things up with a soldering iron and putting a high resistance across one or the other divider resistor. In this case, I used a 24K ohm across the 2.7k ohm, because the 1N753 had only 5.9 V drop at 6 mA. If you put a 1000 ohm pot between the two resistors, with the 2N3904 base hooked to the arm of the pot, it will still work. The capacitor still goes from base to ground.

Connect a capacitor across the reference diode. I've tried everything from 1 to 40 μF, so the size is not critical.

Place ceramic disc bypasses across the diodes (right across—with very short leads) to reduce hash picked up on AM broadcast sets. A number of commercial supplies, as well as automotive alternator diodes, can cause this sort of interference.

In wiring high current supplies, it is good practice to run wires from diodes and the power transformer directly to the big filter capacitors and then run additional wires from the capacitors to the regulator. Run the input leads together to keep the stray field down—those wires are carrying 10 amp pulses. In a ham station, it is also smart to filter all leads for rf; audio rectification in a transmitter power supply can give some strange feedback.

A design-it-yourself program is as follows:

☐ Measure the output voltage. In this case, given 12.6 V, I added 1 V for regulator variations and 15% for power line variation, for a total of 15.6 V minimum instantaneous (dc minus ripple) input to the regulator at 115 V.

☐ Measure the load current and calculate ripple. For 2 amps, given 4200 μF (that was what I had), peak-to-peak voltage = 2 × 1/120 × 1,000,000/4299 = 3.97. On that basis, required dc is 17.6 V and rectifier load R is 8.8 ohms.

☐ Using a vom, check the power transformer for the following values: line voltage = 115 V; secondary, no-load voltage = 39 V (10% higher than nominal); ratio of primary to half secondary = 5.9; dc resistance of primary = 9 ohms. Primary resistance reflected into half secondary is then 9 divided by $(5.9)^2$ = 0.26 ohm. Measured secondary resistance = 1.1 ohm; half of that = 0.55 ohm. Thus the series resistance of the transformer is effectively R_s = 0.55 + 0.26 = 0.81 ohm. At 20 amps instantaneous, the diodes I picked have a 1.2 V drop with a slope equal to 0.01 ohm at that point for a total of 0.82 ohm.

In applications where noise and ripple requirements are moderate, a grounded collector pass transistor is often convenient.

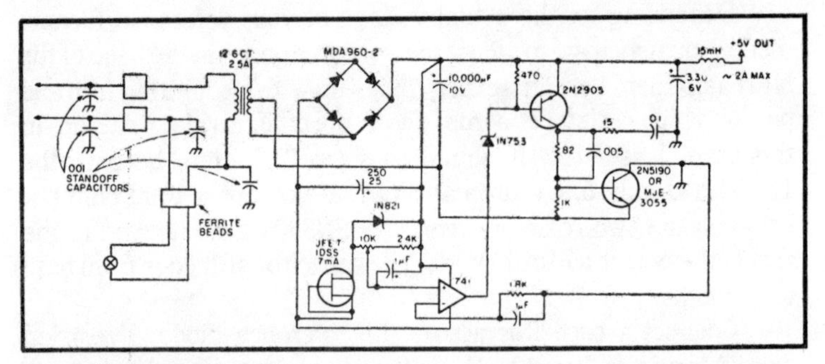

Fig. 1-26. Five-volt supply.

Figure 1-26 shows the 5 V supply I built for a receiver frequency counter that used LED readouts. The display added a strobed 1 amp load to the other drain, putting the requirements out of reach of the usual IC units. The circuit, as shown, fit the parts I had on hand and performed well enough down to below 90 V from the power line. I used a compensated reference diode, but a 1N753 will do as well. The FET current regulator (if you have one with I_{DSS} around 7.5 mA) is an improvement over a 1200 ohm resistor, and of course a 741 is one of the best op amps you can buy.

LOW VOLTAGE POWER SUPPLY

Three power supplies were constructed as shown in Fig. 1-27. The first is a dual tracking supply with variable output voltage 0 to ±20 volts and current to 100 mA on each output (200 mA total current capacity). Also available is a +12, −6 volt option. Current sensing is done in both the positive and negative legs, and when the current exceeds a preset level, a signal is developed to shut down the output from the voltage regulator. This signal latches so that output voltage can only be restored by pressing a reset switch.

The second supply has variable output from 2.6 to 25 volts and current to 1 ampere. Up to 34 volts is available at reduced current. This supply also has adjustable current sensing and, like the first supply, the output voltage shuts down when the current exceeds a preset level. Voltage is restored by pressing the reset switch.

The third supply provides a fixed 5 volt output at currents to 1 ampere for operating TTL circuits. This supply has output voltage sensing and will shut down if the voltage moves outside a preset range from 4.75 to 5.25 volts.

The first supply provides the power for the sensing circuits used in all three supplies. Also, if any one supply shuts down, the other two will shut down also.

All three supplies use voltage regulators that are short circuitproof, an added safety bonus in the event that the current sensing circuits are manually disabled or in the event of the failure of some component in the current sensing networks.

Current Sensing

The current sensing network in Fig. 1-28 operates as follows: Assume that initially no current is drawn from the supply. With R2 set to 500Ω, R2 + R3 = 21k and R4 + R5 = 21k. With the wiper of R4 set closest to R3, the voltage at pin 11 of voltage comparator IC1A will be 14 volts, exactly half the voltage across C1. Assuming for the moment that no current flows in R1, the voltage across R6 and R7 will be 28 volts and the voltage at pin 10 of IC1A will be 14 volts also. When current is drawn from the positive leg of the supply, a voltage drop develops across R1 and the voltage at pin 10 of IC1A drops below 14 volts. This drives pin 13 of IC1A positive and the resulting current in R21 charges C3. Q1 fires, sending a pulse through C4 to SCR1. SCR1 turns on, operating relay K1 and forcing Q2 to switch on. Q2 shorts out R27, thus reducing the output of IC3 to nearly zero volts. K1 interrupts the current to IC6 in Fig. 1-29. Q1 also sends a pulse to C14 in Fig. 1-29. This pulse

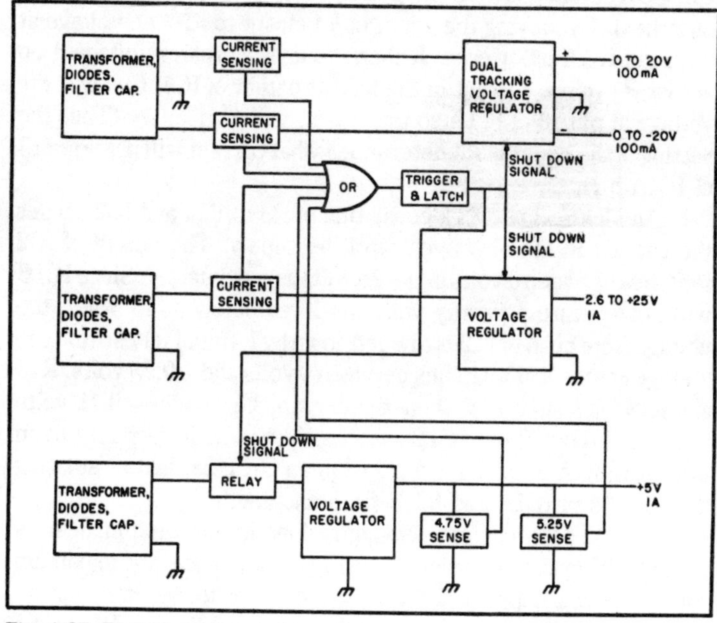

Fig. 1-27. Basic layout.

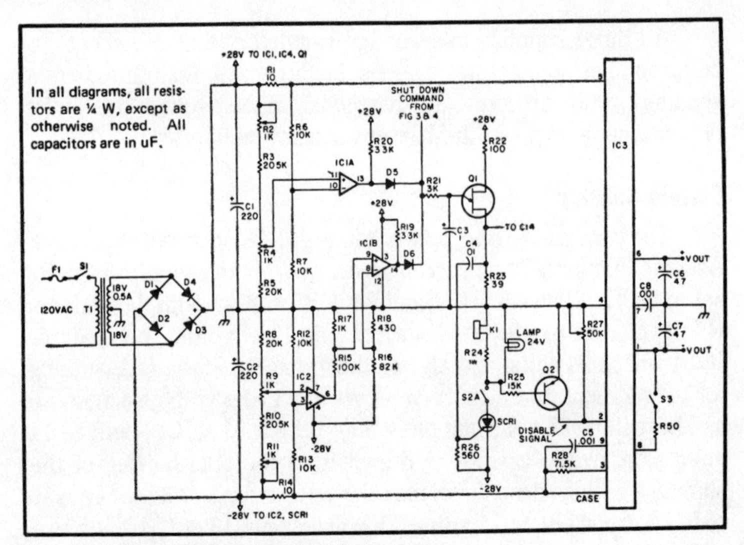

Fig. 1-28. Dual tracking regulated supply.

turns on SCR2, forcing Q4 to switch on; this action reduces the output of IC5 to zero volts.

When the load is removed from the output of IC3, the power can be restored by opening S2A and S2B (normally closed switches). By moving the wiper of R4 closer to R5, the voltage at pin 11 of IC1A is lowered. It then requires a greater voltage drop across R1 (more current in the load at output of IC3) to lower the voltage at pin 10 of IC1A so that pin 13 will go positive. Thus the setting of the wiper of R4 determines what current will drive pin 13 of IC1A high.

An identical network consisting of R8 to R14 and IC2 senses the current in the negative leg of the supply. The output of IC2 switches between 0 volts and −26 volts approximately. Since IC1B will not operate normally with any input below −0.3 volts, the voltage from pin 6 of IC2 is divided down by R15 and R17 so that the voltage across R17 switches between 0 volts and −0.25 volts. R16 and R18 form another voltage divider which provides −0.15 volts to pin 8 of IC1B. Thus IC1B switches like IC1A in response to an overcurrent in R14. D5 and D6 form an OR gate, hence isolating the outputs of IC1A and IC1B from one another.

In Fig. 1-29, current sensing is done in the same manner as described for the positive leg of Fig. 1-28. Since the maximum current for this supply is 10 times greater than for the first supply, resistance values have been adjusted accordingly. D9 forms another part of the OR gate that feeds R21.

Voltage Sensing

For the 5 volt supply in Fig. 1-30, it is more desirable to have output voltage sensing than current sensing. This is because there are wide variations in the current demanded by TTL circuits when they are switching from state to state. The current limit point would always have to be set rather high, and consequently only gross over-currents could be sensed. On the other hand, a circuit that senses when the voltage falls below 4.75 volts, the lower operating limit for 7400 series TTL, is quite useful. Suppose, for example, that you are operating near the 1 ampere limit of IC6; a brief current pulse could exceed this limit and the internal circuit of IC6 would then allow the output voltage to drop. Without voltage sensing this could easily go unnoticed and your circuit would malfunction.

In Fig. 1-30, D14 provides a reference voltage. R41 acts as a voltage divider and is set to 5.25 volts. R42 is another voltage divider and is set to 4.75 volts. IC1C and IC1D compare the output of IC6 to these voltages and, if the output moves outside the window from 4.75 to 5.25 volts, pin 1 or pin 2 will go high. This signal goes to R21 of Fig. 1-28 and eventually shuts down all the supplies.

Response Time

R21 and C3 determine the response time of the circuit. With R21 = 3k and C3 = 1 μF, the circuit responds to an overcurrent,

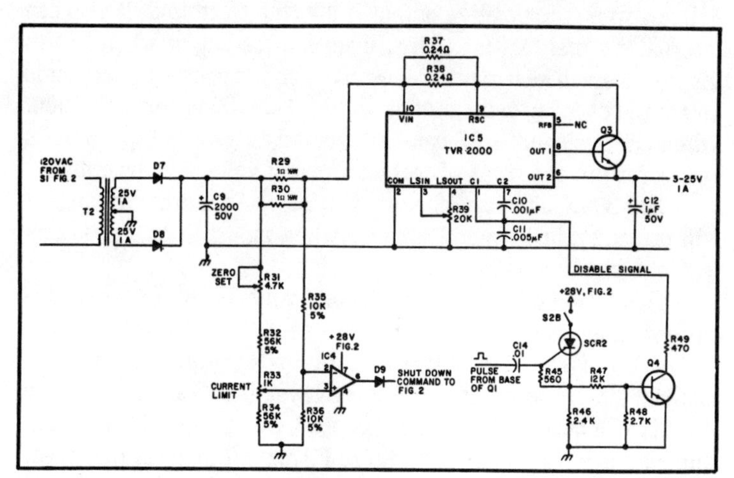

Fig. 1-29. Variable voltage power supply.

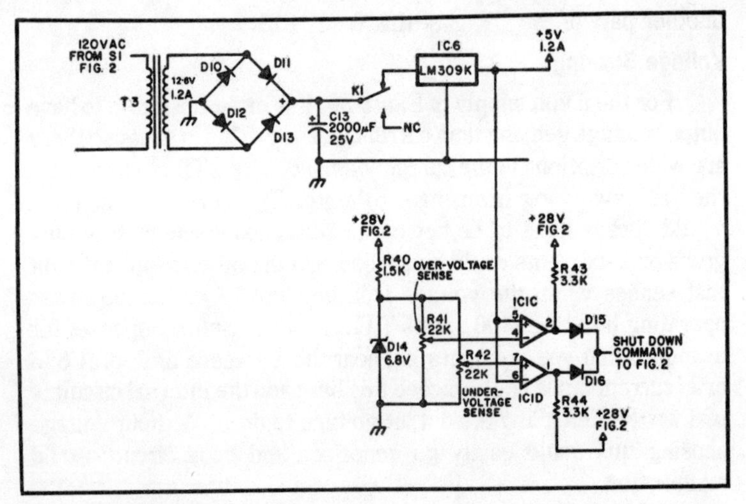

Fig. 1-30. 5 volt power supply.

overvoltage or undervoltage that lasts 3 milliseconds or more. K1 adds an additional 7.5 ms to the time required for the 5 volt supply to shut down. By reducing C3 to 0.1 µF, response time can be made as low as 0.3 ms. R21 can be increased to as much as 10 megohms if desired to lengthen the response time, but should not be reduced below 3k.

The Voltage Regulators

The 4194TK regulator is internally current limited at about 350 mA when the positive output is shorted to ground. It also has internal thermal limiting that will reduce the output when it gets too hot. A small heat sink is required when the operating current is 100 mA in each leg of the output. In Fig. 1-28, S3 is normally open. When S3 is closed, R27 can be adjusted to give +12, −6 volts output for the operation of certain types of voltage comparators.

The 309K also has current limiting and thermal limiting. It will provide a little over 1 ampere when mounted on a heat sink with the circuit shown.

The TVR2000 has been available for a number of years and is quite inexpensive.

In Fig. 1-29, the foldback current limiting option is not used. Instead, simple short circuit sensing is used. R37 and R38 set the short circuit current to a value of about 1.2 amperes. The relationship here is $R_{sc} \cdot I_{out} \approx 0.1$ volt, where R37 and R38 in parallel make up R_{sc}. R39 sets the output voltage. Q3 acts as a current booster and is mounted on a heat sink. C10 stabilizes the current limiting circuitry and C11 stabilizes the regulator section of IC5. Different values from those shown may be required to drive high capacitance loads.

Selecting Resistors

Resistors of 1% tolerance are best for R1, R3, R5 to R8, R10, and R12 to R14. This will make the final adjustments simpler and will keep tracking errors in R4 and R9 to a minimum. In Fig. 1-29, 5% resistors will suffice for R32, R34, R35, and R36, providing you choose them such that R32 ≤ R34 and R36 ≥ R25.

Regarding the tracking of R4 and R9: Since they form a tandem control, it is important that they both exhibit approximately the same resistance between their wipers and their ends for all rotations of the shaft. Failure to do so will mean that the positive and negative legs of the supply will trip at different currents. Several dual controls I bought did not track very well. If you want very good tracking, replace both R4 and R9 with a series of 5% resistors and use a two pole rotary switch to select the current limit you want as shown in Fig. 1-31.

Construction

All three supplies were constructed on a single 4″ × 5″ printed circuit board as shown in Figs. 1-32 and 1-33. Also, see

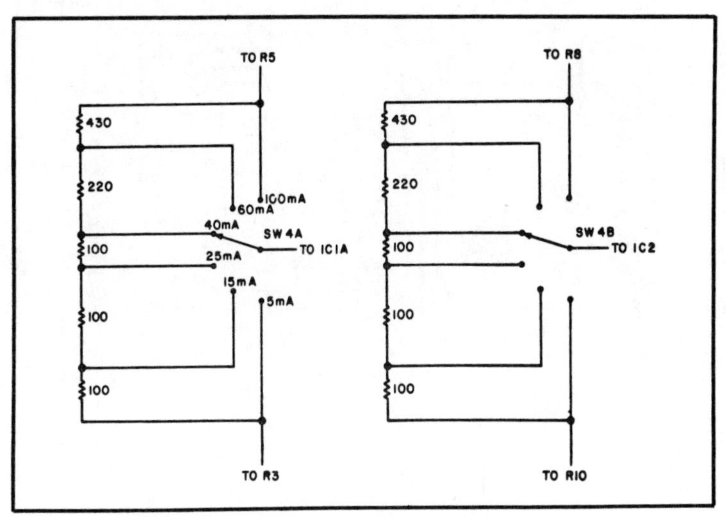

Fig. 1-31. Switch selected resistors replace R4 and R9.

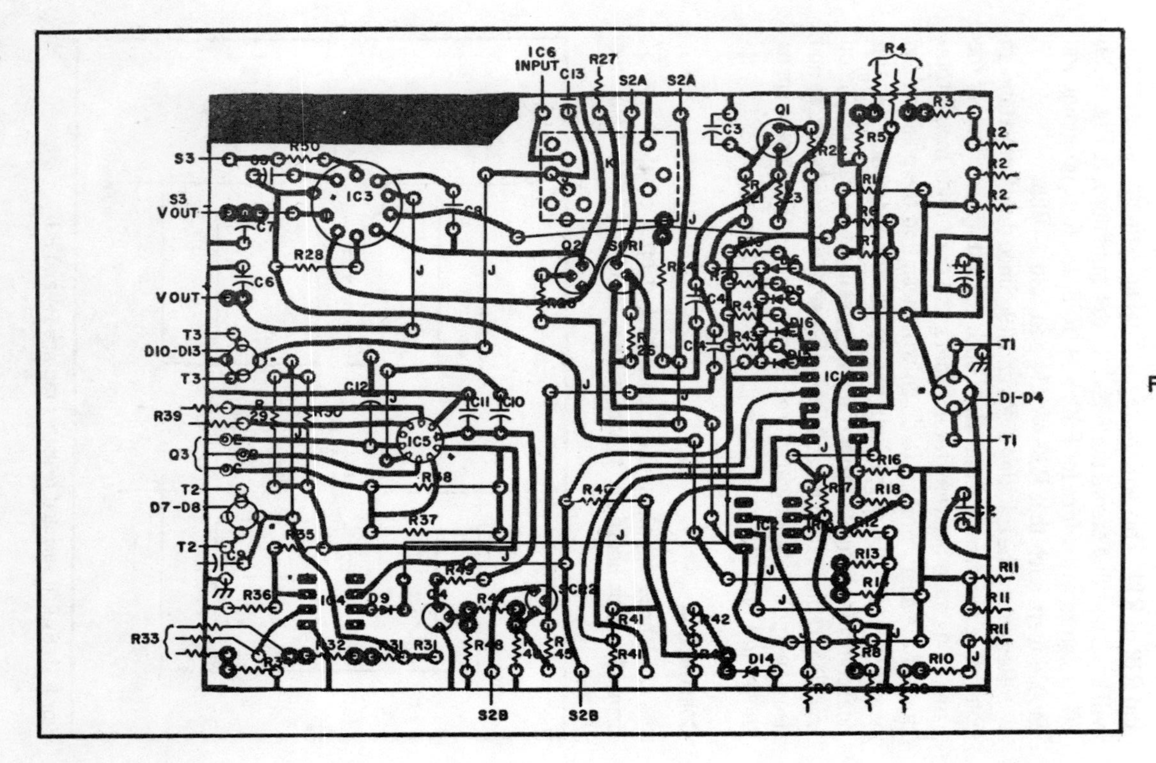

Fig. 1-32. Parts layout.

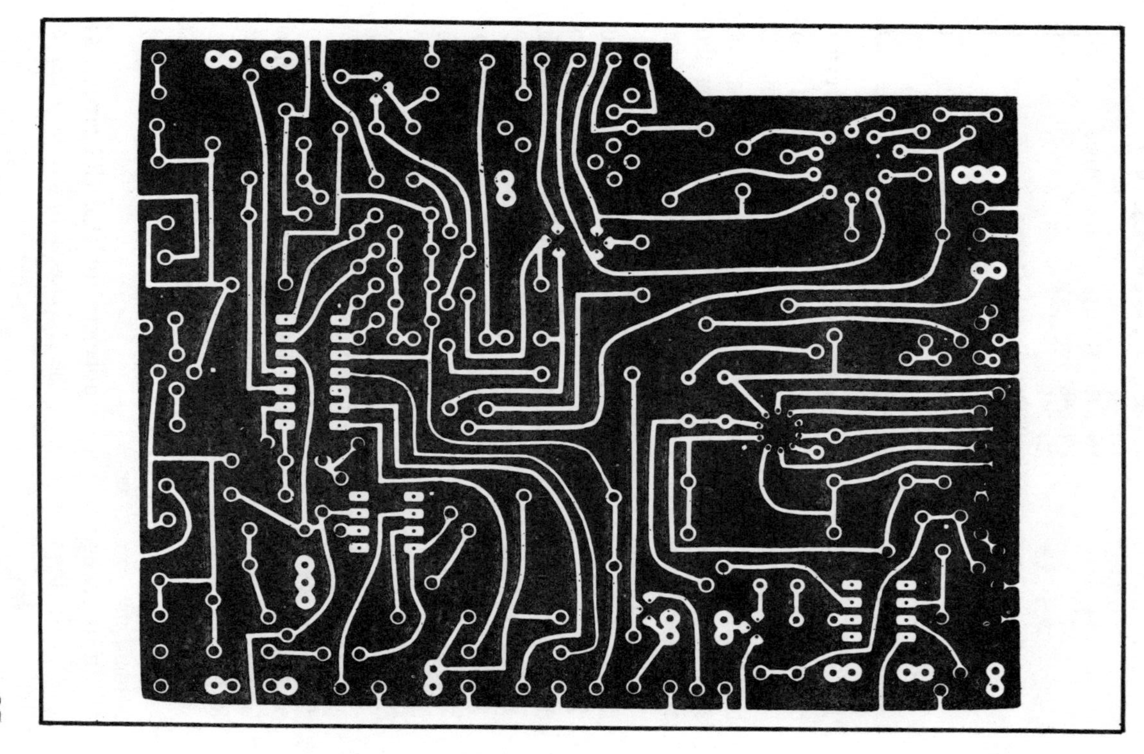

Fig. 1-33. PC board.

Table 1-2. Partial Parts List.

IC1	339
IC2, IC4	741
IC3	4194TK
IC5	TVR2000 (available at Poly Paks)
IC6	309K
D1-D4, D10-D13	2 A 100 piv bridge rectifier
D7-D8	half of 2 A 100 piv bridge rectifier
D5,6,9,15,16	1N4148
D14	1N957B 6.8 V, 0.4 W zener diode
Q1	2N2646
Q2	2N4249
Q3	MJE3055
Q4	2N5550
K1	ITT type 24A02C18A
R4,R9	dual section control; see text
SCR1,SCR2	C103B

Table 1-2. IC3 does not plug directly into the board; the holes in the board have been spaced out to assure clean etching. Solder a short wire to the outside of each pin of IC3; insert the wires into the PC board and solder. A piece of aluminum was bolted to IC3 as a heat sink. There are so many connections to the PC board from the external switches, controls, transformers, etc., that it was not possible to arrange for an edge connector on a board of this size; instead there are about 35 wires soldered at various points around the edge of the board and all are routed to one end of the board so that the board can be hinged outward from the chassis if parts on it need to be replaced in the future.

All components fit nicely on a chassis 10″ × 6″ × 2″ as shown in Figs. 1-34 and 1-35.

Final Adjustments

Switch S2 to reset. Leave S3 open. This disables the shutdown mechanism. Connect a high impedance voltmeter between pin 7 of IC1D and ground. Adjust R42 for a reading of 4.75 volts. Connect the voltmeter between pin 4 of IC1C and ground. Adjust R41 for a reading of 5.25 volts.

Set the wiper of R33 to the end closest to R32. Connect a voltmeter between pin 6 of IC4 and ground. Adjust R31 so that the reading just goes to zero.

If you are using a dual potentiometer for R4 and R9, proceed as follows: Set the wiper of R4 to the end closest to R3; the wiper of

R9 should then be at the end closest to R10. Connect a voltmeter between pin 13 of IC1A and ground. Adjust R2 until the reading just drops to zero. If you run out of adjustment with R2, interchange R6 and R7 and try again.

Connect the voltmeter between pin 14 of IC1B and ground. Adjust R11 until the reading just drops to zero.

If you elect to use the switched resistors in Fig. 1-30, proceed as follows: Set the switch in Fig. 1-30 to the 5 mA position. Connect a load between the positive and negative output terminals of the supply and adjust the output voltage so that the load draws 5 mA. With a voltmeter from pin 13 of IC1A to ground, adjust R2 until the voltage just drops to zero. If you run out of adjustment with R2, interchange R6 and R7 and try again. Connect the voltmeter between pin 14 of IC1B and ground. Adjust R11 until the reading just drops to zero.

ADJUSTABLE BENCH SUPPLY

How about constructing an adjustable voltage power supply that can have up to 1.5 amperes output with good load voltage regulation and full overload protection at minimal cost? Admittedly, a $5.00 estimate depends a lot on what parts are available from one's junk box, but for just a few dollars spent on a new IC, one can have the "heart" of a very versatile power supply.

An LM317 is an adjustable, three-terminal positive voltage regulator. Its simple external connections rather belie the complexity and performance features of the unit. As shown in Fig. 1-36 it has only simple in/out connections and a minimum of three

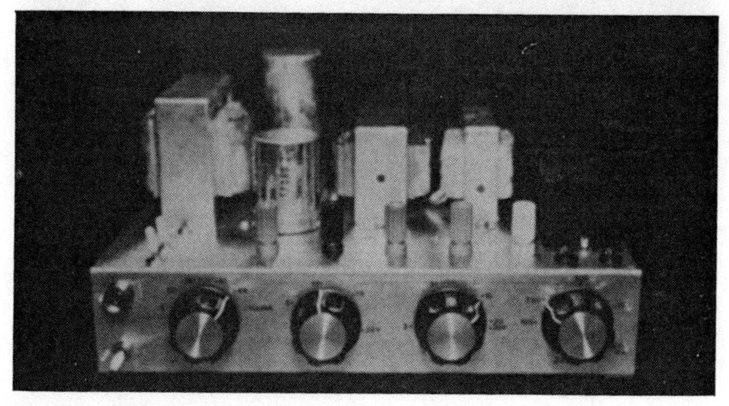

Fig. 1-34. The finished chassis.

Fig. 1-35. Bottom view of power supply.

simple external components are required. The output voltage is set by the ratio of two resistors, R1 and R2. By making R2 variable, one can adjust the output voltage to be any value from a few volts less than the dc input voltage to the regulator down to a minimum of about 1.2 volts output. Thus, if the input dc voltage were 40 volts, the output voltage can be continuously varied from about 37 volts down to 1.2 volts.

Although the output voltage is determined only by a resistor setting, the output voltage is regulated at *any* given setting. The regulation will be about 0.1% going from no load to full load (1.5 amperes, assuming the transformer/rectifier used for the dc input voltage handles this current). The LM317 is also overload and thermally protected. If the current limit is exceeded, such as by a short circuit, the LM317 will simply "shut down." If the regulator gets too hot, either because of excessive load current and/or inadequate heat dissipation, it will also protect itself. Although one can destroy the LM317 like any other IC, it is pretty hard to do with any sort of reasonable care.

The manufacturer suggests two additional capacitors (C2 and C3) be used, which may prove useful in some applications. C2 is used to bypass the adjustment terminal to ground to improve ripple rejection. This bypass prevents ripple from being amplified as the output voltage is increased. About 60 dB ripple rejection is achieved without this capacitor, but it can be improved to about 80 dB by adding it. A 10 mF or greater unit can be used, but values over 10 mF do not offer any significant advantage in further ripple improvement. The manufacturer particularly recommends the use of a solid tantalum capacitor type since they have low impedance even at high frequencies. An alternative is the use of the more readily available and inexpensive aluminum electrolytic, but it takes about 25 mF of the latter type to equal 1 mF of the tantalum type for good high frequency bypassing! C3 is added to prevent instability when the output load presents a load capacitance of between 500 and 5000 pF. By using a 1 mF bypass at the output (solid tantalum again or aluminum electrolytic equivalent), any load capacitance in the 500 to 5000 pF range is swamped and stability is ensured. Both C2 and C3 will not be required for many applications where the LM317 is being used with a specific load

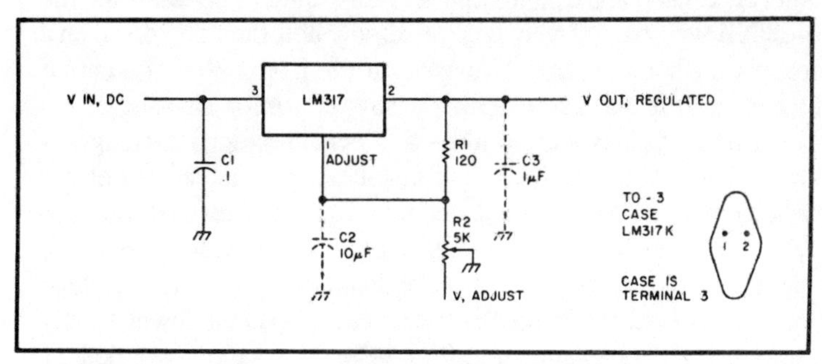

Fig. 1-36. Basic adjustable voltage regulator circuit using an LM317. Normally only three external components are needed, but C2 and C3 may be useful in certain situations as explained in the text.

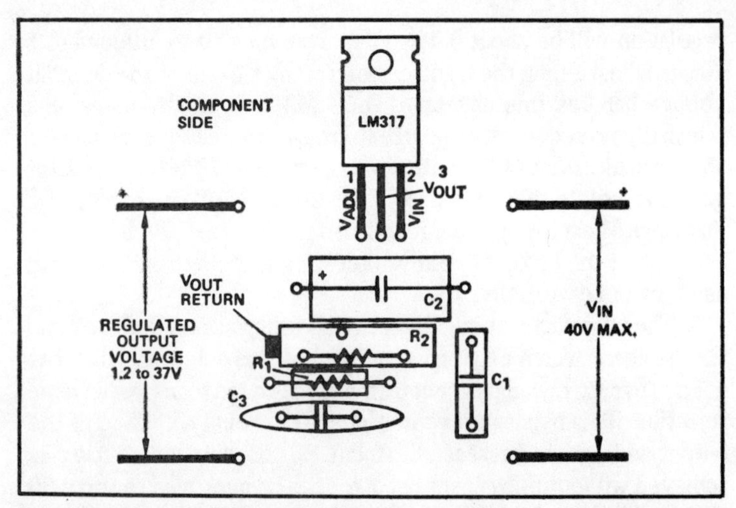

Fig. 1-37. This is a PC board layout for the regulator suggested by the manufacturer. R2 is shown as a multi-turn pot for ease of adjustment. The figure also shows the pin connections for an LM317 if it is obtained in the TO-220 plastic case.

circuit. But if the LM317 is used as the heart of a general purpose bench type power supply, they should be included.

Figure 1-37 shows a PC board layout and component placement diagram. This layout has been suggested by the manufacturer, but there is no need to follow it exactly as long as all of the external components are grouped around the regulator with solid short leads. Figure 1-37 shows the LM317 in a TO-220 plastic case which is designated the LM317T. Most amateurs will probably prefer to buy the LM317 in the familiar TO-3 metal case and, in this case, it is the LM317K. But, when using the unit, note an important difference as compared to the old LM309K. The case on the LM309K was ground so one could simply bolt the thing down on a chassis for heat sinking. The case on the LM317K is the output terminal, so it must be properly insulated from a chassis.

Various power supply ideas and considerations can suggest themselves for the LM317. For instance, R2, instead of being a variable resistor, can be replaced by switchable fixed resistors to obtain some of the commonly used supply voltages such as 6, 9, 12, 15 volts, etc. This idea, plus a continuously variable output voltage position, is featured in the practical realization of a power supply using the LM317 as shown in Fig. 1-38. This supply will deliver fixed output voltages of 6, 9, 12, and 15 volts (depending upon how the trim potentiometers are set), plus a continuously variable output of 1.2 to about 24 volts. All outputs can deliver at least 1.5 amperes with the components specified. The supply is simple to build in any size metal enclosure suitable for the components used. The only precautions to observe are to firmly heat sink the LM317 to one side of the metal enclosure and to keep the 0.1 mF capacitor going from pin 3 to ground, the 10 mF capacitor going from pin 1 to ground, and the 120 ohm resistor going between pins 2 and 1, *all* connected directly at the LM317 terminals. The other components may be mounted wherever it is convenient to do so.

The zener diode/resistor/LED combination at the output of the supply serves as a crude but useful voltage output indicator without having to build a regular voltmeter in the supply. The LED just starts to glow when the output voltage is about 9-10 volts (depending on the tolerances of the components used). The 1k resistor is adjusted so the LED just glows fully when the *maximum* output voltage is reached. So by using the fixed output voltage positions (which are adjusted using a good vom) and watching the LED, one can obtain a fairly good estimate of what the variable output voltage is set for.

VERSATILE POWER SUPPLY

Whether you need a supply to charge batteries, run a portable tape recorder or radio, operate relays, run your mobile FM rig in the house, or power op amps or logic, this circuit definitely de-

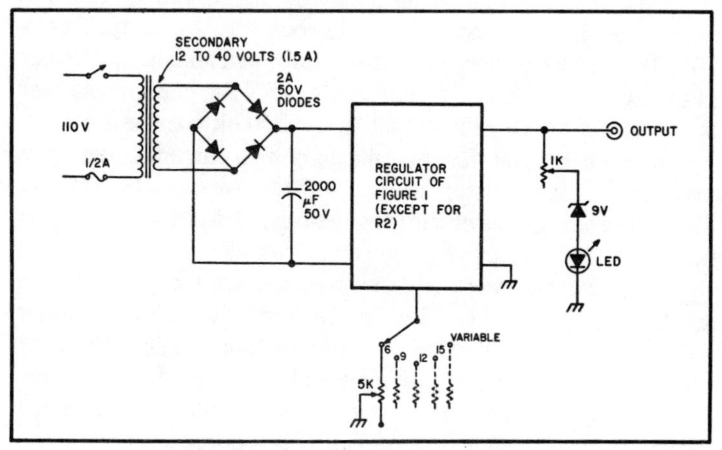

Fig. 1-38. A complete power supply using the LM317. The switch simply selects different 5k ohm pots which are set for 6, 9, 12, 15 and a variable voltage output. The latter 5k pot is front panel mounted. The function of the LED is described in the text.

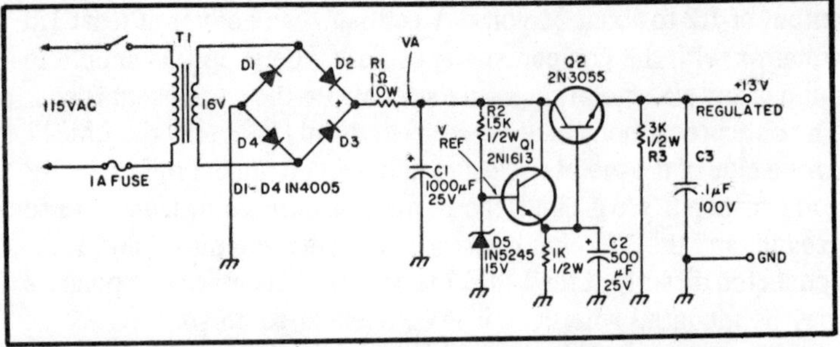

Fig. 1-39. Regulated +13 volts as 2 amps using NPN transistors.

serves a try. It's simple, uses a minimum of easily obtainable parts and, best yet, it works.

The basic circuit (Fig. 1-39) is a simple two transistor regulated supply. One transistor, Q1, acts as a reference voltage source and the other, Q2, acts as a series pass regulating element.

The circuit is the same no matter what the output voltage. Only the transformer, zener diode, resistor R2, and possibly Q1 and Q2 have different values.

Circuit Theory

The circuit consists of three sections—the transformer, rectifier and filter being one, the voltage reference another and the series pass regulator transistor the third.

Pick your transformer for a slightly higher voltage than you wish to regulate. For example: for 12 volts regulated output, use a 16 to 19 volt transformer (a 6 volt and a 12 volt filament transformer in series). For 5 volts regulated output, a 6 V filament transformer is used. And for 15 volts output, use a 24 volt transformer.

If you use a transformer capable of high current output, you may need to put a resistor in series with the output of the rectifier. This prevents the surge current, generated when the supply is turned on, from destroying the rectifier diodes.

Of all the different rectifier circuits in use today, I prefer to use the full wave bridge. The bridge circuit has a higher output voltage than the standard full wave rectifier and a higher frequency ripple than the half wave, making it easier to filter. However, any type rectifier, as well as the conventional voltage doublers and triplers, can be used with no circuit degradation.

For output filtering, I have found that using a 1000 µF capacitor for C1 provides adequate filtering. The more capacitance, the more filtering, so you can increase this value if you wish.

The regulator also acts as a capacitance multiplier so that the total capacitance is the capacitance of C1 plus the capacitance of C2 multiplied by the gain, h_{fe}, of Q2:

$$C_T = C_2 (h_{fe}) + C_1$$

Resistor R2, the zener diode and Q1 form a voltage reference circuit. Set the reference voltage to about 1.5 V above the desired output voltage. The reference voltage is higher because there are two diode voltage drops between this point and the output. To determine the value of the zener current limiting resistor, R2, you must first know the gain of transistors Q1 and Q2.

The dc gain of Q1 and Q2 can be found in a transistor specification manual. Gain, sometimes called Beta or h_{fe}, is the ratio of the collector current to the base current that caused it. For example: for a transistor with a gain of 20, 1 mA of base current causes 20 mA of collector current.

Now, knowing the transistor gains and the desired output current, you pick a value for R2 by calculating how much current must flow from the base of Q1 to make a corresponding amount of current flow from the base of Q2, to cause the desired output current flow into Q2.

For example, we wish to have a regulated output voltage of 13 V at 2.0 amps. The gain of Q1 is 70 and the gain of Q2 is 20. The voltage at point A is 20 V. To cause 2.0 amps of current to flow from the base of Q2 must be 100 mA.

$$I_b = \frac{I\ out}{gain} = \frac{2.0\ A}{20} = 100\ mA$$

And similarly, to cause 100 mA of current flow into Q1, the base current must be

$$Ib = \frac{100\ mA}{70} = 1.4\ mA$$

If you use a 15 V zener diode, the voltage drop across R2 is 5 V, so from Ohm's Law:

$$R = \frac{E}{1} = 5\ V/1.4\ mA = 3.7k$$

Let R2 be 3.3k. The reason for this lower resistance is that most transistors have less gain than that listed in the specification book, so by using a smaller resistor more current will be available, making up for possible low transistor gain. This also gives you a safety margin, in case you need just a bit more current than you thought.

In picking transistors for Q1 and Q2, not only must you pick a transistor with suitable gain but you must also choose it for its type (NPN or PNP), emitter-collector breakdown voltage (BV_{ceo}), collector current (I_c), and power dissipation.

All of these circuits use NPN transistors. If you wish to use PNPs, reverse the polarity of the rectifier output, the filter capacitors and the zener diode. Isolate everything from ground. If you require that the negative lead be grounded, ground the emitter of Q2 (see Fig. 1-40).

BV_{ceo} is the voltage at which the collector to emitter junction breaks down. For Q1 this voltage rating must be high enough to stand the difference in voltage between the rectifier output and ground, and for Q2 the rectifier output and the regulated output.

The collector current (I_c) rating of each transistor is the maximum continuous collector current that the collector to emitter junction can safely pass. For Q2, this is the total output current which you require from the supply.

Power dissipation is the maximum amount of power that the transistor can dissipate before it is destroyed. The power dissipation rating is usually given for an ambient case temperature of +25° C. If you heat sink the transistor (which I recommend for Q2), you can exceed this dissipation by an amount which depends upon how well your heat sink dissipates the power.

The output circuit consists of R3, a bleeder resistor chosen to allow a couple of milliamps of current to flow, and capacitor C3, which acts as a high frequency filter to keep any zener noise or voltage spikes out of solid state equipment.

Construction

The circuit layout is not critical and almost any type of configuration can be used. I have built the supply both on a piece of

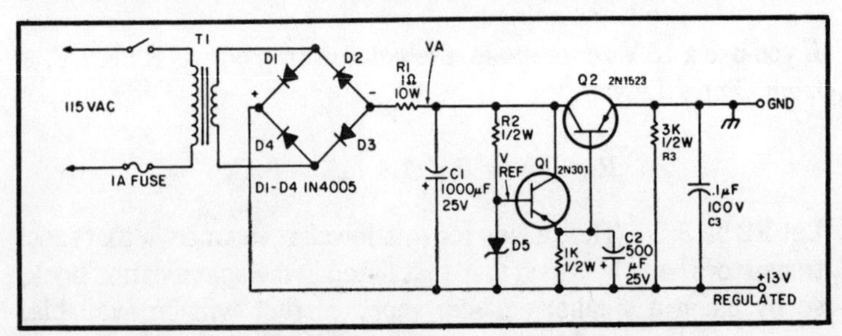

Fig. 1-40. Regulated +13 volts at 2 amps using PNP transistors.

pegboard with the device leads serving as hookup wire, and on a printed circuit board. With the circuit board shown in Fig. 1-41, the electrolytic capacitors are mounted external to the board due to

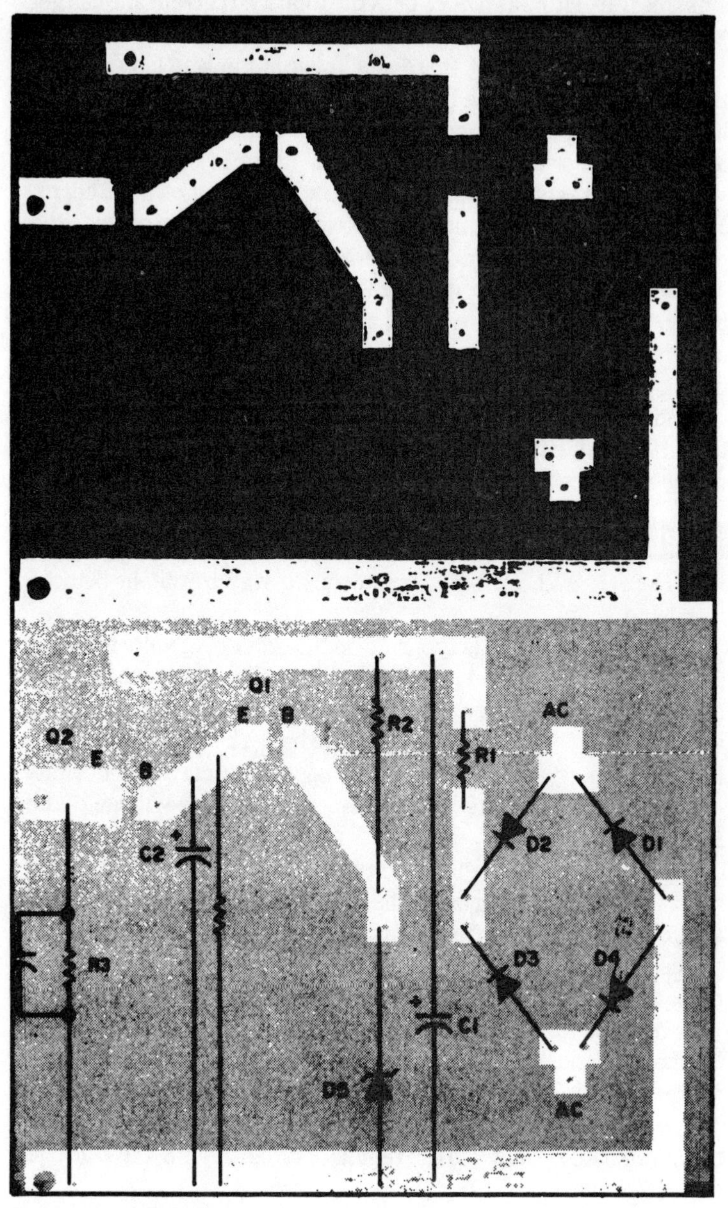

Fig. 1-41. Foil side view.

their large physical size. This figure also shows Q2 mounted on the board. While this will work for low power applications (one example is the supply running my TR22), I would suggest that you heat sink this transistor, as it does pass the total load current and can get warm.

You will notice that there is a break in the land coming from the positive output of the bridge. This is where R1, which must be heat sinked externally from the board, is connected. If you do not use R1, connect a jumper in its place.

I used a resist marking pen to draw the circuit on a PC board. You can do the same or make a photographic negative and use the photoresist method of making a board. The board in Fig. 1-41 is shown full size. I have mounted all of my supplies inside Bud boxes, and, as there are quite a variety of Bud boxes, no two supplies look alike. The only precaution is to mount the heat sinked transistor, Q2, where its case cannot be accidentally shorted to ground. If possible, mounting it on an attachable heat sink and mounting the heat sink inside the box (with a few ventilation holes) will work fine.

INEXPENSIVE POWER SUPPLY

Here is a cheap but very adequate way to obtain 5 to 15 regulated volts.

Output regulation is typically on the order of .1 volts for loads from a few milliamps to 4 amps or so, depending on the components used. This circuit is very noncritical and easy to get going. With it, just about any reference voltage may be had. The only requirement is that the unregulated dc voltage supplied to the regulator circuitry must be few volts greater than the reference voltage of the zener.

The transformers I used were two 6.3-volt filament junk box specials with the primaries connected in parallel for 110-volt primary voltage. 220-volt primary may be used for wiring primaries in series. Caution must be used to keep the phasing correct. If you get no output, change the pairing of wires in the primary.

With the full-wave circuit described in Fig. 1-42, the unregulated output should be about 17 volts or so. This should be adequate for the regulation of 12 volts.

If you want 14 or 15 volts regulated, you may need to increase the unregulated voltage to the regulator circuit. A full-wave bridge should do this.

However, the pass transistor (or Darlington, whichever you

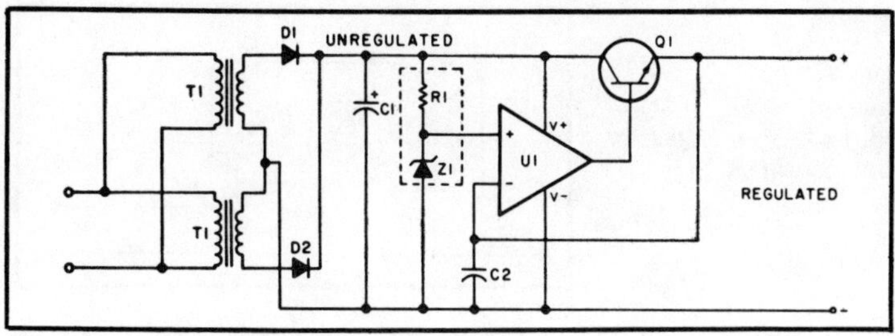

Fig. 1-42. Full wave circuit.

decide to use) will run cooler with the lowest possible unregulated voltage on it to maintain good regulation. To determine what is adequate, try the full wave first and measure the regulated voltage output. If it varies more than a volt on full load from the supply, an increase in the unregulated voltage is needed and a full-wave bridge should be used.

U1, the op amp, can be any type of 709, 741, etc. It is operating with essentially an open-loop gain, and its job is to bias the Darlington or pass transistor. When the unregulated voltage wants to drop, the noninverting input referenced by the zener diode and the inverting input of the op amp "see" a difference voltage and amplify it, supplying the base additional bias which enables Q1 to amplify more and hold the regulated output constant by pulling the voltage back up.

The reference voltage in the noninverting input of the op amp is determined by the zener reference. I use a 12-volt zener and a regular silicon diode, connected as illustrated in Fig. 1-43.

The pair gives about 12.6 volts as a reference—12 volts for the zener and about a .6 drop across the silicon diode. The zener alone could be used for just 12 volts, or more diodes could be used for .6-volt increases for each additional diode used with the zener.

C1 is any old filter capacitor—the more capacity, the better. And, of course, it is rated for at least the unregulated voltage.

Q1 is the main factor in determining the current that can be taken from the supply. The larger the maximum collector current

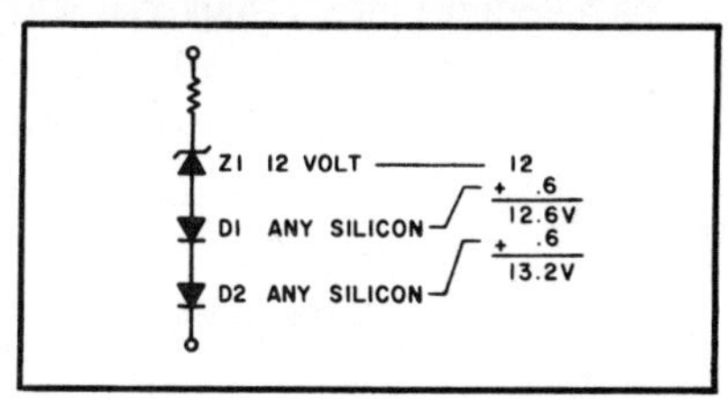

Fig. 1-43. Zener reference circuit.

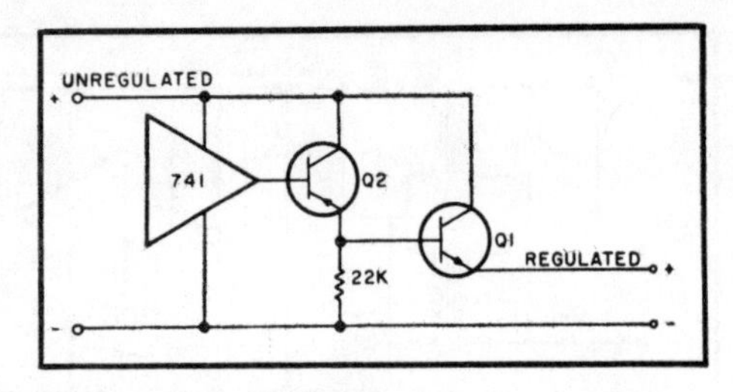

Fig. 1-44. Driver transistor added to circuit.

of Q1 is, the more the current that can be taken from your machine. A Radio Shack power Darlington (TO-3 case) for $1.98 was used, and the supply is good for about 4 amps. With a high-gain power Darlington, the output of a 741 is all that is needed to drive it. If more current demand is required, a larger Darlington or regular power transistor can be used. However, if the beta of the transistor is fairly low, a driver is needed between the 741 and the pass transistor. Just about any PNP transistor will do. It should be connected as shown in Fig. 1-44.

If lots of current is wanted, say 10 amps, a large pass transistor is needed. Pass transistors this large typically have betas of 60 or less, which will require a driving current of about 1/6 amp, which is far too much for the 741 and most other op amps. Therefore, a driver transistor with a high gain (beta) should be used to allow the small output of a 741 or similar op amp to drive anything.

If the op amp heats up to the touch, a drive transistor should be used. Q1 should be well heat sunk, and, of course, the transformer used should be capable of handling whatever current you want at the load.

C2 is a .001 to help keep any rf out of the regulator if the supply is to be used for your 2m radio or the like. If used with digital projects, a 10 μF filter capacitor can be placed across Z1, if noise is a problem, and/or a slightly smaller R1 used to move Z1 from the knee region where noise occurs. See Table 1-3.

There are many other regulators available on the surplus market, but I have chosen to describe only the LM309 and μA7800. However, if you spend a little time studying the published data sheets on other units, I think you will find that most of the circuit design procedures I have outlined here are directly transferable to use with other three terminal regulators. See Fig. 1-45.

Parts Selection

The specifications for the regulators I am describing are shown in Table 1-4.

Table 1-3. Parts List.

T1	6.3 V filament transformer or equivalent
D1, 2	about 50 piv at desired current, plus 50%
C1	20 to 20,000 uF, 25 volts
R1	470 to 820 Ohm, ¼ Watt
Z1	12 V zener, or desired value from 2.4 to 33 volts
U1	741, 709, etc., op amp
Q1	power Darlington or 100-Watt or more pass transistor, NPN
Q2	any PNP: 2sc710, etc.
C2	.001 uF, 50 V

Figure 1-46 shows a practical power supply circuit, using any one of the regulators in Table 1-4. It needs no additional parts. It will work with all of these three terminal devices by simply varying a few component values and ratings.

Figure 1-47 is also practical: If you cannot locate a center-tapped transformer, replace T1 and D1-D2 from Fig. 1-46 with the transformer and bridge rectifier circuit shown in Fig. 1-47. The full wave rectifier is preferred, because of the smaller diode drop of the two diodes in Fig. 1-46 than of the four in Fig. 1-47, but the difference is not that crucial.

Now let's assign some values to the components in those circuits. Use Table 1-4 to determine what regulator to use for IC1.

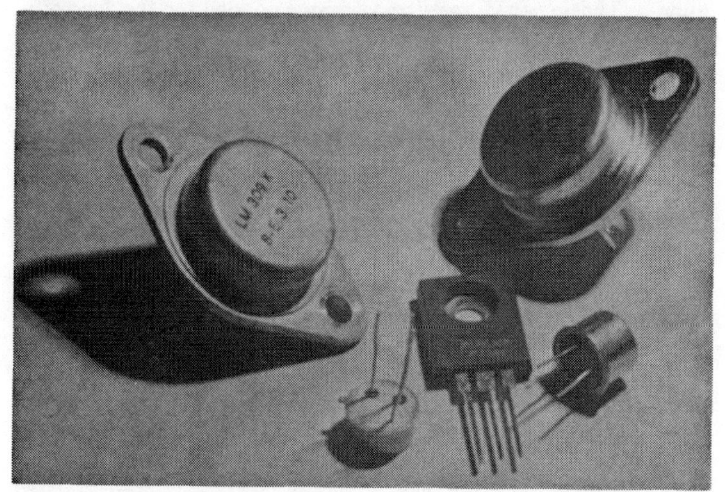

Fig. 1-45. Three terminal IC regulators, available in a wide variety of voltage ratings, current capacities, and package styles, are easily used to make high quality power supplies.

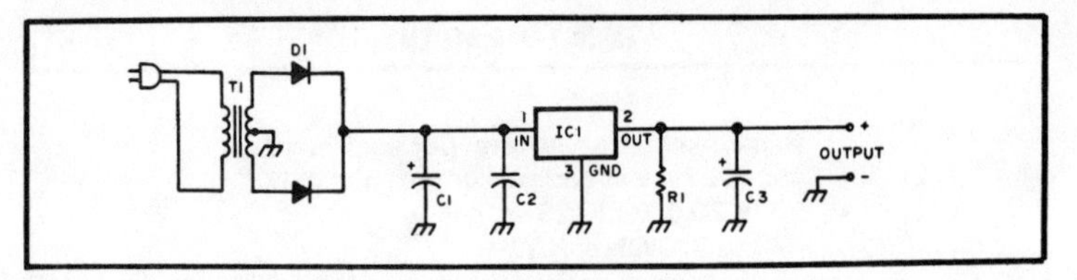

Fig. 1-46. Here is a practical diagram for a power supply of any voltage, needing no other parts. Use this schematic with center-tapped transformers in a full wave rectifier configuration. Regular voltage/R1 value: 5/1.0kΩ, 12/2.2k, 15/2.7k, 18/3.3k, 24/4.7k.

Make your choice on the basis of voltage rating and current capacity.

THREE TERMINAL REGULATORS

Refer to the graphs in Fig. 1-48. There are five graphs for use in determining the T1 transformer voltage and C1 filter capacitance for five separate one amp regulators. The transformer voltage is for a center-tapped transformer with a full wave rectifier, as in Fig. 1-46. If you use the bridge rectifier configuration from Fig. 1-47, halve the transformer voltage but remember to *add one volt* to compensate for the voltage drop of the extra diodes. If you use the schematic from Fig. 1-46, the current capacity of T1 need only be half an amp—or half of what you want the completed supply to deliver. The lower voltage transformer needed for the less efficient bridge of Fig. 1-47, however, must be rated the full amp.

Filter capacitor C1 should be an aluminum computergrade electrolytic. If you have any question concerning interpolating a capacitance to use with any given transformer, round *up*. This will help to make up for component tolerances and line voltage fluctuations. Since readings off the graph are minimum allowable capacitances, you might want to use a larger value capacitor anyway.

Table. 1-4. Regulator Specifications.

Device	Current	Voltage	Package
LM309H	200 mA	+5 volts	T0-5
LM309K	1.0 Amp	+5 volts	T0-3
uA7805	1.0 Amp	+5 volts	T0-220
uA7812	1.0 Amp	+12 volts	T0-220
uA7815	1.0 Amp	+15 volts	T0-220
uA7818	1.0 Amp	+18 volts	T0-220
uA7824	1.0 Amp	+24 volts	T0-220

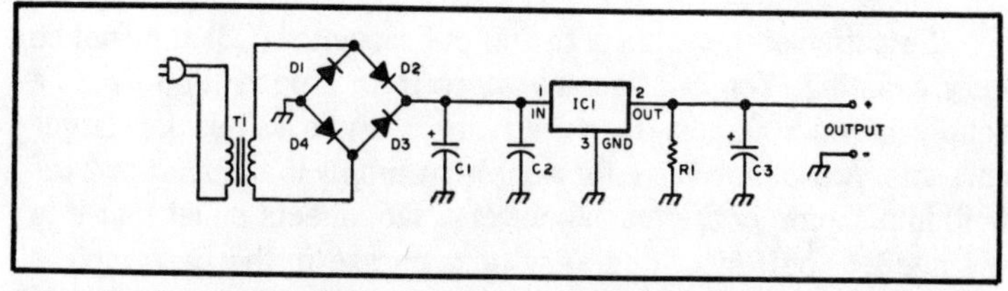

Fig. 1-47. If you cannot locate a center-tapped transformer, substitute this schematic for that of Fig. 1. Keep in mind the additional design considerations explained in the text necessitated by the extra diodes.

Since the graphs were calculated using one ampere of drain, if some fraction of an amp is to be drawn, the C1 value can be reduced by that fraction. For example, if only 100 mA is needed, then only one tenth the capacitance is necessary, or any other such fraction. It works the other way around, also—if you make a ten ampere supply (with a pass transistor), you will need ten times the filter size.

Exercise prudence when calculating capacitor voltage ratings. Use at least one and a half times, but preferably twice, the rms transformer voltage across the capacitor.

Do not omit C2. It is an rf bypass capacitor, especially necessary if the power supply is to be used in an area of strong rf such as a ham shack—but still needed even if used elsewhere. C2 helps to absorb the spikes and glitches on the line that are too sharp for C1 to respond to. It can be anywhere from around .10 μF to .33 μF—.22 μF is a happy medium.

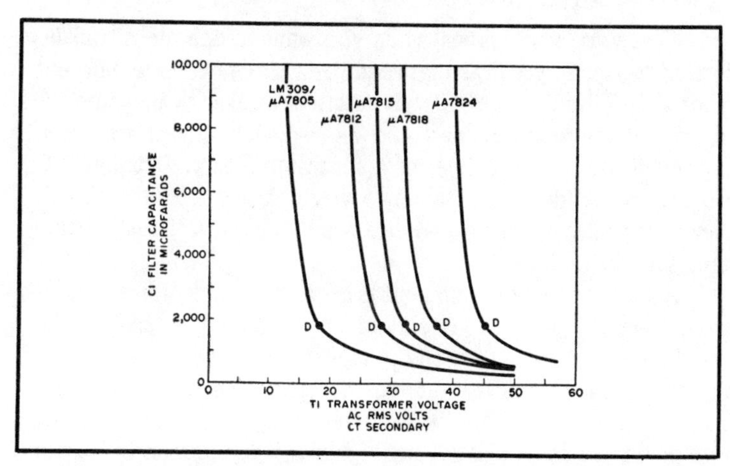

Fig. 1-48. Use this graph for determining C1 capacitance and T1 voltages. The points labeled "D" are the dissipation limits explained in the text.

Selection of the value of the output capacitor (C3) need not be very exacting. You can use anywhere from around .005 μF to a couple of microfarads. Generally, use larger values for larger currents. A common value for a one amp supply is about a 1 or 2 μF tantalum. Some of the manufacturers' data sheets insist that it is not needed, but, since the very next phrase in the paragraph is always that its presence "helps to improve transient response," I would not suggest omitting it.

These regulators do not work well at low current drains—around 5 mA and under. Therefore, R1 has been calculated to draw about 5 mA from the supply. A half or quarter watter is sufficient. A pilot light (LED or incandescent) could be wired in to serve the same purpose. If the supply is being built into a project so that the drain is never allowed to go below 5 mA, R1 could be altogether eliminated.

The choice of rectifiers for D1 through D4 is not at all critical. The graphs were complied using 1.3 volts as the forward diode drop. This figure is based on the 1N4000 series silicon one amp rectifiers. It should be close enough for interchanging with any of their surplus equivalents or replacements. 1N4002 or 1N4003 diodes would be a good choice for any of the transformer voltages on the graphs. Some of the lower voltage units can even get by with 1N4001s. Just make sure that you don't exceed the diodes' piv ratings.

Automotive Applications

There are many times when you want to operate a transistorized device in your car that requires a supply voltage different from that of the standard 12 volt automotive electrical system. A higher voltage requirement would take a power convertor exceeding the scope of this article, or a separate battery. But these IC regulators are ideal for obtaining lower voltages from the 12 volt electrical system. Figure 1-49 is a supply circuit tailored specifically for use in a car.

An automotive environment is however, very prone to transience. Special precautions should be exercised to protect the regulator. Positive spikes on the input line can very easily cause that input terminal to go above the maximum allowable input voltage for the IC. Therefore, the zener diode (Z1) has been put into the circuit to help shunt any higher spikes to ground. You can use whatever zener you have in your junk box in the 24 to 30 volt range.

The rectifiers are absent in this version, but note that the filter capacitor is still there. Keep it there—your car's alternator puts out an awful dc waveform. Filtering is still needed.

All of the other parts are the same as the other circuit diagrams.

Construction Considerations

The construction considerations you need concern yourself with when using these IC regulators are not very numerous. But those that do exist should be carefully observed.

There are definite reasons for the endpoints of the graphs in Fig. 1-48. *Do NOT extrapolate transformer voltages from either end of the graphs.*

These regulators will only work when an input voltage between certain limits is present at the input terminal. The dropout voltage, or level below which the unit will not operate properly, is usually two volts greater than the rated output voltage. Thus the LM309 needs an input of at least 7 volts to work correctly. If the input is too low, it will not regulate at the desired voltage and the ripple will be excessive.

The input level to these regulators cannot be too high, either. The upper limit is almost always 35 volts, except in some of those regulators with a high output voltage. For example, the μA7824 will take an input of up to 40 volts, but the other units in the μA7800 series (and the LM309) are limited to 35 volts. This figure includes ***peak*** ripple voltage. If the input exceeds the absolute maximum input rating, ***permanent damage may result.***

There is another reason for keeping the input voltage as low as possible. The dissipation of the devices in the T0-3 or T0-220 packages must be limited to around 5 to 10 watts, depending upon heat sink efficiency. (The LM309H, in its T0-5 metal can package, is limited to around 1 or 2 watts power dissipation.) In this case, exceeding the junction temperature limit, for whatever reason,

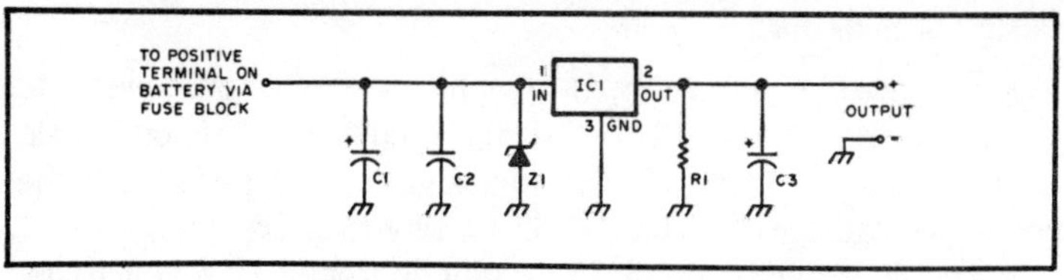

Fig. 1-49. This is a supply tailored for use in the transient-prone environment of a car. Zener Z1 can be anywhere from around 24 to 30 volts. All other parts are as in Figs. 1 and 2.

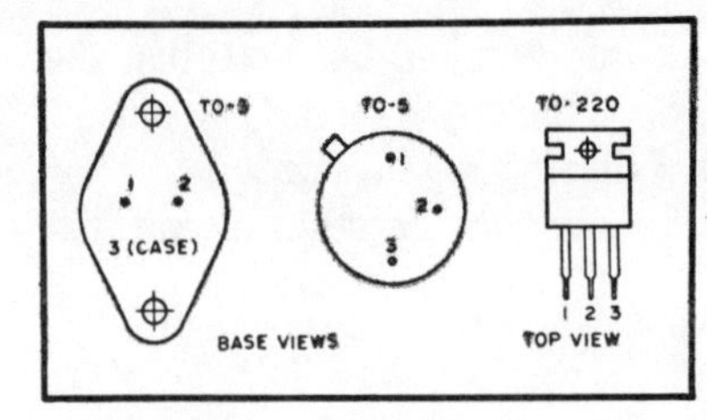

Fig. 1-50. Here are the pin connections for the T0-5 metal can, T0-3 metal power, and T0-220 plastic power packages.

will only trigger the internal protective features—thermal shutdown and current limiting—causing no permanent damage. This is especially important if you want to draw a full ampere from the supply. Although the dissipation capacity is still a function of heat sink efficiency, 20 watts is the absolute maximum dissipation regardless of how much heat you remove from the case.

For these reasons I have included points, labeled D (for dissipation limit), on the graphs in Fig. 1-48. If you want to draw high currents from the IC, use a transformer voltage to the left of (above) points D and attach the regulator to an adequate heat sink with liberally applied silicone grease. The Signetics LM309 data sheet indicates that a typical commercial heat sink, the Wakefield 680-75, should be sufficient to allow the T0-3 package of the LM309K to dissipate from 5 to 7 watts at room temperature before thermal shutdown is triggered. Use this as a guideline for determining your own heat sink requirements. Keep in mind that my placement of the dissipation limits in Fig. 1-48 is in no way absolute. Dissipation capacity is very dependent upon heat sink size and efficiency. Simple experimentation should be used to determine what size is needed for your specific application.

Refer to the base diagrams in Fig. 1-50 for finding the pin connections of the T0-5 metal can, T0-3 metal power, and the T0-220 plastic power packages. Follow normal wiring procedures, with no other special precautions except to make sure that you use large enough wire for the current you intend to draw.

Design Calculations

For those of you who want to apply my design procedures to other three terminal regulators, or just simply want to know a little of the theory behind this shortened design method, here are all the necessary calculations and a little on why they are used:

You do not get exactly 12.6 volts dc from a 12.6 volt transformer. Nor do you get 6.3; in fact, there is no such easy one-to-one relationship. Rather, you get a fluctuating dc waveform whose amplitude varies all the way from zero to a peak value of $\sqrt{2}$ (or about 1.414) times the transformer voltage.

As you increase the size of the filter capacitor, the ripple stabilizes. The waveform approaches an inverse sawtooth—an almost vertical charge time, followed by a nearly steady linear discharge rate, all repeating every 8.333 milliseconds (cycling at 120 Hertz). The peak value of the waveform remains where it was, but the amount of ripple subtracted from it decreases. Approximating a sawtooth will be close enough for our purposes.

Since specific transformer voltages are a lot easier to come by than are capacitor sizes (especially if you rewind them yourself—but that's another whole story), we will calculate the necessary transformer size needed with a given capacitor. Let's start by taking that capacitor size and calculating its ripple per given current drain:

$$V_r = \frac{81}{C}$$

where V_r = peak-to-peak ripple voltage, I = current drain in mA, and C = filter capacitance in μF.

We now have the amount of ripple, the diode drop, and the dropout voltage of our IC regulator. The sum of these voltages equals our minimum necessary peak transformer value. The *minimum* rms voltage that we can use therefore is:

$$V_t = (1.414)\ (V_d + V_r + 1.3)$$

where V_t = transformer voltage, volts center-tapped (rms), V_d = regulator dropout voltage (usually 2 + V_{out}), and V_r = the ripple voltage we calculated above.

Since this equation is for use with a center-tapped transformer and a full wave rectifier, here is a modified formula to use for Fig. 1-47. It is not simply half of the full wave value, so use this modified formula instead of trying to work it out of the other one:

$$V_t = (.707)\ (V_d + V_r + 2.6)$$

where all variable names are the same as before.

If you are using other rectifiers and have more accurate data on them, you can insert your diodes' forward drop in place of the 1.3 and 2.6 that I use for 1N4000 series rectifiers. Remember that, whatever the *average* current, the peak value is much higher at specific instances. Currents above ten amperes can be in a one amp power supply! Keep this in mind if you calculate your own rectifiers' forward drop.

Fig. 1-51. This breadboarded supply was assembled with an LM309K for use in testing TTL circuits.

Round up your final answer. This will incorporate a safety margin to allow for component tolerances and line voltage fluctuations. A ten or twenty percent round up margin could safely be used, providing that you do not approach the absolute maximum input ratings of your device. Your final calculation should be a check to make certain that the instantaneous voltage, with the actual component values you have decided to use, never exceeds the capacitor or regulator input ratings. When you round up, all that you in fact are doing is improving the line regulation at only a slight trade-off in size and efficiency.

The LM309 regulator (both the H and K package styles) has been designated to be worst case 7400 TTL compatible. Barring thermal shutdown or current limiting, their output voltage will never deviate outside the 4.75 to 5.25 volt range that 7400 TTL logic requires—even counting line, load, and thermal regulation summed worst case.

The LM309H is therefore a prime candidate for on-card regulators for logic circuits needing less than 200 mA—and it is in fact somewhat unique in this respect. The larger package styles, although still small enough for PC board mounting, are more conducive to chassis mounting. See Figs. 1-51 and 1-52.

THE 723 VOLTAGE REGULATOR

Whenever you need a regulated power supply for the test bench or for that special IC project you're working on, consider the 723 integrated voltage regulator. This handy little device is inexpensive, simple to use, and offers a lot of capability in a small package. With it, you can build a quality regulated power supply using very few parts. Read on and I'll show you how easy it is to custom design your own regulator around the 723 to suit your particular application.

Before we begin, let's look at a few characteristics of the device itself. The 723 operates with a rectified and filtered input voltage in the range of 9.5 volts to 40 volts. The output voltage is adjustable from 2 volts to 37 volts with .01% line and load regulation.

In case the terms "line regulation" and "load regulation" are unfamiliar to you, line regulation is the percentage change in the regulator output voltage for a change in the input voltage; load regulation is the percentage change in the regulator output voltage for a change in the regulator output current. Ripple regulation, which is defined as the ratio of the peak-to-peak input-ripple voltage to the peak-to-peak output-ripple voltage, is typically 45 dB.

As you can see, the regulation characteristics of the 723 are excellent. Note, however, that the 723 has two disadvantages that

Fig. 1-52. Here is an example of an on-card regulator using the LM309H. The circuit is a TTL heads-tails game, taking its power from a 9 volt battery.

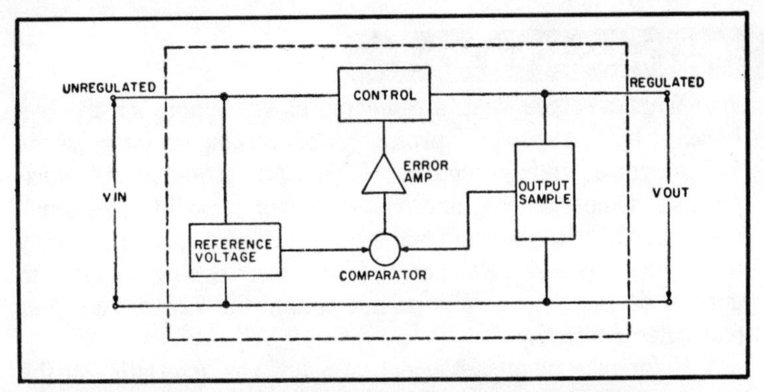

Fig. 1-53. Function block diagram of a series voltage regulator.

we must deal with. The first of these is that this IC can provide a maximum output current of only 150 mA. This limitation can be overcome by adding a single external pass transistor, as we will see later. The second disadvantage pertains to the value of the internal reference voltage of the 723, but I'll discuss this later and show you how to overcome it as well.

To illustrate how easy this device is to use, let's look at the 723 internal structure and then design a few regulator circuits with it. Figure 1-53 shows the essential functions of a typical series voltage regulator. The reference voltage element provides a known stable voltage which is compared with a sample of the regulated output voltage at the comparator. The comparator subtracts the sampled output voltage from the reference voltage. This difference voltage (or error voltage) is then amplified by the error amplifier to provide drive for the control element. The control element behaves similarly to a valve in a water line in that it conducts more or less to adjust its resistance, and hence its voltage drop, to yield the proper output voltage across the load.

Now compare the internal features of the 723 (Fig. 1-54) with the regulator block diagram (Fig. 1-53). The 723 contains all the necessary regulator components except the output sampling element. The reference amplifier, current source, and zener diode comprise a stable temperature-compensated voltage source. Other internal functions include an error amplifier, a series pass transistor (the control element), an adjustable current limiter, and a 6.2-volt zener diode. The input of the error amplifier functions as a comparator by taking the difference between the voltages applied to the inverting and noninverting inputs. Note that the current limiter and the 6.2-volt zener diode are features that are not essential to the operation of the regulator. However, the current limiter is extremely useful for setting the maximum (short circuit) current output from the regulator, and the 6.2-volt zener diode can be used in floating or negative voltage regulator applications. One caution should be noted here—the 6.2-volt zener diode is only accessible in the 14-pin dual-in-line package. The flat pack and the 10-lead metal can packages do not have enough pins to accommodate all the internal functions of the 723, so the Vz output is not accessible. The pin outputs for the various packages are shown in Fig. 1-55.

High-Voltage Regulator (7.1 volts to 37 volts)

A typical 723 regulator circuit is shown in Fig. 1-56. In this figure, the temperature-compensated zener, the reference amplifier, and the current source are represented by an equivalent independent voltage source (a battery) to simplify the diagram. Vr may vary somewhat from device to device (6.6 volts to 7.5 volts), although it is typically 7.1 volts. The value of Vr establishes the lowest possible output voltage obtained from this circuit.

R1 and R2 form a voltage divider network from which the output sample (Vx) is taken. Therefore Vx = VoR1/(R1 + R2). If Vx is greater than Vr, the error voltage (which is Vr – Vx) will be negative and the output of the error amplifier will decrease, causing the series pass transistor to conduct less. This, in turn, causes Vo to decrease until Vx is equal to Vr. At this point, the error voltage is essentially zero and the output voltage (Vo) remains steady at a value equal to Vr(R1 + R2)/R1. Should the output voltage begin to decrease, Vx will begin to decrease proportionally. As Vx attempts to drop lower than Vr, the error voltage

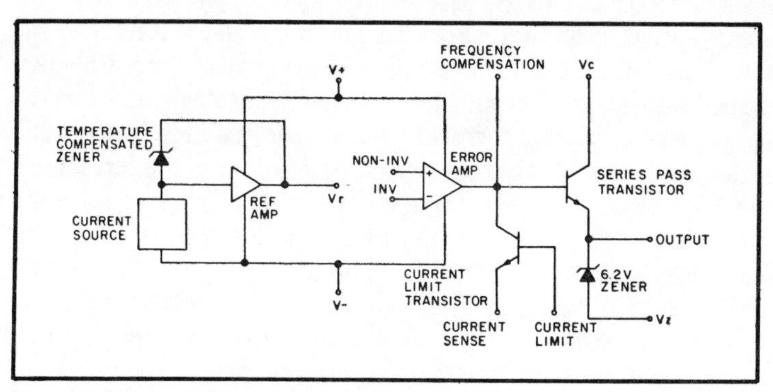

Fig. 1-54. Internal 723 functional block diagram.

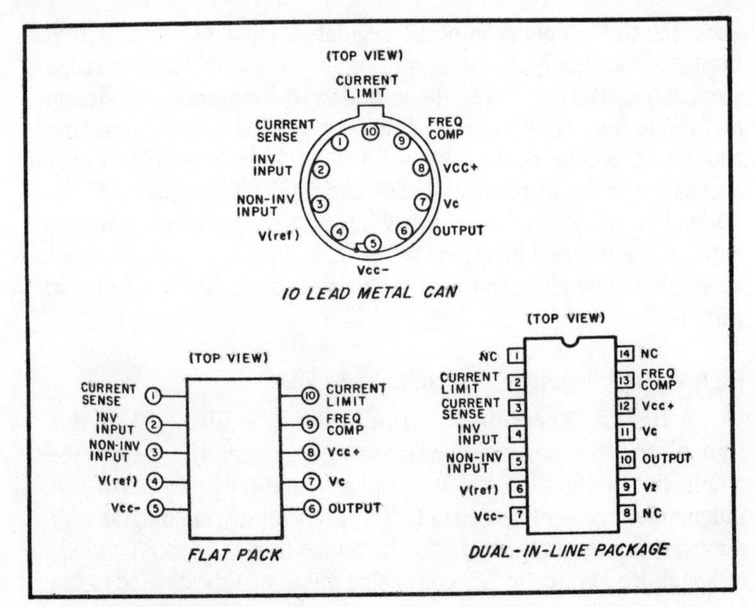

Fig. 1-55. 723 IC pin outputs.

becomes positive, causing the output of the error amplifier to increase. The pass transistor then conducts harder, which causes Vo to increase until Vx = Vr once again. As before, Vo = Vr (R1 + R2)/R1.

Rsc is a series current-limiting resistor which is selected to limit the maximum current the load can draw (Isc). Since this resistor is connected between the base and emitter of the current-limiting transistor, the load current through Rsc forward biases this transistor. Note, however, that the current-limiting transistor does not conduct until its base-emitter junction potential is overcome (approx. 0.65 volts). For example, select Rsc = 0.65/0.10 = 6.5 ohms. When the load attempts to draw more than 100-mA output current, the current limiter conducts and robs the series pass transistor of drive current from the error amplifier. The result is that the output voltage begins to drop off to restrict the current to the limit set by Rsc. Rsc is often selected so that the current capability of the regulator and power supply components is not exceeded. Therefore, you can short the output leads of the power supply without damaging the power supply or regulator.

Follow the steps in Fig. 1-56 in selecting components for this circuit. First, choose R1 so that the current drawn by the voltage-divider network (R1 and R2) is between 0.1 mA and 5 mA, and let's call this current Ib. Suppose we let Ib = 1 mA; then R1 = Vx/Ib. But note that we previously stated that regulator action tends to maintain Vx = Vr, so R1 = Vr/Ib = 7.1/.001 = 7100 ohms. Next, the selection of R2 is dependent upon the output voltage you want; R2 = (Vo — Vr)/Ib. Finally, R3 is chosen to balance the impedances seen by the input of the error amp. This improves error amp stability and accuracy. Therefore, R3 should be equal to the parallel combination of R1 and R2. If you want to keep the parts count to an absolute minimum, then just connect Vr to the noninverting input of the error amp and leave R3 out altogether. In most applications, you'll never notice a difference. The capacitor (C) in most 723 regulator applications should be 100 pF to 500 pF. This capacitor prevents the error amplifier from oscillating.

Low-Voltage Regulator (2 volts to 7.1 volts)

The circuit of Fig. 1-56 has one primary disadvantage in that the lowest output voltage obtainable from the regulator is limited

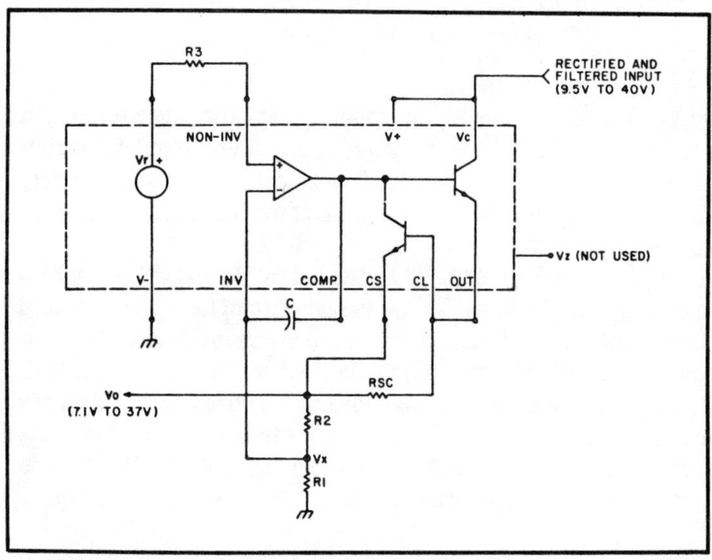

Fig. 1-56. High-voltage regulator (7.1 V to 37 V). To select component values:

1. Choose Vo
2. Measure Vr (or assume Vr = 7.1 V)
3. R1 = Vr/Ib (Ib is between 0.1 mA and 5 mA)
4. R2 = (Vo — Vr)/Ib
5. R3 = R1R2(R1 + R2)
6. Rsc = 0.65 (Isc = max. output current limit)
7. C = 100 pF to 500 pF

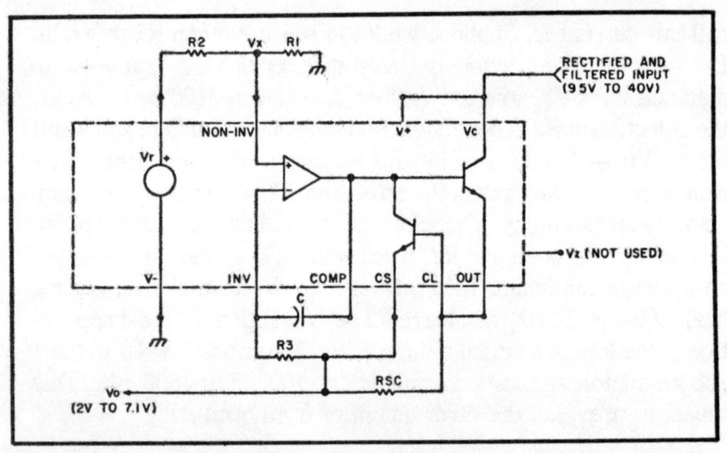

Fig. 1-57. Low-voltage regulator (2 V to 7.1 V). To select component values:

1. Choose Vo
2. Measure Vr (or assume Vr = 7.1 V)
3. R1 = Vr/Ib (Ib is between 0.1 mA and 5 mA)
4. R2 = R1(Vr — Vo)/Vo
5. R3 = R1R2/(R1 + R2)
6. Rsc = 0.65/Isc (Isc = max. output current limit)
7. C = 100 pF to 500 pF

by Vr. Therefore, this circuit cannot produce a regulated output voltage of less than 7.1 volts. If you need a 5-volt regulated output for TTL operation, you're out of luck. All is not lost, however, since we can get less than 7.1 volts out of the regulator by rearranging the components as shown in Fig. 1-57.

In this configuration, Vr is divided by R1 and R2 to obtain a lower reference voltage (Vx) for the error amplifier. The regulated output voltage is sampled via R3 at the inverting input of the error amplifier. Regulator action is similar to that previously described. In this case, the error amplifier adjusts the conduction of the pass transistor until Vo = Vx. First select R1 for a voltage divider bias current (Ib) of from 0.1 mA to 5 mA as before, (R1 = Vo/Ib). Rsc functions as previously described.

Variable-Voltage Regulator (2 volts to 37 volts)

The regulator circuits in Figs. 1-56 and 1-57 have fixed output voltages. Both of these regulators could be made variable by replacing R1 or R2 with a potentiometer, but we still have the disadvantages previously mentioned in that the output of the circuit in Fig. 1-56 is limited to an output voltage range of 7.1 volts to 37 volts, and the regulator of Fig. 1-57 is limited to an output voltage range of 2 volts to 7.1 volts. This is a particular disadvantage when working with differing logic families because TTL circuits require 5 volts, while transistor and MOS circuitry require higher voltages. Therefore, if you're building a variable power supply for your test bench, it is desirable for the supply to be variable over the entire range of the 723, if possible. The regulator configuration in Fig. 1-58 will do this quite nicely.

R1 and R2 form a voltage divider for the reference voltage (Vr), while R3, R4, and R5 form an adjustable network which samples both Vo and Vr. Together, these two networks comprise the input circuitry for the error amplifier. For good bias stability, let R1 = R3 and R2 = R4, so that, when the wiper of R5 is at point A, the voltage from the error amplifier, and hence Vo, is at a minimum (usually around 2 volts).

Now let's determine the component values for this circuit. We can begin by choosing R5 so that the current drawn from Vr is less than 5 mA (Vr is normally capable of supplying 15 mA max.). Once R5 is determined, choose R1 + R2 so that this combination draws

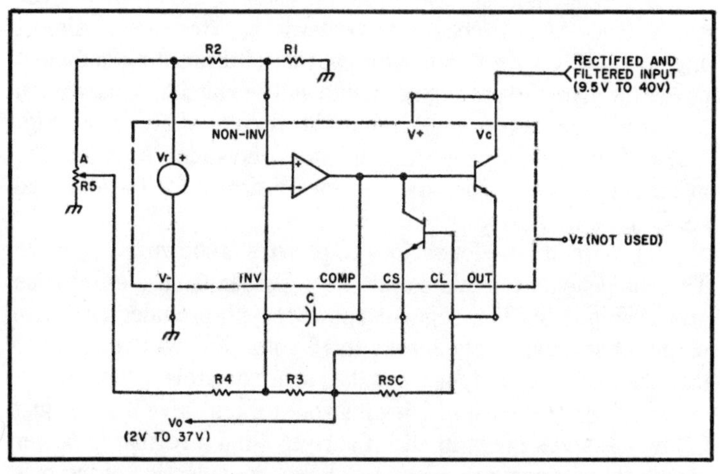

Fig. 1-58. Variable voltage regulator (2 V to 27 V). To select component values:

1. Choose Vo (here, Vo is the max. output voltage)
2. Measure Vr (or assume Vr = 7.1 V)
3. R5 = Vr/Ib (Ib is between 0.1 mA and 5 Ma)
4. R1 = VrVo/Ib (Vr + Vo)(Ib is between 0.1 mA and 5 mA)
5. R2 = R1(Vr/Vo)
6. R1 = R3
7. R2 = R4
8. Rsc = 0.65/Isc (Isc = max. output current limit)
9. C = 100 pF to 500 pF

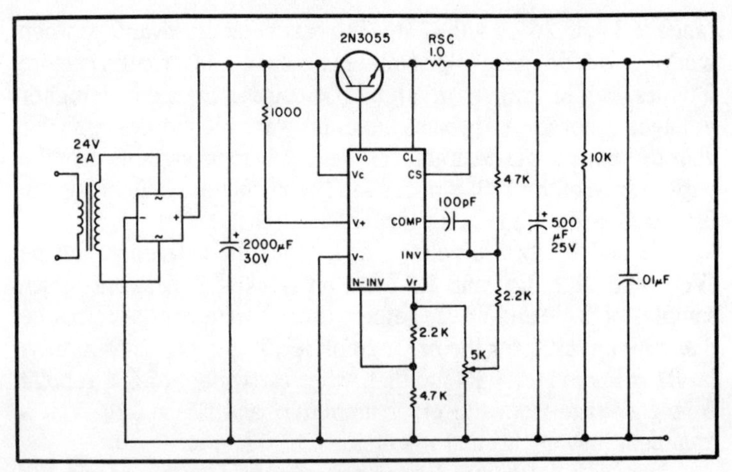

Fig. 1-59. Variable regulated power supply.

from 0.1 mA to 5 mA from Vr. Continue with the steps outlined in Fig. 1-58 to complete the circuit.

Since the internal series pass transistor is only capable of 150 mA, you may want to consider an external pass transistor. Almost any good NPN power transistor capable of dissipating sufficient power for your current requirements will be suitable. An external pass transistor can be employed in the circuits of Fig. 1-56, Fig. 1-57 or Fig. 1-58 by connecting it just as shown in Fig. 1-59. The internal pass transistor now becomes a driver for the external series pass transistor.

The circuit I use for various IC projects is shown in Fig. 1-59. The components for this circuit were chosen from the formulas given in Fig. 1-58. This regulated power supply provides a variable output of approximately 2 volts to 15 volts. Rsc was selected so that the maximum output current is approximately 1.2 amps.

So there you have it! Now it's your turn, so give it a try. Just follow the steps given in each figure to build a regulated power supply to your own specifications. Since the selection of the bias current (lb) is not critical, the component values selected can vary over a rather wide range so that you have plenty of flexibility in using parts you may already have on hand.

ADJUSTABLE ELECTRONIC LOAD

Current drawn by the load is adjustable from 0 to 10 amperes. Input voltages up to 30 volts are permitted at full current; that's 300 watts dissipation! Heat sinking of the pass transistors is sufficient to permit 300-watt operation for about ten minutes, at which point a thermal protection switch shuts the current off. At 100 watts dissipation the heat sink is adequate for indefinite operation. By directing a small blower at the heat sink, I have run the load at 300 watts for hours without difficulty.

Since I desired essentially self-contained operation, I included a small supply in the unit to bias the load transistors. Current is controlled by means of a front panel pot, and reverse polarity protection is included. A front-panel LED indicates operation of thermal shutdown circuitry. I took advantage of extra heat sink area to add two 8-ohm, 50-watt resistors. These are used for testing audio amplifiers or to increase the dissipation capability of the Power Waster circuit.

Circuit Operation

The simplified schemetics in Fig. 1-60 are useful in understanding how the circuit operates. Commercial load boxes use current sensing and feedback to set the load current. My approach is simpler and uses the constant-current collector load-line characteristic of all bipolar transistors.

Figure 1-60A shows the basic idea. The base of Q1 is biased at several volts from voltage source V1. The collector supply (V2) is

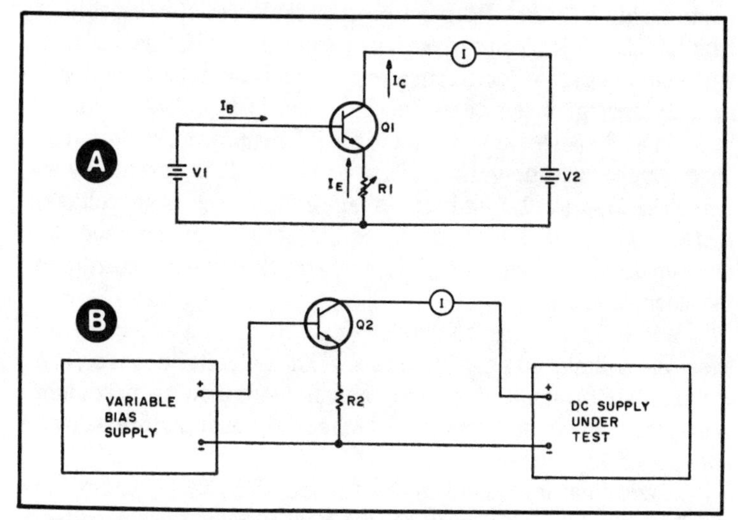

Fig. 1-60. Constant-current bias circuits (a) A transistor with fixed base voltage. Collector current is set by emitter resistor. (b) Basic circuit of the electronic load. Emitter resistor is fixed, and both emitter and collector current are controlled with base-bias supply voltage.

greater than V1 by at least 1 volt. The voltage at the emitter of Q1 is one diode drop (0.7 V) below the base voltage, and the circuit is essentially an emitter follower. Emitter current is set by dividing the emitter voltage by R1. As R1 is decreased, the emitter current increases.

The collector current for any transistor is simply alpha (the common-base current gain) times the emitter current. As alpha is essentially equal to 1 for any modern transistor, we see that setting the emitter current also sets the collector current. And this is the point, the collector current is determined only by the emitter current. Collector voltage has almost no effect on the collector current, provided the transistor is kept from saturation or breakdown. Saturation occurs if the collector voltage becomes less than the base voltage. Breakdown will occur if any excessive voltage is applied.

The constant-current collector load line is a useful property. If the voltage across R1 is small compared to the collector supply, a lot of power can be controlled and will be quite independent of the collector supply voltage. If a relatively low base-bias voltage is used, most of the power will be dissipated in the transistor and relatively little in R1.

As useful as this circuit is, there are some disadvantages. The main problem is that R1 must be a variable resistor capable of handling the entire load current. In a practical circuit, this becomes a 0.2-ohm pot rated for 10 amperes, which is an expensive item.

By arranging the circuit as indicated in Fig. 1-60B, things become a bit easier. Resistor R2 is a fixed highpower resistor. The base supply is made variable. Since the circuit is essentially an emitter follower, the emitter voltage follows the base voltage. Increasing the base bias increases the emitter voltage and the emitter current through R2. This increases the collector current to the desired value.

Within limitations, the collector current is set solely by the base-bias voltage and the value of R2. These limitations are: The collector supply must be at least one volt more than the maximum base-bias voltage, and must be less than the transistor breakdown voltage for the current drawn.

In addition, it is assumed that the base-bias supply is capable of supplying increasing current to the transistor base as the collector current increases. The base current will be the transistor collector current divided by beta, the common emitter current gain. Beta will vary from 10-50 for practical power transistors.

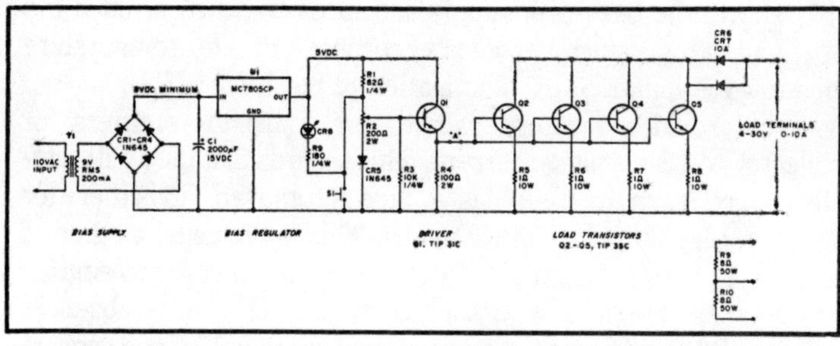

Fig. 1-61. Schematic of the Power Waster electronic load. S1 is a 75° C NO thermal switch. S1 is in thermal contact with the heat sink. The collectors of Q2-Q5 are connected directly to the sink. LED CR8 indicates thermal shutdown. Jumper at point A is opened in test.

To convert these ideas into the final design, I had only to add a base-bias supply, circuit overload, reverse polarity protection, and to increase the number of load transistors to handle the required power. The result is shown in Fig. 1-61.

Power is dissipated in the pass element (consisting of Q2, Q3, Q4, and Q5 connected in parallel). Four TIP 35C transistors are used in order to safely handle the 10-ampere maximum load current. Each pass transistor has a 1-ohm resistor in the emitter which corresponds to R2 of Fig. 1-60. In addition to setting the collector current for a given base voltage, these resistors also equalize the load distribution between the four pass transistors and help to keep the load current constant as the transistor temperature increases.

Variable base voltage for the load transistors is obtained from emitter follower Q1 which operates from a 5-volt regulated supply. This circuit consists of transformer T1, the bridge rectifier (CR1-CR4), C1, and an MC7805CP three-terminal regulator chip (U1).

The base of emitter follower Q1 is connected to a variable voltage divider consisting of R1, the pot (R2), and diode CR5. R1 and CR5 limit the driver base voltage range to approximately 0.8 to 4 volts. An additional 0.8-volt offset in the base-emitter junction of Q1 results in a driver output voltage range of 0 to 3.2 volts. Restricting the drive voltage to 3.2 V sets the maximum current the load will draw to 10 A. CR5 was included to compensate for the change in the base-to-emitter threshold voltage of Q1 as the temperature changes. In use, this simple bias supply has proven completely adequate. There is very little drift in load current as the temperature increases.

Some additional features are worthy of mention. Diodes CR6 and CR7 provide reverse polarity protection. A thermal switch

(S1) shorts the base-bias supply and turns the current off if the heat-sink temperature becomes excessive. An over-temperature shutdown is indicated by illumination of the LED (CR8).

R4 provides a return path to the transistor emitters for collector-to-base leakage current. This assures that they will actually turn off when the base-bias voltage is removed. R3 is there for safety in case the wiper of R2 opens. This is the usual method of failure for pots. Inclusion of R3 assures a path from base to emitter for the collector-to-base leakage current of Q1. In its absence, failure of R2 could cause the load to pull more than 10 A and damage something. With R3 included, the circuit just shuts off.

A series-connected pair of 8-ohm, 50-watt resistors is also mounted on the heat sink of the electronic load. These are provided to increase the dissipation capability of the unit and, in addition, are handy for testing audio amplifiers.

Power Rating

While the nominal input capability of the Power Waster is 10 A at up to 30 V, it may be operated at higher voltages if the proper conditions for safe operation are understood. In this section, I will explain how the nominal power rating is derived.

All high-power transistors have a "safe area" of operation, in which no damage will occur. Figure 1-62 is a safe-area curve for a single TIP 35C. This is a plot of maximum-permitted collector current as a function of collector voltage. Operation at any point in the region below and to the left of the curve will not damage the transistor. Combinations of voltage and current above and to the right of the curve will certainly destroy the transistor.

It is interesting that the power dissipation capability is not constant. At a Vce of 5 V, a current of 25 A is permitted. That's 125 watts. But if the voltage is increased to 50 volts, only 1 ampere is allowed, providing a dissipation capability of only 50 watts! The successful designer of high-power transistor circuits stares long and hard at the safe area curves before picking a final configuration!

When I designed the Power Waster, I tried to be conservative so that it could really take abuse without failure. Thus, four TIP 35C transistors were used. In Fig. 1-63, the composite safe-area curve for the four devices in parallel is shown. This is Curve B. Curve A indicates the "rated" operating envelope for the Power Waster. Only at the 10 A and 30 V point does the "rating" curve approach the safe-area curve. At all other combinations of current and voltage, the power dissipation is comfortably inside of curve B.

This conservative design assures that the unit will never fail as long as the 10 A and 30 V maximums are observed.

Construction

Construction of the electronic load can take almost any form because of the non-critical nature of the circuit. In most cases, the shape of the heat sink will determine the final configuration. In order to dissipate 300 watts, a sink having a thermal resistance of 0.1° C per watt is required. This is a truly massive piece of metal. I elected to depend upon thermal inertia to permit short tests of up to ten minutes duration and to let the thermal cutout act if things got too hot. Continuous operation is obtained by directing a small blower at the heat sink.

I used a 5″ × 11″ piece of heat-sink extrusion having 32 one-inch-high fins as a chassis. To this I fastened a pair of 0.75″ × 2″ pieces of angle to form a U-shaped chassis with the fins on top. All parts are mounted on the underside of the extrusion, and the front and rear panels are formed by the pieces of angle. Vertical fins are more efficient, of course, but since a cooling fan is required to obtain the full power rating, this arrangement is quite convenient.

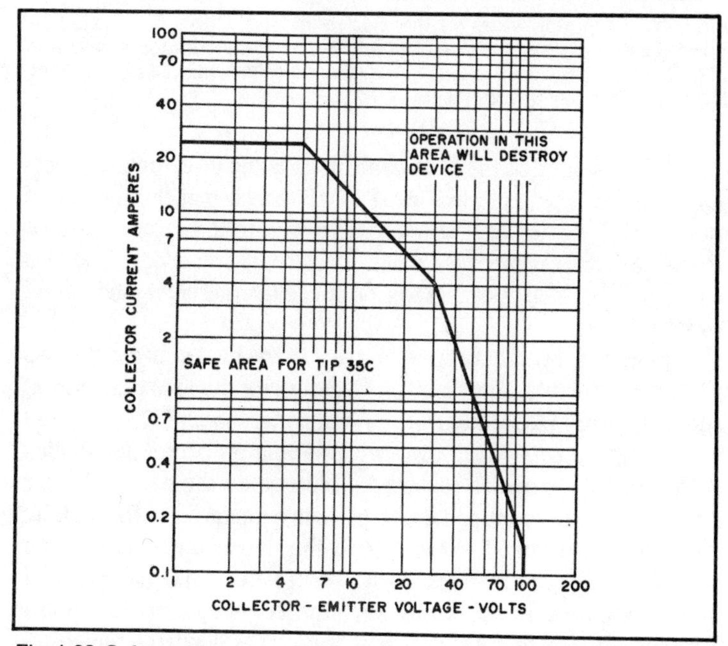

Fig. 1-62. Safe-area curve for TIP 35C. Safe operation is limited to the area below and to the left of the curve. Operation beyond the safe area will destroy the transistor.

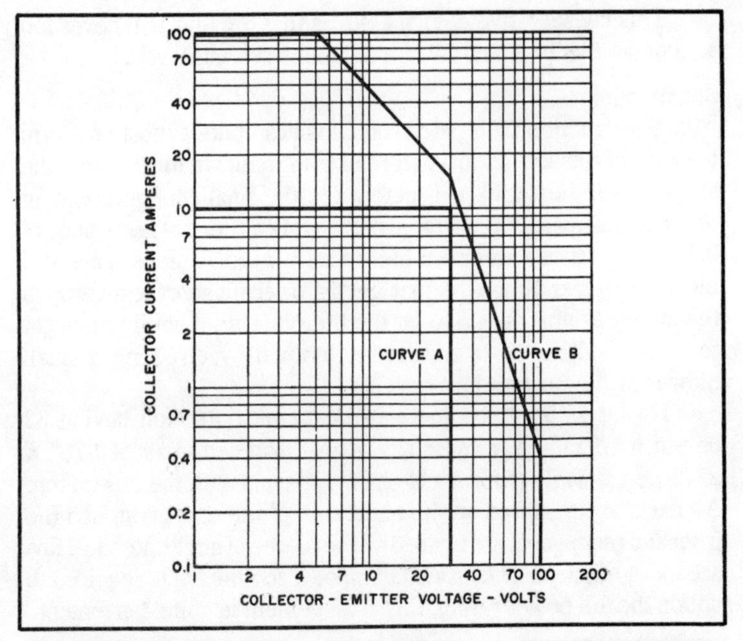

Fig. 1-63. Operating areas for the electronic load: Curve A indicates the rated operating area for the load. Curve B shows the composite safe area of the four TIP 35C transistors used as the load element. The rated operating area (Curve A) is well inside of the safe area.

Parts located on the front panel include the three binding posts for the 8-ohm resistors (R9 and R10). These are at the left end of the panel. Next is the LED over-temperature indicator. Adjacent to the LED are the input terminals for the electronic load. The load current adjustment pot, R2, is on the extreme right end of the panel.

Mounted directly to the underside of the heat sink are the two 8-ohm resistors, the four emitter resistors for the load transistors, and transistors Q2 through Q5. The thermal switch (S1) is located in the center of the sink. Two reverse-polarity protection diodes (CR6 and CR7) are also mounted directly onto the heat sink near the center of the front panel. All parts are mounted to the sink by threaded holes tapped directly into the extrusion. The cathode ends of CR6 and CR7 are connected to the sink via threaded mounting holes. In the same way, the collectors of Q2-Q5 are screwed directly to the sink. Thus, the connection from the polarity protection diodes to the transistor collectors is via the heat sink itself. Fastening the transistors to the sink without insulating wafers puts the entire chassis at the potential of the positive supply input.

Normally, I would have insulated the transistor collectors from the sink but there is a good reason for not doing so. The thermal impedance of the transistor junction to case is 1° C per watt. Most insulating wafers have a thermal resistance of at least 0.5° C per watt. If a wafer were used, the resistances are summed to yield a thermal resistance of 1.5° C per watt. The junction temperature would rise by fifty percent! Instead, I elected to have the positive supply input on the sink and to exercise caution in using the unit. The low voltages involved certainly pose no shock hazard.

On the rear panel are located: the power transformer, bridge rectifier, filter capacitor, 5-volt regulator, and the driver transistor (Q1). The transformer is located at the left end of the rear panel. The filter capacitor is in the center of the panel as is the bridge rectifier, which is hidden by the capacitor in the photograph. Adjacent to the filter capacitor is the 5-V regulator and the driver circuit components. U1 and Q1 are both fastened to the rear panel for cooling. All small parts such as resistors, capacitors, and diodes are mounted with push-in Teflon terminals. Threaded standoffs are used for the larger components.

Not everyone who wishes to build this circuit will be able to find a heat sink similar to the one I used. One approach is to use four smaller sinks and to mount a load transistor on each. A suitable configuration would use four Wakefield Engineering type NC-423 heat sinks. Individually, these units have a thermal resistance of 0.8° C per watt so the resistance of the combination would be 0.2° C per watt. Such an arrangement would premit inputs of about 150 watts without forced-air cooling.

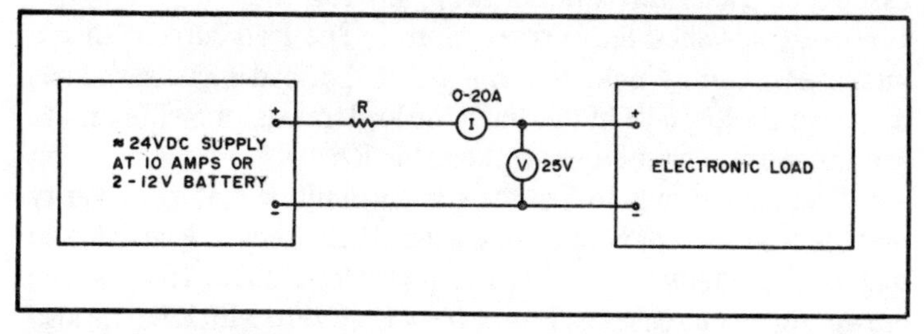

Fig. 1-64. Testing the Power Waster. A high-current supply such as a pair of 12-V batteries is connected to the load via current-limiting resistor R. The resistor should be chosen to limit the current to a few amperes for the initial tests.

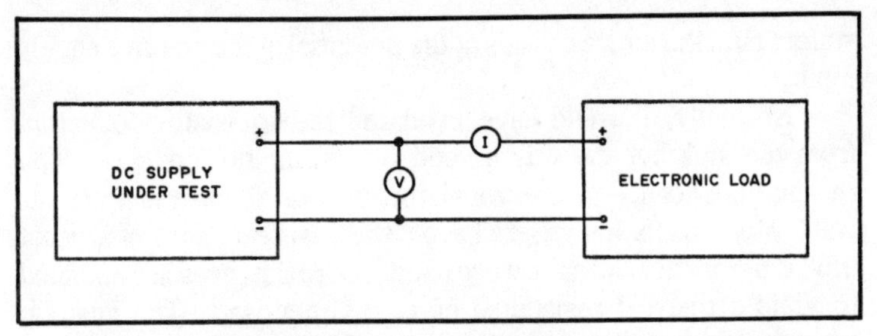

Fig. 1-65. Using the load to evaluate a power supply. Increasing current is drawn from the supply, and the regulator performance is measured. The load also may be used to ensure proper current-limiter operation.

Testing

A few simple tests prior to using the load will prevent damage to it, or to the circuit being tested. First, break the circuit between the driver transistor (Q1) and the load transistors (Q2 through Q5), at point A in Fig. 1-61. Apply 110 V ac to the power transformer and verify that the voltage is variable from about 0 to 3.5 V as R2 is rotated. Zero volts should occur at the extreme counterclockwise position of R2. This is the zero current setting.

Reconnect the circuit at point A. Then arrange a high-current test setup similar to Fig. 1-64. Resistor R is chosen to limit the current to 3 or 4 A under short-circuit conditions. Two automobile headlamps in series may be used for this resistor. Set the current control pot (R2) at minimum (CCW position) and turn on the supply and the bias supply in the Power Waster.

Increase the current setting until 2 A is drawn by the load. Measure the voltage drop across R5 through R8 and assure yourself that the voltages are about equal. About 0.5 V should appear across each resistor for a 2-A load. If no voltage appears, then the associated transistor is not drawing any current.

Next, advance the current control. The load current should increase smoothly until the voltage at the load input terminals drops well below 5 V. At this point the load transistors will saturate and the current will be set by resistor R.

The next step is to test the current-limiting feature to verify that the load current is limited to about 10 A. Reduce R in value so that for the supply voltage present under load, about 15 A may be drawn. Again increase the load current from zero with R2. The load should current limit at 9 to 11 A if the same parts values were used.

If all of the above tests were successful, the electronic load is ready for use. See Figs. 1-66 and 1-67.

Applications

My purpose in building the electronic load was to enable rapid test of power-supply circuits. The 10-A and 30-V input capability will suffice for almost the entire range of solid-state supplies found in amateur equipment. After I built the load, a number of other applications surfaced. Some of these are worthy of mention.

The performance of a voltage regulator is easily plotted by using the load, connected as indicated by Fig. 1-65. The regulator is connected to the load via an ammeter. A voltmeter is connected across the output of the supply. By increasing the load current and noting the meter reading, the internal resistance of the regulator may be found, and the percent regulation as a function of load determined. If the supply has a current limiter, the current-limiting point may be found by slowly increasing the load current until the voltage drops. The smooth control of current afforded by the electronic load makes these tests easy.

At times it is desirable to know if a surplus transformer of questionable ancestry can meet a given requirement. Of particular concern is the "overhead" voltage requirement. This is the input voltage required to maintain a regulator in operation (at full load) subtracted from the output voltage.

Typical three-terminal regulators require a 3-volt overhead to function properly. If all circuit parameters are known, it is easy to

Fig. 1-66. Interior view of the Power Waster. Four TIP 35C load transistors are mounted to the heat sink. The thermal switch is located between the two center transistors. A U-shaped chassis is formed from the heat sink and two pieces of aluminum angle stock.

Fig. 1-67. The Power Waster. This electronic load is adjustable from 0 to 10 A and can dissipate up to 300 watts. The unit is built upon a piece of heat-sink extrusion and has both reverse polarity and thermal overload protection.

determine if a certain transformer, rectifier, and filter capacitor will work. If junkbox parts are used, the calculation is often difficult or impossible. It takes only a few minutes to connect up the unregulated supply parts and apply the expected load current using the Power Waster. The unregulated voltage and ripple may be measured, and the suitability of the components determined.

Battery charging is another application. All batteries must be charged from a current source so the charging current is held constant as the battery terminal voltage increases. To use the load as a battery charger, simply connect it in series with a source having a voltage at least 4 V greater than the full-charge voltage of the battery. Then place the combination in series with the battery and dial up the desired charging current.

One final application is protection of high-power transistors during tune-up. The electronic load is placed in series with the supply and programmed for the maximum current that you feel is safe for the amplifier under test. At currents less than the setting, the load saturates and the voltage drop across it is quite small. This is especially true at the lower currents.

Should the amplifier try to draw excessive current, the maximum will be limited to the set value. Response is very rapid. Current limiting occurs in about a microsecond and is much faster than a normal power-supply current limiter. So effective is this method of protection that I have built it into several solid-state transmitters.

Chapter 2
General Test Equipment

Every electronics experimenter needs good reliable test equipment. Some test equipment you can buy but if you build your own you can save a lot of money and have fun at the same time. Some of the test equipment in this chapter can't be purchased commercially at any price.

METERLESS OHMMETER

This handy device is a low-cost audible continuity tester. It uses both low voltage and low current. The tester is also small enough to put in your shirt pocket because it uses a 35mm film container for an enclosure. Originally I had a need for such a tester during a project that was wire-wrapped and I needed to check a lot of connections in as short a time as possible. The tester will give you an audible indication of resistance up to around 2000 ohms. You can test semiconductor junctions with it and the tester will let you tell the difference between just a few ohms of resistance because different tones will be heard when testing different values of resistance. Since this continuity tester will let you measure small values of resistance, it is nice for testing any sort of wiring or semiconductor components.

The LM3909 used in this tester is almost indestructible provided it isn't fed with more than 1.5 volts. I use an AAA-size 1.5-volt battery in my tester and it has lasted almost a year now. The tester provides enough voltage to turn on transistor and diode junctions and it does so at low current levels. Maximum current levels are obtained when the component being measured has close to zero ohms of resistance. If you use a 1000-ohm earphone with the tester, the current will be approximately 2 mA. If you use an 8-ohm speaker, the current will be around 13 mA. If you're not measuring zero ohms, the current through the component or wire being tested will be in fractions of a milliampere. The enclosure used for my continuity tester was an empty film container and it is just the right size to put in your pocket and get ahold of when you need it. If you don't have a 35mm camera, ask one of your friends that does to give you an empty film container.

Construction of this continuity tester will only take an hour or so if you have all the parts ready. You can buy all the parts at a Radio Shack store. Depending upon what you have in spare parts and your junk box, the total cost will be from five to ten bucks.

The electrical design of the continuity tester is shown in Fig. 2-1. If you look at Fig. 2-2, you can see how the parts are placed on the piece of experimenter circuit board. A completed continuity tester is shown in Fig. 2-3. Looking at Fig. 1-1, you should notice that the earphone or speaker has to be connected for the tester to operate. If you don't use an earphone and jack as I did, you might want to install an on-off switch to turn the tester off in case the test leads touch together while it is waiting to be used.

Before you solder the parts in place on the circuit board, trim

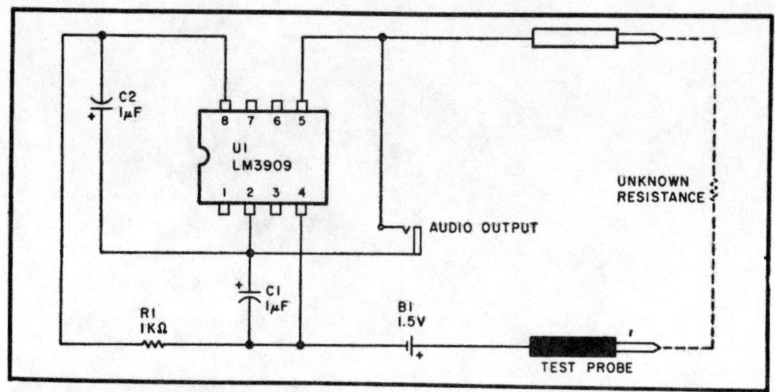

Fig. 2-1. Continuity tester schematic.

it down enough to fit into the film case. Then drill two holes in the lid of the film case and pass the test leads through it. Figure 2-5 is an example of my trimmed down circuit board. The capacitors used in my tester were electrolytics rated at 50 volts, but any rating small enough to fit on the circuit board and into the film case would work just as well. The voltage rating needs to be only a few volts, so tantalum capacitors would work nicely, too. After you have soldered the components to the circuit board, drill a hole in the

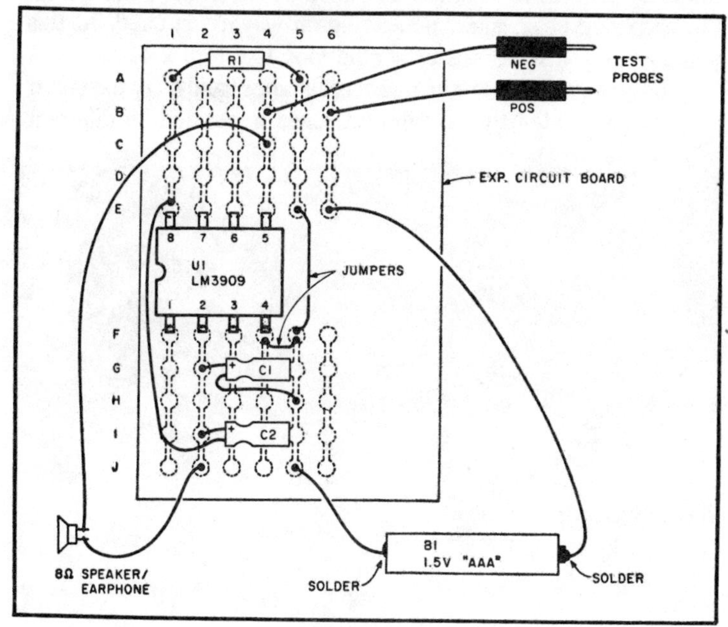

Fig. 2-2. View from top of circuit board.

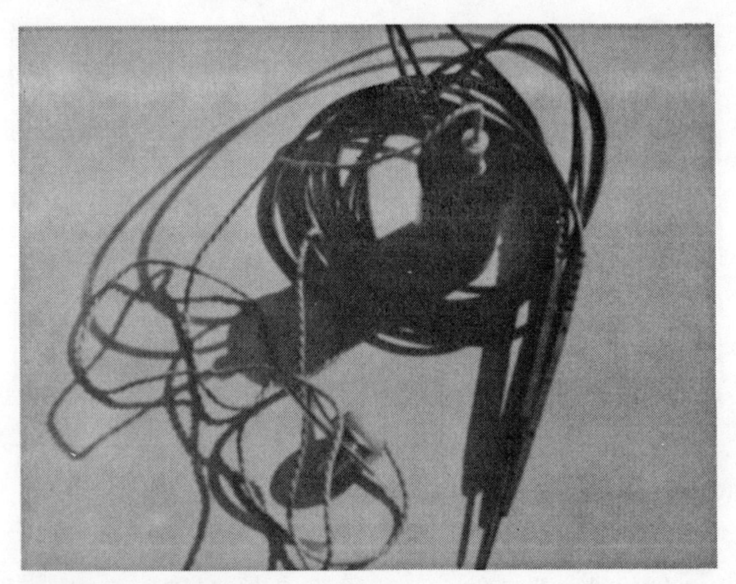

Fig. 2-3. Completed tester.

center of the film container top for the earphone jack (if you use one) and install it. Finally, solder the leads going to the battery and touch the test leads together. You should hear a tone coming from the earphone or speaker, depending on which you used. At this point, your continuity tester should look like Fig. 2-4.

If you have some low values of resistance handy, try the tester on them and listen to the different tones generated by different

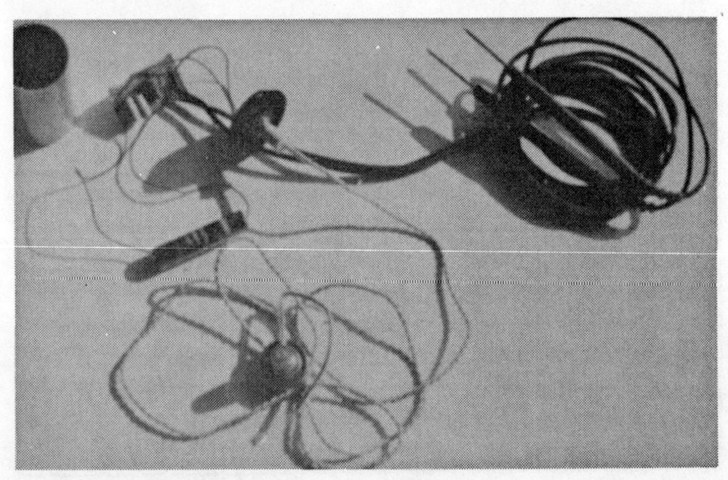

Fig. 2-4. Partially assembled.

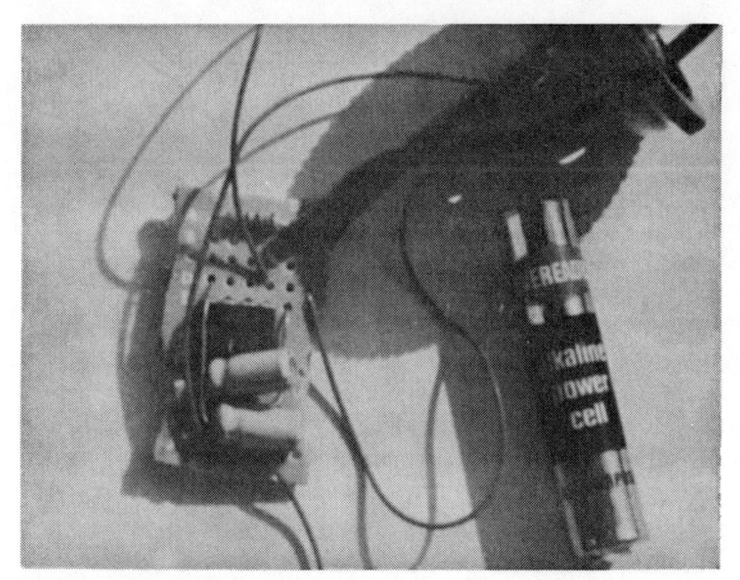

Fig. 2-5. Circuit board.

values of resistance. When you're sure that the tester is working correctly and all the wires are soldered, wrap the circuit board and the battery with electrical tape to prevent things from shorting out once you put everything into the film case. Take a look at Fig. 2-6—you can see what my tester looks like before stuffing everything into the film case. Now that you've got the audible continuity tester put together, you can use it to check wires and semiconduc-

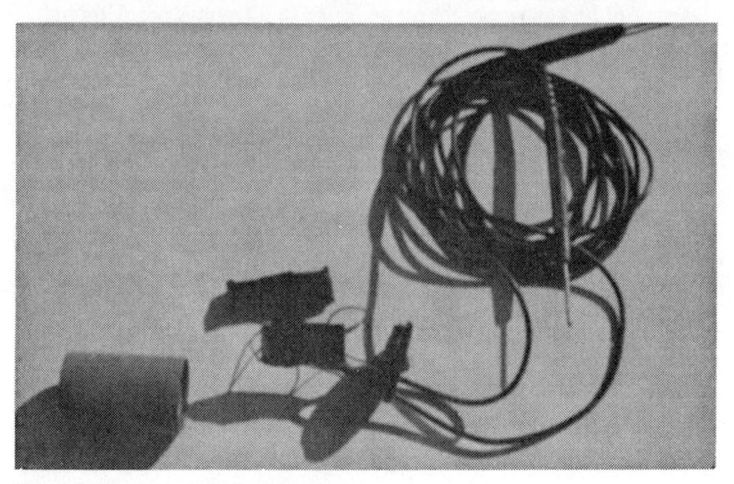

Fig. 2-6. Ready to install in the case.

Table 2-1. Parts List For Meterless Ohmmeter

C1 and C2	1-uF/15-volt electrolytic capacitors
R1	1000-Ohm, 1/4-W resistor
U1	LM3909 flasher oscillator
B1	1.5-volt AAA-size battery
Miscellaneous:	
Circuit board (Radio Shack 276-170, Global Specialties Corp. EXP-300), Test probes, 35mm film container, speaker or earphone and jack, wire, solder.	

tors. By connecting it to a telegraph key, you've got a code practice oscillator. If you replace the earphone or speaker with the correct value of resistor (between 10 and 2000 ohms) and take an output from across it, you have an audio signal generator, the output frequency depending upon the resistance that you use.

AN OHMMETER FOR SOLID-STATE CIRCUITS

Figure 2-7 shows the circuit of the low-voltage ohmmeter. Only about 250 m V are developed across the test prods, far below the value capable of forward-biasing silicon junctions in solid-state devices. But what have we here? A dc current meter is in the collector-base circuit of a transistor connected in the common-base configuration. Emitter-base bias is provided by B1, but look as one may, there is no obvious source of collector bias. Nor is there any subtle or tricky current path for polarizing the collector. How then can the transistor deliver current to the meter? Actually, this is a valid way of utilizing the characteristics of a transistor. Although infrequently encountered, a transistor so employed is capable of providing collector current with zero-applied collector voltage! In

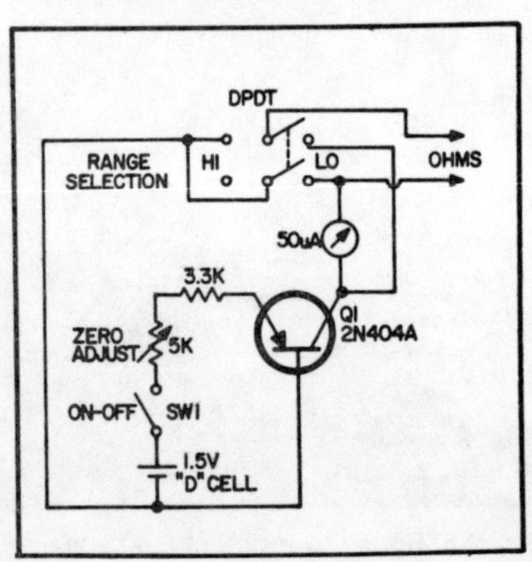

Fig. 2-7. Schematic circuit of the ohmmeter for solid-state circuits. Transistor Q1 need not be a 2N404A. Any small signal PNP germanium type will work.

so doing, the transistor develops about 250 m V between collector and base (this is the maximum voltage available at the test prods). Although this mode of transistor operation is not generally useful, it is just what the doctor ordered for our purpose. It should, however, be realized that a *germanium* transistor such as the 2N404A must be used.

Other than the unique current-source for the microammeter, this ohmmeter operates in the same manner as conventional instruments. Note the range switch, SW2, enables the meter to be used as a shunt-type ohmmeter for the low range, or as a series type ohmmeter for the high range.

Popular 20,000Ω per volt VOM's employ 50 μA meter movements which has an internal resistance in the vicinity of 5000Ω. On many of these instruments, a 50 μA current-measuring function is provided. However, rather than clutter up the already congested scales of such meters with additional markings, it would appear desirable to make a conversion table relating microamperes to ohms. Inasmuch as this introduces an inconvenience during test procedures, the best bet is probably to obtain a 50 μA dc current meter. Then one is free to inscribe high and low ohms calibrations as shown in Fig. 2-8. This drawing is intended as a general guide and is very approximate. So-called 5000Ω meters will vary considerably in actual resistance and it would not be practical to provide a universal template for transferring scale markings to meters. Meters also vary in linerity and accuracy. It is much better to calibrate the individual meter even though accuracy may not be the primary goal in a meter for generalized trouble-shooting.

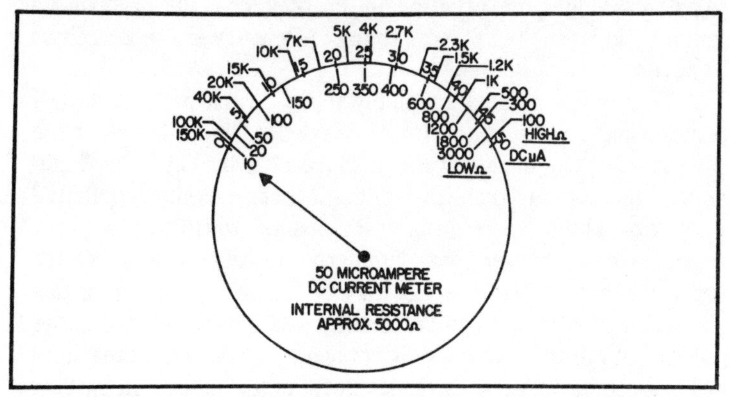

Fig. 2-8. Approximate appearance of high and low range ohms scales. To be used only as a guide. Actual calibration is made with the use of known resistances.

With regard to scale calibration of the meter face, the ensuing procedure is a straightforward way to get a reasonable start. Acquire *pairs* of the following 5%, ½W, composition resistors: 10, 100, 1000, 10K and 100KΩ. According to the way each pair is used (parallel, singly or series) we then have *at least* the following calibration values: 5, 10, 20, 50, 100, 200, 500, 1000, 2000, 5K, 10K, 20K, 50K, 100K, 200K. By appropriate combinations of different values, one readily comes up with such values at 70, 700, 7000Ω, etc., or approximately 14, 140, 1400, etc. Additional resistors can be obtained for more extensive calibration. Decade values are the most useful. Caution should be exercised in any attempt to interpolate calibration markings—this just is not easy to do on a scale as nonlinear as that of an ohmmeter. Two scales are calibrated on the meter face, one for the high ohms-range, and the other for the low-ohms range. The extremes of the high range are 100Ω and 150K. The 150K marking is close to the zero of the 50 μA scale. The extremes of the low range are 10Ω and 3K. The 3K marking is close to the "50" of the 50 μA scale. When calibrating, frequently check the zeroing of the meter. This is accomplished by means of R1. On the high range, an exact full-scale meter-deflection must exist with the test prods *shorted*. On the low range, an exact-full-scale meter-deflection must exist with the test prods *apart*.

The total measurement range of this ohmmeter is ten to one-hundred thousand ohms with useful estimates possible somewhat beyond this range. It happens that such a range is adequate for the majority of tests in solid-state circuits. Although megohm resistance values are occasionally encountered, the pronounced tendency is for the resistances to range from several tens of ohms to several tens of kilo-ohms.

To use the ohmmeter simply place switch SW1 in its ON position and zero the meter (full-scale deflection, or 50 μA on the current scale) by means of R1. If range switch SW1 is on "high-ohms," zeroing is accomplished with the test prods shorted. If SW2 is on "low-ohms," zeroing is accomplished with the test prods apart. As with a conventional ohmmeter, ascertain that no voltage sources are active in the circuitry to be tested. In using this ohmmeter, the circuit can be investigated without regard to the polarity of the test prods. This applies to electrolytic capacitors and to all semiconductor devices, except tunnel diodes. In the vast majority of test procedures using this ohmmeter, it will be unnecessary to remove solid-state devices in order to make meaning-

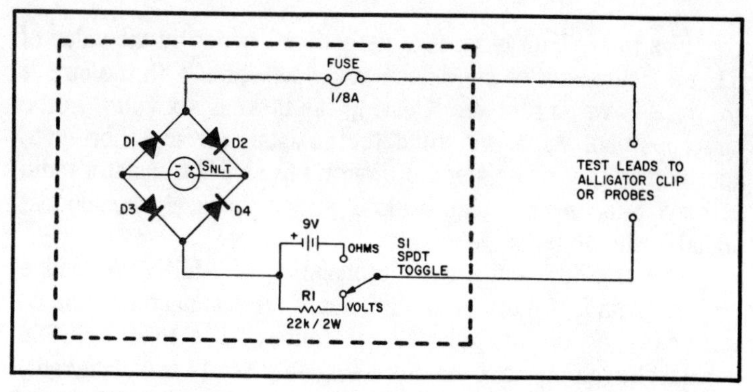

Fig. 2-9. Simple audio volt-Ohm detector. SNLT—Mallory SC628 Sonalert.® D_1 through D_4—1N5061 or equivalent. (Any general-purpose diode or package bridge rectifier with a piv rating of 500 V or more should work.)

ful resistance tests of the associated passive-circuitry. Possible exceptions can occur with germanium devices. In the case of germanium transistors, one can revert to the technique of reversing the polarity of the test prods. Germanium tunnel-diodes, however, should have one lead disconnected in order to free circuit tests from the effect of their conductivity. Most modern circuit-boards tend to have silicon devices. In addition to ignoring the junctions of bipolar transistors, this ohmmeter will ignore the junctions of common signal-diodes, rectifier diodes, zeners and varactors. A similar statement applies to the gates of FET's, the emitters of UJT's, and the entire family of SCR devices, including TRIACS. Insofar as I have been able to determine, one should likewise be able to ignore the presence of most IC modules. Because of the great variety and the rapid evolution of IC's, some reservation is purposely held here.

BUILD AN AUDIO VOM

The simple circuit described here will produce an audio output in proportion to the level of voltage (ac or dc) and will check continuity. See Fig. 2-9.

This unit will detect ac or dc voltage to at least 300 volts and down to as low as approximately 6 volts. It will distinguish between ac and dc, with the audio from ac sounding a bit raspy and the dc producing a more pure note. Checking for dc voltage is further simplified by not having to worry about the polarity of the test leads. The sound intensity is proportional to the level of voltage applied.

The lower limit of voltage detection depends on the value of R1, the sensitivity of the device, and your ears. With the circuit shown, I have no problem "hearing" as little as six volts. In the ohms position, the circuit will detect resistance from a short up to approximately 40k ohms or more, with a fresh 9-V transistor radio battery. Shorting the test leads together produces the loudest signal in the ohms mode.

The circuit is quite basic, consisting of a full-wave bridge rectifier with a Mallory Sonalert® transducer connected to the dc terminals, observing polarity, of course. The model SC628 Sonalert has a range of 6-28 V dc using only 3 to 14 mA of current, so the device is fairly sensitive. With the series resistor, R1, or the battery as shown in the circuit, the range of input voltage, or resistance that produces a sound output, will surprise you. Switch S1 selects either an internal 9-V battery for ohms or series resistor R1 for voltage.

The value of the resistor was determined experimentally to permit a range of voltage to be checked that would most likely be encountered by the average person, and limit the voltage drop across the device to a safe level. At 300 volts, the Sonalert has about 20 volts across it which is still within its range. Finally, a ⅛-Amp fuse is included in series with one of the leads, in case someone goofs and tries to hear voltage with the switch in the ohms position.

The unit was built to fit into a small plastic instrument case, with the test leads brought out directly to alligator clips for maximum convenience and economy. Just about any small box or enclosure that will hold the parts should work nicely, however.

One word of caution when using this tester—which holds true for any VOM. Always make sure the circuit is de-energized before checking resistance or continuity either by disconnecting all sources of voltage and/or checking for no voltage first. Always return the instrument to the voltage position after using the ohms position, and you should never have to replace the fuse inside. Also, it is not recommended for voltages in excess of 300.

VOM DESIGN

The most expensive part of a VOM (if bought new) is the meter movement. What to look for is the most sensitive (lowest mA at full scale) milli-or micro-amp meter. A 500 μA will do very nicely, but anything up to about 10 mA will do.

After obtaining a meter, measure its internal resistance using the resistance scale of a friends VOM.

For voltage measurement, resistors are wired in series with the meter to obtain different voltage scales. Decide what voltage scales you would like and see Table 2-2 for computing series resistor values. It is shown using 1, 5, 10, 50, 100 and 500 volt scales. Notice that none of the series resistors are standard values, but two or more resistors could be wired in series to obtain the proper value. To enable the meter to measure ac in addition to dc a rectifier and capacitor must also be wired into the circuit. Fig. 2-10 shows the ac-dc voltmeter.

For the addition of current scales different resistors must be wired in parallel with the meter. For a scale two times that of the meter, a resistor, equal to the meter's internal resistance, is wired in parallel so that ½ of the current goes through the meter and ½ through the resistor. For a scale three times that of the meter two resistors are wired across the meter so that ⅓ of the current goes through the meter and ⅔ through the resistors. The number of parallel resistors that are equal to the internal resistance is found in the following formula (A+B) – 1 where A equals the desired current scale in mA and B equals the full scale current reading of the meter. The parallel resistors for any one scale can be combined into one by using Ohm's law for parallel resistance: Example:

I want 500 μA, 1mA, 10m mA, and 010 mA current scales using a 500 μA 150 Ω meter. For the 500 μA scale no resistor is needed. For the 1 mA scale one 150 Ω resistor is used. For the 10 mA scale 19-150 Ω resistors are used in parallel which is equal to about 7.9 Ω. Two 16 Ω resistors in parallel could be used. For the 100 mA scale 199-150 Ω resistors are needed, equal to about .85 Ω. Just use two .43 Ω resistors in series.

The next step is ohms scale. Figure 2-11 shows the basic circuit for an ohmmeter. Calibration of the ohms scale will have to be found experimentally. Different values of resistors should be

Table 2-2. Resistor Values for VOM.

Desired Voltage Scale	Current Rating of Meter in Amps	Necessary Resistance	Internal Resistance of Meter	Series Resistor
1	.0005	2000Ω	150Ω	1850
5	.0005	10K	150Ω	9850
10	,0005	20K	150Ω	19850
50	.0005	100K	150Ω	99850
100	.0005	200K	150Ω	199850
500	.0005	1 meg.	150Ω	999850

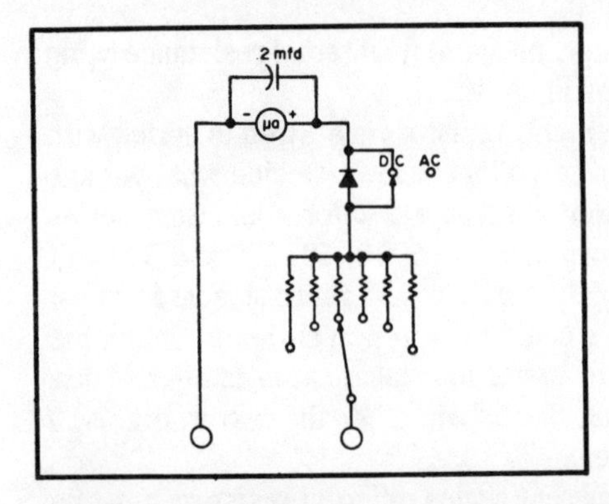

Fig. 2-10. Ac-dc voltmeter.

tested in the circuit to obtain approximate calibration. In my circuit using a 500 μA 150 Ω meter indicated about 100K at little deflection of the meter and about 1K at mid-scale.

Figure 2-12 is the circuit for the combined volt-ohm-milliampmeter using a 500 μA 150 Ω meter. Five or ten % resistors can be used but 1% are better if they are available at a low price.

R_X and C_X SUBSTITUTION BOXES

Every electronics experimenter should have both a resistance and capacitance substitution box.

Resistor Substitution Box

Construction of this unit is simple. The resistors are mounted across the switches' terminals. By opening a switch, *that* resistor is connected in circuit.

The switches are arranged in rows of seven across and four down (See Fig. 2-13 and Table 2-3). Then the resistor/switch combinations are connected in series. I used slide switches in my unit.

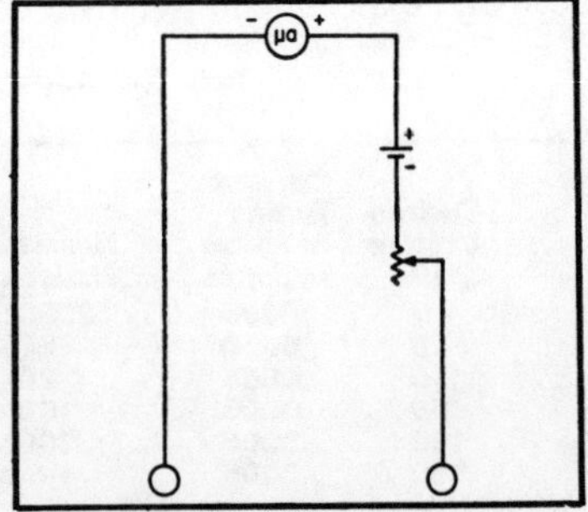

Fig. 2-11. Basic circuit for ohmmeter.

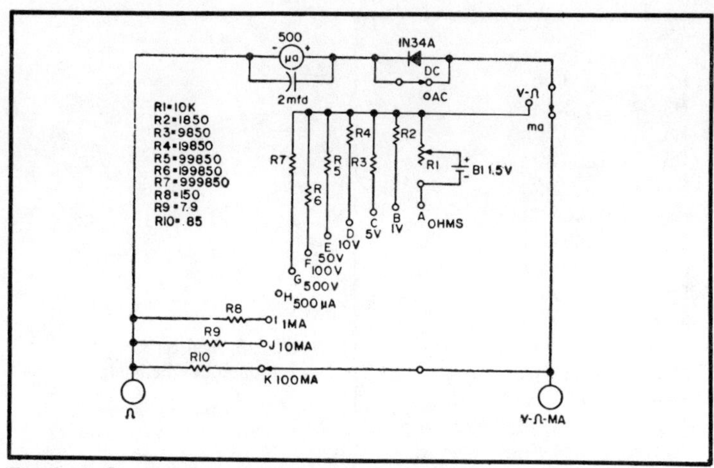

Fig. 2-12. Circuit for combined VOM with a 500 μA 150Ω meter.

The rectangular openings were cut out with a nibbling tool and the switches were mounted to the box panel with pop rivets.

With the use of 1% resistors, there is a possible error of ± 100k (that's with all resistors in circuit for a total of 9,999,999 ohms).

With this circuit, there is a possible monetary advantage over conventional resistance substitution boxes which usually require

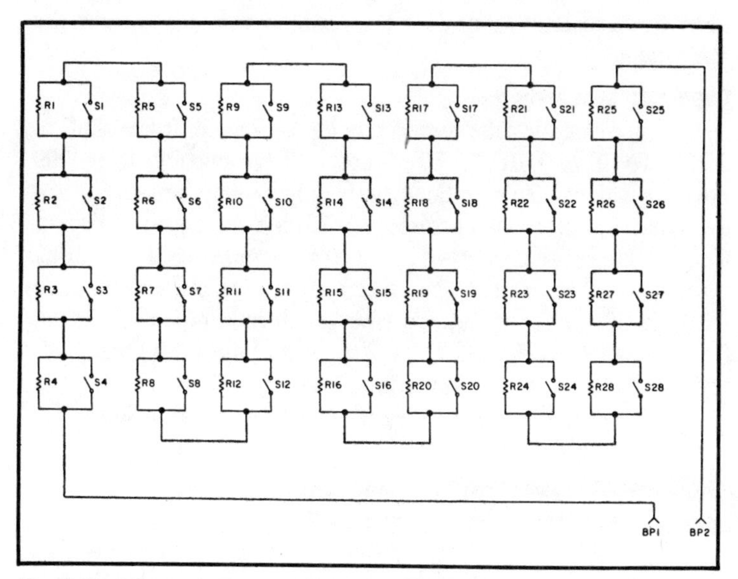

Fig. 2-13. Schematic for resistor substitution box.

R1	1 Ohm
R2	2 Ohm
R3-R4	3 Ohm
R5	10 Ohm
R6	20 Ohm
R7-R8	30 Ohm
R9	100 Ohm
R10	200 Ohm
R11-R12	300 Ohm
R13	1k Ohm
R14	2k Ohm
R15-R16	3k Ohm
R17	10k Ohm
R18	20k Ohm
R19-R20	30k Ohm
R21	100k Ohm
R22	200k Ohm
R23-R24	300k Ohm
R25	1 Megohm
R26	2 megohm
R27-R28	3 megohm

S1 through S28—SPST slide or toggle switches
BPI, BP2—5-way binding posts
Misc.—wire, cabinet, rub-on letters and numbers

Table 2-3. Resistor Substitution Box Parts List.

sixty-three resistors and seven ten-position switches to cover the same range.

Capacitor Substitution Box

This capacitor substitution box has a range of from 10 pF to within 10 pF of 1 μF, in 10-pF steps. That amounts to 99,999 possible values. This is done with only twenty capacitors and switches. Construction of this unit is simple and straightforward. The capacitors are connected between a common line (B1) and one terminal on each switch. The other terminals on the switches are wired to B2. The switches are arranged in rows of four down and five across (Fig. 2-14 and Table 2-4). I used mica (5%) and polystyrene (2%) capacitors in my unit.

Of course, the tighter the tolerance on the capacitors, the more accurate the unit.

WIDE-RANGE Rf RESISTANCE BRIDGE

Figure 2-15 shows the wide-range bridge principle explained by a simple numerical example. The two legs of the bridge, R_A and

R_B, are made up of a single variable 100-ohm two-watt linear potentiometer. R_C is a half-watt 50-ohm carbon resistance. R_X symbolizes the unknown resistance we wish to measure. E is a voltage source, and I is the meter I will measure our null with when the bridge is in balance.

Without going into a lot of bridge theory and proofs that can be found in any textbook, all we need to know is the basic bridge equation for a balanced condition, which is: $R_X/R_B = 50/R_A$. Solving for R_X, we find that $R_X = (R_B)(50)/R_A$. Now, as I've already said that R_A and R_B are nothing more than a 100-ohm potentiometer, let us set the potentiometer near one end of its travel so that R_A is equal to ten ohms. Then R_B will be equal to 100–10, or 90 ohms.

Now, if we set the potentiometer near the end of its travel so that R_A is equal to 90 ohms, we find that R_B will be equal to 10 ohms. And, solving for R_x, we now find that $R_x = (50)(10)/90 = 5.56$ ohms. These calculations show that a wide-range bridge design can be achieved quite easily.

I know that already some of you will be saying "Why don't you run the potentiometer to zero for both calculations just shown and come up with a bridge with a range of from zero to infinity? That is a very good thought, but we do run into some practical limitations. We would find that our bridge sensitivity would have dropped to zero, for one thing. And secondly, as there is no such thing as zero

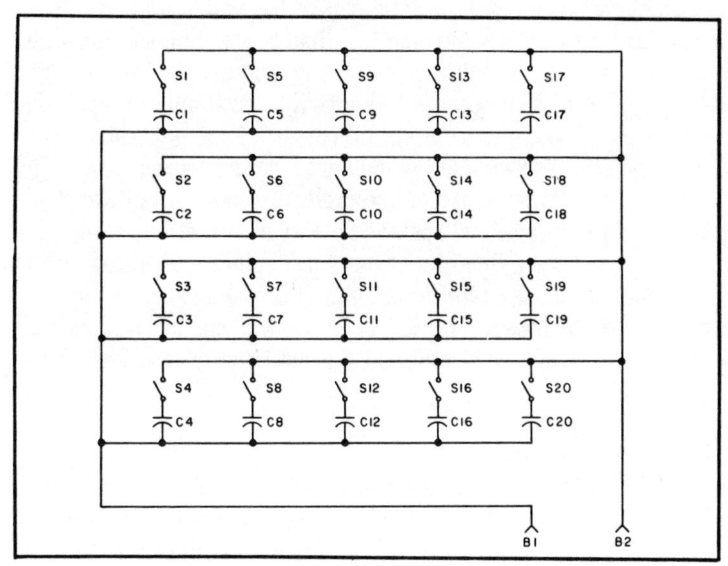

Fig. 2-14. Schematic for capacitor substitution box.

C1	10 pF
C2	20 pF
C3	30 pF
C4	30 pF
C5	100 pF
C6	200 pF
C7	300 pF
C8	300 pF
C9	0.001 uF
C10	0.002 uF
C11	0.003 uF
C12	0.003 uF
C13	0.01 uF
C14	0.02 uF
C15	0.03 uF
C16	0.03 uF
C17	0.1 uF
C18	0.2 uF
C19	0.3 uF
C20	0.3 uF

S1 through S20—SPST switches (slide or toggle type)
B1, B2—5-way binding posts
Misc.—wire, cabinet, rub-on letters and numbers

Table 2-4. Capacitor Substitution Box Parts List.

resistance when the potentiometer is at the end of its travel, our actual bridge range is somewhat less than zero to infinity. But with a small transistor amplifier we can easily increase the basic bridge sensitivity. And this increased sensitivity also allows our exciting voltage to be anything from an ordinary rf signal generator, to a grid-dip meter, to a simple low-powered variable-frequency oscillator. If a grid-dip meter is used, a single turn coupling coil attached to the bridge input jack will provide plenty of excitation voltage. In my own case, I use a Heathkit® Model LG-1 signal generator with a maximum output of a tenth of a volt. This is adequate for proper operation of the bridge. So, if care is taken in our calibration, we can come up with a wide-range bridge sufficiently accurate for all amateur usage.

The Actual Bridge Circuit

The basic bridge circuit just described is used in the final circuit of Fig. 2-16. As in all rf circuitry, every effort should be made to keep the internal bridge leads as short as possible to avoid excessive lead inductance and capacity to ground. The input and

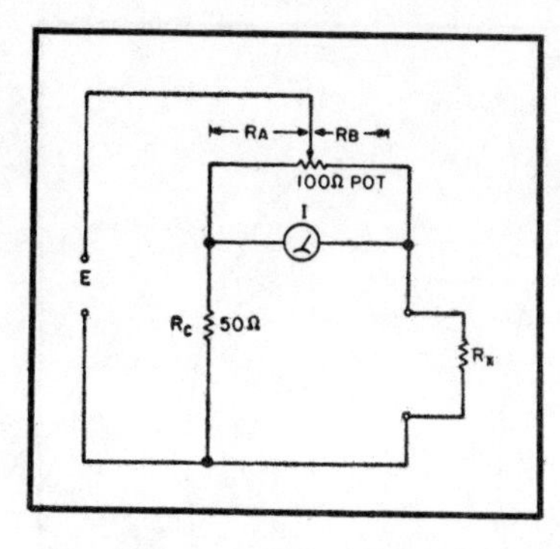

Fig. 2-15. Basic wide-range bridge.

output coax jacks are about an inch and a half apart on the chassis. This allowed very short leads to be used.

A pair of binding posts, A and B, are in parallel with the output jack J2 so that calibration resistors and wire-connected leads can be easily connected without having to use clip leads or other methods of connection. The 1N34 diode detector rectifies the nulling voltage and feeds it to the transistor amplifier through a 4.7k resistor, along with a couple of .01- μF bypass condensers. The transistor amplifier circuit was about the simplest one I know; it is a standard differential amplifier.

The transistors used were a couple of surplus PNP 2N396As. Any standard general-purpose transistor will work as well, although it may be necessary to use another value of resistance of R4 of Fig. 2-16. In my own case, I temporarily used a variable 25k resistor for R4 and adjusted its value until the amplifier gave the gain and stability I desired. Then the variable resistance was measured with an ohmmeter, and a fixed value used in its place in the circuit. It's a quick and easy way to optimize a circuit. The

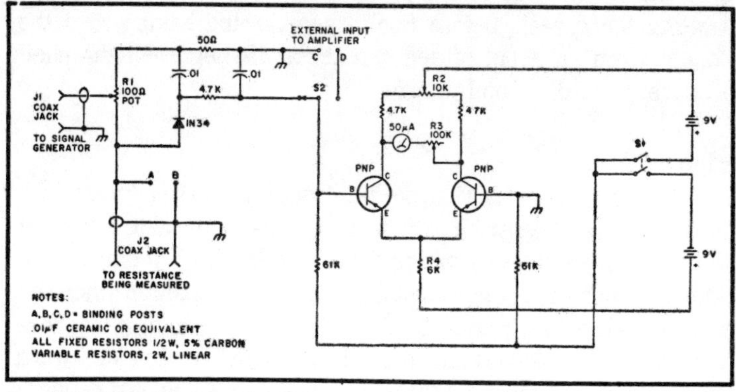

Fig. 2-16. Wide-range rf bridge.

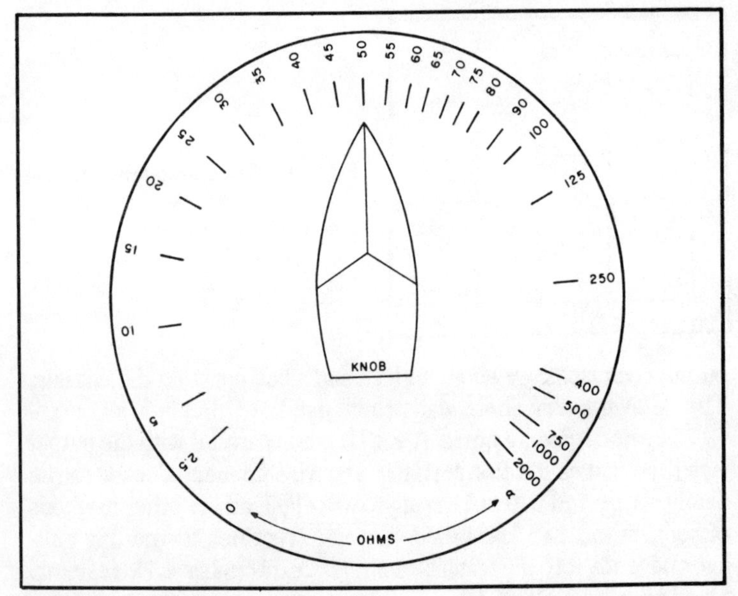

Fig. 2-17. Direct-reading bridge scale.

variable 10k potentiometer R2 used for balancing the meter current to zero during the initial null was an available 2-watt potentiometer, although a small trimpot can be used if desired. R3 is a 2-watt potentiometer-type variable resistance used as a sensitivity control that limits the current through the 50- μA microammeter when setting for initial null. Switch S1 turns the transistor amplifier "on" or "off". Switch S2 allows the transistor amplifier to be used for other purposes, as will be described later on. The entire bridge is self-contained in a 7″ × 5″ × 2″ chassis. A coat of gray enamel paint gives it a professional look. I have found that painting the chassis before final assembly and baking it in the kitchen oven for about fifteen minutes at 250 degrees Fahrenheit produces a hard, smooth finish.

Calibration

The simplest way to make the bridge direct-reading, which avoids the use of separate calibration curves, is to paste a piece of white cardboard directly on the chassis under the 100-ohm potentiometer knob. This allows actual values to be inked in when the calibration is made. Figure 2-17 is a theoretical calibration sheet illustrating the method just described. Theoretical values were shown, as every individual calibration will vary somewhat depending upon the linearity of the 100-ohm potentiometer and the actual values of the resistance of the potentiometer at the ends of its travel.

The calibration standards used were ordinary 5% tolerance small fixed carbon resistors. The calibration technique is very simple. Connect the calibration resistance selected across binding posts A and B. Set the balance control to about the midway position, and set the sensitivity control to minimum sensitivity (maximum resistance). Turn the power switch to the "on" position. Now adjust the balance control until the meter reads zero. Apply an rf voltage from the signal generator to J1, the input jack. The signal generator frequency I used was 3.75 MHz, although the frequency is not important since the basic bridge is not frequency-sensitive. The meter will read some current. Adjust the 100-ohm bridge potentiometer until a null is obtained. Now the sensitivity control can be increased to maximum sensitivity, and the output of the signal generator can be increased as desired. The bridge potentiometer is now rotated until the best null is obtained. The dial scale can now be marked for the calibration resistance value. As many calibration points as desired can be obtained in this way. The zero resistance point can be obtained by shorting the binding posts and observing the null. The infinite resistance calibration point is found in the same manner except that the binding posts are left open. In both of these instances, the null will be poorer than those obtained at other points on the scale.

Actual Uses of the Bridge

Although the bridge is normally used to measure rf resistance, its wide range allows it to establish other measurements where the actual resistance values are of secondary importance. Therefore, I am going to include those items which make the bridge so useful under these circumstances. These secondary uses are of great value, not only for obtaining specific data, but also in tying many theoretical concepts we read about to simple observations that we are now equipped to make. In almost all rf measurements, the circuit being checked is frequency-sensitive so that at resonance the circuit becomes a pure resistance. Knowledge of the resonant frequency is read from the signal generator. The resistance value read from the bridge will provide sufficient information for almost any problem that the amateur may want to solve.

Antenna Resistance at Resonance. Connect your antenna coax line to J2 and apply rf voltage from the signal generator.

Establish a first approximate null by rotating the 100-ohm potentiometer knob on the bridge. Then vary the frequency of the signal generator until the null is more pronounced. The final nulling consists of adjusting both the frequency of the signal generator and the variable control on the bridge. This is necessary to obtain the best null. At the lowest null obtained, the resistance will be the antenna system resistance, and the frequency will be its resonant frequency. I was careful to say antenna system, as this is defined as your antenna with the normal coax feedline attached. If for some reason you want to know the actual antenna resistance, then your feedline will have to be an electrical half wavelength long.

Tuning Your Matching Unit. This bridge is a wonderful device to allow you to load your transmitter to exactly 50 ohms without even putting a test signal on the air. You no longer have the worry of inadvertently loading up into a high swr condition and possibly damaging your final. At the same time, you can't create any unnecessary QRM on our already crowded bands. And once you've done this for any frequencies, just note the dial settings of your transmitter and matching unit, and you can quickly put your transmitter on and be assured that you will be perfectly matched. First off, load up your transmitter at the desired frequency into a dummy load. (Normally, this is 50 ohms.) Then connect J2 to the input of your matching unit, and connect the output of your matching unit to your antenna system. Select the frequency you just tuned up your transmitter to on your signal generator. Then set the variable 100-ohm potentiometer on your bridge to 50 ohms on the calibrated scale. Turn up your bridge, and adjust your matching unit until the best null is obtained. You are now completely tuned up. You don't have to touch your transmitter tuning or loading at all. Just reconnect your transmitter to your matching unit input and you are finished. Actually, I can do the above more quickly than I've written about it. And in addition, when you are done, you can't even see your swr meter or reflected power meter wiggle when you put your transmitter on the air. It's really fun to amaze your friends by obtaining such a perfect match, quickly and scientifically, and without any QRM on the air. It's tuning up your station in a really engineering fashion.

How Good is Your Matching Unit? From the preceding step it is easy to go a bit further and establish the resistive matching limits of your tuning unit. It's a quick method of comparing one kind of a unit against another. I'm sure we all agree that to define the limits of a matching unit as "being able to load into a random wire" is not a very exact technical description. This is particularly true since random could mean anything from a couple of feet to a couple of wavelengths. I've found in my own case that my bridge has allowed me to design a wide-range matching unit, and quickly establish which variables are important to obtain the wide range I desired for my specific antenna needs.

All you do is see what the resistance limits are that your matching will load into and still retain 50 ohms at the input jack. The technique is very easy. You can use your 5% calibrating resistors for this purpose. Just connect a resistance to your matching unit output and jack J2 of the bridge to the matching unit input. Set the bridge resistance knob to read 50 ohms on the scale. Now adjust your matching unit until you can achieve a bridge null at the frequency of operation you have selected on the signal generator. In this way you can quickly determine the upper and lower resistance loading limits of your matching unit. If the resistances you select are too high or too low, you will not be able to obtain a satisfactory null on the bridge. My home-built matching unit matches from about 10 to 200 ohms without any difficulty. Generally speaking, the wider the resistance range, the wider the range of the matching unit for varying impedances. And you'll be surprised to find the different ranges obtained by different kinds of matching units. It's an education in itself.

Resonant Frequency of Tuned Circuits. The ease with which the bridge can determine the resonant frequency of a totally enclosed tuned circuit is quite amazing. And it will even tell you whether or not it is a series-tuned circuit or a parallel-tuned circuit. All you need is two leads from the circuit being tested. Connect the leads to output binding posts A and B. We know that if a circuit is a series one, its electrical resistance will be low. So, just to obtain the best null, vary the signal generator frequency and the 100-ohm control at the low resistance end of the scale. At the best null obtained, the signal generator frequency will be the resonant frequency of the tuned circuit. The resistance reading will be its equivalent series resistance.

The parallel circuit is measured in the same way, except that the bridge potentiometer is at the high end of the scale. The resistance value measured will be the equivalent parallel resistance of the tuned circuit. And, as before, the signal generator will indicate the resonant frequency of the tuned circuit. I mentioned that the tuned circuit could be completely enclosed. This bridge can measure resonant frequencies where it would be impossible to inductively couple in a grid-dip meter to make a similar measurement.

Measurement of Velocity Factor of Coax Cable. The technique of measuring the velocity factor of coaxial cable is simple when it is realized that an electrical quarter wavelength of line (or odd numbered multiples thereof) open at the far end acts like a series-tuned circuit at the near, or measured, end. Just connect your piece of coax cable to J2 and null the bridge at the low resistance end of the scale, at the same time varying the signal generator frequency for the lowest frequency for the best null obtainable. The equation for velocity factor in this case is: Velocity factor = (length in feet) (F_{MHz})/246 for a quarter-wave section of line. And now, as an experiment, if you triple your signal frequency, your line will again null the bridge. It is now acting as a three-quarter-wavelength line. In this way, it is very easy to demonstrate basic transmission line theory.

The same principle is used in determining the electrical half of a line open at the far end. This condition is equivalent to a parallel-tuned circuit. Just adjust your bridge null at the high resistance end of its scale, and at the same time adjust your signal generator for the lowest frequency that will give you the best null. The velocity factor can be checked and the formula will be velocity factor = (length in feet) (F_{MHz})/492 for a half-wave section of line. You can also use this method for setting up for a half-wave section of line if desired.

Characteristic Impedance of Coax Cable. The characteristic impedance of a coax cable can easily be determined with the rf bridge. All you do is connect one end of the cable to J2 and connect a selected resistor across the other end of the cable. The selection of this resistance is such that when the signal generator frequency is varied there is no change in the bridge reading. At that point, the resistance selected is equal to the characteristic impedance of the coaxial cable. It is an easy matter to try several different values of resistance to terminate the cable, so that a variation of signal generator frequency will show no change in the meter reading. The null can now be obtained, and the resistance reading will be the characteristic impedance of the cable. The value read will be equal to the terminating resistance at the end of the line. This illustrates the basic principle that a properly terminated line presents a constant pure resistance to a varying input frequency.

Reflected Resistance of Tuned Coupled Circuits. Although it is possible to wade through a great deal of circuit theory to prove that resistance can be reflected from one circuit to another, this easily performed experiment will show it in a manner in which, if once performed, it will never be forgotten. An actual experiment is such a wonderful way of firming up any theoretical proof in anyone's mind. The principle of reflected resistance can be demonstrated as shown in Fig. 2-18 where a parallel-tuned circuit is inductively coupled to a series circuit. Loading down the secondary with a resistance generates a reflected resistance in the primary whose value can be determined by measuring the change of resistance with the bridge. Close-coupling effects can also be demonstrated by observing the re-tuning necessary to reestablish resonance when the secondary is coupled into the primary. This item described is an easy way to demonstrate rather complex principles, particularly when teaching basics to students studying for an amateur license.

The Transistor Amplifier. In addition to the many uses just described, I found that the addition of an SPDT switch, S2, and a couple of binding posts, C and D, would allow me to use the transistor amplifier for other duties in my amateur station. One valuable use is as a field-strength-meter amplifier. My regular simple field-strength meter diode circuit is now far more sensitive, which allows me to make my field-strength measurements further away from the antenna and avoid close-field effects. This makes my field-strength measurements more accurate and meaningful. A second use is as a small experimental capacity bridge, which I built using the same principles as the bridge previously described.

AC WATTMETER

The basic circuitry is as shown in Fig. 2-19. The formula I used to determine the milliammeter range is: watts ÷ line voltage ÷ 1.414 ÷ ratio = meter current in amps.

To construct your unit, you will need to decide what are the minimum and maximum wattages you will want to measure. Since

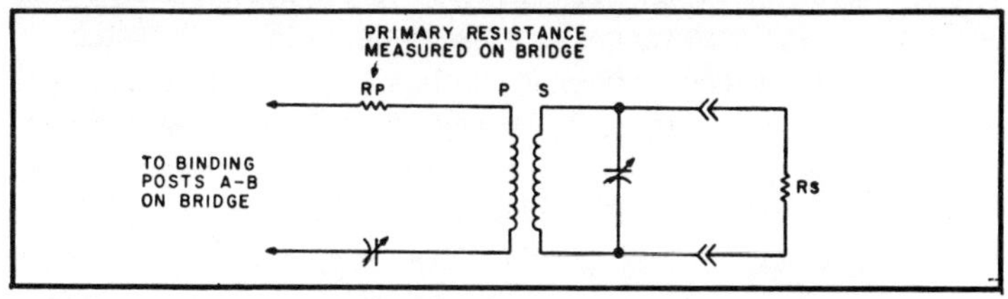

Fig. 2-18. The effect of secondary resistance being reflected in primary circuit.

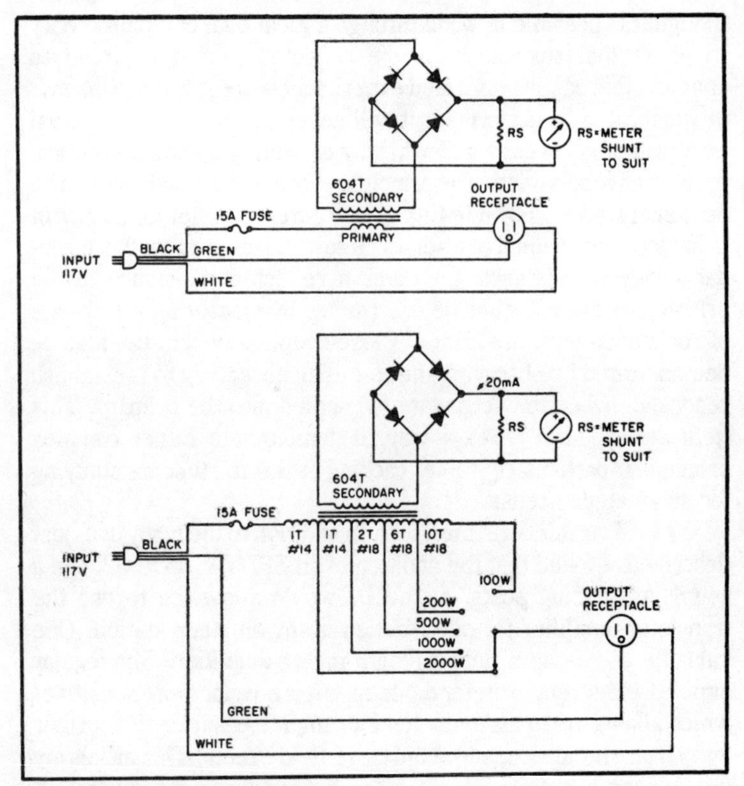

Fig. 2-19. Multirange circuitry.

the meter scale is not a linear function, the meter scale will be compressed at the lower end. I settled on a multirange unit.

The fuse should be of the household variety, since the 15-amp size of type 3AB is difficult to find, and the type 3AG will vaporize on the interior of the glass and provide no protection.

The bridge rectifiers can be made of type 1N4001 diodes, since they only supply the meter movement current.

To have a multirange instrument, you can use either multiple outlets, one for each range, or a heavy-duty (15-amp) switch.

The current transformer is the heart of the instrument and must be wound to suit your desires, as it is not commercially available unless you want it custom made. To wind one yourself is not a big undertaking and only requires some wire, tape, and an old core. I used old speaker output transformers for my cores. The size is not too critical, since there is only miniscule wattage requirement of the meter movement. You should ascertain that you will have ample winding space. I wound my transformers with 604 turns of #36 on the secondary for 117 volts line voltage. If you have 120 volts, use 590 turns, and for 115 volts, use 615 turns. The primary was wound as follows: 10 turns #18 tap, 6 turns #18, tap, 2 turns #18, tap, 1 turn #14, tap, 1 turn #14. This gave me a total of 20 primary turns which, with the secondary of 604 turns, resulted in the following ratios: 100 watts = 604:20 or 30.2, 200 watts = 604:10 or 60.4, 500 watts = 604:4 or 151, 1000 watts = 604.2 or 302, 2000 watts = 604:1 or 604.

The meter used had a fullscale value of 20 millamperes. A recalculation of turns ratio may be made to accommodate another meter range. A simple solution for the utilization of, say, a 0-1 milliammeter is to put a 100-ohm trimmer pot across the meter movement and adjust it to read 100 watts for full scale deflection.

To calibrate my units, I used several 25-50-100 watt light globes to provide me with loads to calibrate the meter scale. By using them in parallel, I was able to acquire enough plot points to make a new meter scale. I made my meter scale on white paper, then inked the marks to suit and reinstalled it. I pasted the scale on the reverse side of the original scale using rubber cement.

An error of ± 5% can be expected, since the line voltage will vary by this amount. This error should not be objectionable, since the equipment will be subject to the varying line voltage in operation anyway. In order to lower the error rate, it would be necessary to utilize an electrodynamic meter movement, whereby the electromagnetic field of the meter could vary with the varying line voltages.

RF WATTMETER

As the schematic (Fig. 2-20) shows, the circuit is a basic rectifying type rf voltmeter. The prototype was built in a small can of the plug-in-module variety that was scrounged from the junkbox. About the only critical part is the series resistor. The capacitors in the prototype were mica, but ceramic disks would work as well. The diode can be a 1N34A, 1N270, 1N52, 1N38A, or just about

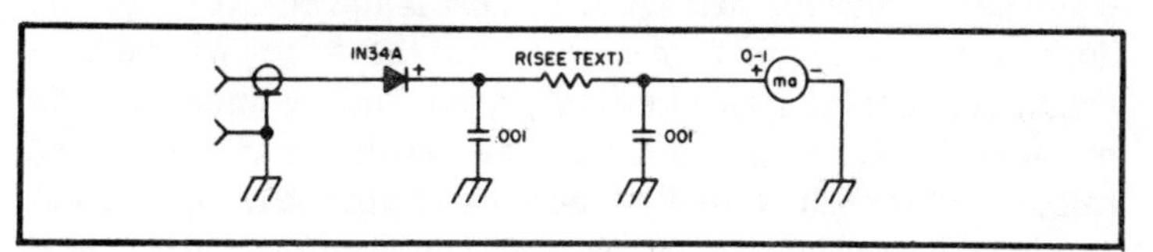

Fig. 2-20. Schematic diagram of the simple, accurate, and easy-to-build rf wattmeter.

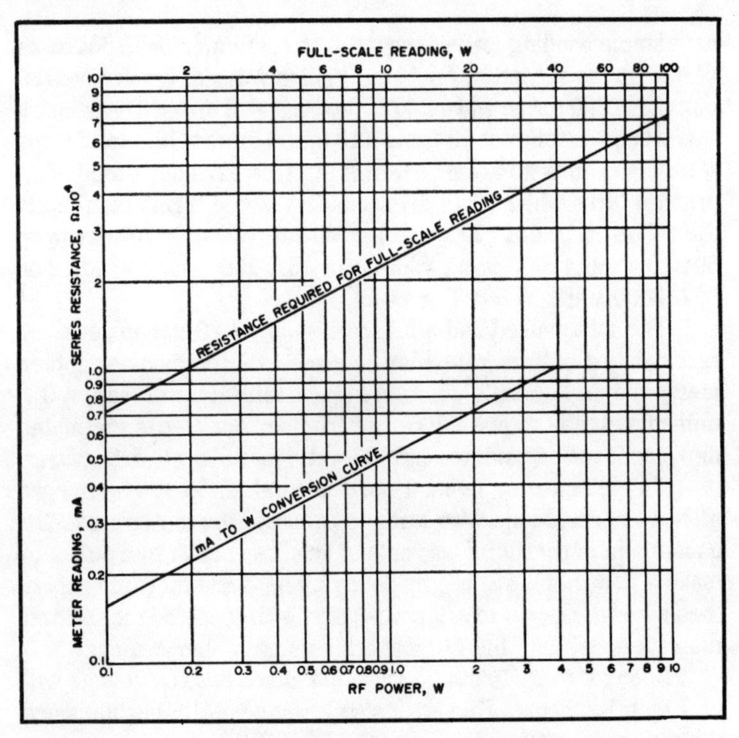

Fig. 2-21. Logarithmic plots for determining power. The upper curve gives resistance values for determining what the full-scale meter deflection will be (remember to multiply the series resistance value shown on the chart by 12 kΩ). The lower curve will allow you to determine your precise power out if you use a 0—1 mA meter.

anything else. Use the old ham's rule of thumb: "When in doubt, try it out!"

Two sockets might prove more convenient rather than one with a coaxial tee as shown. Conventional minibox construction or building into a new or existing rig will be more than adequate. Point-to-point wiring is used to permit compactness and reduce lead length.

"Fine," you say, "but I don't have a huge mound of test equipment. How do I calibrate it?" That is the beauty of it—you don't! If the series resistor is accurate, the meter will be self-calibrating to a log scale. Remember, you know R and Z, and the full-scale W. Now assume a half-scale reading, $I = 0.0005$, and calculate W for half-scale. Plot these two points at 1.0 and 0.5 mA on Fig. 2-21, and connect by a straight line, which you may extend the length of the graph.

Install the meter through a coaxial tee at your antenna connector, or through some other predetermined means, and terminate with a dummy load. Apply power and read the meter. That's it! The meter can be used with an antenna if your swr is below about 1.2:1.

CAPACITY METER

The generator is a 555 oscillating at approximately 200 kHz. The frequency is adjusted by R2 when calibrating prior to use or during use, if a reading is doubtful.

The output of IC1 is used for the 0-to-100 pF range and is also used to clock the first 74LS90. The output of IC2 (pin 11) is the 555 frequency divided by 10. This frequency is used for the .001 full-scale range. The remaining 74LS90s operate similarly. Each division by 10 is the frequency for the next larger decade of capacity. The 74LS00 was added as a buffer. Since it would be wasteful with the other gates doing nothing, the LED circuit was added to give a visual indication that the clock and all dividers are

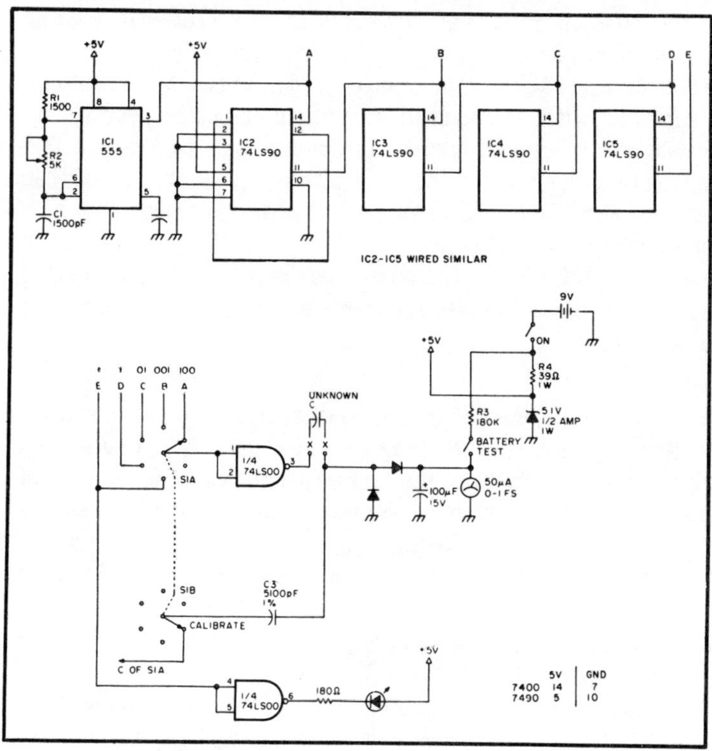

Fig. 2-22. All resistors are ¼ watt except as noted.

working. There is a definite pulsating of the LED, so the user is sure that the circuit is working.

The accuracy of the instrument is very dependent upon the tolerance of the calibration capacitor, C3. Prior to use, C3 is switched into the circuit and the frequency of IC1 is adjusted by R2 until the meter reading is equal to the value of C3. Figure 2-22 shows C3 as being 5100 pF, 1% tolerance. Actually, any convenient value could be used, but the tap on the range switch would have to be changed to make sure the correct frequency is being applied through C3.

The condition of the battery can be checked by pressing S3. When making a battery test, ensure that there isn't a signal path through a capacitor to the meter, as this would give a higher battery voltage reading than really exists. The multiplier resistor will give a reading of .9 on a full scale of 1.

Construction

I like sockets. It makes troubleshooting easier if a chip must be removed, but sockets are not necessary. All 74LS90s are wired the same as IC2. The range switch is wired to make 5 ranges available. The additional position is used as the calibrate position. A separate switch can be used for calibration, but the user has the chance of damaging his meter if the calibration capacitor is a .005 and the range switch is in the 0-to-100 range position. The ICs can be replaced by 7490s and 7400s, but the current drain is rather high for a 9-volt battery. If a power supply is used, there is no problem using the TTL chips. The meter rectifiers are run-of-the-mill diodes. The meter is a 50 μA meter with a 0-to-1 scale marked off in .1 readings.

Use

Prior to use, the meter should be calibrated to compensate for those variables that cause errors. After calibration, the unknown capacitor is placed across the test terminals and read. A word of caution is offered: Use this meter as you would a voltmeter in a strange circuit. Start on the high range and work down until you get a reading. This prevents pegging the meter and possibly damaging it.

AUDIBLE TRANSISTOR TESTER

If this sad experience has happened to you, it may be because you are like many experimenters who routinely use an ohmmeter to test transistors and diodes. Though this method is effective, it can also cause the instant death of low-current semiconductor devices. The typical ohmmeter can, on some scales, apply a rather high voltage and current across the test leads, which may cause the destruction of some semiconductor devices. Another fault of the ohmmeter method is that the instrument must be checked visually; this can lead to erroneous results or damage to equipment if a test lead slips while making continuity checks. What is needed is a low current (under 1 mA) device which gives an audible indication of continuity. Such a device should also be low in cost (for those of us who are cheapskates or poor) and simple to assemble. Figure 2-23 shows such a device using a single 555 IC timer as a square wave oscillator and an NPN general purpose transistor for switching. Any method of construction can be used, as there is nothing critical about the circuit, but the most convenient method is perfboard and a socket for the IC. The checker can be assembled as shown in Fig. 2-23 for use as a continuity tester or, if desired, the device can also be used as a code practice oscillator. If it is desired to use the device as a CPO, it would be advantageous to be able to vary both the tone and volume, although this will run the cost up a couple of dollars unless the builder has the required parts in the junk box. The two potentiometers can be added to the circuit as indicated in Fig. 2-24.

VERSATILE TRANSISTOR TESTER

Many of the new transistor testers appearing on the market today provide two unique features:

1. Random lead connection.
2. Identification of transistor leads.

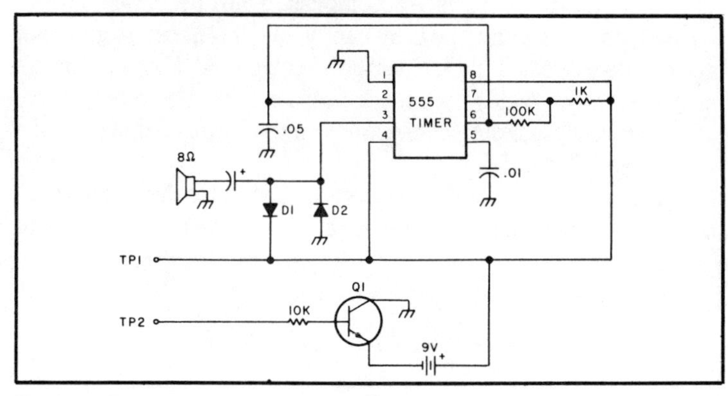

Fig. 2-23. Test oscillator (low current).

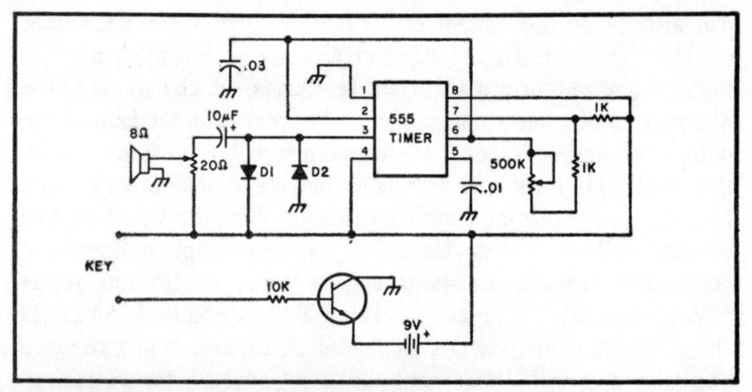

Fig. 2-24. Code practice oscillator.

With these features, the test procedure is greatly simplified, and the person performing the test is assured that the test leads are *always* connected correctly to the transistor under test. The advantages become immediately obvious when you are culling through a large assortment of surplus transistors or when performing in-circuit tests on a crowded circuit board where it is often difficult to identify the leads on a transistor.

The principle on which this circuit operates is simple. There are only six possible ways in which the leads can be arranged on the base of a transistor. See Fig. 2-25.

By the simple expedient of interposing a 3-pole, 6-position switch between the test leads and the transistor, it is possible to present to the input of the tester, one by one, all six of the possible lead arrangements. The circuit for this switch is shown in Fig. 2-26.

Up to this point we have accomplished the goal of random lead connection. To identify the transistor leads, it is necessary to color code the wires from the switch to the transistor. Next, prepare a chart which correlates the switch position with the colored wires and their corresponding connections to the input terminals of the tester. See Table 2-5.

For convenience, this chart can be attached to the box in which the switch is mounted, or it can be affixed to the transistor tester

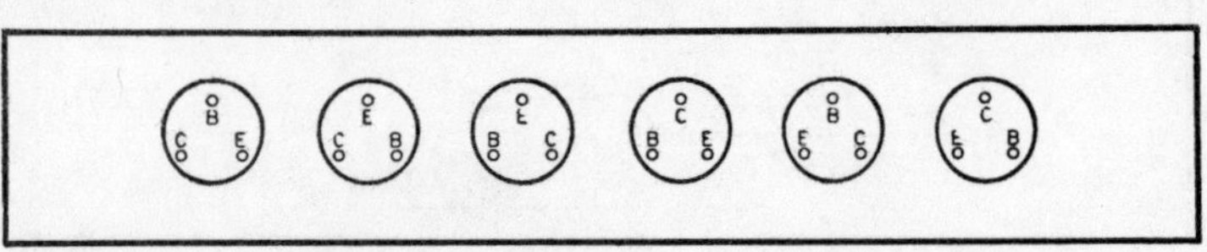

Fig. 2-25. Possible transistor leads.

itself. One thing which should be mentioned is that the leads from the switch to the input of the tester should be connected exactly as shown on the schematic in Fig. 2-26. If these leads are changed around, you will still retain the random lead connection feature, but you will no longer be able to identify the transistor leads.

When using this switch for the first time, you will notice that there is more than one position which will give an up-scale reading on the meter of your tester. This presents no particular problem in the interpretation of the test results, once you understand why this happens.

To illustrate this, select a good transistor on which you can positively identify the leads, and connect it to your tester. After you have made the initial adjustments on your tester, press the gain button and observe the beta. Now reverse the emitter and collector leads on the transistor, and repeat the test. You should observe that the transistor still has a forward gain, but the beta will be lower than that obtained previously. Why does this occur?

Actually, nothing strange or contrary to the laws of transistor physics has occurred here. The emitter and collector of any transistor are made of the same type of material and, from a theoretical standpoint at least, we could say that the designation of the emitter and collector are purely arbitrary.

In the real world of transistor design, however, the physical structure of the emitter and collector are different. These differences are dictated by design considerations such as input-output capacitance, collector heat dissipation, and forward gain requirements. From a practical standpoint, all of these factors add up to one thing when testing a transistor. The connection of the emitter and collector test leads which give the highest beta reading on the meter positively identifies the emitter and collector leads on the transistor.

On some testers, a collector-base reversal will give an indication on the meter during the initial setup and adjustment of the

Table 2-5. Color Codes.

Switch position	Collector	Base	Emitter
#1	red	yellow	black
#2	red	black	yellow
#3	yellow	red	black
#4	black	red	yellow
#5	black	yellow	red
#6	yellow	black	red

tester, but the transistor will not have any forward gain. In this case, the emitter-base junction is forward biased, as it should be, and it is this current that you see indicated on the meter in the first part of the test. The collector circuit, however, is also forward biased, and, therefore, the transistor will not have any forward gain.

SEMICONDUCTOR TEST GADGET—USE WITH YOUR SCOPE

Transistor testers come in many assorted kits and variations providing confusing data such as BV_{cbo}, BV_{ces}, BV_{evo}, R2D2, $C3_{po}$, and so on. All I want to know is if the transistor or diode is good or bad.

This transistor/diode tester tells you just that. With a glance at an oscilloscope face, you can see shorts, opens, and leakage between collector and emitter and determine the overall quality of the transistor in a matter of seconds.

All you need is an oscilloscope that works reasonably well, a 110/6.3-volt transformer, 2 resistors, an SPST switch, and a set of probes. See Fig. 2-27.

The transformer provides low power to the transistor or diode to be tested and is read directly on the oscilloscope face.

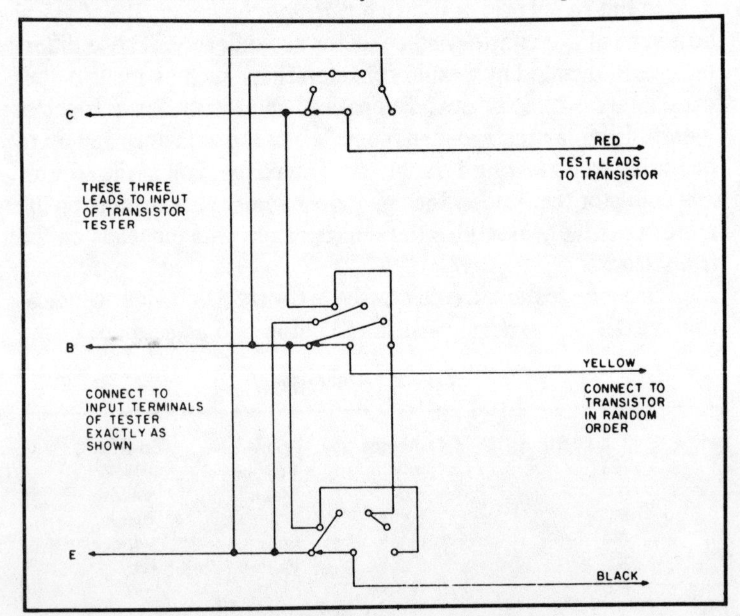

Fig. 2-26. 3-pole, 6-position switch (Mallory 3236J).

An in-circuit or out-of-circuit switch is provided so that the transistor or diode does not have to be removed from the circuit to be checked. However, I feel the transistor is checked best if removed; this is a matter of personal preference.

At the in-circuit position, both resistors are in series providing the lowest current applied to the circuit. This low current, 1 mA or less, should not harm surrounding components associated with the transistor or diode being tested.

This transistor/diode tester is to be used on deenergized circuits only. I have not used it on an energized circuit, but I feel it will be of no real value there.

After the tester is installed, touch the probes together. The scope will go from its normal horizontal line to a vertical line. Adjust the horizontal and vertical gains so that the line will be the same length both vertically and horizontally. Adjust the centering of both vertical and horizontal positioning so that the line will be in the center of the scope. You are now ready to check transistors and diodes.

Select a transistor from the junk box; determine if it is silicon or germanium and hold one probe on the base and touch the other probe to the collector or emitter. If the transistor is good, the screen should show a right angle. (If it is inverted, it doesn't matter; just turn your probes around.) Now read from collector to emitter. A straight line should show. This is a good silicon transistor.

Germanium transistors give a reading that will be slightly different. From collector to emitter only, they will give some type of right angle. With germanium transistors, I recommend that you compare them with others of the same number out of the circuit.

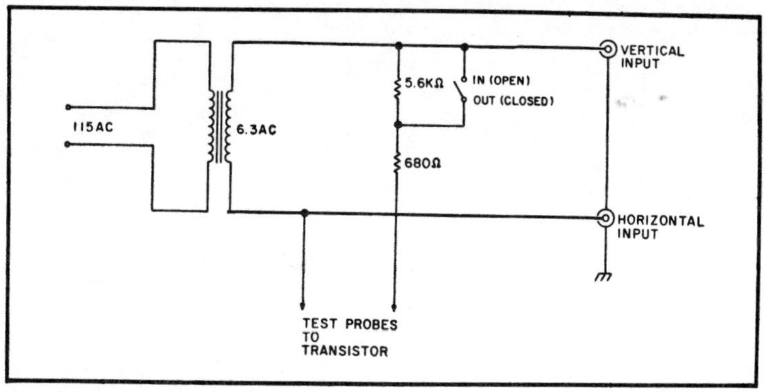

Fig. 2-27. Test gadget schematic.

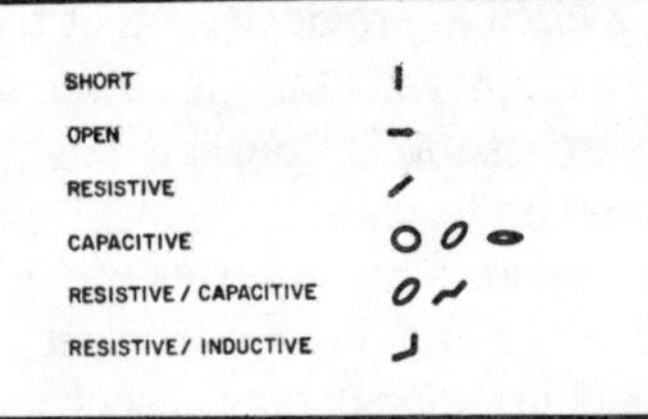

Fig. 2-28. In-Circuit readings.

Figures 2-28 and 2-29 will give you an idea of what to expect on your oscilloscope. Some of them can be confusing when checking transistors or diodes at the in-circuit position; this is why I feel that out-of-circuit is best used to check transistors and diodes. However, after some experience using this device and a little circuit tracing to find resistors and capacitors used in conjunction with the transistor or diode, you should be able to determine with some degree of accuracy if the circuit is operational.

BARGAIN ZENER CLASSIFIER

The circuit is as simple as they come. Follow Fig. 2-30 and you should have no problem building a useful gadget which comes in very handy almost 90% of the time you build a new project. Looking at the schematic, you'll notice the power transformer is very important as it isolates the circuit from the ac line for safety. Next, the diode rectifies the ac into pulsating dc. The following 120k resistor drops the voltage to the potentiometer. The electrolytic capacitor smoothes out the dc ripple to provide a dc voltage of about 30 volts to the alligator clips. (It's the next thing to an instant project.)

Place the cathode (the end with the band) of the zener that you wish to classify in the plus voltage alligator clip, connect the vom,

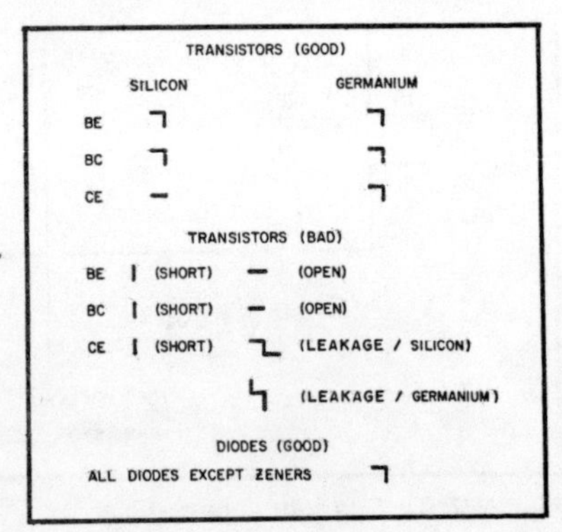

Fig. 2-29. Out-of-Circuit readings.

as shown in Fig. 2-31, run the voltage up until the vom needle stops, and that's the zener regulating, or breakdown, voltage. If you connect the zener backwards in the circuit, you should get very little, if any, indication on the vom meter because the diode is conducting and presents a short circuit to the tester. This circuit passes very little current through the zener under test, so you shouldn't have any problems with zener burnouts.

If, after classifying all your diodes, you still don't have the correct value, try putting two or three of them in series. When I built my power supply, I needed a 12-volt to 13-volt zener. I put an 8.2-volt and a 4.3-volt zener together in series to obtain 12.5 volts and, because of this, I could also tap at the junction of the zeners to obtain a regulated 8.2 volts.

You'll find that regular diodes and transistors have a zener action as well. Although voltage values will be random and current capabilities small, they can often be used in place of a regular zener diode. When the transistor is bad anyway, it beats throwing it away.

You'll notice all parts have a wide tolerance, so you can use any of the old parts you may have in your junk box. See Fig. 2-32.

IC AUDIO FREQUENCY METER

The 555 timer integrated circuit can easily be made into an inexpensive linear frequency meter covering the audio spectrum. The 555 is used in a monostable multivibrator circuit. The monostable puts out a fixed time-width pulse, which is triggered by the unknown input frequency.

Referring to Fig. 2-33, transistors Q1 and Q2 are used as an input Schmitt trigger. The unknown frequency input is clipped between 9 volts and ground by these transistors. Positive feedback is used to insure the waveforms have fast, clean edges. The output of Q2 is a square pulse with the same frequency as the input signal. The output of Q2 is differentiated by C2 and R2 to provide a short pulse for the 555. A small signal diode is connected across the

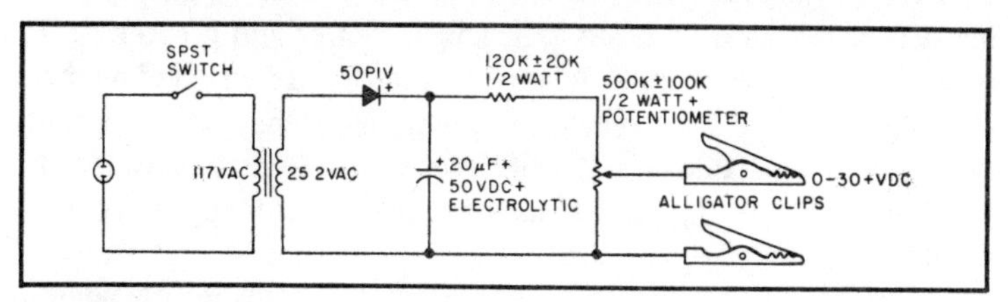

Fig. 2-30. Zener tester.

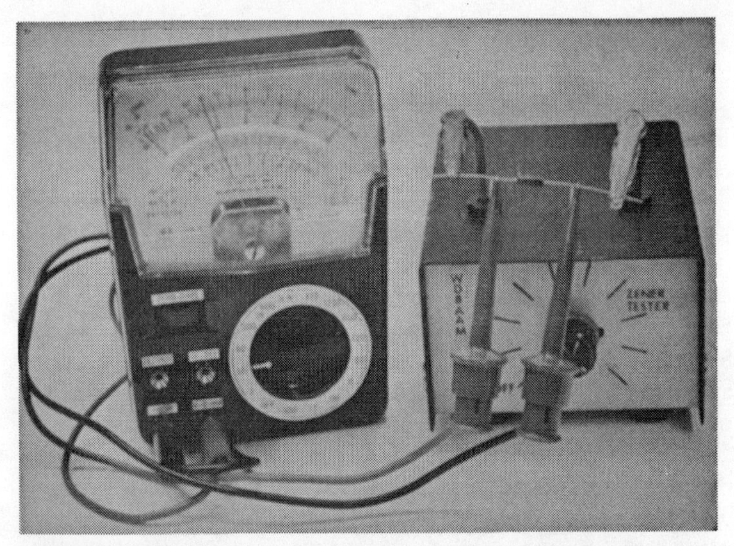

Fig. 2-31. Unit with a zener diode being tested and showing proper hookup to the VOM. Note that the VOM is indicating 8.6 volts even though the tester voltage output is 28 volts. Don't forget to set the VOM to measure dc voltage. It doesn't show up very well in the photo, but I used red wire for the plus alligator clip and black for the negative. These wires were then inserted into 2 phone tip jacks on the top of the case.

differentiator to insure that the 555 input never exceeds 9 volts. The 555 is connected in the standard monostable circuit. Since a Schmitt is used to trigger the monostable, square, sine and ramp type waveforms may be used at the input to the frequency meter. A nominal voltage of 1 volt rms is required to trigger the Schmitt circuit.

The range scale timing resistors R3, which determine the monostable pulsewidth, are small potentiometers mounted directly on the circuit board. These pots are used to calibrate each frequency scale.

The output of the monostable is a fixed width pulse. Every time a zero crossing of the unknown frequency occurs, the monostable is triggered. Thus, as the frequency of the trigger pulse increases, the monostable output has a greater and greater duty cycle. The frequency limit on any one range is determined by the R3C3 time constant chosen. As the input frequency becomes too great, the monostable output will never return to zero, because of constant re-triggering, and a constant 9 volts will appear at the output.

The monostable output is a pulse with a duty cycle dependent upon the input frequency. Thus, by intergrating or averaging the output waveform, a dc voltage is developed. This voltage is directly related to frequency. Resistor and capacitor R4C4 is used as a pulse averager, important on the lower range setting. As the input frequency increases, the panel meter itself can act as a waveform averager. Input frequencies greater than 50 Hz will be averaged by the meter fairly well; however, at lower frequencies the meter will respond to each cycle of the unknown frequency input. The meter is used as a high impedance voltmeter. A 1 mA meter could be substituted with a change of resistor R5. For a 1 mA meter, R5 should be about 9.1k and R4C4 should be changed appropriately. R4 should be a factor of ten less than R5, and the same R4C4 time constant should be kept. This would make C4 a very large value, so R4C4 could be left out if the lower frequencies are of little interest (less than 50 Hz).

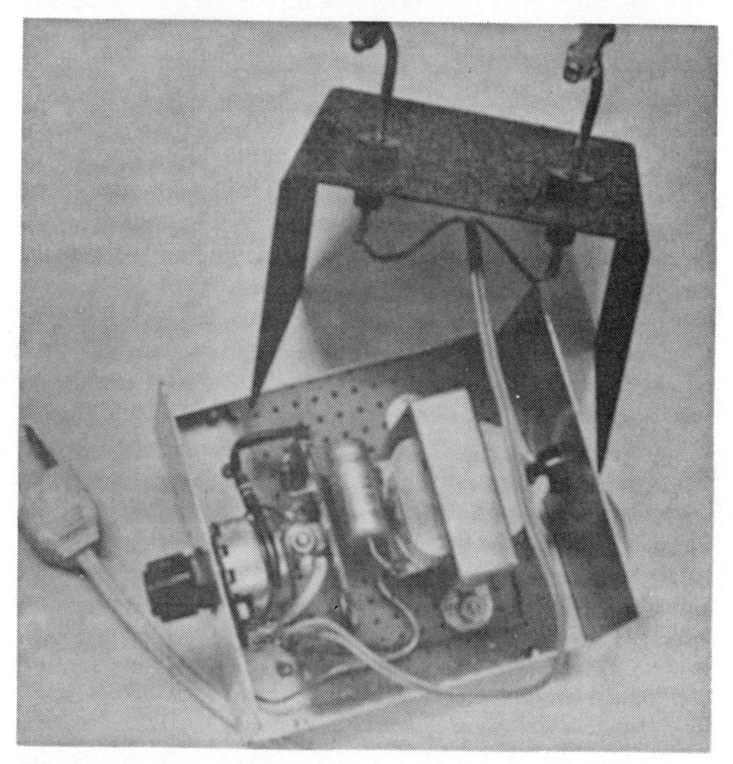

Fig. 2-32. Interior view. All the parts came from my junk box with the exception of the case and power transformer which are Radio Shack stock items. A small piece of perfboard was used to mount the parts, and spacers were used to lift the perfboard off the metal chassis.

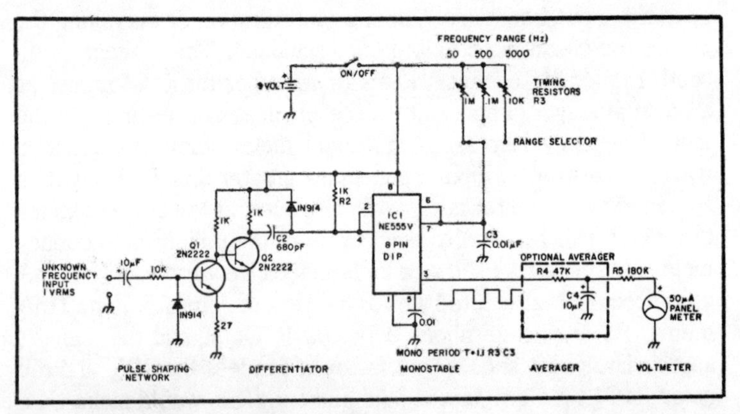

Fig. 2-33. 555 timer frequency meter.

The range scales are set up in decades. To calibrate each scale, a standard input frequency is connected to the input. About 1 volt rms is needed to trigger the first stage. The monostable output pulse period must be less than the maximum input frequency to be measured on each scale. With the maximum input frequency applied, each range potentiometer should be adjusted until the value of the input frequency corresponds to the full scale meter reading. The duty cycle of the monostable should be roughly 90% at the maximum frequency input for each scale. This will give the maximum dynamic range on each scale. This will give the maximum dynamic range on each scale setting. If, for a 90% duty cycle, the meter will not read full scale, meter resistor R5 should be lowered accordingly. Each scale setting should be calibrated by adjusting the respective R3 potentiometer.

During operation, when the scale reads off scale on any range, the scale should be changed to the next higher setting. Once calibrated, this frequency meter should read within 5% of full scale. The useful frequency range of the meter is from tens of hertz to well over 50 kHz. Although decade ranges are shown, the ranges between the decades can easily be added to give as many frequency ranges as deemed necessary.

EFFECTIVE RADIATED FIELD METER

The construction of the meters is very basic and quite flexible. As you design rf meter for higher frequencies such as the VHF bands, you should place components so that lead lengths in the rf circuitry are as short as possible. The enclosures I used were

Universal Meter Cases but most any type chassis or small cabinet will be adequate. The enclosure ought to be metal as the circuitry should be shielded. To shield the Universal Meter Case, I used some roof valley aluminum to enclose the rear opening. It can be easily cut to size with hand scissors. You will find this available at most lumber yards or hardware stores.

Instrument Sensitivity

Sensitivity of the instrument is determined by the full scale deflection of the meter chosen. The smaller the full scale deflection, the greater the sensitivity; for example, a 0-50 μA meter will be more sensitive than a 0-1 mA meter. I used a 0-1 mA meter for the untuned meter and a 0-200 μA for the tuned meter.

The length of your reference antenna will also affect sensitivity: the longer the antenna, the greater the sensitivity. Also, dc resistance of the rfc will affect sensitivity. Choose an rfc with low dc resistance for maximum sensitivity.

Instrument Application

Once you have your field strength meter constructed and operating, you can begin making some reference measurements for future use.

What I do is install a small reference antenna outside the immediate area of the transmitter; it does not have to be a great distance from it. The important item is not to change the location or length of the reference antenna once you start basic measurements.

For measurements with a beam, I note and record on paper degrees of rotation "direction" and power input into antenna transmission line. With a dipole or fixed array, direction need not be recorded, but you must record input. Next, I proceed to measure field strength with the meter. I record in my log the relative reading for future reference, and if I suspect a problem antenna or otherwise, I immediately go back to my original readings and make a measurement. If you have significant energy loss in your ERF, it will show.

You will also find the "ERF" meter a valuable instrument for mobile or portable operation See Figs. 2-34 and 2-35.

SIMPLE BRIDGE FOR MEASURING METER RESISTANCE

There comes a time in every ham's life when he must seek that unknown meter resistance. Here's a simple solution to that age-old

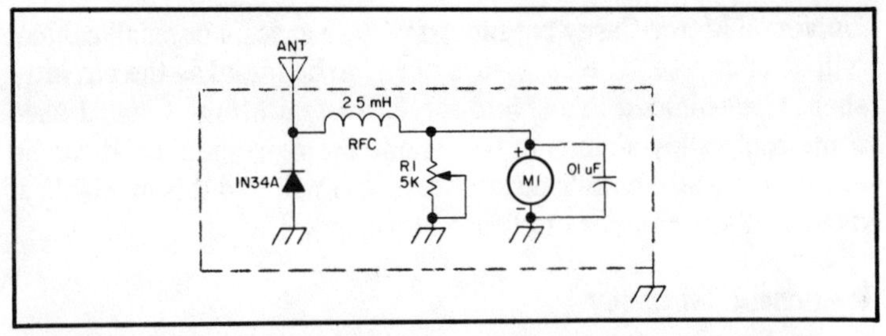

Fig. 2-34. Untuned circuit. M1 — Dc meter (see text). R1 — Sensitivity control.

problem. The schematic is shown in Fig. 2-36. It's equivalent circuit is shown in Figs. 2-37A and 2-37B.

In Fig. 2-37B, R2 is equal to R2, and R_{BP} is the equivalent parallel resistance of branch 1 and branch 2. Neglecting R_{BP}, the current through R2 would be 1.5 (E)/1500(R) = .001 A or 1 mA, the full-scale reading of most meters. Thus, when we reinsert R_{BP}, we know that the current is less than 1 mA. This keeps the current through each branch (Fig. 2-36) less than 1 mA, protecting both meters.

In Figs. 2-36 and 2-37B, when the resistance of branch one is equal to the resistance of branch two, the currents through both are equal. Thus, you know that when the reading on the meter under test and the current reading on your meter are equal, the resistances of the two branches are equal.

The resistance of branch one is equal to the resistance of M1 (which *must* be known) plus R1, a potentiometer with a calibrated dial. If we select Rx so that it is equal to R_{M1}, then, when R1 is equal to $R_{M\text{-}test}$, the resistances of the branches are equal. If the resis-

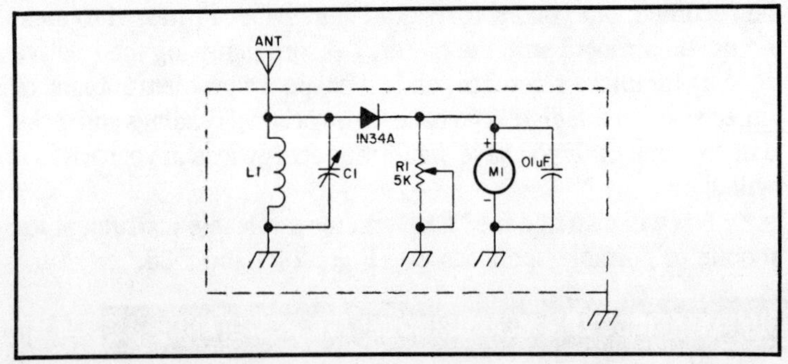

Fig. 2-35. Tuned circuit. M1 — Dc meter (see text). L1 and C1 — Resonant combination to cover frequencies you desire.

tances of each branch are equal, the currents through them are equal.

To find the meter resistance, one must plug in the meter under test and rotate R1 until the currents through both meters are equal. Then we know that R1 = 1 $R_{M\text{-test}}$ and its resistance can be read directly off the calibrated dial.

The smaller the value of potentiometer R1, the more accurate is the measurement of $R_{M\text{-test}}$. This is because the dial can be calibrated in smaller units.

As an option, a more accurate circuit is shown in Fig. 2-38. A rotary switch can select different values of resistance to be added to R1. Thus, R1 can be made as small as you wish. $R_{M\text{-test}}$ is now equal to R1 plus the switched-in resistance.

Let's say you wanted to measure a meter's resistance using only R1 (Fig. 2-36). If your dial was calibrated with 100 notches, the result would be 5 ohms per notch. If we use the circuit in Fig. 2-38, the potentiometer is only 200 ohms, leaving 2 ohms per notch on the same calibrated dial. Thus, we see how there is more accuracy in a circuit such as the one shown in Fig. 2-38.

I would suggest that you choose a meter with a low resistance. Also, if you prefer, you can use an ohmmeter to read the resistance of R1, thus saving yourself the trouble of finding a calibrated dial.

As you can see, the circuit is a flexible one and can be customized by the builder. All that is needed is a pen, paper and E = IR.

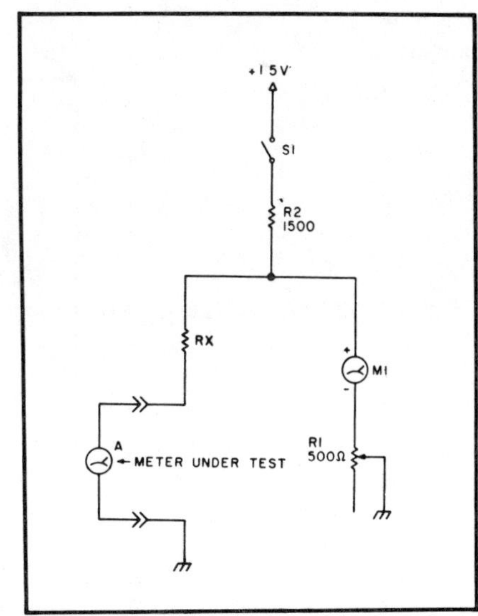

Fig. 2-36. Bridge schematic.

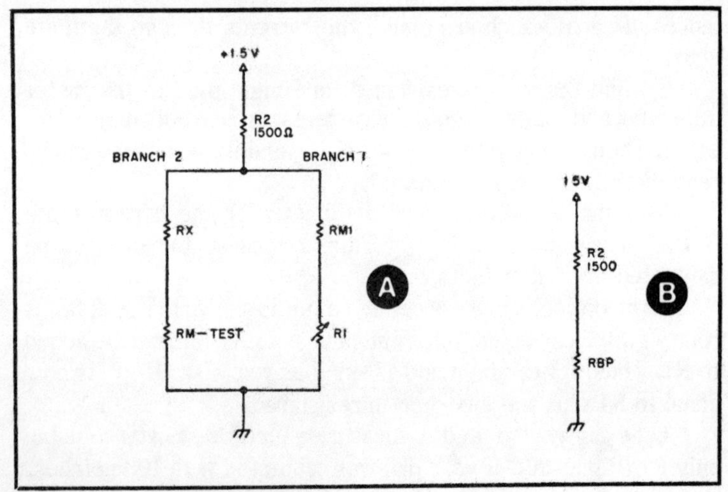

Fig. 2-37. Equivalent circuits.

FIELD STRENGTH METER

The manual for my rig suggests that a "better procedure" for loading it up is to use an swr bridge on forward or a field strength meter to peak plate and loading controls for maximum output power. The field strength meter seemed the easy way out, since a bridge costs fifteen dollars. The meter would cost me next to nothing to build considering the present status of my junk box. Therefore, I decided on the junk box approach to measure output power.

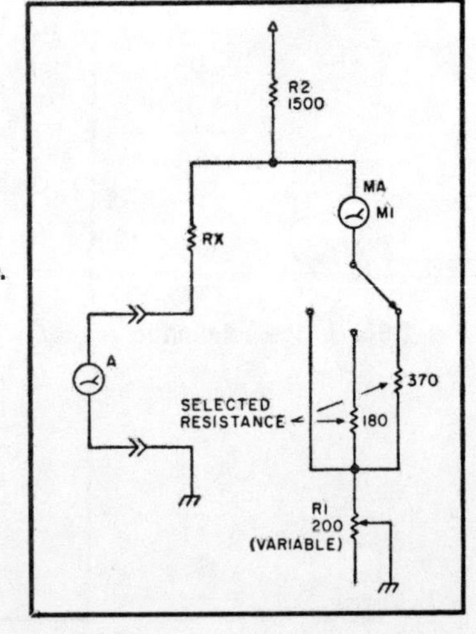

Fig. 2-38. A more accurate bridge.

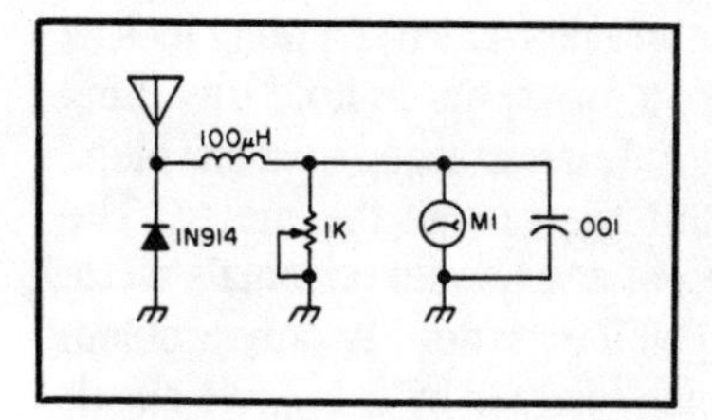

Fig. 2-39. Field strength meter schematic.

While building the field strength meter, I consulted an article in the July, 1976, issue of 73. The article dealt with building the world's smallest field strength meter. The design given was standard and allowed wide margins for parts substitution. Recently, another ham wrote 73 about an rf choke that cost four dollars, adding that amount to the cost of his otherwise free field strength meter. I just thought I'd mention that all parts in my field strength meter came from junked PC boards, with the exception of the meter movement.

When you go scounging parts for this project, remember about the choke. Anything that will choke rf works. So, substitute freely, but don't change from a diode to a resistor.

As I said, the design is standard, but the antenna is quite unconventional. Instead of using a piece of twelve gauge bare copper or coat hanger wire, I used twenty-two gauge tinned copper wire bent into the form of my initials.

Select the amount of wire you feel necessary to complete your initials or whatever. Straighten the wire with a vise and a pair of pliers. Starting at the top of the left initial, make a bend to form the beginning of the antenna. Try to use one continuous piece of wire to make the antenna. To insure this, draw your idea on a piece of paper first. If you use one piece of wire the antenna will look a lot better, and you will eliminate having to solder on extra pieces of

Table 2-6. Parts List for Field Strength Meter.

1 diode, 1N914 or just about anything else
1 50 uA meter
1 rf choke
1 pot, 1k or thereabouts (I have a 500 Ohm)
1 .001 uF capacitor
1 mini box
1 banana plug
1 banana jack
2 shouldered washers
Plenty of twenty-two gauge wire

wire. The joints make bulges in the antenna which just don't look as nice on the finished meter. Solder a few spots to hold this thing together. While the iron was still hot, I also attached a banana plug.

With the antenna done, I built the rest of the meter. The easiest way to go is to mount the pot and use its terminals as tie points. The wiring is the easiest you'll ever do. My only problem was mounting the banana jack, which I solved by using two shouldered washers. See Fig. 2-39 and Table 2-6.

Taking a lead from modern art, you can make just about anything for an antenna on your FSM. You've got to admit, now that we have an alternative, that bare copper does look a bit uncouth! This meter may not be the best or the smallest, but it does have class!

Chapter 3

Special Test Equipment

Not everyone is going to need each and every piece of special test equipment in this chapter, but it sure would be great if you had them all! Even if you only build one of these projects you can save a bundle, not only on the piece of test equipment itself, but also every time you use it.

The electrical circuit of the generator is shown in Fig. 3-1. One FET is used as the basic oscillator in a Hartley-type circuit. The second FET is lightly coupled through the 5 pF capacitor in its gate lead to the oscillator. This stage functions as source follower isolation stage. The last stage, the 2N3866, is designed to boost the signal level up to about 1 volt output on most bands. This level is far more than what is required for most receiver-type work, but the increased level comes in handy when doing transmitter exciter work, where the generator might substitute for a vfo. The output of the 2N3866 stage can be regulated by the 500 ohm carbon potentiometer, which will provide about a 30 dB variation in output level. A 47 ohm resistor can also be switched in across the output, so a true nominal 50 ohm generator source impedance can be simulated for tests such as receiver sensitivity. The switching in of this resistor also serves as a high-low output level selector for the generator.

Although the generator is not complicated electrically, its true potential will not be achieved unless it is carefully constructed. Fortunately, no elaborate construction work is required, but attention should be paid to the few details mentioned here. All of the circuitry is mounted on a single-side copperclad board measuring about 2½ by 2½ inches, as shown in Fig. 3-2. The board is wired point-to-point, using the isolated pad technique, starting with the oscillator stage toward the back panel of the enclosure and progressing forward to the 2N3866 stage towards the front panel. There is nothing critical about the wiring, whatever technique is used, but the circuitry should just be stretched out to provide maximum separation between the oscillator and 2N3866 output stage.

The heart of the signal generator lies in the band-switched oscillator coil assembly and the variable tuning capacitor. The tuning capacitor is readily available broadcast receiver type, which contains a single section AM section of about 300 pF and a single section FM section of about 25 pF. Such capacitors can often be found with built-in tuning shaft drive reductions of 3:1 to 6:1. A simple alternative to the AM/FM type is the even more readily available standard dual section AM type, where one section, designed for local oscillator usage, has fewer plates. Remove more plates from the oscillator section, so it is left with 4 stator plates and 3 rotor plates.

The coils for the six bands can either be purchased or constructed from a mixture of home brew and commercial coils. As a completely purchased set, one can use the Conar CO-69 through CP-74 series. These are replacement coils for an old-fashioned National Radio Institute tube-type signal generator, but they work just fine in the FET oscillator described in this article. The coils are available from Conar, National Radio Institute, Washington, D.C. 20016.

Another alternative is to just purchase the coils for the lower frequency bands, which would be almost impossible to home brew, and wind the other coils. In this case, for the first three frequency ranges one can use prewound J.C. Miller-type coils, which have the necessary tapped windings. The types are 9015, 9013, and 9013. For the highest three frequency ranges, one can self-wind the necessary coils on ⅜" diameter slug-tuned forms. The windings necessary and the tap points for each of the three coils are as follows: 15 turns tapped at 4 turns from the ground end; 7 turns tapped at 3 turns; and, 4½ turns tapped at 2 turns. The latter coil is wound using #18 wire while the other two coils use #24 wire.

The coils can be mounted directly on the 3P6T rotary bandswitch, as shown in Fig. 3-3. The coils are secured to the bandswitch with epoxy cement and wired in place. In order to ensure a good ground connection for the coils, a piece of sheet copper was cut out to resemble a 6-legged starfish and placed over the bandswitch shaft, so one of the "legs" could be soldered to each

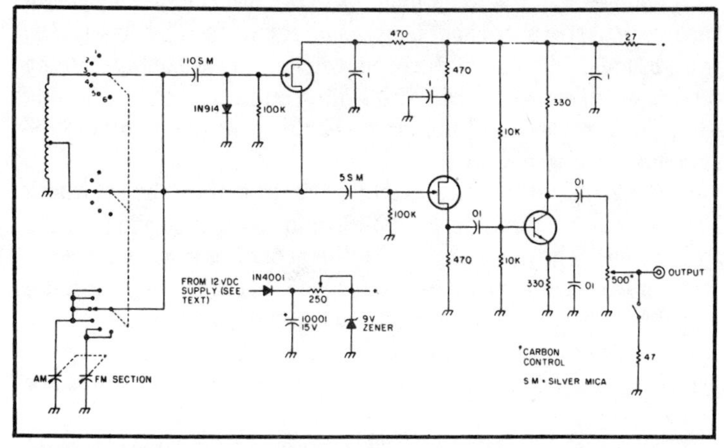

Fig. 3-1. Schematic of the generator. Only one of the six oscillator coils is shown for clarity. The FETs are HEP802 or MPF102 types. The output transistor is a 2N3866 or 2N706. Other details are covered in the text.

Fig. 3-2. The oscillator circuitry is mounted on a 2½-inch square circuit board, and the components associated with the zener regulator are on a terminal strip on the rear panel of the enclosure. At this point, the bandswitched coil assembly has not been installed.

coil. This arrangement is probably a bit overdone. Ground connections from the coils to several ground lugs, equally spaced around the shaft of the bandswitch, should suffice just as well.

The generator is assembled in a standard commercial enclosure, which measures about 5 inches on each side. The dimensions were based on the size of the bandswitched coil assembly, turning capacitor and circuit board. With a bit of effort, one should be able to fit the generator into the more readily available 4″ × 5″ × 6″ aluminum enclosure.

The generator can be powered either from the ac line or from a 12-volt battery source, making it ideal for both fixed and portable applications. The power supply for the generator is *not* included in the same enclosure as the generator, and this seems to contribute significantly to the total lack of ac hum on the output signal. The ac power supply is an ac wall plug-mounted 12-volt dc battery replacement supply of the type commonly sold to power transistor radios.

Within the generator enclosure there is only a diode to protect against reverse voltage polarity, a 1000 mF filter capacitor, and a 9 V zener regulator. For portable application, 12 V from a battery pack, or even 9 V from a transistor radio-type battery, can be used. The 250 ohm variable resistor before the zener is adjusted for the maximum resistance value that still allows the zener to maintain a constant 9 V output.

Because of the lack of a frequency readout scale, there is not the usual need to try to adjust the low and high frequency range excursions on each band. However, they should be checked with a counter to see that sufficient overlap exists between ranges. The slug tuning of the coils suffices to correct the tuning on each range. Although one has some latitude to adjust the frequency coverage on each band to suit individual preferences, Table 3-1 shows a typical arrangement, starting with the lowest frequency band.

The stability of the oscillator proved to be good enough on all ranges except the highest, so that temperature compensation was not needed. This is probably due to the low power operating requirements of the circuit and to the fact that the power supply is mounted externally to the generator. Since the highest band was not used extensively, it was not temperature compensated. But, by selection of small value NPO capacitors placed across the small section of the tuning capacitor and by watching the output fre-

Fig. 3-3. The bandswitch coil assembly is preassembled and then mounted in place. Individual coils are fastened to the switch by epoxy cement and wired in place.

Table 3-1. Frequency Coverage.

Band	Low end	High end
1	100	570 kHz
2	400	1400 kHz
3	1.2	4.5 MHz
4	4.1	17.0 MHz
5	15.0	39.0 MHz
6	25.0	75-80 MHz

quency change on a counter, it should be possible to achieve excellent frequency stability on the highest range also.

Tone modulation can be added to the generator by the circuit shown in Fig. 3-4. The circuit provides a single frequency tone modulation, which is useful to identify the presence of the rf signal when working with a receiver having only an envelope (AM) detector. By placing the output of the audio oscillator on the gate of the second FET (instead of on the base of the 2N3866 as Fig. 3-4 indicates), a slight FM modulation of the oscillator will occur. So, the signal generator can be utilized with SSB/CW, AM or FM receivers.

A sweep frequency capability can also be added to the generator, by means of a varactor diode connected across the gate terminal of the oscillator FET to ground, and driven by a suitable sawtooth of triangular waveform.

SIGNAL TRACER

For this project you'll need a small transistor radio, a resistor and a capacitor. The modification to the pocket radio involves four steps. First, find the earphone jack and the earpiece and cord that plug into it. Inside, the jack will have three wires connected to it: a ground, a lead to the speaker, and another one trailing off to the innards of the radio somewhere. This last wire actually goes to the secondary winding of the audio output transformer (see Fig. 3-4).

Leave the ground wire undisturbed. Unsolder the wire to the speaker and the one to the innards, both at the jack, and note which went where. Solder the ends of these two together and tape them. Now the radio is permanently connected to its built-in speaker.

The second step involves finding the point where the diode detector connects to the volume control. This can be found by tracing back from the center pin of the volume control along the foil until you find the glass diode. Unsolder the end of this diode which

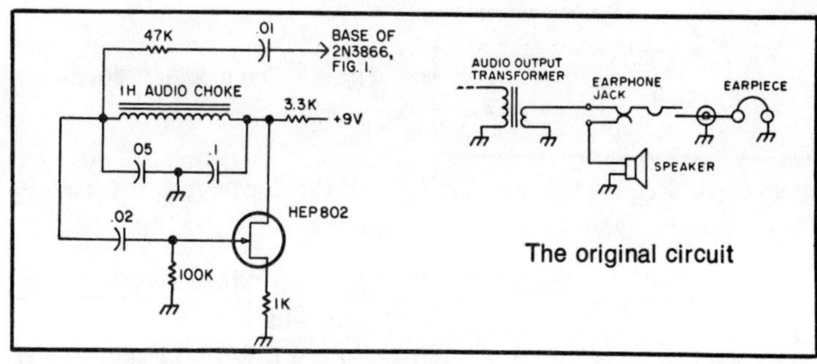

Fig. 3-4. Audio oscillator which can be added to the basic signal generator to aid in identification of the rf signal.

goes to the volume control, but leave the other end connected. Solder a piece of insulated hookup wire to the free end of the diode. The other end of this wire is soldered to the pin on the earphone jack that was formerly connected to the speaker. Solder another piece of insulated wire to the point on the circuit board where the free end of the diode used to be. The other end of this wire is connected to the remaining pin on the earphone jack that used to be connected to the innards. Now, without a plug in the earphone jack, the pocket radio will play normally, since the diode detector is connected to the volume control once again, although now through the contacts of the earphone jack (see Fig. 3-5).

For the third step, cut the earpiece off the end of its cord. Strip the ends of the wires, and with an ohmmeter or continuity checker, find out which of the wires goes to the inner pin of the jack, and mark it. The other lead is the ground connection, which can be connected to an alligator clip. Solder one lead of a 1 μF capacitor to the "hot" lead. This capacitor will keep stray dc voltages out of your pocket radio, thus preventing premature trauma. The free end of the capacitor is the probe tip, and is to be connected to the equipment under test, wherever you suspect audio should be. With the earphone plug inserted in the jack, and the probe connected to the circuit under test, you should now hear the desired signal. The lead with the capacitor can be built into the plastic end of a discarded ballpoint pen, to make a neater probe tip. The voltage rating of this capacitor must be higher than any voltage you have in the equipment under test. For tube type receivers, 600 volts is usually adequate, while a 50 volt capacitor is adequate for transistor receivers and hi-fi gear.

If you're going to run the pocket radio from its own battery, this step may be omitted. If you would like to run the pocket radio

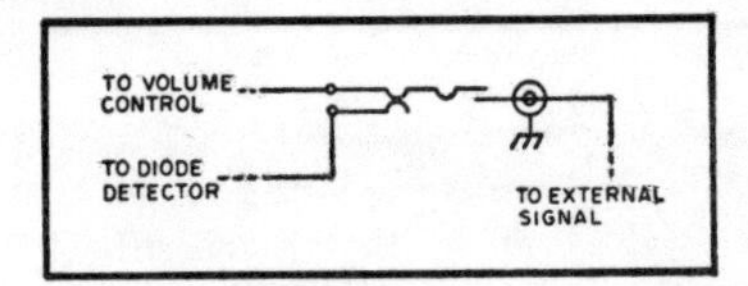

Fig. 3-5. The modified circuit.

from the voltage in the gear under test, this formula can be used to find the right value of dropping resistor:

$$\text{Resistance} = \frac{(\text{Available Voltage}) - (\text{Voltage Needed})}{\text{Receiver Current}}$$

For example, if your pocket radio needs 9 V to operate, and 12 V is available, and the pocket radio draws an average of about .010 A, then by plugging the numbers in:

$$\frac{12\text{-}9}{.01} = \frac{3}{.01} = 300\Omega \text{ resistance needed}$$

The wattage needed for the resistor can be figured by the formula $I^2R = P$; that is, the current multiplied by itself, times the resistance, gives the needed power rating in watts. In the example above, it would be (.01) × (.01) × 300 = .03 watts. A ¼ watt resistor would give a more than adequate safety margin. If you're planning to use a 150 V supply to run the pocket radio, a 14.1 kΩ at 1.4 watts is the calculated value, and 15 kΩ at 5 watts is adequate and a practical common value. This resistor is connected between the positive terminal of the battery holder and the supply voltage point, as shown in Figs. 3-6 and 3-7. Also, see Fig. 3-8.

Once completed, the pocket radio can be used normally, and, by just plugging in the earphone plug/test probe, it becomes a signal tracer or audio amplifier.

AN AUDIBLE LOGIC PROBE

The most serious short-coming of conventional logic probes is the need to watch for signs of a change. It's not always convenient to be watching the probe during a test.

This unit gives an LED indication of the static state of the line under test. When the LED is lit, the line is logic state "1." Any logic transition which lasts at least 50 nanoseconds is detected, and an audio beep is generated. Both a "1" to "0" and a "0" to "1" transition are detected.

Construction

My unit is built on a perforated epoxy board. Discrete components are mounted on a 16-pin header which plugs into an IC

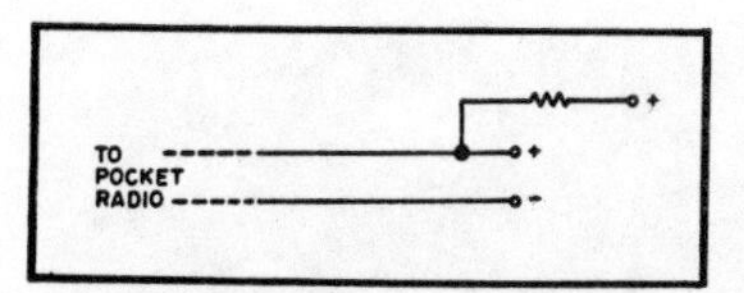

Fig. 3-6. Battery terminals with added dropping resistor.

socket. Sockets are mounted by installing printed circuit board eyelets under pins 7 and 14. Each eyelet is peened in its hole by using opposing automatic center punches. The sockets are mounted, and pins 7 and 14 are soldered to the eyelets. The rest is a simple wirewrap job.

It is possible to run the device from batteries, although the current is a bit high (about 70 mA). Rather than use batteries or steal power from the circuit under test, I decided to include a small power supply. Everything is mounted in a small phenolic box.

Operation

There is no special procedure. Simply connect the input to the line to be monitored, and a ground between this device and the one under test. Once connected, an audible beep will be heard whenever the line pulses or changes state. The LED indicates the line's static state.

Theory

The input is squared by (Schmitt) IC1A, which is also the LED driver. IC1B and IC1C provide a delay before applying the signal to

Fig. 3-7. This photo shows the dropping resistor connected to the positive battery terminal, and the lead from the volume control to the external jack.

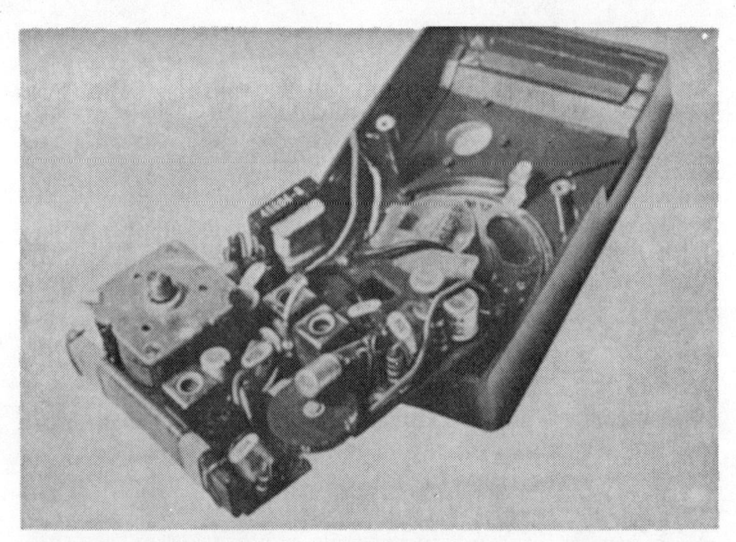

Fig. 3-8. This photo shows the diode detector (nestled snugly between two i-f cans) and its lead (going to the jack).

the exclusive OR gate, IC2. When the input changes state, the two signals applied to the exclusive OR gate will be different for a period equal to the delay through IC1B and IC1C. The gate will pulse low for this time. After inversion by IC1D, the positive pulse is applied to the first half of the NE556. (IC3). This half is wired as a one shot with a duration of approximately one half second. The output of this one shot is applied to the second half of IC3, which is wired as an audio oscillator. When the one shot is active, it turns on the oscillator, which in turn drives the speaker. Changing C1 will adjust the duration of the beep; C2 determines the output tone.

Troubleshooting

During initial wiring, leave out the jumper between pin 5 of IC3 and pin 10 of IC3. Check the power supply to be sure it is +5 V. Plug in IC3. The oscillator should be heard running. This is a good time to vary C2 if you would prefer a different tone. After the tone is working properly, connect the jumper between pin 5 and pin 10 of IC3. Plug in IC1—but not IC2. Touch pin 12 of IC1 momentarily to ground. The oscillator should beep for approximately one half second. C1 can be adjusted to vary the beep duration.

The LED should be lit. Touching pin 2 of IC1 to ground should make the LED go out. Insert IC2 and test the complete unit. Short the input (pin 2) of IC1 to ground. The LED should go out and the oscillator should beep. Remove the short and another beep will occur.

If you want to test its ability to capture a pulse, I suggest wiring a one shot (such as an SN74121) as a switch deglitcher. This device will capture the pulse every time the switch is pushed. See Figs. 3-9 and 3-10.

TV TEST UNIT

This is to describe a device of considerable value to the many hams that service TVs either in the shop or at home. It has not, so far as I am aware, been used or suggested by anyone else. It is simply a regenerative receiver set on 15.75 kHz, the TV horizontal oscillator frequency.

With it the horizontal oscillator frequency can be set correctly without a signal and the adjustment made in seconds without any doubt as to whether to increase or decrease the oscillator frequency. Many of us have spent valuable time in the shop blindly turning the slug in and out without the slightest idea where it should be for the correct frequency. If the oscillator is not working, there will be no signal regardless of adjustment. Adjustment of the horizontal oscillator in the usual manner only brings the frequency near enough that the sync pulse can lock it in step and thus is no assurance that in the free-running state it is on frequency.

This receiver may be built into any small case such as that from a defunct transistor radio. Its variable condenser, audio amplifier and speaker may also be used if good. It may be made small enough to carry in a shirt pocket.

To conserve space and avoid hand capacity effect, an 88 mH toroid coil was used rather than a regular horizontal oscillator coil which would have required some shielding. See Fig. 3-11.

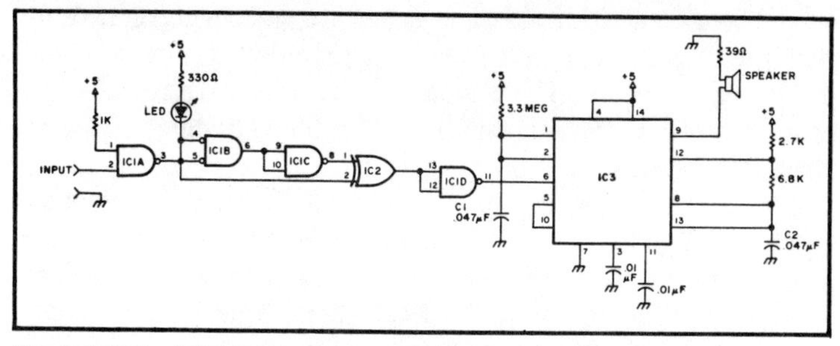

Fig. 3-9. IC1: SN74132N. IC2: SN7486N. IC3: NE556. All resistors are ¼ watt, except for the one which is ½ watt, 39 Ohms. Power supply current is approximately 70 mA.

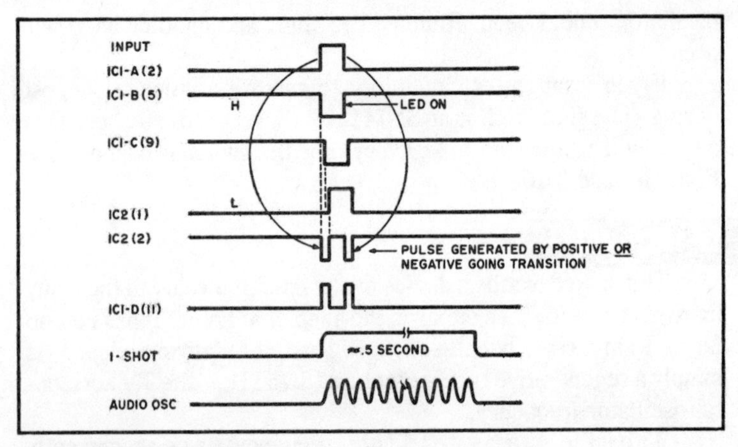

Fig. 3-10. Timing diagram.

This circuit using the collector at ground potential for rf was chosen to simplify the audio takeoff. R5C5 provides additional filtering to keep rf out of the audio output. Use the audio amplifier in the original case, if convenient. If you must build your own, a small IC is suggested for compactness.

A regenerative receiver is most sensitive when not oscillating at full strength, so it is a good idea to use variable resistors to determine the best values for reliable but not excessive feedback, replacing them with the nearest fixed small resistors. While I used a 2N706 transistor, it is safe to assume that at least a hundred other types, requiring different bias resistors, etc., may do as well or even better.

C1 is made up of one or more fixed mica condensers in parallel with a small variable condenser or mica compression trimmer; the latter is definitely second choice. The total capacity required should be around .0018. Marked values are seldom correct. Temperature sensitive condensers are to be avoided for tuning. Silver micas are preferred. The temporary use of an external variable condenser of considerable capacity will expedite finding the proper value and frequency.

With this receiver's antenna near the horizontal area of an operating TV, listen for the 15.75 kHz signal when you are sure the receiver is oscillating. When zero-beat is obtained, with final tuning condensers in place in the final assembly, no further adjustment is required and it is ready for use.

With no signal or antenna on the TV to be serviced and with front panel control, if any, set at midrange, adjust the horizontal slug for zero-beat with the receiver and you are finished with that part of the job—no guesswork! If the sync doesn't take control, then that is a different problem, and there is no need to twiddle with the horizontal oscillator.

POWER SUPPLY TESTER, USING A LOAD BANK

The first question asked might be, "What is a load bank?" The term is taken from the electrical power industry to explain a device to simulate a load on various power sources in order to check the load performance of those sources.

Using the basic ideas presented here, load banks for most any type of low voltage power supply can be built.

In order to properly evaluate the performance of a power supply (whether home built or commercially built), the supply must be tested at various degrees of loading—that is, no load, half load, full load, etc. The major requirement for a suitable load bank is that it be capable of providing these various load conditions. Thus, several load elements are required, or at least one large variable element is required, in order to cover several different load conditions. These load elements would typically be large fixed or variable power resistors. If the power supply is of any size, the load elements have to be of high power rating (10, 25, or even 50 watt). A look at a parts catalog will show that power resistors get expensive as the power rating goes up and the different ohmic values available go down. Also, power resistors present the problem of how to get rid of the heat and how to conveniently mount the resistors.

Instead of using the regular power resistor in the design of these load banks, automotive bulbs and pilot lamps were used.

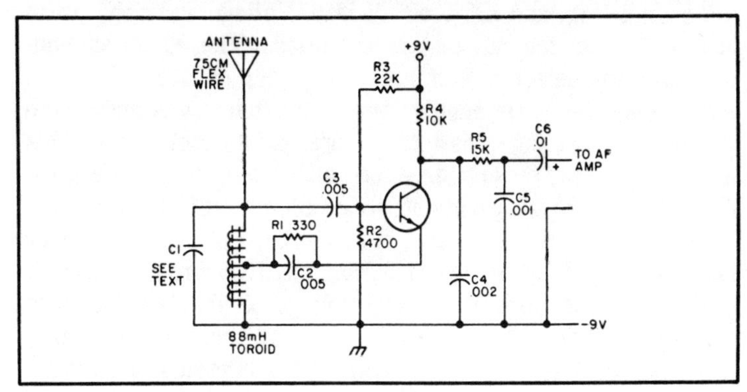

Fig. 3-11. 15.75 kHz oscillator schematic.

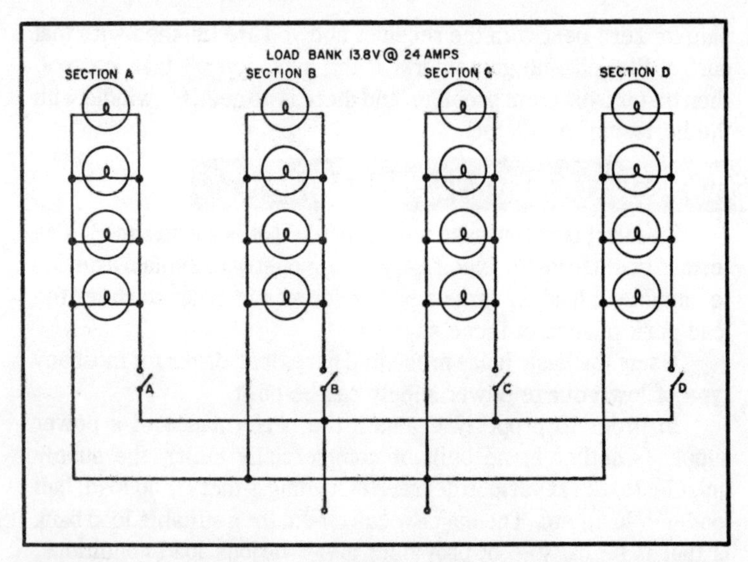

Fig. 3-12. All bulbs #1073. All wire #14 TW. A, B, C, D—10 amp relay contacts or 10 amp switches (optional since connections can be made directly).

These are readily available at auto parts stores, electronic parts houses, and even department store electrical parts sections. Bulbs are selected on the basis of voltage and current requirements and typically work out very nicely in regard to both values. The problem of power dissipation is very minimal and mounting can be very simple. With the load elements decided, the actual construction of the load bank can begin.

The first load bank requirement was for a rather large 13.8 volt dc 24 amp dc supply. It was decided to check one third, two thirds, full load, and 133% of full load. Voltage and ripple were observed under the various load conditions. The #1073 automotive bulb was selected with a nominal rating of 12.8 volts @ 1.8 amps. Since the power supply voltage was to be a volt higher, the current at that voltage was approximated to be about 2 amps. This is a fine feature of these bulbs—the voltage can be increased up to 20% of nominal rating without any problem in this application.

The bulbs were arranged in four groups or sections of four each. See Fig. 3-12 for the circuit diagram. Each group presents a load of about 8 amps at nominal voltage to simulate four load conditions. In order to keep costs down and construction simple, the wires consisting of #14 TW wire were soldered directly to the bulb bases. The start and end of each section wire were wrapped around a nail which was driven into the wooden base board. This eliminated expensive sockets and, after all, a load bank is not a device that is normally on display in the shack.

Since this load bank was to be used for a considerable amount of testing and since four surplus relays were available, each load section was wired through relay contacts. Any combination of loading can be quickly selected. This extra feature can be eliminated with just the use of solder connections or switches with proper current rating. The switching feature is certainly handy if considerable testing is to be done, but adds considerably to the parts cost if the junk box is not well stocked.

The second load bank was for a much smaller dc power supply with a rating of 5 volts @1 amp. Following the same design features of the previous load bank, a #502 bulb was selected as load element in four sections with two bulbs in each section, giving 30%, 60%, 90%, and 120% of full load. Since the current requirements were much less, about 0.3 amps per section, regular #22 solid hookup wire was directly soldered to the bulb bases and toggle switches (or even slide switches) used for controlling the load sections. A small wood base board was again used for mounting the bulbs and wiring with a small ⅛ inch pressed board panel (nailed to the edge) to support the switches. This supply is most useful for checking logic power supplies used with ICs. Overload or current foldover characteristics can be checked on supplies designed with that feature. See Fig. 3-13 for the circuit.

There are many advantages to these simple load banks besides the cost and ease of construction. The bulbs present a large amount of light under full load to leave no doubt that the supply is working. Also, there are no burn marks from power resistors on the workbench or, even worse, the dining room table. Buying the

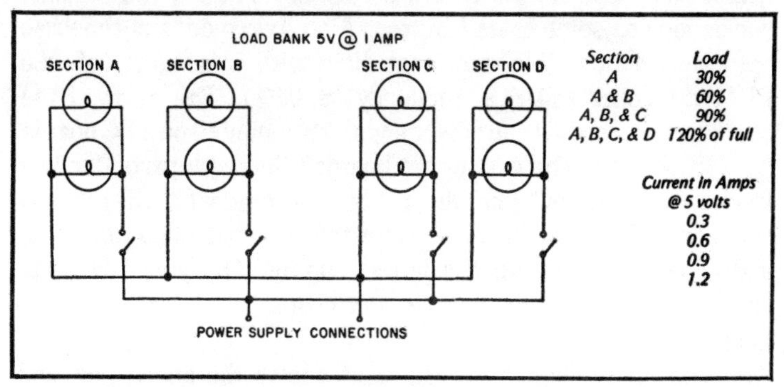

Fig. 3-13. All bulbs #502. All wire #22 solid. A, B, C, D—0.3 amp switches.

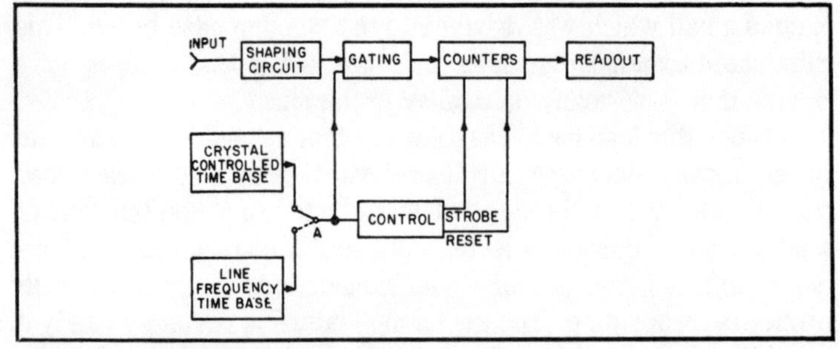

Fig. 3-14. Block diagram.

bulbs by the box reduces cost, and automotive type bulbs are readily available, even at gas stations.

The purists may argue that the bulb is not a constant load resistance due to the filament characteristics. This sudden heavier than normal load when voltage is first applied to the bulb is very short in duration and provides an even stricter load test of the supply being tested.

The load bank ideas presented here can be extended to other sizes and types of load banks by changing the bulb types and the number of bulbs in a section, and even the number of load sections.

These load banks have been used to check voltage and ripple conditions on a variety of home brew and commercial supplies. The cost is certainly cheap if proper shopping around is done on the choice and source of bulbs.

FREQUENCY COUNTER

My original goal in designing this counter was to produce an inexpensive, no frills, reasonably accurate, minimum parts design.

High impedance FET input, .2 to 30 MHz frequency range, sensitivity ≤ 60 mV RMS across most of its range, and compact packaging are some of its features. Also, two modes of timebase operation are possible: xtal controlled with an accuracy of ±20 PPM ±1 count and line frequency (±.05%). The heart of the counter (less readout and power supply) is mounted on 2 PC boards (≅ 3″ × 5″). The cost of the boards filled with parts (for line frequency operation) plus the 6 Minitron readouts will run you about $40. An additional $8 pays for the xtal controlled timebase, and of course you'll need a power supply. Any junk box parts substitute will naturally reduce your costs.

Operation

The block diagram of Fig. 3-14 shows the signal flow and control. Referring to Fig. 3-15 Q1 to Q4 shape the incoming signal to produce TTL compatible levels. A precise 100 ms gate (at IC19 pin 2) is then "or"ed together with the signal (IC19 pin 3). This gated signal is counted down and displayed on the readouts. The trailing edge of the 100 ms gate triggers the one shot IC24. A pulse is generated that strobes the quad latches IC7 to IC12. The stored data is transferred to the decoders and readouts. The trailing edge of the one shot triggers another one shot IC25 to produce the reset pulse to counters IC13 to 18. All counters are reset to zero until the next gated input is counted. Thus the sequence is: gate, count, transfer, reset.

Construction

As mentioned earlier, the heart of the counter is mounted on 3″ × 5″ PC boards (actually 3″ × 5-3/16″). You can package these boards any way you like. Just allow sufficient ventilation in any enclosure you use. One of the drawbacks of laying out small boards is that you can't get all the circuit connections on them that you'd like. So we're going to have to add some jumpers. Refer to Fig. 3-15 and Figs. 3-16A & B. After all the components are mounted, jumper IC19 pin 2 to IC24 pin 3. Also jumper point "A" to the pad right above it (for line base operation). That's all the jumping we need for the Input and Control board.

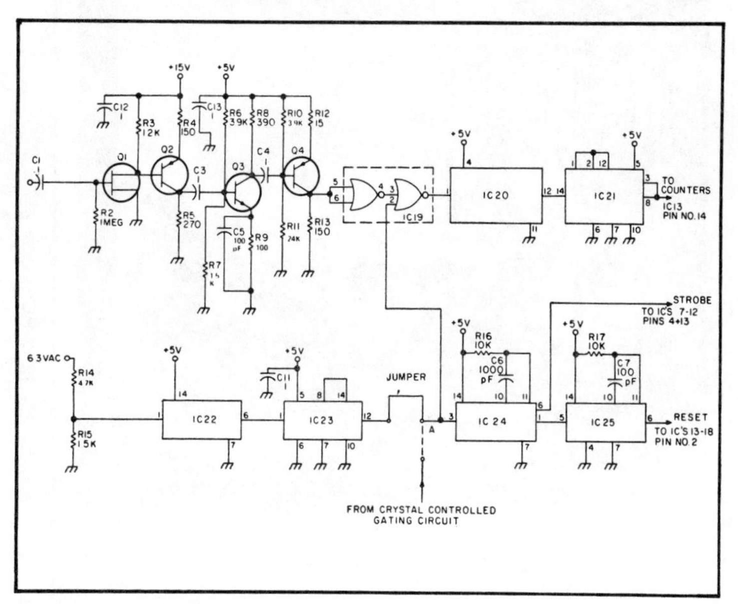

Fig. 3-15. Input and Control board.

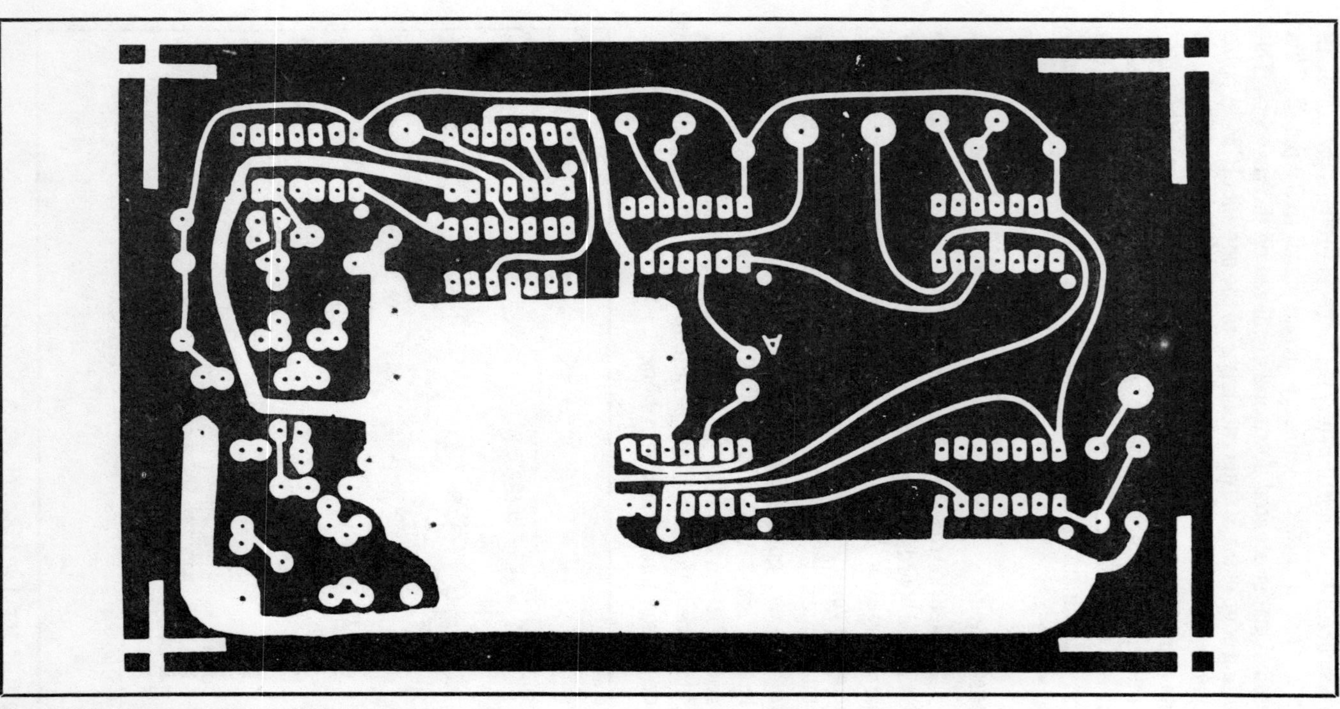

Fig. 3-16A. Input and Control PC board.

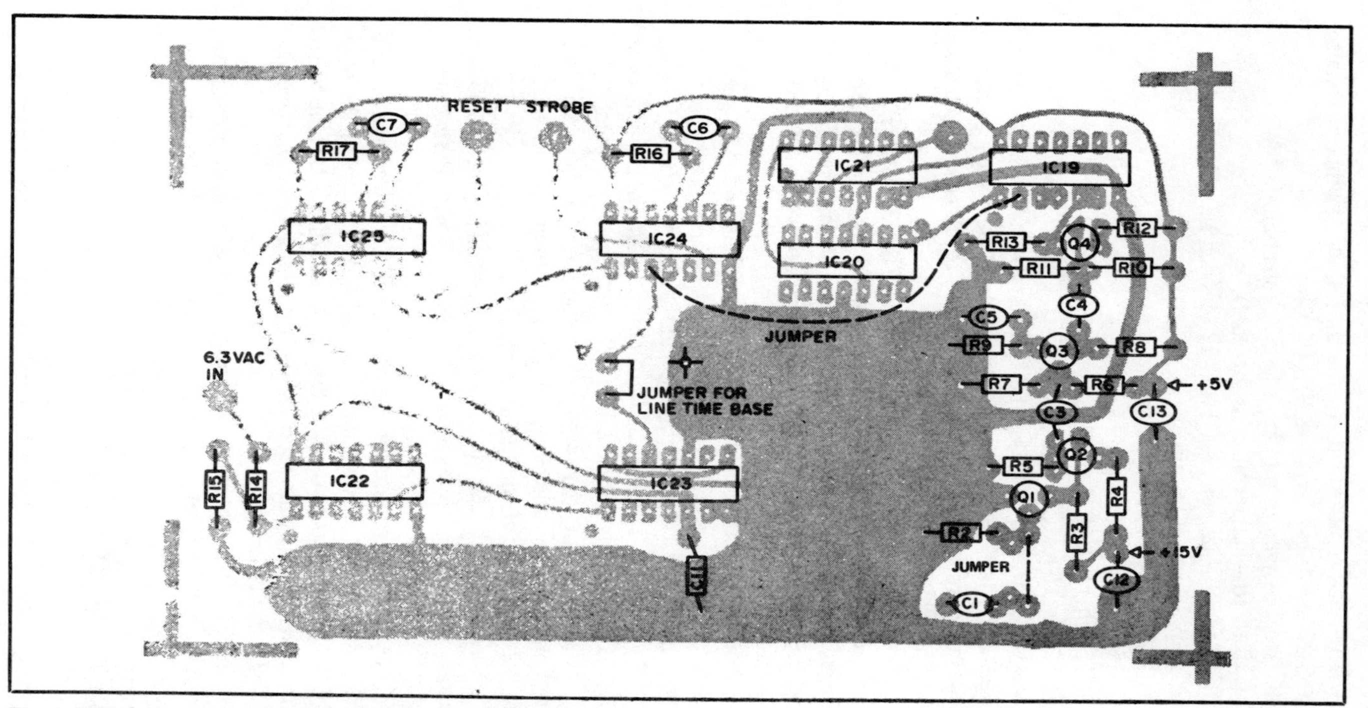

Fig. 3-16B. Component layout, Input and control board.

The Readout board will require the following jumpers (refer to Fig. 3-17 and Figs. 3-18A & B):

1. IC7 to 12—jumper all pins 4 and 13 (strobe).
2. IC13 to 18—jumper all pins 2 (reset).
3. IC13 to 18—jumper pins 1 and 12 on each IC only.
4. IC13 pin 11 to IC14 pin 14.
5. IC14 pin 11 to IC15 pin 14.
6. IC15 pin 11 to IC16 pin 14.
7. IC16 pin 11 to IC17 pin 14.
8. IC17 pin 11 to IC18 pin 14.

Also, power to all ICs and all power grounds will have to be added. Refer to Fig. 3-17. Jumper grounds to the ground plane and between boards. I suggest you use Molex terminals to mount all ICs. The added cost is worth it.

The Input and Control board artwork (Figs. 3-16A & B) is laid out for 60 Hz operation. If you're satisfied with an accuracy of approximately 500 PPM on your counter, use it as is. You won't need the LM340-5T in the power supply, so that's an additional saving.

The crystal controlled timebase should be added to get the most accuracy from your counter Fig. 3-19 is a suggested circuit. Layout is not critical and could be hand-wired in no time. If used, you won't need IC22 and IC23. To connect, remove the jumper at point "A" and connect the output on the last counter to point "A".

That's about it. If you got all the jumpers in right and your power is connected right (double check), you should be finished. The dot near each IC on the artwork is pin 1.

Make sure all ICs are installed correctly. That's another reason for using the Molex terminals. It's a lot easier to reverse an incorrectly mounted IC with them.

There is no calibration procedure. If it's wired right, it'll work. One word of caution. If the counter fails to count all the way to 30 MHz, IC21 could be the cause. It has to divide by 5 up to 15 MHz. Substitute it with other 7490s until you get one that goes to 30 MHz on the input.

Finally, your counter is compatible with any of the prescalers available on the market, extending its range to 300 MHz. Also, see Fig. 3-20.

FREQUENCY STANDARD

With the proliferation of subbands and interference-obscured net frequencies, the use of an accurate secondary frequency standard is both good operating procedure and also helps fill our legal requirements for frequency measurement capability under FCC regulations. In addition, such a standard can be used as a timebase or signal injector to test digital logic circuits.

This frequency standard was designed to be a low cost answer to the need for a good secondary frequency standard. It generates marker signals of 1000, 500, 100, 50, 25, 10, 5, and 1 kHz, and 100, 10 and 1 Hz. With harmonics usable well beyond 30 MHz, markers available to denote subband edges, align receiver dials, find net frequencies, and measure the frequency of unknown signals. Should your latest logic circuit not perk properly, two TTL level outputs are available as substitute clocks or signal injectors.

Short term accuracy is approximately 1 part in 10^6. The unit is easily aligned to WWV with a short wave receiver. An attenuator is included to permit matching signal strengths with the received signal so that a zero beat can be easily and accurately identified. Only one trimmer need be adjusted to align the standard.

With the standard still attached to the receiver and providing an audible measure of its accuracy, a counter or other TTL-compatible device may be attached to either TTL output and that device's accuracy checked.

A frequency burst mode is provided to allow identification of the standard's markers in a crowded receiver passband. Enabling the burst turns the output on and off ten times per second, resulting

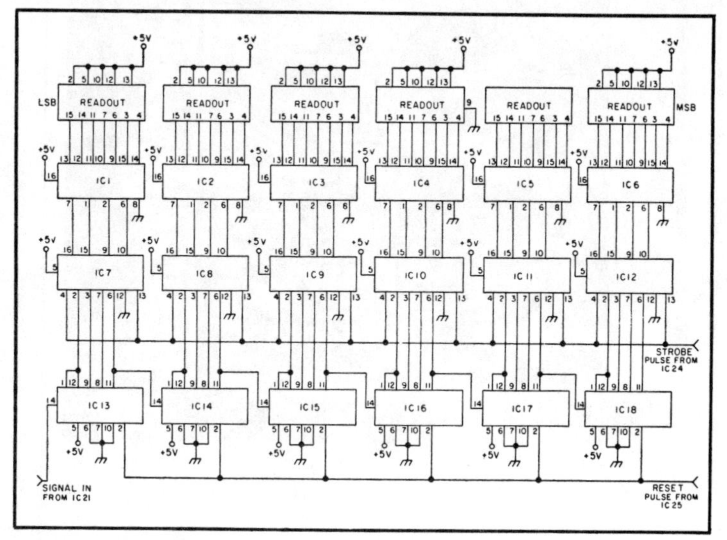

Fig. 3-17. Display board.

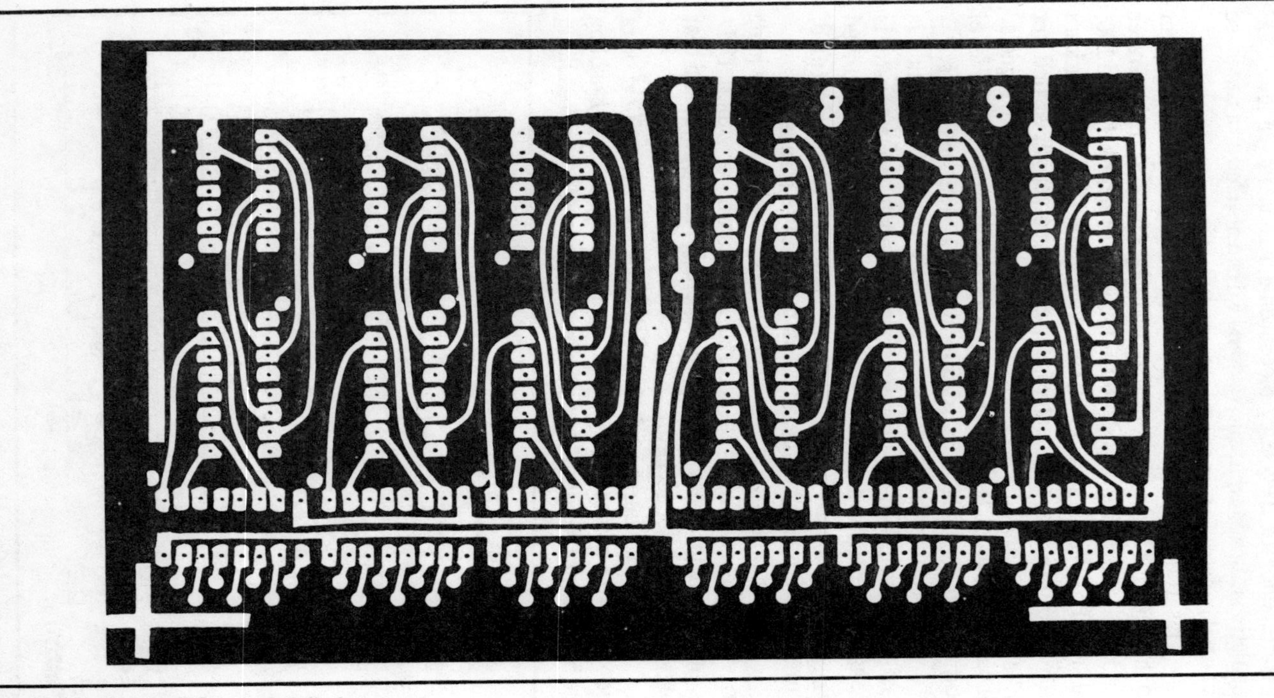

Fig. 3-18A. Display PC board, full size.

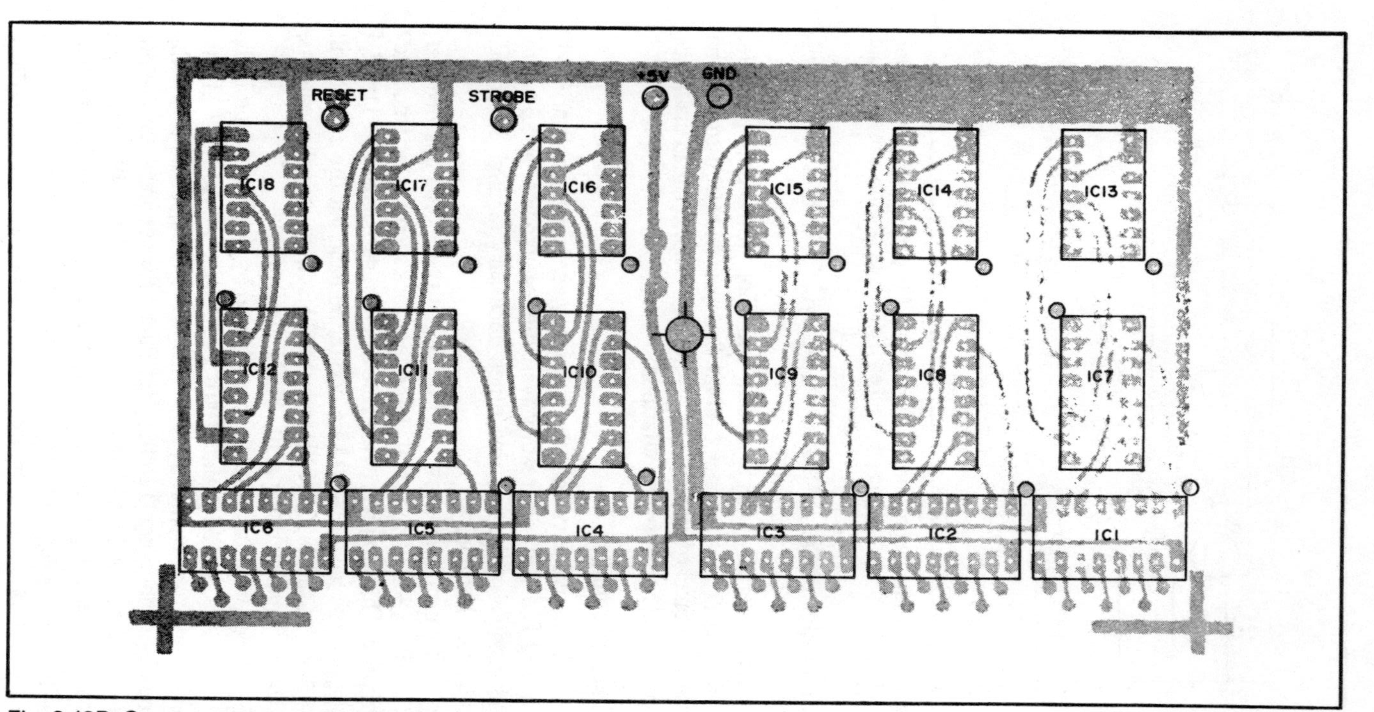

Fig. 3-18B. Component layout, Display board.

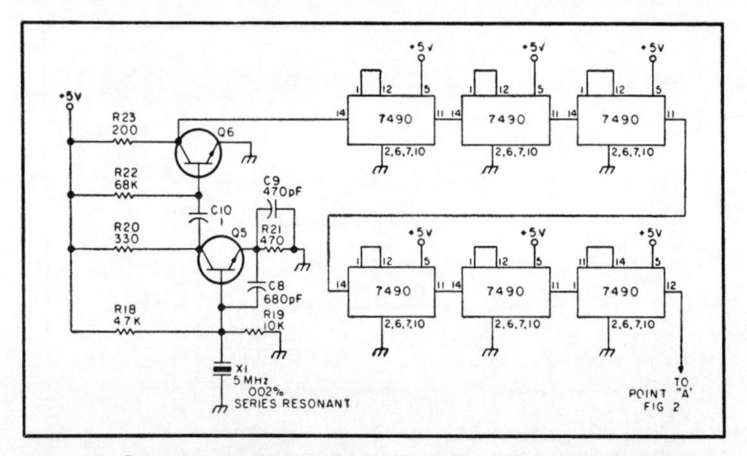

Fig. 3-19. Crystal controlled timebase. Note: For a more versatile counter, and to prevent overloading the front end, the input circuitry shown should be added.

in an easily, recognized "beep-beep-beep." This burst also is available on one TTL output.

An external dc input makes operation in the field possible. 9-15 volts at 250 milliamps is all that you will need to have an accurate standard available for Field Day. This is cheap insurance against FCC out-of-band citations.

Circuit Description

The active devices in the frequency generation chain are 7400 series TTL. They are readily available, easy to use, inexpensive, and capable of the fast rise times necessary for high level high frequency harmonics.

The oscillator shown in Fig. 3-21 uses a 7404 hex inverter, A1, with a 200 kHz crystal as the feedback element for frequency stability. This circuit provided the cleanest output of those I tried. I happened to have a 2000 kHz crystal on hand but the circuit will work with a surplus 1000 kHz crystal. In fact, you can eliminate one 7474 package by using a 1000 kHz oscillator.

Frequency division is accomplished by 7474 dual type D flip flops A2, A9, and A10, and 7490 decade counters. A3-A6, A11, and A12 wired for division by ten. The 7474s are used as toggle flip flops by connecting Q to the data input. Preset and clear are not used and are tied high through 1.8k resistors. The 7490s are ripple counters and are prone to spikes and level changes in their output. Proper bypassing of all ICs is necessary to prevent these devices from putting spikes on the power buses. The .1 μF bypass

Table 3-2. Parts List for Counter.

IC1 to IC6	7447	R12	15
IC7 to IC12	7475	R14	4.7k
IC13 to IC18, IC21	7490	R15	1.5k
IC24, IC25	74121	C1, C3, C4, C11	.1 mF 20 V
IC22	7413	C5	100 pF
IC23	7492	C6	.001 mF 20 V
IC19	7402	C7	100 pF
IC20	74S73	Q1	MPF 102
R16, R17	10k ¼ W	Q2, Q4	2N3638A
R2	1 meg	Q3	2N2222
R3	1.2k	Q5, Q6	2N2222
R4, R13	150		
R5	270		
R6	3.9k		
R7	1.5k		
R8	390		
R9	100		
R10	3.9k		
R11	24k		

IC terminals Molex Soldercon
Series 90 Minitron Display (6)

Note: Transistors, ICs, terminals and display are available from Digi-Key Corp., P.O. Box 126, Thief River Falls MN 56701.

capacitors are not superfluous—use one at each IC.

The various frequency outputs are selected by a rotary switch and fed to A8c, one section of a 7400 quad gate. The selected frequency can either be passed without change or gated with the 10 Hz output of A7, an NE555 astable oscillator, producing an easily identified frequency burst at the output terminals.

Should you not have a switch with enough positions to suit your needs or just wish something different in the way of frequency selection, try the electronic switch of Fig. 3-22.

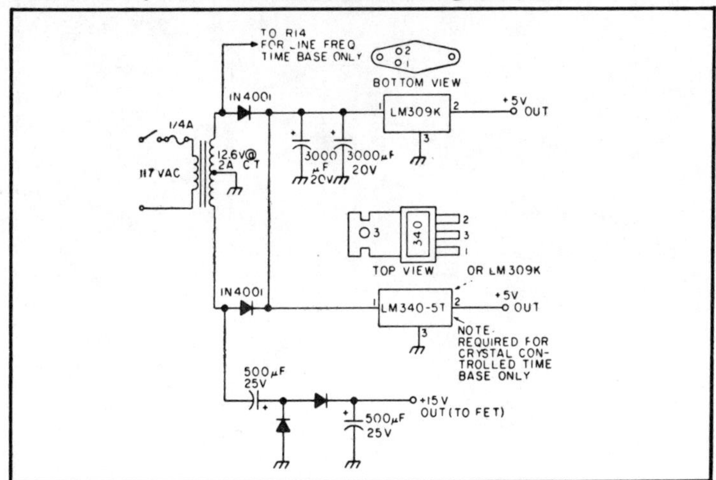

Fig. 3-20. Power supply.

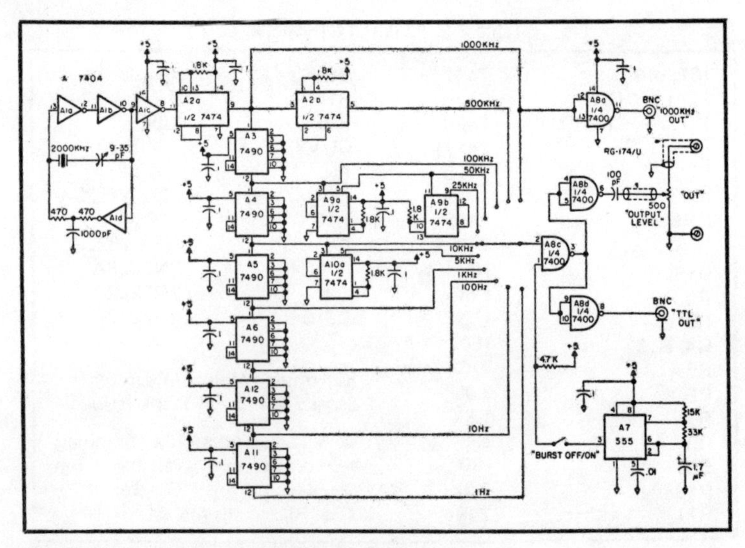

Fig. 3-21. Frequency standard schematic.

The NE555 timer A14 is used as an astable oscillator whose output is a train of 24 millisecond low-going pulses with a period of .81 seconds. A15a inverts this to a train of positive-going pulses. A15b and A16a gate the pulse train under control of the STEP push-button switch.

On each low-going pulse edge of pin 14, the 7493 binary counter A17 increments by one. A16b forces a reset on a counter

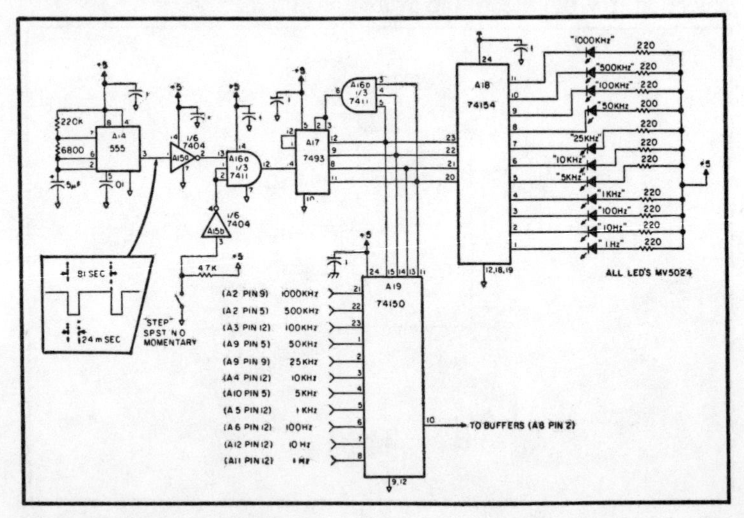

Fig. 3-22. Electronic switch.

output of 1011, permitting outputs from 0000 through 1010 to select the eleven frequencies of the standard.

The binary output of A17 causes the 74150 multiplexer A19 to select a signal from its inputs to be fed to the standard's output buffers. A18, a 74154 binary to one of sixteen decoder, enables the corresponding LED.

To change frequencies on the standard, depress the STEP push-button. The standard will step through its eleven outputs, one every .8 seconds, until the button is released. When the desired frequency is reached, as indicated by its LED, you have three-quarters of a second in which to release the button before the standard steps again.

Because of the additional current required by the electronic switch, the 2200 μF filter capacitor should be changed to 4700 μF if this circuit is added.

The remaining gates of the 7400 are used as output buffers. The two used for TTL outputs will drive ten TTL loads apiece. One of these gates buffers the 1000 kHz output of A2a and makes it available at a BNC jack on the rear panel. The other takes the output of A8c, which is controlled by the frequency selector and the BURST switch, and makes it available at a BNC jack.

The output to a receiver is from A8b through a 100 pF capacitor and a 500 ohm pot used as a signal level attenuator. Connection of the pot as shown prevents the receiver sensitivity from being affected by the attenuator setting.

The power supply uses the ubiquitous LM309K +5 volt regulator. Since the circuit draws only 250 milliamperes, the project case can be used as the heat sink. Dissipation of the 309 is only 0.7 watts.

Substitution for the surplus 7 volt power transformer I used is easy. Use a 12.6 volt center-tapped filament transformer in the configuration of Fig. 3-23.

In order to use the unit in the field, provision is made for an external dc input. The .01 μF capacitor removes stray rf from the power lines and the series diodes prevent damage from polarity

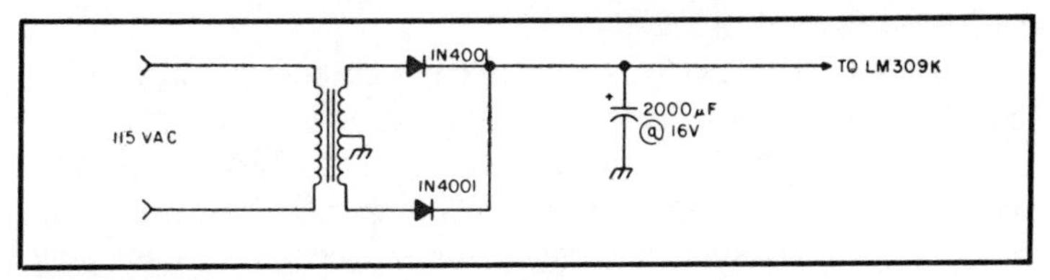

Fig. 3-23. Center-tapped filament transformer circuit.

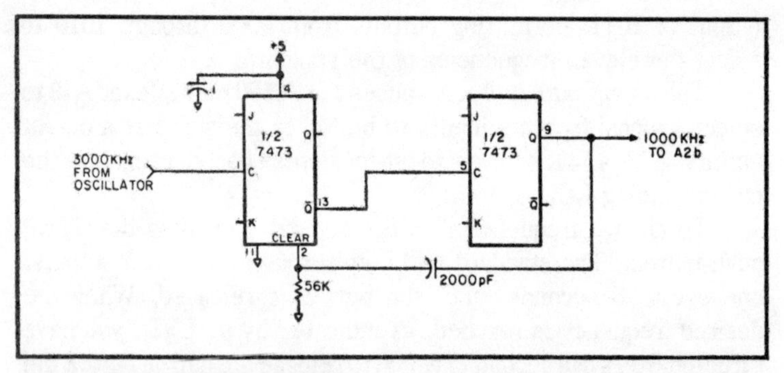

Fig. 3-24. Divide by three circuit.

reversal. Two diodes were used because I feel that the 14.5 volts of a car battery under charge come perilously close to the filter capacitor's 16 volt rating. With the two diodes shown, applicable dc input voltage is 9-15 volts. With the appropriate filter capacitor voltage rating and a single diode, voltages in the range of 8 to 25 can be used.

Do not leave out the .22 μF bypass capacitor on the 309 input. It prevents oscillation of the device should a remote battery be used as a power source.

Construction

Because of the high speed switching characteristics of TTL and the high frequency harmonic content of the output waveforms, each IC is a transmitter and each interconnecting wire is an antenna. A thoughtful layout and careful construction are important to minimize unwanted radiation.

Switching transients appearing on the V_{CC} and ground lines can add unwanted noise to the output. I used a prototyping board with V_{CC} and ground planes to minimize glitches. A printed circuit board would be even better. If you wish to build the circuit with wire wrap sockets on vectorboard, use bus wire for the power leads to the sockets. An effective technique is to interleave V_{CC} and ground bus wires for each row of IC sockets, with a row not containing more than a half-dozen ICs.

Bus wire should also be used to connect the board directly to the 309. Do not invite problems by grounding the board to the chassis. Connect the chassis to the 309 case, the board to the 309 case, and the rectifier ground to the 309 case. This prevents the board ground from rising above the power ground and developing noise problems.

Asynchronous TTL devices such as the 7490 generate plenty of switching transients. Put a .1 μF bypass capacitor between V_{CC} and ground at each device socket and another at the power input to the board. Bypass the 309 input and output as shown to prevent spikes and oscillation.

The leads connecting the frequency divider ICs and the selector switch make excellent antennas, so shield the output from the board to the attenuator pot and from the pot to the output terminals with RG-174 coax to reduce unwanted pickup. Long unshielded leads will reduce the effective control range of the pot.

Mechanical stability will be reflected in electrical stability. Rigidly mount the crystal and trimmer capacitor close to the 7404 oscillator.

Changes

For those who would like accurate channel markers for 2 meter FM, try a 3000 kHz crystal in the oscillator. This frequency can be divided by two 7490s to 30 kHz for standard channels and then by half of a 7474 to cover splinter channels. The divide by three circuit of Fig. 3-24 replaces A2a to retain the outputs of the present standard.

With CMOS prices falling and use of these devices increasing, you may wish to provide an output compatible with CMOS logic circuits. A 7406 open collector hex inverter package with a 2k pullup resistor could be added to interface with CMOS logic levels. Such an arrangement will have a deleterious effect on high frequency harmonic content, but will be fine for signal injection to a CMOS circuit.

AUDIO FUNCTION GENERATOR

The generator features a frequency range of .05 Hz to 300 kHz, digital frequency display, and a sweep range of 1000:1 or better. The unit produces sine, triangle, and square waves as well as left- or right-sloped ramps and pulses with an adjustable duty cycle of 1% to 99%. The output may be amplitude- or frequency-modulated by an external signal and adjusted in amplitude from six volts peak-to-peak down to millivolts. The sine, triangle, and square waves may be swept in frequency by a built-in linear sweep circuit or by an external signal. The cost is in the $50.00 range if you have a moderately stocked junk box. The majority of the parts are stocked by Radio Shack. About half of the cost is in the digital frequency display, which may be easily replaced with a frequency counter or an analog scale.

The entire unit consists of the function generator circuits, the sweep circuits, the digital display circuits, and the power supply. The function generator circuits actually contain two function generators, designated the primary and secondary and labeled F1 and F2 in Fig. 3-25. The two generators may be set independently of each other and it is possible to shift between the two merely by changing the logic level at the FSK jack. The amplitude of the output is absolutely constant from the lowest frequency up to 300 kHz with a usable signal generated up to about 1 MHz. Distortion of the sine wave is adjustable and quite good at approximately .5% THD.

Construction

The heart of the function generator is an XR 2206 IC. I purchased my chip from Jameco Electronics, 1021 Howard Ave. San Carlos, CA 94070. Be sure to ask for the spec sheet; the additional 25 cents is well worth the price. I used the manufacturer's recommendations on wiring the 2206 except for several modifications that are peculiar to my function generator. If you do not want to build the sweep circuit, then just leave those components off of the circuit board; there are no components that are used for both the function generator and the sweep circuits. See Figs. 3-26 through 3-30.

I used ¼-watt, 5% resistors except for R12, which is not critical but should be at least 1 watt. Capacitors C1 through C5 are

Fig. 3-25. The completed swept function generator. The knob labeled F1 is the primary frequency control and is mounted on a 3:1 gear drive. The SIN/TRI output jack also provides a ramp function. All unlabeled jacks are grounds, except the jack, far left, which is the variable dc output.

frequency-determining components and should be 5% or better, and for best results should be made of some stable material such as polystyrene. Other types of capacitors could be substituted, but the generator will be slightly less stable. Wherever possible, I have tried to provide large pads on the circuit board to accommodate capacitors of different sizes.

Trimmer pots R1, R2, R3, R6, R7, and R8 all are mounted on the PC board. I have provided space on the circuit board so that either the stand-up or the lay-down type of trimmer may be used. Front-panel pots R4 and R5 are the other frequency-determining components. During normal use, R4 is the main frequency control, and I have mounted mine using a 3:1 gear drive so that the frequency can be varied in small increments. Potentiometer R5 is used to set the frequency of the secondary function generator and to adjust the shape of the pulse and ramp functions. I found it sufficient to mount R5 directly on the front panel without a gear drive.

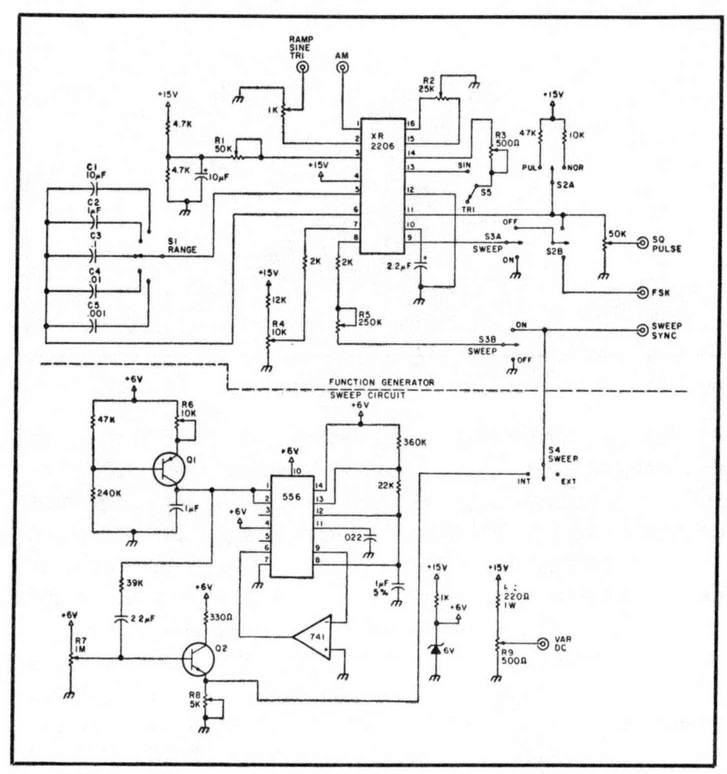

Fig. 3-26. Schematic of function generator and sweep generator.

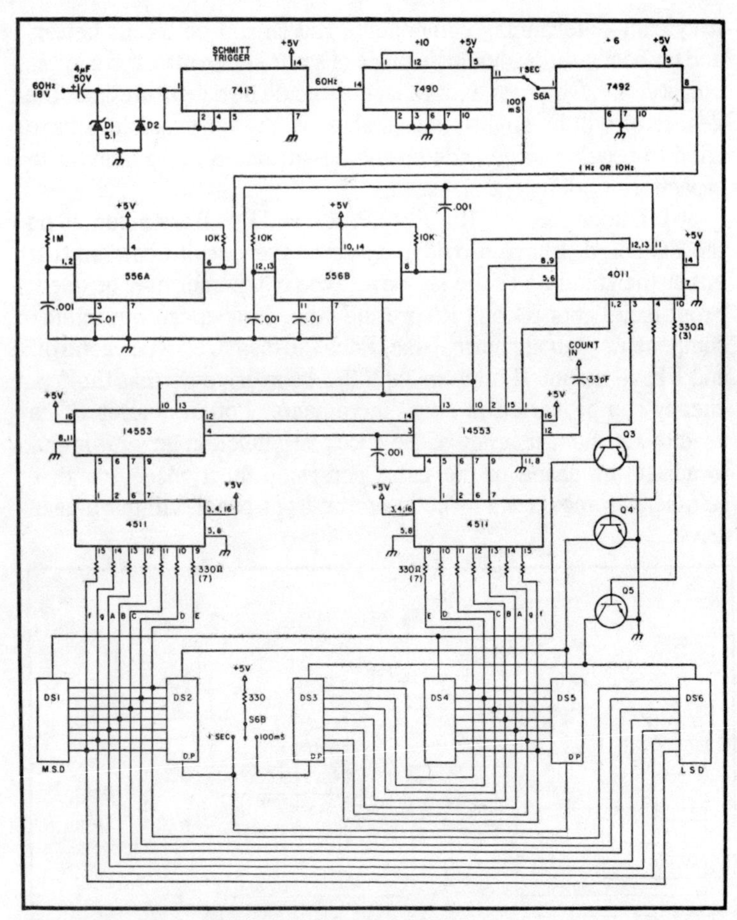

Fig. 3-27. Schematic of digital frequency display.

The function generator may be swept in frequency by applying a varying voltage to R5. For maximum frequency sweep, 1000:1 or better, this voltage should vary from 0 V to +3 V, with the highest frequency being generated at 0 V. I have arranged the front-panel switches so that the function generator may be swept or frequency-modulated by applying a varying voltage to the sweep/sync jack from an external source or from internal circuitry. The internal sweep circuit is a linear ramp generator which provides a 3-V ramp at a frequency which allows one sweep to fill my scope screen when set to 2 ms/cm. When the internal sweep circuit is used, the ramp signal is applied to the sweep/sync jack so that it may be used to trigger your scope.

Operation of the sweep circuit is not overly complex. One half of the 556 dual timer is wired as a monostable multivibrator which is used for a time base. The 741 op amp inverts the output of the multivibrator to provide a good trigger signal. The actual ramp is generated by the other half of the 556 and the two transistors. Transistor Q1 is a 2N4250 or similar and provides a constant current source to capacitor C6. This constant current allows C6 to charge linearly. The other half of the 556 timer acts as a switch which grounds the positive side of C6, when triggered by the op amp. The voltage across C6 almost instantly goes to zero and then begins to recharge. Transistor Q2 (a 2N2222) is a buffer amplifier. By adjusting the dc bias on this transistor, the amplitude and position of the ramp relative to ground may be precisely set. The output of this emitter-follower type circuit is the final sweep signal.

The digital frequency display is a multiplexed unit and is based on the Radio Shack three-digit BCD counter IC, part 276-2489. The entire display was point-to-point wired on a multipurpose 22-pin edge connector circuit board which will just accommodate the 9 ICs and three 2N2222 transistors. The timebase is derived from the 60-cycle line. I have found the accuracy to be quite good at audio frequencies, often showing the same frequency as my lab grade counter, up through 50 kHz or so. Even at much higher frequencies, the difference is usually under 100 Hz and the error will always be on the low side. The frequency generated by the function generator is sampled at pin 11 of the XR 2206 and is coupled to the digital display through a 33 pF capacitor. This signal is then counted and displayed by the two sets of 14553 counter and 4511 driver ICs.

The wiring of the LED displays is a little unusual and requires an explanation. If the counter is placed in the 100-ms timebase

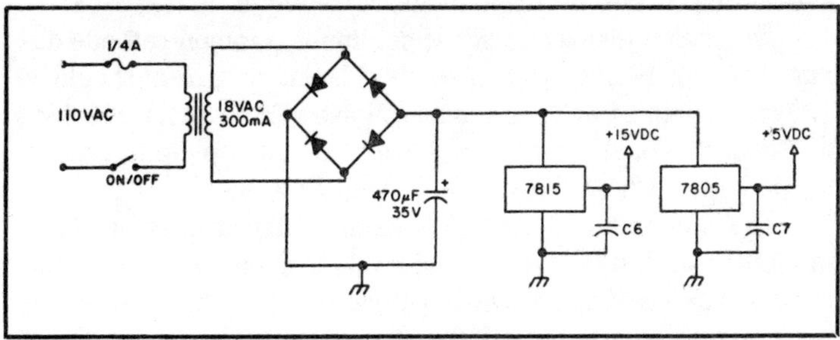

Fig. 3-28. Power supply. C6 and C7 are 2.2-µF tantalum and are mounted on the circuit board. All other components are mounted on the chassis.

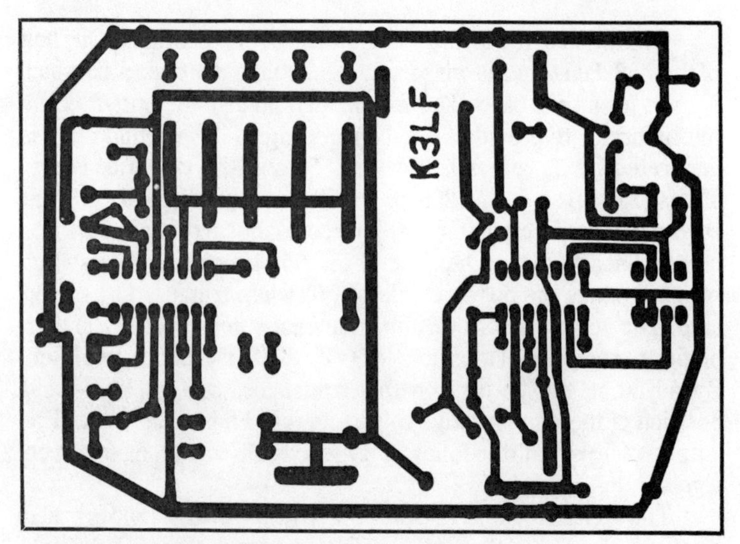

Fig. 3-29. PC board layout.

position, it will count and display 1/10 of the actual frequency being generated, so that 4552 Hz would appear as 455. I found this to be distracting, so I added another LED display in the "ones" position and set it to display a "0". Now the frequency display would appear as 4550. In order to get the newly-added display to have the same brightness as the other digits, I wired it as if it were in the "millionth" position. There is no leading-zero blanking, so the new digit will always show a zero unless a frequency of one MHz or higher is counted. In the one-second timebase mode, this newly-added digit will appear in the "tenths" position and will be zero unless a frequency of 100 kHz or higher is counted. In this timebase, the display would appear as 4,552.0. The decimal points are wired to the timebase switch so they will always correctly separate the hundredths and thousandths digits.

The digital display circuit is set up for common-cathode displays with right-hand decimal points. Maximum current should be limited to about 25 mA per segment. I used Radio Shack 276-1644 displays but would recommend some of the multiple-digit, multiplexed units such as those sold by Digi-Key, Inc.

The timebase and control circuits for the display are fairly straightforward. A 60-cycle signal is taken off the low voltage side of the power transformer. Diode D2 and zener D1 limit the voltage excursions to −0.7 V and ±5.1 V. This 60-cycle signal is applied to a Schmitt trigger to produce an accurate square wave, and then passes through the 7492 IC wired to divide by six. The resulting 10-Hz square wave triggers the two one-shot multivibrators which form the latch and counter reset pulses. In the one-second timebase, the 60-Hz signal is divided by ten in the 7490 IC before going through the 7492, so that a 1-Hz square wave is applied to the one-shots. Again, this is a surprisingly accurate system, and I have been very pleased with the results.

The power supply is fairly conventional, using an 18- or 24-V ac transformer, a bridge rectifier, and two regulators, a 15-volt and a 5-volt. If you use a 24-V ac transformer, be sure you do not exceed the maximum voltage which may be applied to the regulators; mine were rated for 35 V dc. All of the power supply components were mounted on the chassis. Note that a 2-wire power cord is used so that the chassis ground is floating and is not connected to the house wiring ground. This allows the function generator ground to be floating and set in relation to other pieces of test gear.

Calibration and Operation

Calibration is not difficult, but does require a dc-coupled oscilloscope. The function generator circuit has three adjustments. Trimmer R1 adjusts the amplitude of the signal with a maximum of six volts for the triangle waveform. Symmetry is adjusted by R2, and R3 adjusts distortion of the sine wave and may be replaced with

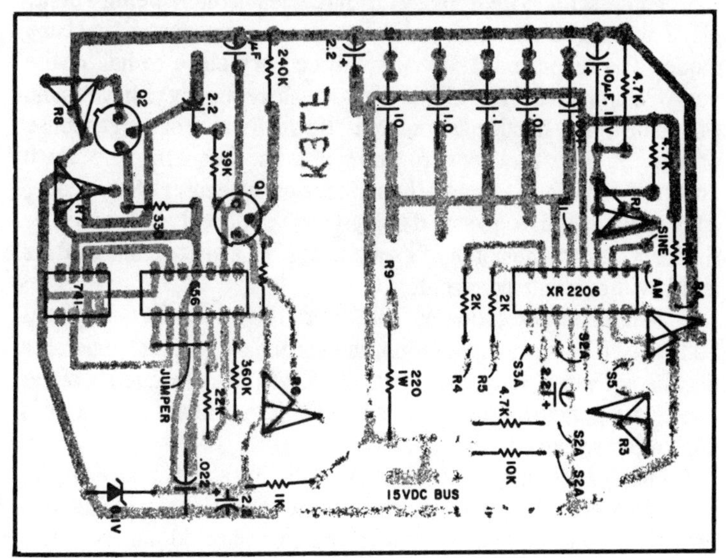

Fig. 3-30. PC board component location.

a 200-ohm resistor with fairly good results. The three adjustments will interact to some degree; the object is to maximize amplitude and minimize distortion. The sweep circuit is adjusted using R6, R7, and R8. The collector of Q1 should be monitored while adjusting R6, which should be set to a position giving the longest ramp. Too little resistance here will make the ramp generator operate at twice the trigger frequency. With an oscilloscope connected to the output of the sweep circuit, adjust R7 so that the ramp just touches the 0-volt baseline. The amplitude of the ramp is adjusted by R8 to a value of three volts p-p. There is some interaction between R7 and R8, and these adjustments are somewhat critical—so be patient.

Operation of the swept function generator should be fairly obvious, but here are some helpful hints. When operating in the sweep mode, S3 should be placed in the ON position and S4 placed in the INT. This will apply the ramp signal from the sweep circuit to the function generator and to the sweep/sync output jack for synchronizing or triggering your oscilloscope. You will find the signal appears to sweep from right to left as the lower frequencies are to the right. If S4 is placed in the EXT position, then an external device may be used to sweep or frequency-modulate the function generator. Whenever S3 is ON, the average frequency will be controlled by R5 and the range switch.

To determine the range of frequencies actually being swept, place the digital frequency display in the one-second timebase mode. The displayed frequency will be very close to half of the peak frequency and 500 times the lowest frequency. I have found this to be a very handy phenomenon. If I am looking at the response of a notch filter and want to know at what frequency the notch occurs, I can adjust R5 until the notch is at the center of my scope. At that point, the frequency displayed on the digital readout will be the frequency of the notch. If I want to apply a swept signal of 10 Hz and I will be right on target.

When operating the generator with the sweep turned off and S2 in the NOR (normal) position, the secondary function generator may be activated by grounding the FSK jack. The frequency of the secondary generator is controlled by R5 (the primary generator is controlled by R4).

One last item: Any amplitude-modulating signal must have the same dc bias as the output. Potentiometer R9 has been included to provide an adjustable dc reference voltage. The 220-ohm resistor, R12, limits the current if the VAR dc jack is grounded.

Chapter 4
Amplifiers

Here are some easy-to-build amplifier projects. Preamps are some of the easiest projects and the results are really dramatic. Linear amps are for hams that want to really get their signals out. All of these projects can be modified to suit your particular needs.

Referring to Fig. 4-1, the circuit can be seen as a direct coupled pair of 2N3904 transistors. This transistor is cheap, high gain, fairly low noise, and very easily obtained. The Q2 transistor is hooked up like any ordinary amplifier stage, but the base resistor that normally goes from its base to ground has been replaced with another transistor, Q1. This Q1 transistor varies the bias on Q2, so the circuit is immune to heat effects. The way it's hooked up, if Q2 draws more current, the voltage on R2 rises, turns on Q1 harder via the 100k resistor connected to its base, and cancels out the increased current in Q2. The result is almost no change in current due to temperature variations. The capacitor C2 prevents the ac signal from being fed back and reducing the overall gain. By placing the capacitor as shown, a very small value, which is also small in physical size and cheaper, will permit the amplifier to keep its full gain to low frequencies as well as would be the case for a very large C placed across R2. The values in Fig. 4-1 will amplify down to about 10 cycles using a physically small capacitor. To make the amplifier roll off at a higher frequency on the low side, reduce C2 to about 1 μF or less, or, alternately, you could reduce the 100k resistor to about 10k. This would make the frequency roll off around 100 cycles and turn the circuit into a speech amplifier rather than a hi-fi type.

The circuit shown in Fig. 4-1 performs best when driven by a moderate impedance source from 500 ohms to 3k ohms impedance. With this kind of source, the gain will be about 250, and the output noise with no signal in will be about 2 millivolts. This is equivalent to an input noise of only 8 microvolts, so the noise is quite low for all but extraordinary uses.

If you wish to drive the circuit with a low impedance source, such as a speaker of 4 to 16 ohms or a telephone earphone (which makes an excellent high output mike), use the circuit in Fig. 4-2. Here, the base is tied to ground via a capacitor, and the signal is fed to the emitter of Q1 through a capacitor. This circuit will perform very similarly to Fig. 4-1, but will have slightly highly gain reaching perhaps 500 and about the same low noise performance.

Ten microfarad capacitors are used throughout because they are small and cheap, and are more than enough to do the job here.

This simple circuit can be made up in a ball smaller than an acorn and put into mikes to give you more gain than you need to drive even the worst transmitter. It also works well when driven by a speaker put out in the yard to let you listen for prowlers at night, when you don't care to get out of a warm bed, but the dog barks like he's on to something. Fed into any hi-fi input, such a preamp will let you hear better than if you were out in the yard. There are many other uses, and most of them will please you because the low noise of this preamp lets you really hear clean audio.

BARGAIN PREAMP

Here is the world's best deal in an audio preamp. It has high input impedance for crystal and other high-impedance mikes. It has low output impedance so it will feed into just about any following amplifier, or even through a long shielded cable to an amplifier. It uses a minimum of parts and a bargain op amp IC.

The circuit (Fig. 4-3) is based on the popular 741 op amp, the μA741. While this number is a Fairchild origination, there are now many manufacturers and it is also part of the HEP line by Motorola, available almost anywhere. The one used in this preamp was purchased on sale from Radio Shack, marked 741C, for only 39¢. The HEP equivalent is Motorola number C6052G, and it performed identically to the Radio Shack special. I tried the RCA CA3160, but it oscillated in this circuit until I added the 150 pF capacitor across the 1-megohm feedback resistor. The RCA CA3160 is a premium-grade op amp meeting military specifications which is cheap enough for this application. The 741 used is in a round 8-pin socket. They are also available in an 8-pin mini-DIP form.

The use of the 741 was ideal for my application because it works so well into a low-impedance load. I am replacing a carbon

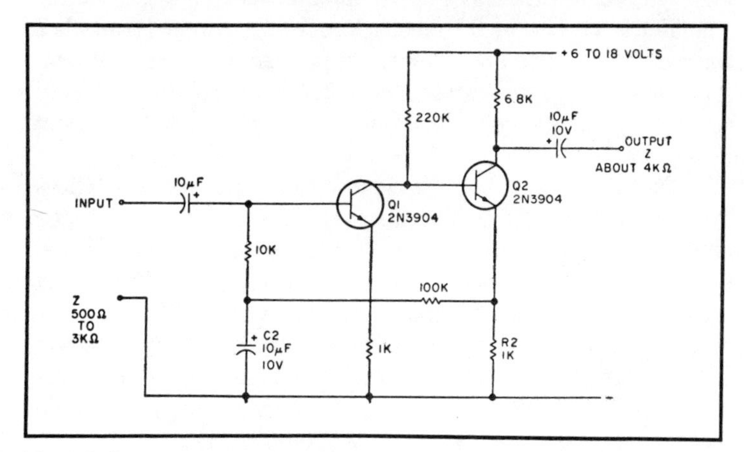

Fig. 4-1. General purpose preamp.

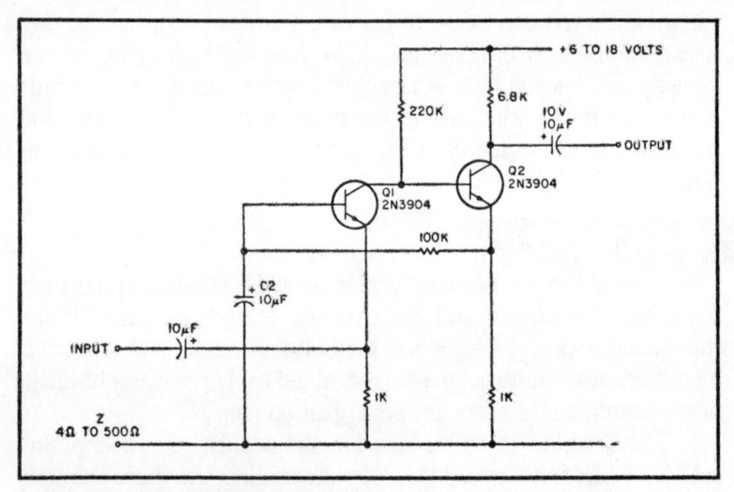

Fig. 4-2. Low impedance preamp.

mike used in an RCA Carfone converted to a 2 meter base station. The input impedance to the converted RCA rig is 1,000 ohms. I could have removed that part of the circuit and fed a preamp directly into the grid of the first audio stage, but I wanted to retain the carbon mike jack and circuit for a secondary mike input because plans call for the use of the crystal mike in a different mobile rig.

Measured frequency response of this circuit is 750 hertz to 9 kHz at the 3 dB down points. This response is the same whether working into an infinite load or 1,000 ohms, but gain measurements do differ with load. Voltage gain is 150 (43.4 dB) into a 1,000 ohm load, but somewhat less into an infinite load. The gain of the amplifier is affected by the feedback resistor between terminals 2 and 6 on the IC. The lower the resistor value, the lower the gain. Without any feedback resistor, the gain can be as high as 100,000, which is obviously more than one needs or can handle without instability.

The 150 pF capacitor across the 1-megohm resistor stabilizes the IC against self-oscillation when using the CA3160, and may be omitted for other 741-type ICs. With the capacitor, the CA3160 replaces the 741 directly. The only apparent adverse effect is that it extends the frequency response to well over 100 kHz. There is no apparent effect on performance as, obviously, any response above about 5 kHz is lost in the associated circuitry of the equipment with which this preamp is used. On-the-air reports indicate that the voice is clean and crisp, with no distortion.

Because the preamp was designed for use on a 2 meter rig,

the input rfc is only a 10 μH (Miller #4612) choke. For lower frequency use, a 1 mH rfc is recommended. Also, a carbon 2,000 ohm resistor might be just as good, but it was not tried.

The rectifier diode may be any silicon diode except small-signal types. The lowest voltage- and current-rated axial-lead diode will do. Other values of electrolytic capacitors than those lised may be used as long as the capacitance values are reasonably high and the voltage rating exceeds the voltage source by a factor of two.

An ac tap across one of the tubes in the rig is the source of the 12.6 V ac. This was convenient. If 9 to 12 V dc is available from other points of the associated circuit, use it. Omit the rectifier diode, but retain the filter to reduce hum.

Current drain is less than .5 mA. A 9 V battery could be used instead of the ac source, and filtering could be left out. The battery would last a long time.

All small parts, including those of the rectifier-filter system, fit easily on a 2″ × 1½″ PC board.

The PC board is not an etched-copper board, but a "PC-style" board using Circuitstik stick-on wiring elements. The system consists of pre-made circuit elements such as for transistors, ICs, multi-element connections, wire strips, etc. These have adhesive backs and a protective backing over that. Their board is glass epoxy with perforated holes in a grid with .1″ spacing. The hole spacing matches that of the circuit elements. A circuit is made by stripping the protective backing from the circuit pieces and apply-

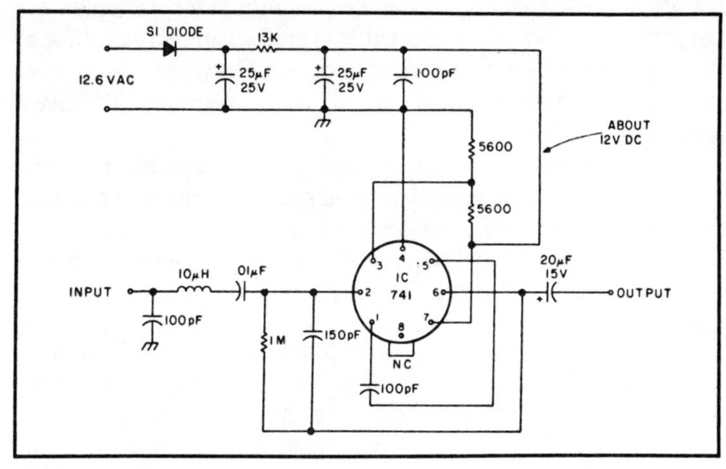

Fig. 4-3. Amplifier schematic diagram.

Fig. 4-4. Completed amplifier with cover off 4"×2"×1½" aluminum case. The glass-epoxy board measures only 2"×1½".

ing the circuits to the board. The same circuit planning and layout as for printed circuitry is required, but the need to etch copper from a copperclad board is eliminated. Circuitstik is available at most large electronic supply stores. Of course, the usual etched-type printed circuit may be used, or even point-to-point wiring. For the latter, you may require a larger board and case.

Parts are mounted to one side of the board in the same manner as with any PC board. The leads are soldered on the other side to the metal circuits, which are solder plated for easier soldering. A few jumper wires are located on the bottom side for convenience. Layout of parts is not critical, but it is pretty hard not to follow a straightforward layout from input to output. The three-circuit jack shown in the photo matches the plug on the mike for PTT relay operation.

Two 6-32 screws hold the board to the bottom of the case. Two nuts, or spacers, should be used to support the board above the case bottom to clear the solder hills.

Shielded wire should be used between the output and the input of the associated circuitry, if the distance is more than about six inches.

While the 4"×2"×1½" aluminum case into which this amplifier was built is not essential, it does provide good shielding against rf pickup. If the amplifier is made part of its associated circuitry, it may be necessary to add some form of shielding. Also, see Figs. 4-4 & 4-5.

The popularity of push pull circuits in rf amplifiers dropped off shortly after World War II. The use of the pi network pretty well solved the harmonic problem, worked well with a single tube or two tubes, in parallel, lent itself to bandswitching, and used a few less components. However, with the advent of TV, everybody started putting low pass filters in the output line anyway, so we are just about back to where we started from. So let's take a look at push pull again.

The number one objection to push pull seems to be the necessity of an input tuned circuit to get the grids 180° out of phase. As will be seen later, this problem has been eliminated. Secondly, the push pull circuit does not lend itself to bandswitching. This is true, but there are people who are interested in one band only. Also, with a small compromise in L/C ratio, you can cover two bands with one coil. This is especially true if the two bands happen to be 15 and 20 or 10 and 15. You can even cover 10, 15 and 20 if you prune very carefully and use a fairly sizable maximum capacity tuning capacitor. Another possiblity, is the National all band tuner, which does away with band-switching entirely.

Of recent date the grounded grid circuit has come to the forefront. It requires no neutralizing and is ideal for the exciter that

Fig. 4-5. Bottom side of perfboard. The .1"-spaced holes match the Circuitstik wiring elements. It looks and works like etched circuitry, but the wiring elements are stick-on and eliminate etching.

Table 4-1. Amplifier Parts List.

Qty	Part
1	741 IC
1	Silicon diode, HEP R0050 or equivalent
3	100 pF capacitors
1	150 pF capacitor
1	.01 uF capacitor
1	20 uF, 15V dc electrolytic
2	25 uF, 25 V dc electrolytic
2	5600 Ohm, 1/4 W resistors
1	13k Ohm, 1/4 W resistor
1	1 megohm, 1/4 W resistor
1	10 uH rfc
1	8-pin IC socket
	Mike jack, PC board, and parts based on your choice.

already has considerable power output. How about a push pull grounded grid? It ought to work fine, but we have all of that circuitry mentioned above. It occurred to me that since this is a low impedance circuit to the input, a 4:1 balun could be used to get the push pull excitation. Accordingly a two inch toroid core was ordered from Amidon. Upon receipt of same, we wound 8 turns of 72 ohm ribbon (receiving type) on it, and discovered that the drive to the grids was about the same as that derived from a tuned circuit. And it works on all bands.

The circuit in Fig. 4-6 will work at any power level but in this design three restraints were imposed. First of all, the all band tuner is rated at about 1500 volts, and it was desired to voltage double an old TV power transformer for the dc supply which also comes out about 1500 volts. Secondly, the box to be used was of limited size, so the number and size of components had to be kept at a minimum. Last but not least, operation without a fan was desired.

Grounded grid with a filamentary cathode is a headache because you need to have big, bulky, hard to mount, filament rf chokes which are expensive if you buy then ready-made. An indirectly heated cathode is much to be desired. A pair of 7094s was on hand. These were ideal except at 1500 volts they don't operate at maximum ratings. Several power tubes with indirectly heated cathodes and higher dissipation ratings have come out recently. Obviously many of the TV tubes could be used in the circuit, operated at lower voltage and power levels.

Figure 4-6, using the all band tuner, was selected. Trial one did not include the ferrite beads and trouble with parasitics was experienced. Beads were added to the grid circuit primarily because the grids were easy to get at. Beads in the plate circuit might be better. Ordinary parasitic chokes with shunt resistors will do the same thing. Beads seem to be the cheap and easy way to do it. At any rate, the parasitic problem was solved. The plate circuit-pickup coil was tapped and the taps connected to a rotary switch. This gives adjustable loading ability. If you have a set of BC 610 plug-in coils with a swinging link, you get the same effect, only it is continuously variable.

The unit shown in Fig. 4-7 uses a box left over from some former project. It is in the vertical position shown because of the National all band tuner which fitted best this way. All of the power supply is in the bottom. The rectifiers are actually five 1000 V piv diodes on each side, without any voltage dividing resistors or capacitors.

At 10¢ each, it's too much trouble to add resistors and capacitors. Just put in twice as many diodes as required for the voltage. It saves space. The front panel is a piece of aluminum sheet from an off-set printing system common to many newspapers. They are pretty thin but, as in this case, will cover an old panel with the holes in the wrong places. The switches are ordinary toggle switches mounted on the grounded front panel. There has been no voltage breakdown to date. You could mount them insulated but then touching them might be an electrical hazard.

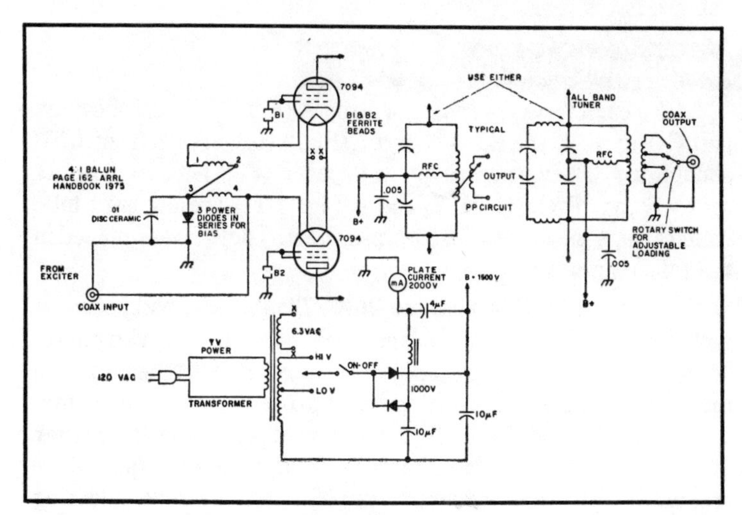

Fig. 4-6. Push pull grounded grid linear.

Fig. 4-7. The completed amplifier, ready to use.

If you use the all band tuner, you should carefully check the output frequency with a wave meter (grid dipper on diode position). Make sure you have the right band and mark same on the dial. For example, the 40 meter and the 15 meter points are close together. You could be tuned to 15 when you think you are on 40 if you are not careful.

50 WATT AMPLIFIER FOR 1296 MHz

The cavity bypass capacitor, part A (Fig. 4-8), top of cavity, part C (Fig. 4-9), and chassis, part E (Fig. 4-10), are cut from .016" sheet brass. The vertical cavity wall, part D (Fig. 4-9), is cut from a cat-food can. Parts G, H, I, and J (Fig. 4-11) are press-fit tube contact rings made out of .010' brass sheet. They are assembled in the following sequence:

1. Wrap part G around the 2C39/7289 anode, and secure it with 2 pieces of #18 bare hookup wire twisted tightly with pliers. Make sure the ends of part G do not quite touch each other. Slip the tube with part G wired on just barely into part A so that the bottom of parts G and A are flush. Tack solder G to A about every quarter inch on the upper surface before smoothly flowing the solder around the full circumference. Use plenty of Nokorode soldering paste and an Ungar #4033 50-watt soldering element to make the

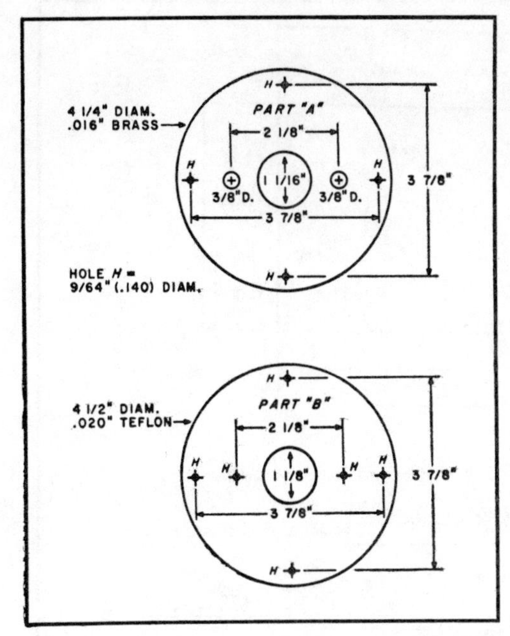

Fig. 4-8. Cavity bypass capacitor.

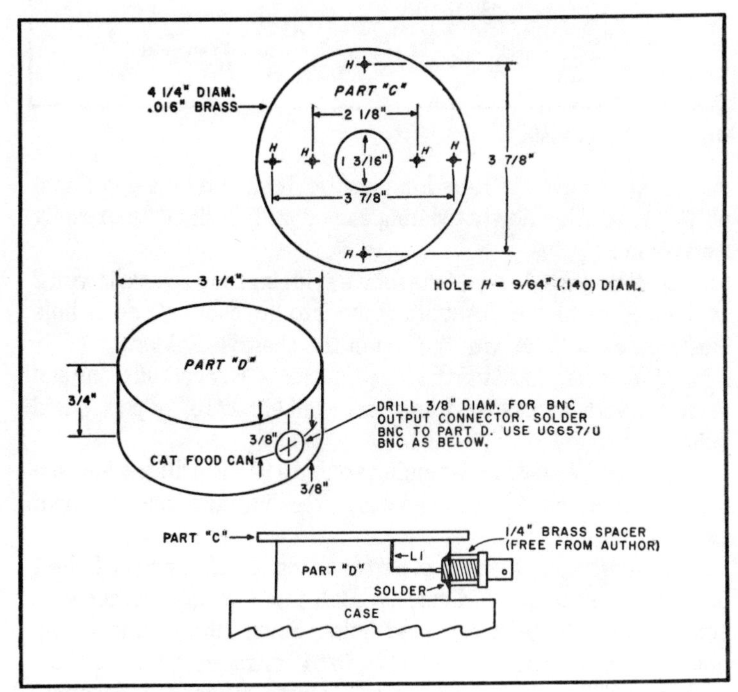

Fig. 4-9. The top of the cavity.

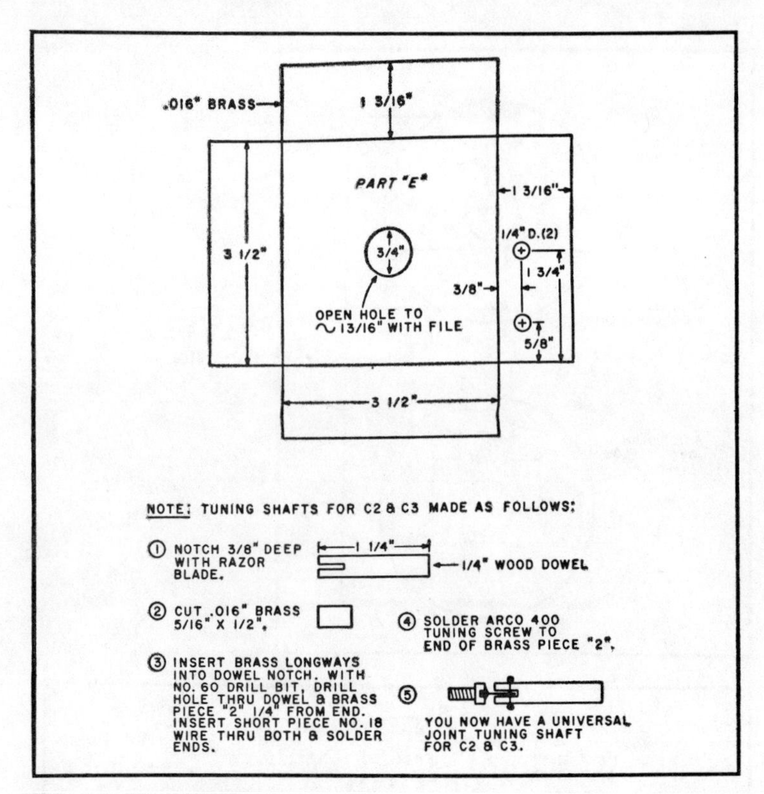

Fig. 4-10. The chassis.

job easy. Remove the tube from part A. It should be a good tight press-fit, but, by slowly twisting as you pull, it should be readily removable.

2. Wrap part H around the tube's grid ring, and secure it with 2 or 3 wires, as above. Insert the tube into the chassis' center hole until part H is flush with the top of the chassis. Solder.

3. Wrap part I around the tube's bottom outer cathode ring and secure it with wire. Insert into part F all but 1/16″ of part I, and solder.

4. Part J is made by wrapping around a 9/16″ drill and squeezing with pliers until it makes a snug press into the inner filament ring in the tube base.

5. Center part A on top of part B on top of part C and drill the 4 outer holes each 9/64″ diameter. Bolt the 3 parts together with conventional ¼″ 6-32 nuts and bolts. Solder the 4 nuts to the bottom of part C. Also drill the two 9/64″ diameter holes for the 2 tuning capacitors, and disassemble. Run two ordinary ¼″ 6-32 nuts

and bolts through the holes in part C, and solder the nuts to the top of part C. Drill ⅜" diameter holes through part A's tuning capacitor holes, and then reassemble parts A, B, and C with four ¼" 6-32 nuts and bolts each.

6. With scissors cut a ¾"-wide section of the cat-food can of your cat's choice. File it smooth, or, better yet, sand it smooth on a belt sander. With parts A, B, and C still bolted together, turn them upside down on a flat surface and tack solder part D to the bottom of part C before flowing the solder smoothly around the entire circumference.

7. Install and solder the output link and BNC connector as shown. Thread in the two tuning capacitors in the bottom of part C.

8. Install the tube in part G and A so that the top of part G is flush with the bottom of the protuding anode ring (tube is in as far as it will go). Carefully insert the tube into the chassis and part H until the bottom of the cavity is flush with the top of the chassis. There should be no radial or axial side loads on the tube if you have

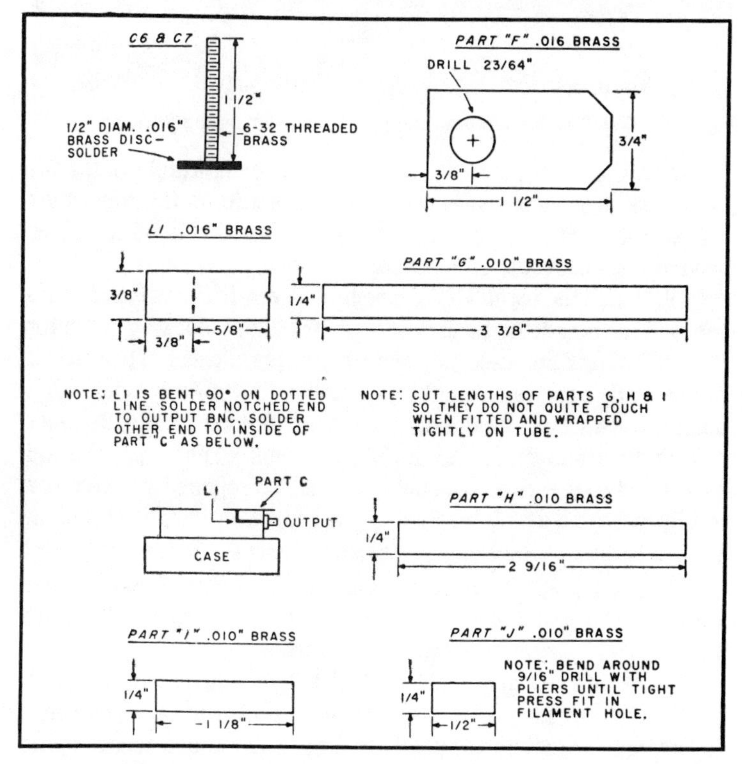

Fig. 4-11. Tube contact rings.

Fig. 4-12. These knobs should be replaced with insulated knobs.

followed these steps in sequence—just a comfortable press-fit. Now, maintaining a gentle downward pressure on the tube, tack solder the cavity to the chassis before smooth flowing solder around the entire circumference.

9. Install the input BNC connector. Press-fit Parts F and I onto the cathode ring as far as they will go. The input blocking capacitor is modified by filing each side of a 500 pF ceramic disc cap until half of the leads have been filed away. Carefully tin each side with a small (25 watt or less) soldering iron. Solder in the 2 pi-net capacitors as shown. Solder the 500 pF modified disc cap to the end part F and to the inner conductor of the BNC connector. This is a fragile part and easily broken when removing and/or installing tubes. To further mechanically stabilize part E, take two burned-out ¾″-long glass fuses and solder them to part F on the side opposite the two pi-net capacitors. They are structurally quite strong, good insulators, and they are free.

If you are like me, unable to resist buying those one buck each 2C39 tubes at hamfests, you are most probably, also like me, throwing your money away. Of the first twelve tubes acquired this way, one was good; the others gave only marginal gain or were

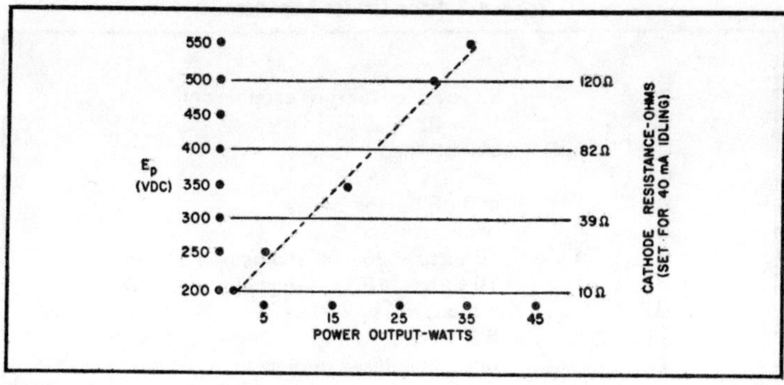

Fig. 4-13. Input held at 4 watts.

worthless. Jaro Electronics offers factory-new 2C39s or 7289s in quantities of 5 each that cannot be beat. I prefer the 7289 tubes with the more heat resistant ceramic seals.

It is best to tune up this amplifier at a reduced plate voltage of, say, 300 volts. Use W1HDQs favorite microwave 50-ohm load, consisting of 100 feet of RG-58 coax. If you are an optimist, by all means terminate the coax with 10 each 1000-ohm, 1-watt resistors in parallel. Be assured that you will not burn them out.

Using a Bird Thruline™ wattmeter with plate power off, and a 24-watt full-scale 1.0-1.8 GHz slug, adjust the input pi-net capacitors for zero reflected power with approximately 5 watts input. If you are using a varactor tripler, you must have a good 1296 MHz filter between the tripler and amplifier input, otherwise your measurement will be totally meaningless as you will probably be

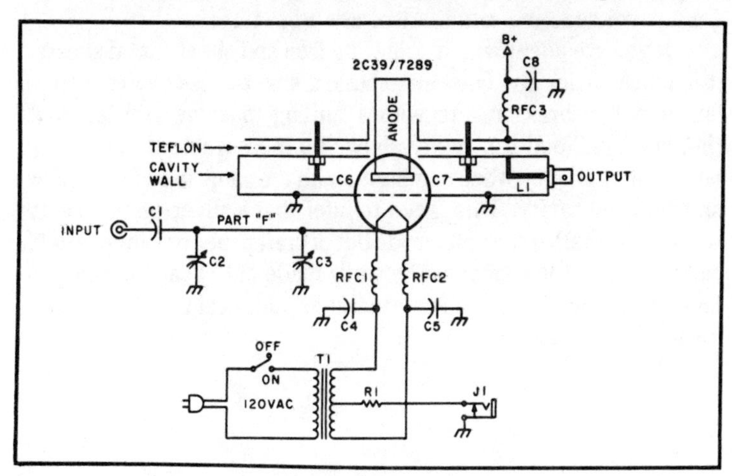

Fig. 4-14. Amplifier schematic.

Table 4-2. Parts List for Amplifier.

C1	500 pF (see text) or ceramic chip
C2, C3	Arco 400 (see Fig. 1)
C4, C5	500 pF feedthrough
C6, C7	see Figs. 1 and 5
C8	500 pF button bypass
L1	see Figs. 1 and 5
RFC1, RFC2	10 turns #26 1/8" diameter x 1" long
RFC3	10 turns #18 1/8" diameter x 1" long
R1	50 to 270 Ω, ½ Watt
T1	6.3 V c-t @ 1 Amp
J1	normally-closed mini jack

measuring as much power at frequencies not on 1296 MHz as you are on 1296 MHz. With a milliammeter in series with the cathode resistors, you should obtain 30 to 50 mA of grid current with the plate supply off. Remove the milliammeter.

With the Bird Thruline™ wattmeter in the cavity's output line and your dummy load attached, you should indicate about 1-watt output, still with no voltage on the plate, when the cavity is properly tuned. Normal position of the tuning capacitors is about ⅛" from full IN. Though the aluminum knobs make a pretty photo, it is best to remove them and use 2½" × ¼" diameter wood dowels drilled to screw on the 6-32 tuning capacitors for adjustment when the plate voltage is ON, or you will be in for a "shocking" surprise. See Fig. 4-12.

With a new tube, you should obtain results as shown in Fig. 4-13 at voltages indicated. Also, see Fig. 4-14.

If you are adept with a soldering iron and sheet-metal shears, it is much easier and quicker to build a sheet-brass cavity than to hog it out of brass ingots with a milling machine and lathe. A tin-plated cat-food can cavity shows no measurable difference in output or efficiency when compared with the same tube in a similar silver-plated cavity. This is not to infer that a silver-plated cavity would be no better than one made out of kraft paper or Plexiglass™, just that the IR losses of cavity walls made out of cat-food cans is less than expected and not measurable with ordinary test equipment.

Chapter 5

Radio And TV Receivers

Some of the most rewarding electronic projects are those that involve building your own receivers. The joy and satisfaction that you feel when you've finished assembling a receiver and you hear (or see) those signals coming in is one that cannot be described, it has to be experienced. So dig right in and build your own!

There are 3 coils or transformers to wind and one i-f transformer to be modified before construction begins. I use standard ⅜″-square 455 kHz transistor i-f transformers for a lot of my construction work. In this case, T1 and T2 are wound on stripped-down i-f transformers using salvaged wire. Refer to Fig. 5-1 for winding instructions. The vfo tank coil is pie-wound on a slug-tuned form. The bfo tank consists of a standard transistor i-f transformer whose secondary has been modified to one turn. This can be accomplished by gently breaking the secondary leads right close to the core with tweezers and unsoldering the remaining wire at each pin. Wind a new one-turn link over the existing windings, and solder the ends to the same pins used by the original winding.

Since it was all being redone, I thought I'd make this version smaller by mounting the resistors and diodes hairpin fashion. The resulting layout is 2.1″ wide by 3.9″ long.

All resistors are ¼-watt, 5% units with one lead bent around a full 180 degrees so that they can be mounted vertically in closely spaced holes. Diodes are mounted similarly. All polarized capacitors are dipped tantalum. All other capacitors are low-voltage discs, except where silver micas are indicated. See Fig. 5-3.

The two 5-30 pF trimmers are very small 5 mm types. These, as well as some of the other parts, may give you trouble when searching for sources.

After all parts have been mounted, you'll have quite a few empty holes left over. These are meant for connecting leads to external controls and for dc power. The PC layout shows where everything goes (see Fig. 5-2). Some holes will not necessarily be used, such as all grounds located around the copper border. These are for convenience and not specifically assigned. You will also find spare pads for +12 volts and +7 volts regulated.

With a 12-volt supply, the receiver should draw between 50 and 60 mA with no signal. A meter in series with the power lead or a metered supply can be used for the initial smoke test in order to determine that there are no shorts or faulty parts. If drain is normal, the following procedure should have you receiving signals in 15 minutes or so.

Using a voltmeter, adjust the trimmer resistor for an output of 7 volts from the regulator. Make sure the regulator is working by varying the voltage from the power supply up to 14 volts and down to 9 volts. You should lose regulation at about 10 volts.

Next, turn the rf and af gain controls to maximum and the bfo to mid position. Slowly run the slug in the bfo transformer out towards the top of the can. When you hit 455 kHz, you'll hear a rushing noise with two peaks and a null in between as the core is rotated. Set the slug to the null point which is zero beat. Make sure turning the bfo tuning control varies the frequency on both sides of zero beat.

The remaining alignment steps should be done with 180-degree operating range of the tuning pot centered so that the excess travel is equally divided at both ends. Some sort of vernier drive is a necessity, and a scale should be mated with the drive so calibration points can be marked off.

Set the tuning pot to the low end starting point and check the vfo frequency. It should be at approximately 3.0 MHz. I use my scope, but a counter can be used. If you have no convenient way of reading this frequency, use brute force. Feed in a hefty signal at 3.5

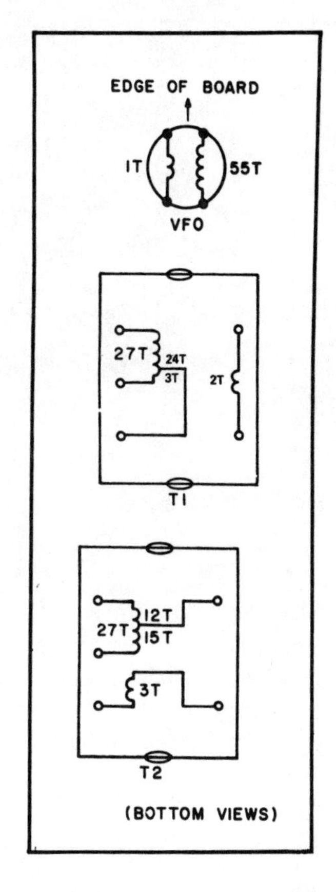

Fig. 5-1. T1 and T2 are wound on stripped 455 kHz transistor i-f transformers using salvaged wire. The vfo tank coil is pie-wound with 7/44 litz wire on a Gowanda series 7 coil form with carbonyl E core. The 1-turn link is wound over the pie.

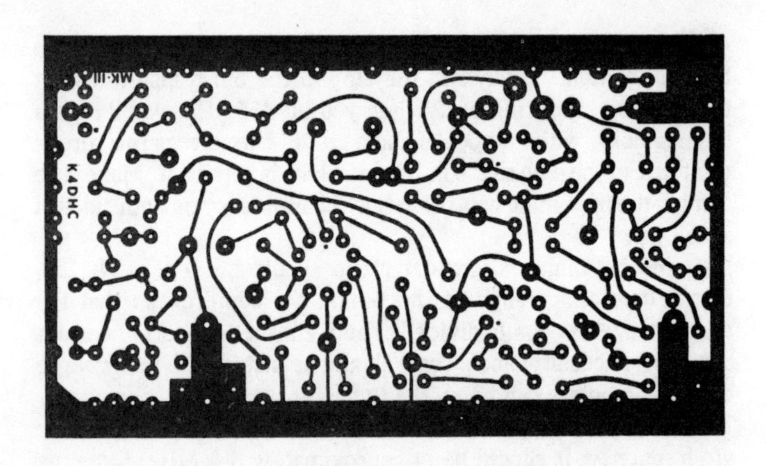

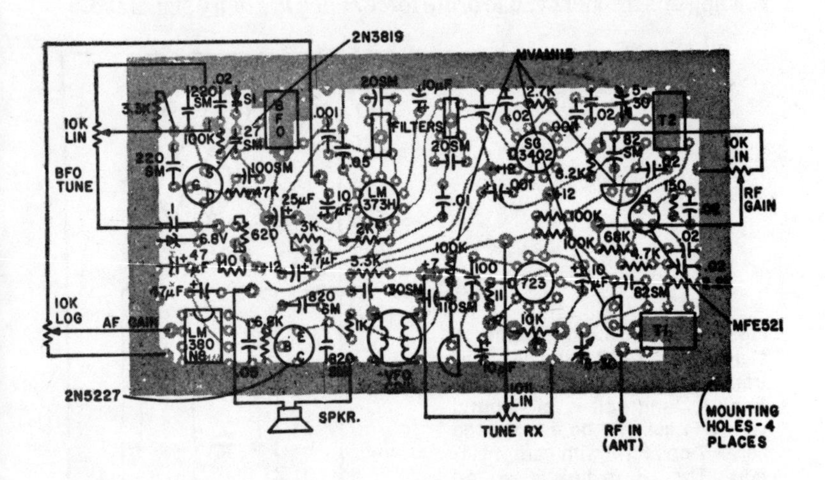

Fig. 5-2. PC board and component layout. Mount the ceramic filters so the circle on top of the case is towards the 20 pF coupling capacitors.

MHz and rotate the slug in the vfo tank coil until you pick up the signal. The gain controls should still be at maximum and the vfo at zero beat. Adjust the slugs in T1 and T2 for maximum response. At this point, you can turn down the af gain to a comfortable level and reduce the rf signal to prevent overloading of the receiver front end.

Once the low end has been set, run the signal generator up to 4.0 MHz, and rotate the tuning pot up towards the high end until

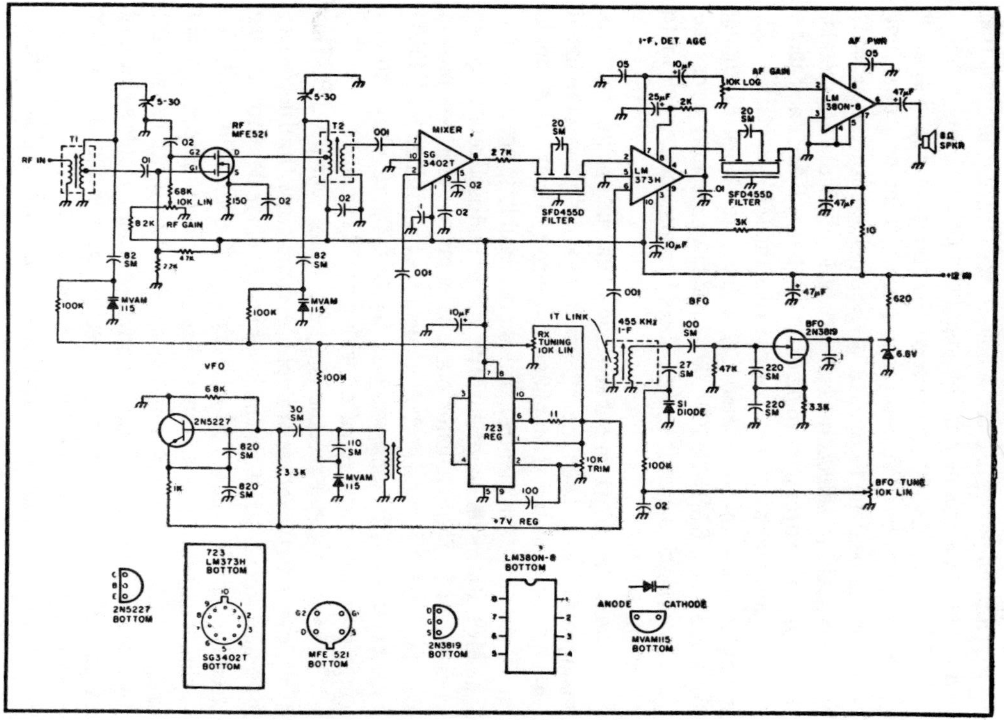

Fig. 5-3. Complete receiver schematic. All resistors are 1/4 watt, 5%. All polarized capacitors are dipped tantalum. All SM capacitors are silver mica. Remaining capacitors are low-voltage discs. Whole numbers are pF unless marked otherwise. Decimal values are in μF.

you find the signal. The two ceramic trimmers across T1 and T2 should be peaked with the generator level set at a low level. Repeat these steps several times until good tracking is achieved.

Finally, mark off intermediate frequencies on your dial with whatever spacing you desire.

HF RECEIVER

The receiver is a single conversion, superheterodyne type, with an FET front end, and is crystal-controlled. No bandswitching is required when it is used over the 6-15 MHz range. Coil usage has been held to a minimum to simplify construction. Construction is also facilitated by the use of a single IC for all audio amplification and the use of a commercial i-f amplifier module.

The schematic for the receiver is shown in Fig. 5-4, as it would be used for WWV reception. Note that the only switching which has to be done to receive WWV on different frequencies is that necessary to select the appropriate local oscillator crystals. The frequency coverage can be extended below 6 MHz and above 15 MHz, by using a different coil between the MPF 102 (HEP 802) rf amplifier and mixer stages. Or, in the case of just extending coverage below 6 MHz, a 100-200 pF padding capacitor, across the 210 pF variable capacitor shown, should extend coverage down to the 80 meter band.

The MPF 102 rf amplifier stage is untuned at its input. Its main purpose is to keep the antenna from loading down the tuned circuits between the rf amplifier and mixer stages. This single tuned circuit is sufficient to provide reasonable image rejection. The MPF 102 mixer stage and MPF 102 crystal oscillator stage are conventional. The oscillator stage is untuned. This has proven satisfactory for general reception, using regular miniature HC6/U type crystals. With some sluggish crystals, the rfc shown in this stage may have to be replaced with a tuned circuit.

The i-f amplifier module is a J.W. Miller type 8902-B. This module is just a two-stage i-f amplifier, complete with all necessary i-f transformers, and it also includes an AM diode detector. Its use greatly simplifies construction. If one can't find it readily available, a simple substitute is to cannibalize the i-f section from a small transistor portable radio. But, use an i-f section which has at least two stages. The really cheap $5 portables often use only a single i-f stage, and this will not provide sufficient gain for any sort of reasonably sensitive reception.

The audio amplifier IC is a Motorola MC1306P. This is a neat,

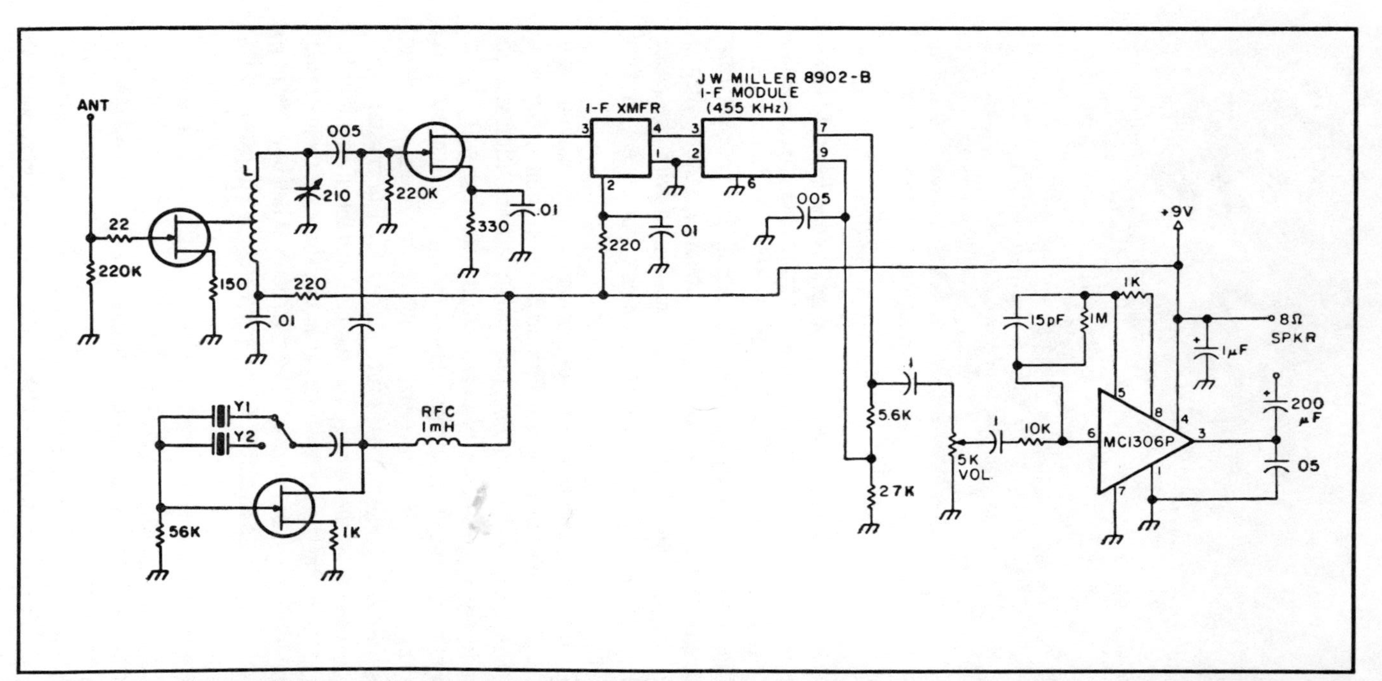

Fig. 5-4. Complete diagram of the receiver. All transistors are MPF 102 or HEP 802. The i-f transformer comes as part of the J. W. Miller i-f module. L = 26 turns #26 on ¼" form. Tap at 13 turns (for 6-15 MHz). Y1 = 9,545 kHz (10 MHz WWV). Y2 = 14,545 kHz (15 MHz WWV).

Fig. 5-5. This is the complete receiver, as assembled on an approximately 4″× 2″ piece of perforated board stock.

inexpensive IC, which combines a preamplifier and ½ watt output in one package. A minimum of external components are needed to make it function. If you did "borrow" the i-f strip from a cheap AM portable to build this receiver, *don't* be tempted to "borrow" the audio section of the AM portable, also. Generally, the quality of such audio sections is horrible, when compared with the clean sound of the MC1306P used with any small, but decent, 8 ohm speaker.

Figure 5-5 shows how the receiver was initially laid out on a piece of perforated board stock. Simple point-to-point wiring was used. The layout wasn't planned, but, rather, construction started on a slightly larger piece of board stock. Starting with the rf amplifier stage, the components were simply grouped together as closely as possible, as I worked from left to right. The rf and mixer stages were grouped around the interstage coil. The crystal oscillator stage is below the i-f amplifier module, and the af amplifier IC is just to the left of the electrolytic capacitor, shown at the extreme right middle side of the board. When the receiver had been assembled, the oversize perforated board was carefully cut down to its final size.

The tuning capacitor used is a regular BC type and is temporarily shown attached at the left side of the board. The receiver should be mounted in a metal enclosure, and the ground leads used in the receiver should be carefully grounded to the enclosure at several points. Although the receiver did work fine wired as shown in the photo, it probably would be safer, from the viewpoint of avoiding possible spurious oscillations, to utilize an isolated pad type of component mounting/soldering technique.

To use the receiver to monitor SSB transmissions, a product detector and bfo have to be added. The circuit for a suitable product detector/bfo is shown in Fig. 5-6. It is relatively simple and inexpensive. If the product detector circuit is added to the receiver using the J. W. Miller i-f module, you have to remove the shield can from the module and take the i-f signal off the first 1N67A *before* the diode detector is built into the module. This operation is fairly simple and obvious, if one uses the module, since a diagram comes with it, illustrating the modification. The diode AM detector need not be disconnected, however. So, one can, if desired, add a switch at the volume control to choose either the output of the product detector or the output of the AM diode detector.

With a mixture of some parts from one's junk box and newly-bought main components, the receiver can be constructed for about $20. This represents a rather modest cost for utility-type HF receiver, for which one can find many applications around the shack or in portable use.

FIVE BAND RECEIVER

I am sure many radio amateurs who have home brew rigs would love to have a matching receiver. Deciding to do something positive about this emptiness in the shack, I came up with a plan that made the home brew receiver not only a possibility, but a reality.

The plan centered on reducing the complexity and time of construction dramatically by using a drugstore transistor AM broadcast radio set as the main building block. Even if you never

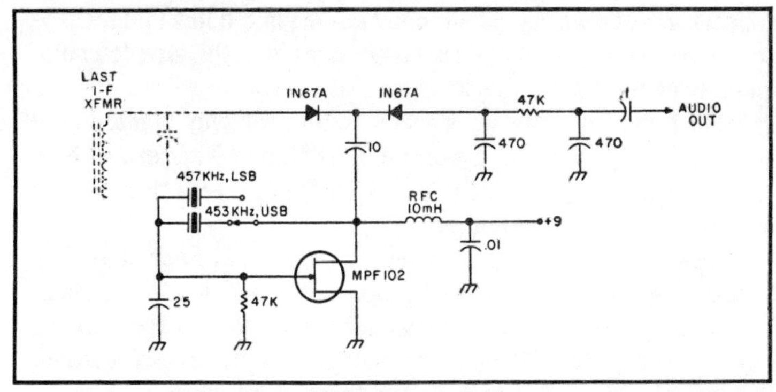

Fig. 5-6. Product detector/bfo, which can be added for SSB reception.

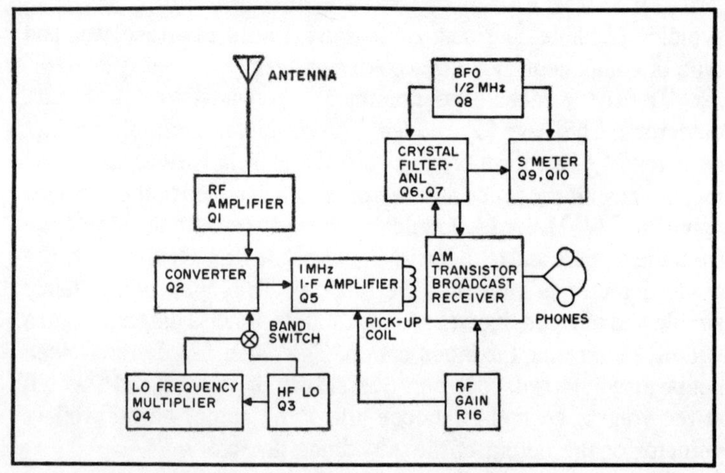

Fig. 5-7. Block diagram.

start to build this receiver, I am sure that you will find it comforting to know that if it ever became necessary, you could do it.

This receiver, being a dual conversion type, has two local oscillators (LO) and two intermediate frequency amplifiers (IF). In this circuit we have to make our own HF LO, while both IF amplifiers and the MF LO are parts of the AM broadcast set.

We also have to make six other circuits to support our BC set to make certain that our project winds up a real communication receiver. These are the HF radio frequency amplifier, 1 MHz amplifier, beat frequency oscillator (BFO), S-meter, crystal filter, and automatic noise limiter (ANL).

When all the circuits are working together, their operation is spectacular for such a simple design. When the rf gain control is two thirds up, a 0.2 microvolt 7.1 MHz signal at the antenna connector will read S9. The receiver noise is too low under these conditions for me to make a measurement with the simple equipment available to me.

All I can say about it is that I could hear only signal in the phones, and I just finished working F6ARC on 40 meters with no trouble at all. Any dual conversion birdies are less than S2 and located so they are no bother.

Drift and broadcast station feedthrough is nil. Each of the five bands can be selected by a front panel control and is 0.9 MHz wide. The 6 dB down bandwidth signal selectivity is 300 Hz with the phase control in the CW position, and 1.2 kHz in the SSB position. It is powered by a 9 V battery and the current drain is 30 mA.

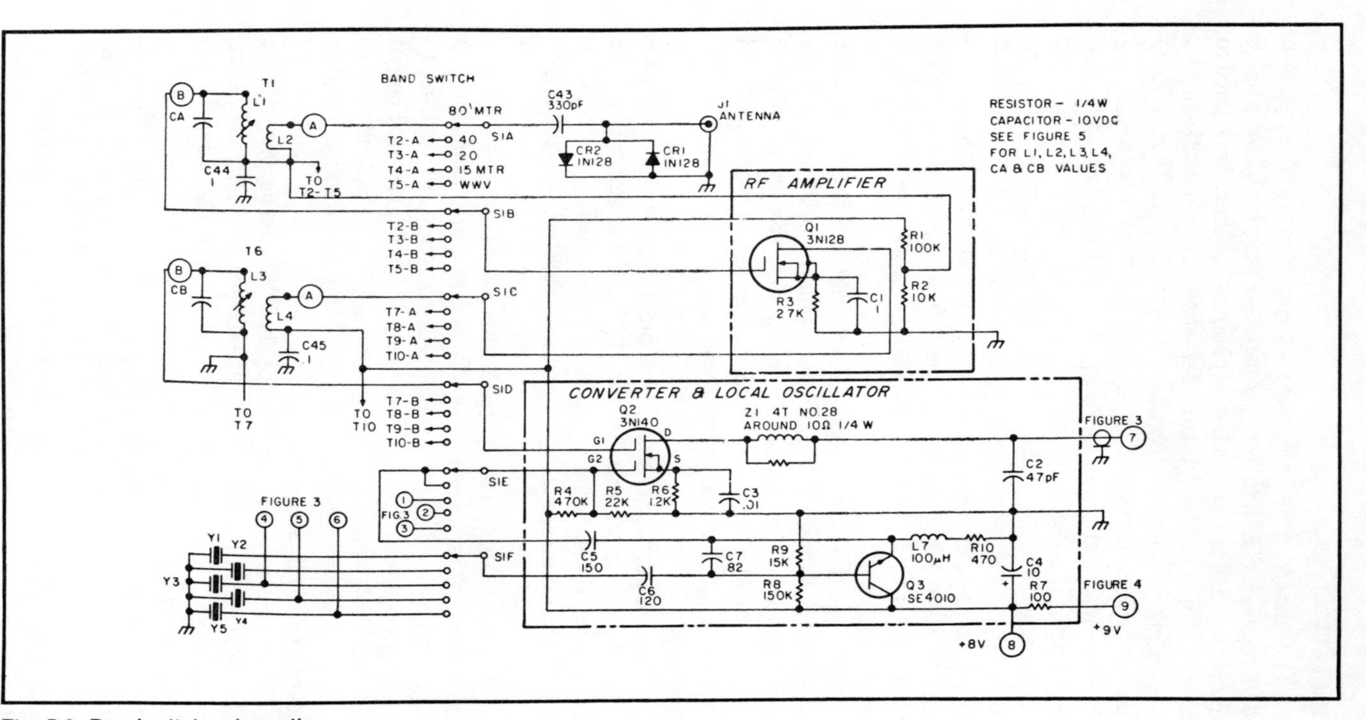

Fig. 5-8. Bandswitch schematic.

Circuit Description

The transistor AM broadcast set just keeps on doing what it did before we bolted it to the front panel—changing .55 MHz to 1.6 MHz rf to sound at the speaker or phones—so there is no need to describe it any further. Figure 5-7 shows how it works in our receiver and is supported by the outboard circuits. These will be described in detail because each is unique in this receiver.

Starting from the antenna connector, Fig. 5-8, the bandswitch, S1, selects one of the rf transformers. T1-T5. They are broadband-tuned to the center of the desired frequency range. Therefore, all the signals in the frequency range selected appear at the gate of rf amplifier Q1. Here they are amplified as much as possible without adding noise to the output. By using a low noise MOSFET for this amplifier, the receiver signal-to-noise ratio is greatly improved.

To prove this point without a lot of math, it is logical that if the rf signal is made greater, the following gain controls will have to be turned down to yield the same output that was present before amplification. If noise was not added in the amplifying process, all the frying sounds generated by these turned down stages will be much less.

The output of the rf amplifier is inductive coupled to the gate coil of the converter transformer, T6-T10, and selected by the bandswitch, which, also through other poles, applies this signal to gate 1 of the HF converter Q2 (along with the HF LO output to gate 2).

The HF LO is crystal controlled for stability and uses FT-243 type crystals. The 20 meter, WWV, and 15 meter bands are at a frequency higher than that at which these crystals will oscillate, so a multiplier is used to double or triple their fundamental output when the bandswitch is in these positions. This multiplier is a class C amplifier whose output is tuned to the selected frequency with rf chokes and fixed capacitors.

Now things really start to happen. While the converter Q2 is doing what is natural, its output is a real mess of signals, and we are only interested in the ones that are the difference between the LO and rf frequencies.

The unwanted signal that will cause the most harm is the very strong one at the LO frequency. If it gets into the BC set loop stick, overloading will take place and there will be birdies all over the bands. To stop this LO feedthrough, the converter output is filtered by using a well shielded oscillator coil T11 (Fig. 5-9), removed from another BC set, and tuning it with a fixed capacitor to about 1 MHz.

Because of a long coaxial cable run to this improvised transformer, Z1 was fabricated to swamp any VHF parasitics that might develop. The base of the 1 MHz amplifier Q5 is connected to the pick-up coil in T11, resulting in a clean converted signal being amplified. It produces a strong field around rf choke L8, which is tuned to about 0.8 MHz with fixed capacitors.

This choke is mounted close to the BC receiver loop stick so its field will be picked up with little attenuation. Strong spurious signals (birdies) are unacceptable. Therefore, the importance of keeping the HF LO signal out of the BC set, and the BC LO signal out of the HF rf amplifier cannot be overemphasized. Most of the receiver shielding and parts placement was made to achieve this isolation.

The broadcast receiver is now able to tune and detect the different HF signals that have been converted to frequencies that are within its range. It is still not ready to be used for a reliable contact, because it needs at least a beat frequency oscillator and more selectivity.

The BFO is a series-tuned Colpitts type. It uses a transistor BC set IF transformer for the frequency controlling element and a front panel controlled capacitor to vary the pitch. Its output is taken from the small untuned winding in the IF transformer.

The receiver's fine selectivity is achieved by connecting a crystal filter between the collector of the BC set's first IF amplifier transistor and its output transformer. To implement this, the collector lead is disconnected from its original place, and reconnected through a coaxial cable to another identical IF transformer located on the ½ MHz crystal filter and ANL circuit board (Fig. 5-10).

This transformer, T12 provides the input for the FT-241 low frequency crystal Y6 and the 180 degree out-of-phase signal for the phase control C20. When C20 is critically adjusted from the front panel, stray signals shunted around Y6 are canceled and the filter output has an extremely narrow bandwidth.

When it is closed, it sends a strong signal around Y6 and the bandwidth is useful for SSB communication. The output of Y6 is

Fig. 5-9. Shielded oscillator coil.

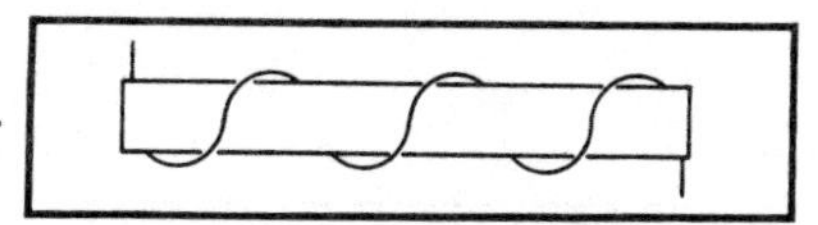

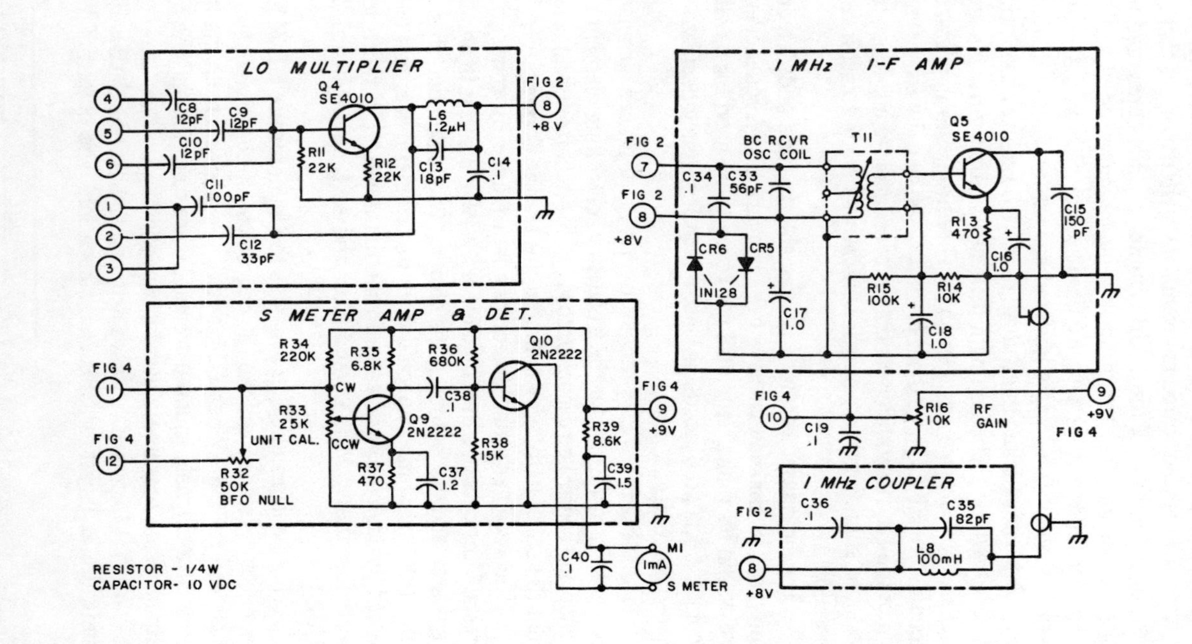

LO MULTIPLIER
Q4 SE4010
C8 12pF
C9 12pF
C10 12pF
C11 100pF
C12 33pF
R11 22K
R12 22K
L6 1.2μH
C13 18pF
C14 .1
FIG 2
+8 V
1 MHz I-F AMP
FIG 2
FIG 2
+8V
BC RCVR OSC COIL
T11
Q5 SE4010
C34 .1
C33 56pF
CR6
CR5
IN128
C17 1.0
R13 470
C16 1.0
C15 150 pF
R15 100K
R14 10K
C18 1.0
S METER AMP & DET.
FIG 4
FIG 4
R34 220K
R35 6.8K
R36 680K
Q10 2N2222
CW
C38 .1
R33 25K UNIT CAL.
CCW
Q9 2N2222
R38 15K
R39 8.6K
+9V
C39 1.5
R37 470
C37 1.2
R32 50K BFO NULL
FIG 4
C19 .1
R16 10K
RF GAIN
+9V
FIG 4
1 MHz COUPLER
C36 .1
C35 82pF
FIG 2
L8 100mH
+8V
C40 .1
M1
1mA
S METER
RESISTOR - 1/4W
CAPACITOR- 10 VDC

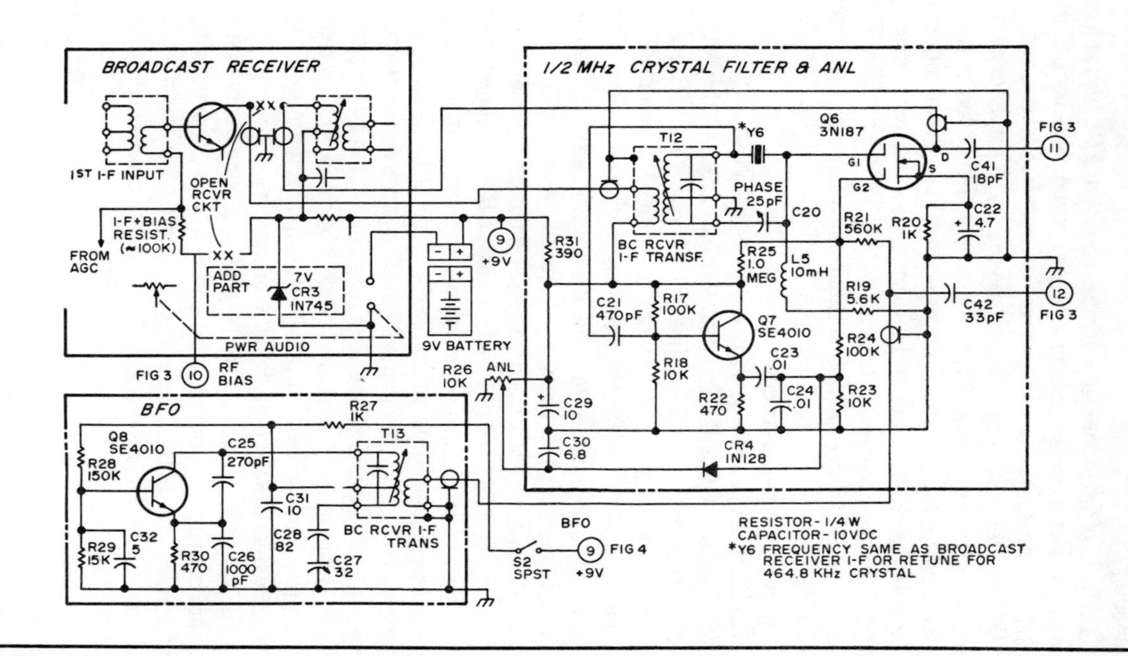

Fig. 5-10. Radio schematic.

kept at a very high impedance and connected to gate 1 of Q6. Gate 2 has the BFO output and the ANL bias feed to it. The BFO is mixed with the IF in this manner to prevent strong signals from pulling its frequency.

The gain of Q6 is regulated by the amount of ANL bias at gate 2. Its drain is connected through another coaxial cable back to the BC set IF transformer at the original collector connection of the first IF amplifier transistor. This completes the IF amplifier circuit again, but with the crystal filter, ANL, and BFO added to it.

To develop the automatic noise limited bias, the IF signal at the input of the crystal filter is transformed to a low impedance by Q7 and diode CR4 changes it to negative dc, filtered by C24. This diode is biased to different values above cut-off by the front panel control R26.

When the signal exceeds this bias, the negative voltage is developed which is subsequently fed to gate 2 of Q6.

The gain of Q6 will vary with a noise pulse all the way to cut-off, depending upon the setting of the ANL pot R26. The diode limiters are also part of the ANL but they are not adjustable. The main function of CR1-CR2 is to prevent serious overloads from damaging any components when the transmitter is keyed, and that of CR5-CR6 is to prevent audio distortion.

The S-meter circuit has an unusual input network that nulls out of the BFO component of the IF signal so it will not deflect the meter. This is accomplished by adding the exact amount of 180 degree out-of-phase BFO power to the input of Q9 (Fig. 5-10).

It might look like a marginal balance, but I have not had to change the original adjustment of R32, and a year has passed without the meter being slightly deflected by the BFO. The rest of the circuit is conventional with a voltage amplifier Q9 followed by a collector detector Q10 that deflects the meter.

The final two modifications require soldering inside the BC set. One is to add manual IF gain control to prevent overloading, and the other to stabilize the collector voltage that feeds the MF LO to prevent modulation and drift. To locate the proper place to do both jobs will take some looking around.

A zener diode, CR3 (Fig. 5-10), is connected across the large capacitor on the load side of the decoupling resistor feeding the collector power to the rf circuits. The resistor is about 100 ohms and the large can type capacitor makes it fairly easy to locate.

Next you will have to find the forward biasing resistor of the IF amplifier base bias divider. My receiver has only one IF stage and it was no trouble to find. It will be about 150k and it feeds the power we just stabilized through the AM detector diode, which also doubles as the AGC generator, on to the cold side of the IF transformer base winding. This resistor is disconnected from the stabilized voltage and reconnected to the wiper of the front panel controlled rf gain pot, R16 (Fig. 5-10).

Construction

I assembled my receiver on a 15 cm × 10 cm × 5 cm (6 × 4 × 2 inch) chassis having a 18.4 cm × 12.7 cm (7¼ × 5 inch) front panel. The transistor AM broadcast receiver was selected because of its tuning dial and volume control layout. The negative side of its battery was connected to the ground plane, and the speaker opening was covered with a gold metal screen that would make a pretty good shield.

I found later that it had only one IF stage, but this certainly did not affect its sensitivity or degrade the project. The BC set must have extension shafts epoxied to its tuning and volume control dials so they can be operated outside the front panel. The new tuning dial is a vernier type and had to be mounted on a 1.1 cm (7/16 inch) homemade spacer so it would fit on the capacitor shaft.

The front panel controls, S-meter, speaker, and phone jack are located so that they are easily accessible. After the BC set has had its IF retuned to match the filter crystal, it is fastened to the front panel with two #2-56 bolts. One of the bolts has a solder lug under its nut so the BC set ground plane can be connected to the metal front panel through it.

The outboard circuits that convert the BC set into our communication receiver are made on pieces of "vector" breadboard

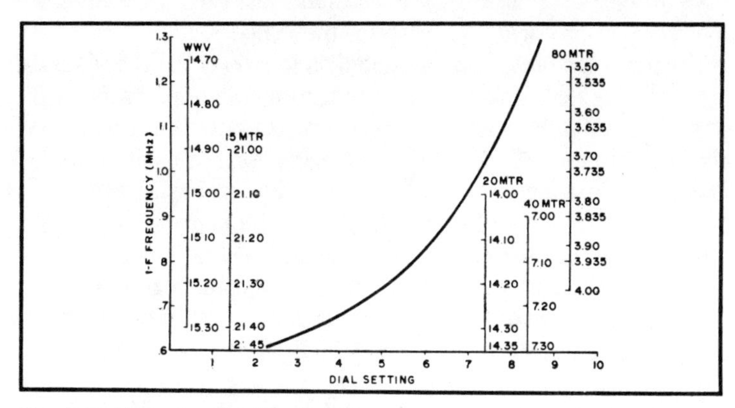

Fig. 5-11. Typical calibration chart.

Fig. 5-12. Looking toward front panel with top half of cabinet removed. Top, left to right: S-meter, 1 MHz coupler next to loop stick antenna, AM broadcast set. Bottom: HF LO crystal shield, S-meter amplifier over the rf transformer shield, crystal filter and ANL circuit board.

material, and the components are soldered to press-in terminals. They were all made as small as possible and tested before they were mounted, using spacers, in the main chassis.

The bandswitch has two levels with three poles on each. All rf amplifier connections are kept on one level, with converter and HF LO connections on the other. A lot of effort was spent trying to keep the leads short and separated from each other, but it still turned out a mess. However, it works better than anyone could have imagined. I used #22 AWG solid insulated wire for the interconnections, and bare wire for the jumpers.

All rf transformers are mounted on the top of the chassis and are well shielded. The converter transformers are preassembled on a plate that fastens to a flange around a cut-out in one end of the chassis. These coils project into the underside of the chassis, isolating them from the rf amplifier input. The tuning is broadened by the heavy loading of both sets of coil with more primary (untuned) turns than would be used for high Q operation.

I used as much shielding as I could make without getting sick. It is very important to keep the outputs of the two local oscillators out of each other's converters, and the BFO harmonics out of the HF rf amplifier.

The cabinet is fabricated out of aluminum sheet and provides the shielding needed to keep out broadcast station signals. The front and rear panels are marked, after painting, with Datak dry letter transfers, and then sprayed with clear plastic to prevent them from being worn off. When the wrap-around top, bottom, and side piece is buffed carefully, the receiver has a professional appearance.

My semiconductors were selected because they were readily available to me. The SE 4010 transistors came from an old printed wiring board bought from a mail order house. When I ran out of these, I used 2N2222 transistors from another board. Both are replaceable with Motorola HEP55, a NPN rf transistor.

The MOSFETs, Q1, Q2 and Q6, are the contents of a Radio Shack Archer Pack #276-628 called "Three MOSFET N Channel Transistors." You must watch how you solder these units in place. Keep all the leads shorted together during the process or the gates will surely be ruined. After they are in place nothing seems to be able to keep them from working. See Figs. 5-12 through 5-14 and Table 5-1.

Alignment

I aligned the tuned circuits of my receiver using equipment commonly found in the ham shack. Operation one is to tune the rf and converter transformers to the center frequency of their bands using a grid dip meter for an indicator.

Fig. 5-13. Looking toward front panel with all shields removed, along with cabinet. Left to right: converter transformers on loose panel, HF LO crystals on home brew holder, 1 MHz coupler, rf transformers, rf amplifier, BC set, S-meter amplifier, crystal filter, ANL board.

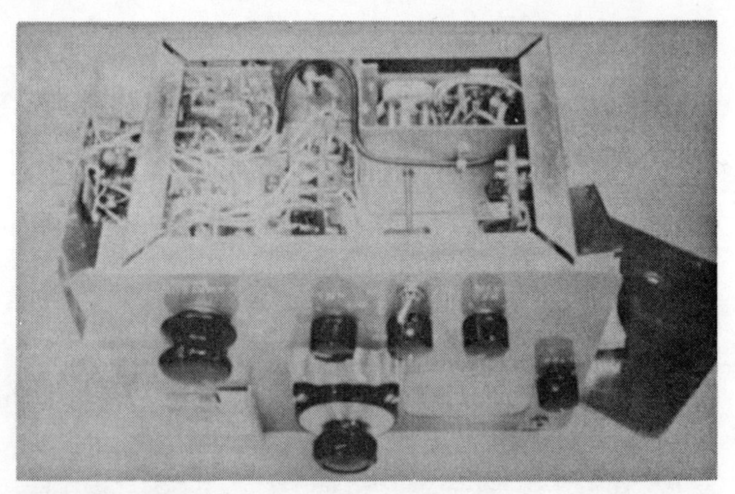

Fig. 5-14. Underside of chassis. Left to right: converter transformers on loosened panel, HF LO converter circuit board in back of bandswitch, LO multiplier, BFO in back of shield, 1 MHz amplifier.

Next, you must have the shields fastened into place, power switch placed on, rf gain turned fully on, audio gain one quarter up, ANL off, BFO off, phase capacitor fully closed, and the antenna input supplied with a signal from a VFO. The coupling must be very loose to the VFO.

The coupling recommended is two 50 ohm resistors side by side, one fed by the VFO and the other across the coax connected to the receiver antenna terminal. There is no hard electrical connection between the resistors (only the rf field), and the spacing between the two should be variable.

Back on the receiver, an oscilloscope is connected to the input of the filter crystal, Y6, using a high Z probe. The slugs in T11 and T12 are centered, and the bandswitch set to the VFO frequency range. The VFO or the receiver dial is varied until the rf is picked up and a ½ MHz IF signal is seen on the scope. T11, T12, and the rf converter transformer combination, when switched in, are adjusted until the IF signal is maximum amplitude.

The rf gain should be reduced along with the coupling to the VFO during these adjustments, to keep the scope presentation at an amplitude easy to see but not overloading the circuits (about a volt peak).

Connect the scope probe to the drain of Q6 and retune the receiver dial for a maximum display. Set the phase capacitor to its minimum bandwidth position, which is found by moving the VFO

Table 5-1. Values for Coils and Crystals.

BAND (METERS)	RECEIVER FREQUENCY RANGE (MHz)	RF AMPLIFIER					CONVERTER					CRYSTAL HF LO	
		TRANSF. SYM.	TURNS		WIRE (AWG)	CA (pF)	TRANSF. SYM.	TURNS		WIRE (AWG)	CB (pF)	SYM.	FREQ. (KHz)
			L1	L2				L3	L4				
80	3.13 – 4.18	T1	65	4	28	33	T6	65	7	28	47	Y1	4735
40	6.30 – 7.35	T2	42	2	28	47	T7	42	4	28	56	Y2	7900
20	13.35 – 14.40	T3	24	2	24	15	T8	24	3	24	47	Y3	7475
15	20.50 – 21.50	T4	14	1	24	6.8	T9	14	2	24	33	Y4	7350
WWV	14.40 – 15.40	T5	24	2	24	12	T10	24	3	24	39	Y5	7975

COIL DIAMETER - 1/4 in. (.635cm.). CRYSTALS - FT-243 PRESSURE MOUNT TYPE.
L2≈1mm. FROM L1 RF GROUND SIDE. L4 & L3 SAME SPACING: CA, CB, L1, L2, L3, & L4 SHOWN ON FIG. 2

dial around the detected frequency. Move the scope probe back to the input of Y6 and tweak up T12 and the first BC set IF transformer so that their bandpass is centered on Y6's frequency.

Connect the scope to the output of the last BC set IF transformer and tweak up the remaining BC set IF transformers so they are also center tuned to Y6's frequency. The BFO is switched on, the pitch capacitor centered, and the slug in T13 adjusted for a zero beat, noted at the scope and the tone at the speaker.

The last and simplest adjustments are made to the S-meter calibration pots, R32 and R33. Without an rf signal being applied to the receiver input, rotate R34 fully clockwise, place the BFO switch on, and adjust R33 to a position where M1 indicates zero. When an rf signal is present, M1 will deflect to a value proportional to its power. There is no clear-cut amount of rf power per S unit, so set R33 to a place where what you believe is a S9 signal in the phones reads S9 on the meter.

If you use a dial marked 1 through 100 like I did, a calibration chart will have to be made. One curve, and only one, is needed for all bands, because the BC set does the tuning each time. Figure 5-11 shows how an easy-to-read chart may be laid out. The points for the curve are located by picking up the output of a 100 kHz crystal calibrator, and knowing the frequency of the converter crystals. Subtracting the HF rf frequency from it will locate the band scales on the chart.

Conclusion

This whole project was a very satisfying success. However, you could always do better if you had a second chance. The next time, I would replace the FT-243 style crystals with smaller devices, even though it would run up the cost.

Also, their frequencies would be such that the LO multiplier could be eliminated, reducing the battery drain by 5 mA. I believe that I would make the front panel larger to accommodate a different type of dial. I cannot find any fault with the semiconductor devices or the BC set, so I would stick with them.

LOW-COST RECEIVER FOR SATELLITE TV

Unlike the vestigial-sideband AM video standard used for terrestrial TV broadcast, DOMSAT video incorporates a wideband FM format, with audio multiplexed onto a subcarrier prior to modulating the composite. The resulting wideband channel (see Table 5-2) affords considerable "FM advantage" (signal-to-noise enhancement for a given carrier-to-noise ratio); however, the bandwidth and format tend to complicate the receiver design task.

Were a signal consisting of vestigial-sideband AM video with intercarrier narrowband FM audio available from the satellites, receive processing would involve merely heterodyning the selected channel in the 4-GHz transmission band against a stable microwave local oscillator (LO), and applying the VHF difference signal directly into the tuner of a conventional TV set. Unfortunately, with DOMSAT signals as they are currently formatted, such a down-conversion process would merely spread unintelligible sidebands across six adjacent TV channels. Thus, it becomes necessary to design a complete receiver, including heterodyne conversion stages, demodulators, and video and audio processing circuitry, to recover and display satellite TV.

Frequency Agility

It will be noted from Table 5-2 that the downlink band used by most North American DOMSATs is 500 MHz wide, and that for a given antenna polarization there will be present up to twelve video

Table 5-2. Typical DOMSAT Signal Characteristics.

Video Carrier	
Channels	24
Adjacent channel spacing	40 MHz
Orthogonal channel spacing	20 MHz
Frequency band	3.7-4.2 MHz
Peak deviation	10.25 MHz
Max. video frequency	4.2 MHz
Pre-emphasis curve	CCIR 405-1
Audio Subcarrier	
Frequency	6.8 MHz
Peak deviation	75 kHz
Max. audio frequency	15 kHz
Pre-emphasis time const.	75 usec
Energy Dispersal	
Waveform	Triangular
Frequency	30 Hz
Peak deviation	750 kHz
Composite	
EIRP	+65 dBm
Path loss	−196 dB
99% power bandwidth	36 MHz
Received spectral density	−206 dBm/Hz

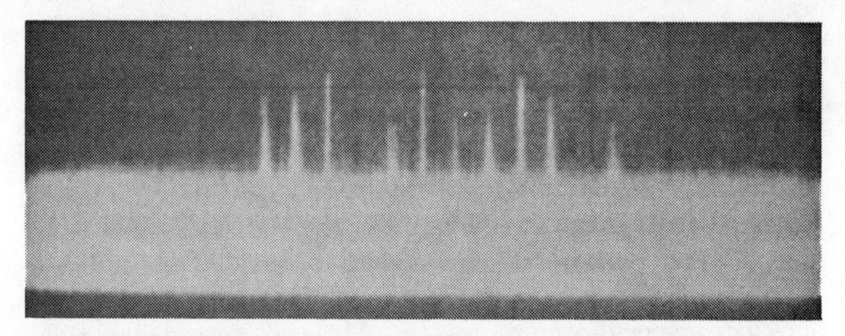

Fig. 5-15. Spectral display of a 4-GHz DOMSAT downlink recovered on a 4.7-meter antenna and amplified by a GaAs FET Low-Noise Amplifer (LNA). Horizontal deflection is 100 MHz/div, and vertical sensitivity is 10 dB/div. Eleven video carriers, along with their associated FM sidebands, are visible. Note that the fourth channel above the bottom of the band is vacant. Otherwise, channel spacing is 40 MHz and carrier-to-noise ratio appears to be on the order of 10 dB.

carriers, spaced 40 MHz apart (see Fig. 5-15). That these signals are of extremely low amplitude complicates the design of the Earth station's antenna and low noise preamplifier, but we will assume for the moment that an adequate signal-to-noise ratio exists at the input of the receiver to permit signal recovery. The problem at hand, then, is to select a particular 40-MHz wide channel from among 12 such signals in a 500-MHz wide band, while adequately attenuating the adjacent channels.

A Tuned Radio Frequency (TRF) approach, with detection occurring directly at the downlink frequency, would require readily-tunable bandpass filters of high Q (to accommodate the 1% or so channel bandwidth) and skirts steep enough to reject adjacent channels. Tuning requirements rule out both LC and resonant cavity filters, suggesting the use of Yttrium-Iron-Garnett (YIG) sphere resonators for channel selection.

Although YIG filters can readily be bias-tuned, their cost and the complexity of the required driving circuitry tend to rule them out for private terminal applications. Furthermore, it is far easier to tune a single oscillator than a bank of filters. This suggests heterodyne-downconverting a selecting channel into a fixed intermediate frequency, at which demodulation may take place.

I-f Selection

The selection of intermediate frequencies for superheterodyne receivers involves careful attention to the required and realizable mixer bandwidths, image rejection criteria, demodulator circuit capabilities, and tuning constraints. These various considerations tend to be mutually exclusive, but it has been shown that for narrowband systems, a reasonable compromise is achieved by selecting an intermediate frequency approximately one-tenth the frequency of the incoming the frequency of the incoming signal. Although DOMSAT video hardly qualifies as a narrowband service, we can use the one-tenth rule of thumb to establish a starting point. For a 4-GHz input signal, this suggests a UHF i-f. However, the various demodulator circuits compatible with wideband FM video (quadrature detector, ratio detector, Foster-Seely discriminator, phase-locked loop and the like) are all the most readily realized in the lower portion of the VHF spectrum. An obvious solution is to utilize dual downconversion, with first and second i-fs near 400 and 40 MHz, respectively.

In fact, numerous experimental DOMSAT video terminals have adopted the above frequency scheme, many employing UHF TV tuners for the second downconversion. The drawbacks to such an approach include the typical UHF tuner's restricted channel bandwidth, relatively high noise figure, and poor local oscillator stability. Nevertheless, when cost is the primary design constraint, these problems can be circumvented.

Not so readily resolved is the input filtering requirement which such a frequency scheme imposes. Assuming low-side first LO injection and top-channel reception, the first conversion will generate an image frequency which falls a mere 300 MHz below the bottom edge of the down-link passband. An input filter capable of providing adequate passband flatness and minimal insertion loss over the 3.7- to 4.2-GHz band is unlikely to provide adequate image rejection if a 400-MHz first i-f is utilized. One may wish to raise the first i-f high enough to separate the image frequency band well away from the downlink passband, thus simplifying input filtering.

In fact, if the Low-Noise Amplifier (LNA) which precedes the receiver utilizes a waveguide input, then an image filter already exists. Rectangular waveguide is a high-pass transmission line. If low-side first LO injection is used and the first i-f is carefully selected, the LNA's waveguide input will itself reject the image frequency.

Most commercial LNA's utilize an EIA standard WR-229 waveguide input. This guide has a lower $TE_{1,0}$ cutoff frequency near 2.5 GHz. This cutoff frequency is about 1.2 GHz below the bottom edge of the receiver's required passband, so input losses will be minimal. But a first i-f of, say, 1.2 GHz, will place the image frequency as far *below* cutoff as the input passband is *above* cutoff. The image thus ends up quite far down the waveguide high-pass filter's skirts, and may effectively be ignored.

True, the fixed 1.2-GHz first i-f requires that the first LO be tunable, but we mentioned earlier that it's far easier to tune a single oscillator for channel selection than a bank of filters. And even at the top of the down-link passband, where the first LO must be tuned up to 3 GHz, the image at 1.8 GHz is sufficiently far below cutoff so that a 12-cm long input waveguide will afford on the order of 60 dB of image rejection.

Another signpost pointing to the selection of 1.2 GHz as a first i-f is realizable amplifier Q. The 3-dB bandwidth of the i-f amplifier string must be greater than or equal to the 20-dB channel bandwidth in order to avoid unduly attenuating significant sidebands. Assuming a channel bandwidth of 40 MHz and an i-f of 1.2 GHz, this dictates an effective first i-f Q of 30. This value is readily realized with microstripline circuitry.

Despite the obvious economic advantages of the modified UHF TV tuner conversion scheme, it was decided to employ a 1.2-GHz first i-f in the Microcomm DOMSAT video receiver. But what of the second i-f—is it similarly constrained by the wide downlink passband? Actually not. With channel selection occurring in the first downconversion, the second i-f need only be wide enough to accommodate a single video channel. Downconverting the 1.2-GHz first i-f to any desired VHF frequency will allow ample second-conversion image rejection with simple i-f filtering while providing adequate bandwidth to pass the 40-MHz composite.

Since the communications industry has long utilized 70 MHz as a standard i-f for microwave links, it was decided to employ a 70-MHz second i-f in the DOMSAT video receiver. This makes it possible to utilize any of the readily available 70-MHz wideband FM i-f strips to demodulate the video information.

Gain Distribution

Gain partitioning for the DOMSAT video receiver depends upon the available power from the satellite, the threshold sensitivity of the demodulator circuit selected, and the gain of the receive antenna utilized. It has been shown that for the illumination contours typical of most North American DOMSATs, an optimum private-terminal antenna will exhibit on the order of +41-dBi gain. Given the EIRP and path loss numbers listed in Table 5-2, it appears that the signal level available to the LNA will be on the order of −90 dBm.

The input threshold for a typical phase-locked loop (PLL) integrated circuit operating as an FM demodulator at 70 MHz is on the order of −20 dBm. This suggests that between the antenna and the demodulator, roughly 70 dB of conversion gain is required.

There are three sources of gain available between the antenna and the PLL. These include the LNA and first and second i-f amplifiers. There are, similarly, three sources of loss in the system: the insertion loss of the transmission line which connects the LNA to the receiver, and the conversion loss of the first and second mixers. For a typical home installation, the feedline insertion loss may be on the order of 6 dB, and if double-balanced diode mixers are used for the two frequency conversions, it is safe to assume that the conversion loss of each will be on the order of 7 dB. This suggests that the overall gain of the LNA, first, and second i-f amplifiers will need to total 90 dB for adequate DOMSAT video reception.

In the interest of maximizing system stability and dynamic range, it is desirable to distribute the required 90 dB of gain uniformly between the rf and two i-f frequencies. A 30-dB gain LNA is clearly feasible at 4 GHz and would require three stages of GaAs FET amplification. This amount of LNA gain is sufficient to adequately mask the noise temperature contribution of the feedline and receiver, allowing the low-noise temperature of the FETs to predominate. Similarly, it is practical to achieve the desired 30 dB of 1.2-GHz gain by cascading two stages of ion-implanted silicon bipolar transistor amplification. At 70 MHz, the required gain is readily available from thin-film wideband gain blocks produced by a number of different vendors.

A block diagram for the dual downconversion portion of the DOMSAT video receiver, partitioned in accordance with the foregoing discussion, is shown in Fig. 5-16.

Construction of Conversion Circuitry

During the initial system-development phase of any new product, it is common practice to build a number of different amplifier, mixer, filter, and oscillator circuits, each connectorized for coaxial input and output and with each circuit separately boxed and shielded. A modular developmental system provides the engineer with the flexibility of changing one or more circuits without having to disrupt the rest of the system.

But modularization has advantages for a production system as well. If every function represented by a block in Fig. 5-16 is implemented in a separate, shielded module, then isolation between stages is maximized and the crosstalk and stability problems associated with stray rf coupling can be eliminated entirely. Further advantages are realized in the area of maintainability. Should a receiver fail, fault isolation by module substitution becomes a viable troubleshooting technique. And, of course, a modular system maximizes user flexibility by allowing customers to assemble from standard modules a custom system designed to meet their precise needs.

The specifications of the modules developed to implement Fig. 5-16 appear here as Table 5-3. Each of these modules employs microstripline construction, as shown in Fig. 5-17, to minimize component count and assure duplicability.

Baseband Processing

Before the wideband FM composite shown in Fig. 5-18 can be displayed, several processing steps are necessary. The 70-MHz i-f signal will, of course, be demodulated first, and this may be accomplished readily by using a monolithic PLL in a standard circuit. The output waveform from the PLL will contain both the video waveform and the modulated audio subcarrier, but superimposed on these will be found the 30-Hz triangular energy dispersal waveform added to all DOMSAT downlink signals as an interference reduction technique. This waveform is evident in the oscilloscope display in Fig. 5-19.

Prior to attempting to remove the energy dispersal waveform, it is desirable to amplify the rather feeble video level available from the PLL demodulator, and this may be accomplished using a single monolithic TV video-amplifier IC. Next an emitter-follower permits splitting off the 6.8-MHz audio subcarrier for demodulation in a standard TV sound i-f microcircuit, whose associated circuitry is modified slightly for compatibility with the higher carrier frequency and peak deviation used on satellite audio.

After passing through a de-emphasis filter and passive video low-pass filter, the video waveform may finally be applied to a diode clamp circuit which will remove the energy dispersal waveform (see Fig. 5-20). An emitter-follower then establishes the desired 75-ohm video output impedance to drive recording or display circuitry, as required.

A block diagram for a complete baseband processing subsystem is shown in Fig. 5-21. This circuit can be constructed on a

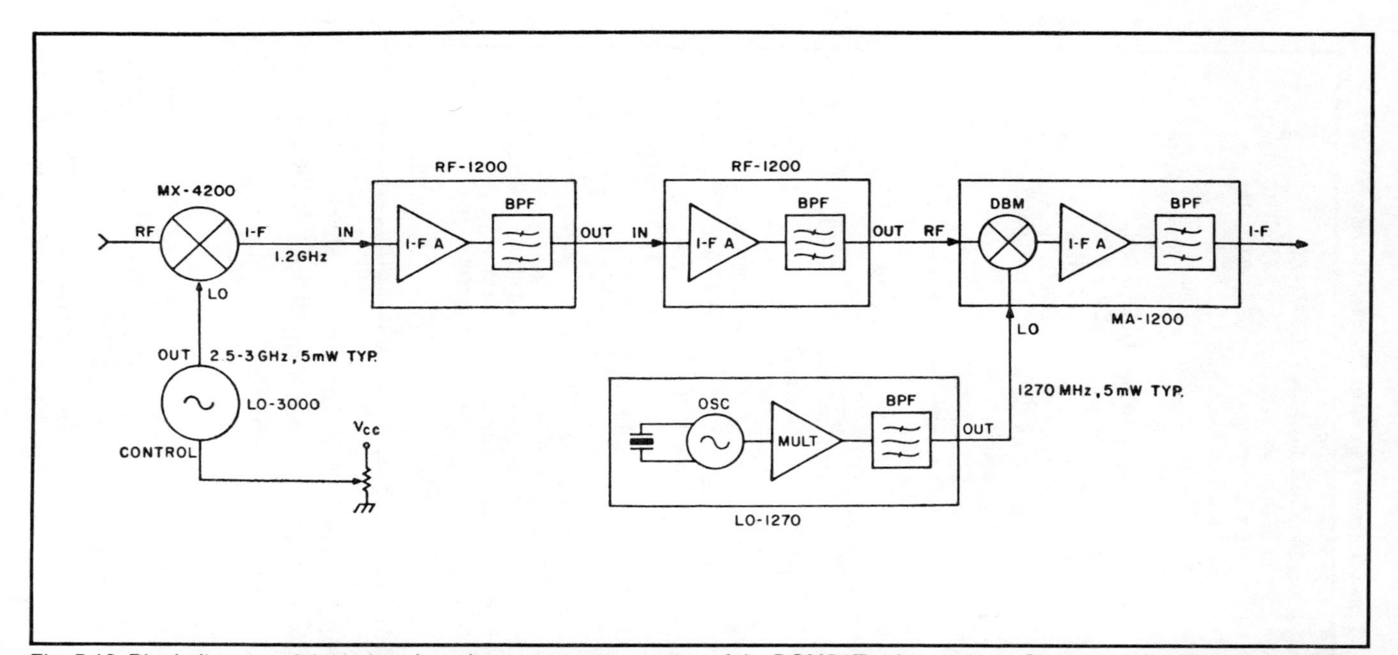

Fig. 5-16. Block diagram of the heterodyne downconversion portion of the DOMSAT video receiver. Shown at left is the 3.7- to 4.2-GHz input terminal from the LNA and feedline. The 70-MHz i-f output at right feeds the baseband demodulator and processing circuitry (see Fig. 7). The potentiometer shown at the lower left represents the resistive voltage divider which tunes the first LO for channel selection. Gain partitioning of the various blocks is discussed in the text.

Table 5-3. Typical Parameters for Conversion Modules.

Parameter	Value
MX-4200 DOUBLE-BALANCED MIXER	
Input frequency	3.7 – 4.2 GHz
LO frequency	2.5 – 3.0 GHz
Intermediate frequency	1.2 GHz
Isolation	20 dB
Conversion loss	7 dB
LO-3000 VOLTAGE CONTROLLED OSC	
Output frequency	2.5 – 3 GHz
Output power	+7 dBm
Spurious rejection	20 dB
Tuning voltage range	3 – 10 V dc
Supply potential	+13.5 V dc
RF-1200 AMPLI-FILTER	
Center frequency	1.2 GHz
3-dB bandwidth	50 MHz
Gain	15 dB
Noise figure	2 dB
Supply potential	+13.5 V dc
LO-1270 LOCAL OSCILLATOR	
Output frequency	1270 MHz
Stability	±0.001%
Power out	+7 dBm
Spurious rejection	40 dB
Supply potential	+13.5 V dc
MA-1200 MIXER-AMPLIFIER	
Input frequency	1200 MHz
LO frequency	1270 MHz
Intermediate frequency	70 MHz
Conversion gain	20 dB
3-dB bandwidth	40 MHz
Supply potential	+13.5 V dc
Isolation	20 dB

single printed circuit board and incorporated into a complete DOMSAT video receiver by simply interfacing it to the downconversion circuitry shown in Fig. 5-16.

Display Options

An ideal DOMSAT video receiver for the home Earth-station market would provide an intercarrier audio, vestigial sideband rf output for direct interface to the user's VHF TV receiver. Such an rf output may readily be realized by using any of the available video modulator microcircuits developed for the TV game and home computer industries. In fact, the video and audio levels available from the baseband unit shown here are entirely compatible with such modulators.

Unfortunately, incorporating an rf modulator in commercial DOMSAT video receiver would subject the entire receiver to FCC type acceptance in United States. As more than one home computer manufacturer has discovered, the type-acceptance procedure is burdensome in the extreme, with bureaucratic delays often precluding a timely market entry. In addition, the resolution and clarity of most of the available low-cost video modulators leave quite a bit to be desired, and rf modulation would tend to degrade overall video quality noticeably.

A possible solution would be to provide the user with simply a video and audio output from the DOMSAT receiver and allow him to display the receiver's output on a studio-quality TV monitor. However, few videophiles possess such a monitor, and the cost is prohibitive.

Fortunately, most videophiles do possess a video-tape recorder (in fact, the owner of a home satellite Earth station would most likely find it impossible to function without one!) and the average video recorder contains an extremely high quality rf mod-

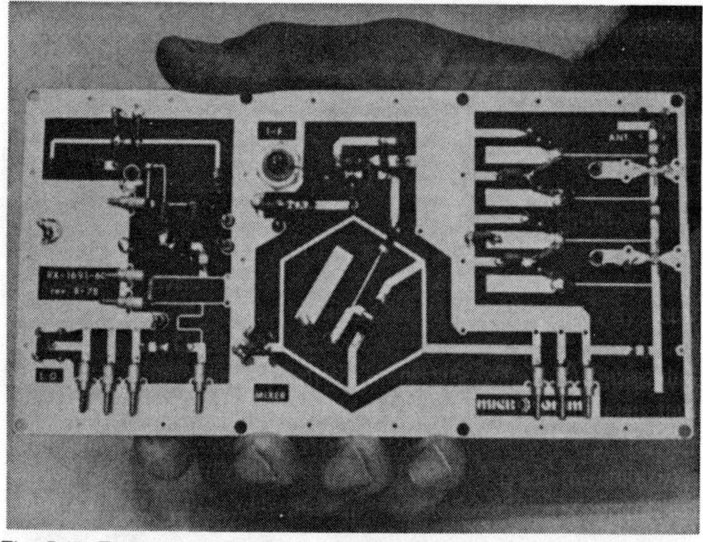

Fig. 5-17. Typical microstripline circuit module used for satellite video downconversion. Each module in the receive system is mounted in its own shielded enclosure, and all are interconnected via coaxial cable as discussed in the text.

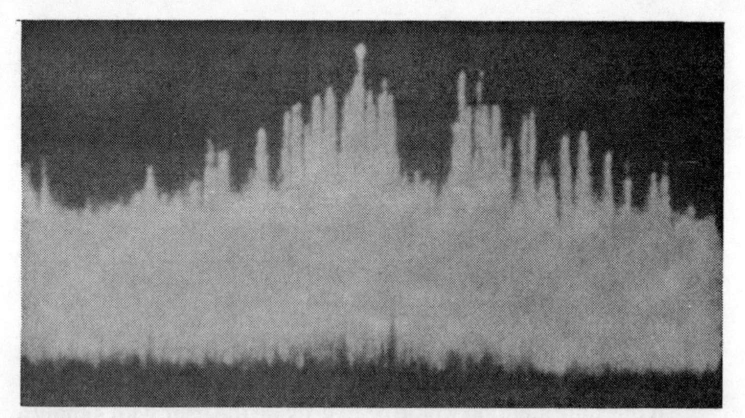

Fig. 5-18. Spectral display of a single wideband FM video channel after dual-downconversion to a 70-MHz second i-f. Horizontal deflection is 3 MHz/div. and vertical sensitivity is 10 dB/div. This is the composite FM signal from which video and audio are to be demodulated.

ulator. Allowing the user to interface his DOMSAT receiver to the TV via a video recorder provides an ideal solution to the type-acceptance dilemma. And those users who have no recorder are, of course, free to add an external rf modulator, any number of which are available in kit or assembled forms.

Equipment Availability

Once priced in the tens of thousands of dollars, DOMSAT video receivers are now being brought within the reach of the

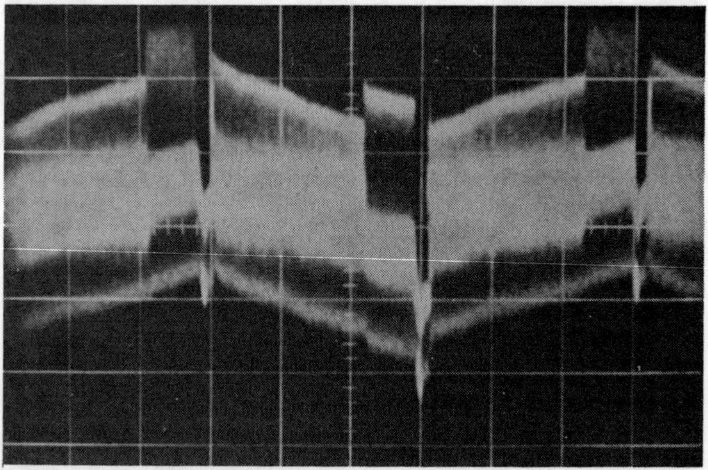

Fig. 5-19. Baseband output of the PLL demodulator. Presence of the energy dispersal waveform on the video composite is evident.

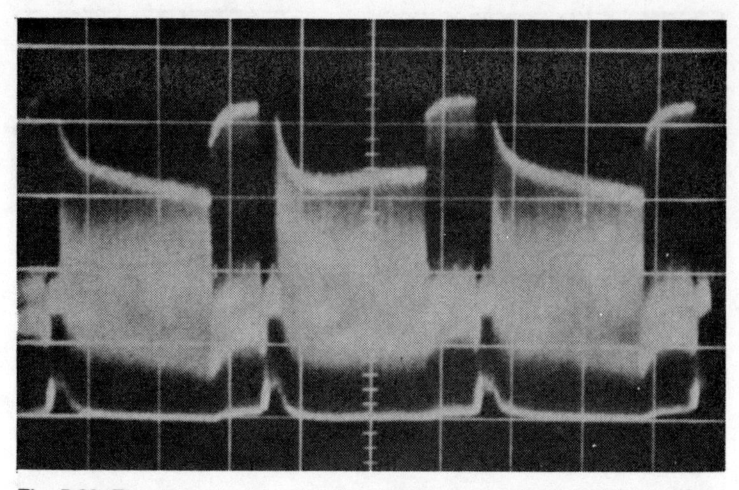

Fig. 5-20. The dc restorer is simply a diode clamp circuit which removes from the video waveform any vestiges of the energy dispersal waveform seen in Fig. 5.

American consumer. The conversion modules shown in Fig. 5-16, for example, have, since late 1978, been available to the experimenter at $1000 complete, and to the OEM at significantly lower prices in production quantities.

WWV-TO-80-METER CONVERTER

This simple inexpensive frequency converter will place the WWV 10-MHz signal anywhere within the 80-meter band.

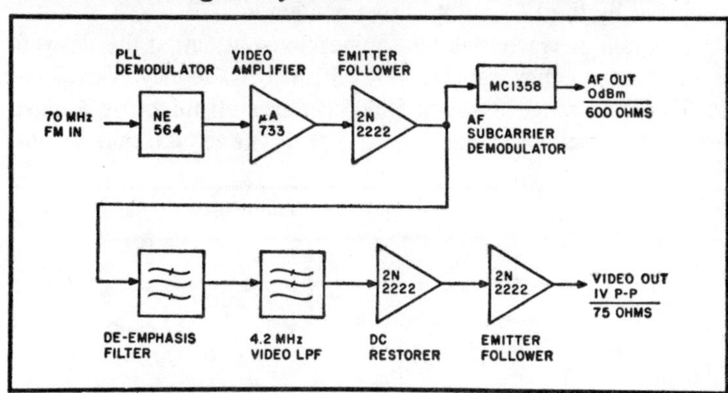

Fig. 5-21. Block diagram of the complete baseband processing portion of the DOMSAT video receiver. The above circuitry is driven by the output of the downconversion module set (see Fig. 3), and provides standard 1-volt video and 0-dBm audio outputs to an external modulator, studio monitor, or video tape deck.

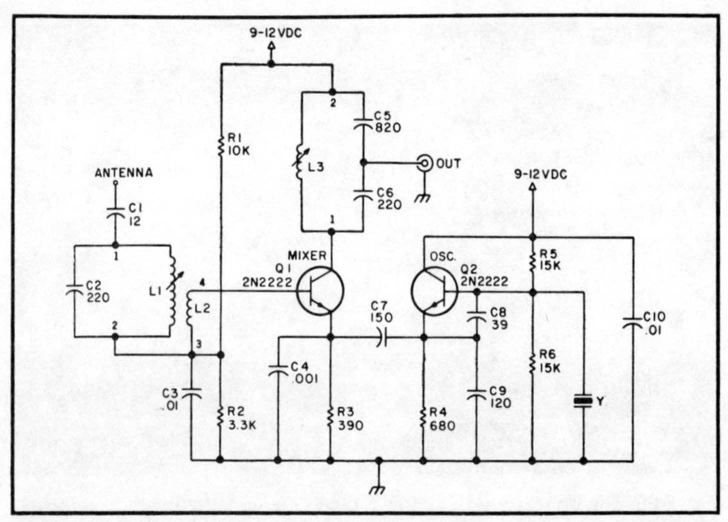

Fig. 5-22. Schematic diagram of the WWV converter. Resistors are ¼ or ½ watt. Capacitors C3 and C10 are Mylar™, with all others being disc ceramic.

Referring to the schematic in Fig. 5-22. L1-C2 are tuned to the WWV 10-MHz signal. This signal is coupled to the base of Q1 by L2. Oscillator Q2 operates at any selected crystal frequency between 6 and 6.5 MHz, and is coupled to the emitter of Q1 by C7. Q1 mixes the two frequencies. The L3-C5-C6 combination is tuned to the 3.5-to-4 MHz difference frequency which appears at the collector of Q1. Impedance matching to the 50-ohm receiver antenna is provided by the C5-C6 capacitive divider.

Crystal frequency is determined by subtracting the desired 80-meter frequency from the WWV 10-MHz frequency. The 3750- to 3800-kHz range (6250-to-6200-kHz crystal) might be a good choice for minimum signal interference. The crystal may be ob-

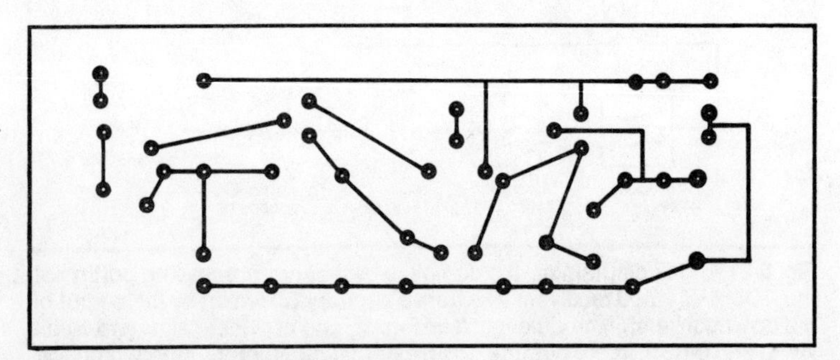

Fig. 5-23. Foil side of the circuit board.

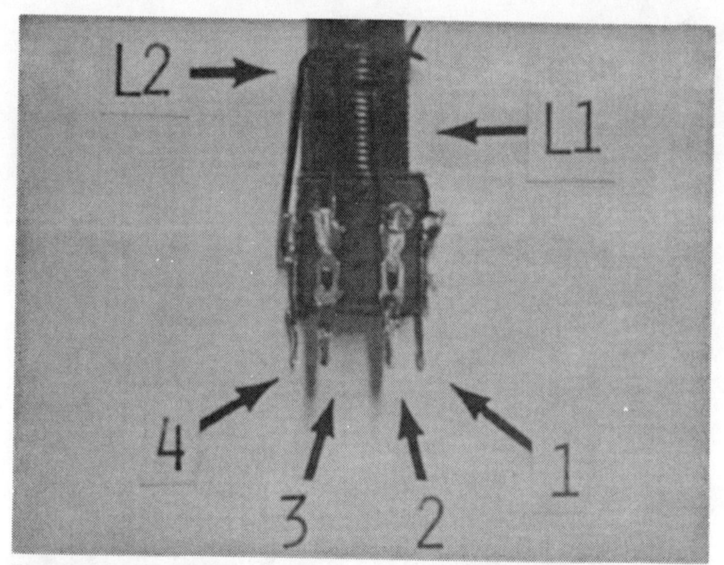

Fig. 5-24. Close-up of the L1-L2 inductor. L1 is 12 ccw turns and L2 is 4 ccw turns of #24 enameled wire. L3 is 35 cw turns of #32 enameled wire. All inductors are wound on a ¼-inch slug-tuned coil form.

tained from Jan Crystals, 2400 Crystal Drive, Ft. Myers FL 33906. Specify type FT-243 holder and desired crystal frequency. This crystal will be .005% tolerance.

The circuit board can be quickly and easily made by first positioning and securing the copper face of a 1-¾" × 3-⅜" board under the circuit pattern in Fig. 5-23. Next, mark through the pattern at each hole location and then drill a #60 hole at each mark. The inductor pin and crystal socket holes may require pattern adjustment and larger holes. Also, check the lead spacing of your

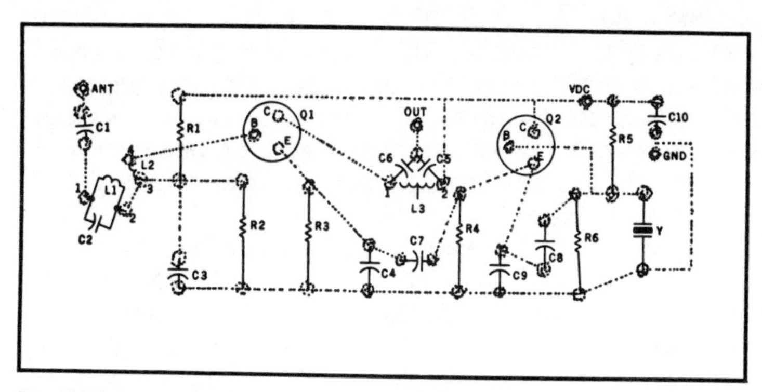

Fig. 5-25. Parts placement guide (foil side shown).

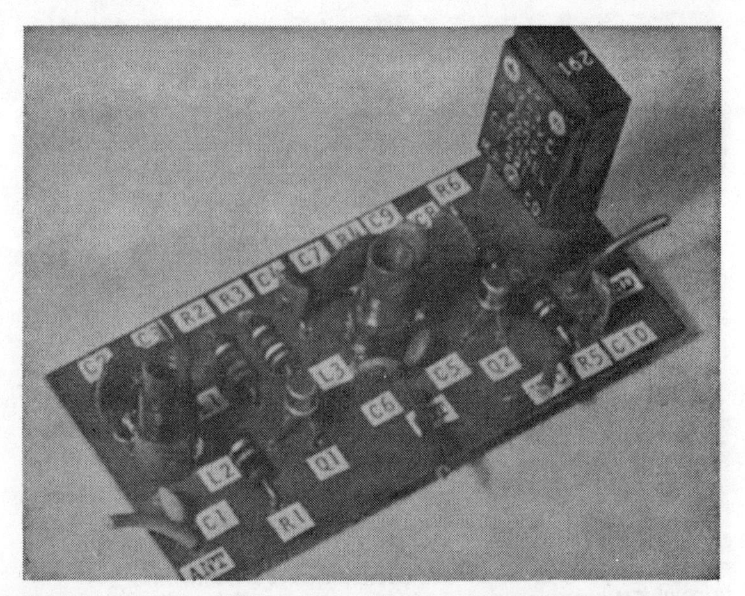

Fig. 5-26. Top view of the completed WWV-to-80-meter frequency converter.

capacitors. The layout is for ¼-inch spacing but room is available for the ⅜-inch variety.

Finally, carefully connect the related holes with ⅛-inch strips of art or masking tape. Place masking tape over the component side of the board to prevent acid from entering the holes. Thoroughly clean the copper surface after etching. Using this method, I easily etched and assembled a checkout board in one afternoon.

The inductors are wound on a ¼-inch diameter slug-tuned coil form as shown in Fig. 5-24. These may be found in most junked TV sets and radios. As viewed from the base, coils L1 and L2 are wound counterclockwise with #24 enameled wire, and L3 is wound clockwise with #32 enameled wire. All three inductors are wound with no space between turns. L2 begins at the end of L1 with no space between the end of L1 and the start of L2. Secure the coil ends with thread or tape and apply two or three coats of varnish to hold the coil in place.

Capacitors C2, C5, and C6 are soldered directly to the inductor pins. I tried several sets of transistors, both NPN and PNP types, and they all worked. Just reverse the voltage polarity for PNP types. To align, connect a short antenna and set L1 and L3 for maximum S-meter reading with a nonmetallic tool. Use shielded cable for hookup to the receiver. Also, see Figs. 5-25 and 5-26.

Chapter 6
Transmitters, Transceivers And Accessories

Transmitters are easy to build and also a lot of fun. This chapter is intended for ham radio operators who want to build "home-brew" rigs. There are also projects for some pretty interesting accessories that can be easily built for a fraction of the cost of a commercial unit.

The CW keyboard keyer described here should appeal to any amateur who wants to increase his code speed and send letter-perfect, space-perfect CW—or to anyone seriously interested in teaching or learning the code.

Although there are a lot of articles and letters in ham magazines about amateur radio clubs teaching code, the use of a keyboard keyer as a teaching aid seems to have been neglected, possibly because most such keyers are quite expensive. Keycoder I, makes an excellent club project. When used as a code teacher, the keyboard keyer has the following advantages to recommend it:

☐ Perfect code—The learner always hears correctly formed characters

☐ Variable speed—Character rate is continuously variable from about 5 wpm to over 50 wpm (unlike code records or cassettes)

☐ Repetition—The student can practice by himself the characters that he is having difficulty with (again unlike recorded code lessons)

☐ Instructor need not know CW!—A unique feature—XYLs, harmonics, and other non-hams can help the novice learner without themselves knowing the code.

Other technical advantages of Keycoder I are:

☐ TTL Logic—9 inexpensive and readily available ICs, plus about 2 dozen discrete components, make up most of the circuit; 7 toroidal cores and another couple of dozen discretes provide an encoding matrix that would require over 100 diodes.

☐ RF Insensitive—With TTL ICs and only minimal bypassing, no rf problems have been experienced with the two units illustrated.

☐ Self-contained—Power supply, logic, monitor (with tone and volume controls), and output circuit are all contained within the keyboard case

☐ Simple construction—A commercially available IC board makes wiring the logic easy (simple point-to-point wiring with no critical steps)

Figure 6-1 is a close-up photo of the unit built by Charles H. Haut W9UBA. Figure 6-2 shows my keyboard with a few control variations.

Keycoder I is a TTL adaptation of the Touchcoder II keyboard keyer described in the July 1969 issue of *QST*. That article contains a very good discussion of the circuit theory and of the toroidal core transformers used, so this information will not be repeated here. Our theory discussion will take you on a tour of the TTL-version schematic diagram (Fig. 6-3). For convenience of discussion, the schematic is separated by dashed lines into the following six areas representing functional blocks: (1) dot generator, (2) dash generator, (3) shift register, (4) encoding matrix, (5) monitor, and (6) output stage. The discussion that follows will be so divided also, with some necessary overlapping. In Fig. 6-3 the logic state H ("1") or L ("0") is shown for the output of each IC. These states are for static conditions, when no character is being generated.

Dot Generator

IC1 is a 555 integrated circuit timer set up as a triggered multivibrator. The frequency of the multivibrator (and hence the code speed) is determined by the values of R1-R3 and C1, and is adjustable by means of R2. The duty cycle is 50% for all practical purposes, thus providing dots and character interelement spaces that are equal. (Of course other R and C values can be substituted. Those on the schematic are the ones in W9UBA's unit; I used R1 = 1k, R2 = 50k, R3 = 15k, and C1 = 2.2 μF, with equally satisfactory results.)

Fig. 6-1. Keycoder 1 built by W9UBA.

Fig. 6-2. The author's version of Keycoder 1.

R4, R5, and CR1 are bias components that are necessary to make the first dot the same length as following ones. This is because normal timer operation charges and discharges C1 between ⅓ and ⅔ vcc. However, on the very first cycle, C1 is discharged, and has to charge from 0 volts to ⅔ vcc. The resistors, which are merely a voltage divider across the power supply, maintain a bias voltage on C1; CR1 prevents interference with the normal triggering of IC1.

The output of the dot generator (pin 3) is routed to both the output circuit and monitor through two sections of IC4 (NOR gates); the first section is a summing gate (discussed later), and the second section disables the output (and monitor) for the duration of 3 dot lengths at the completion of each character, to provide automatic character spacing.

Dots are generated when pin 6 of IC1 goes low, which occurs only when pin 6 of IC3 goes low.

Dash Generator

As with other keyers (both keyboard and manual), Keycoder I forms dashes by filling in the space between two successive dots, thus producing dashes exactly 3 units long. The circuit action is as follows: FF8 is a "D" type flip flop that clocks on the rising edge of pulses applied to pin 11. When no dashes are called for, pin 3 of IC3 applies a "1" to the data input (pin 12) of FF8, and the Q output of

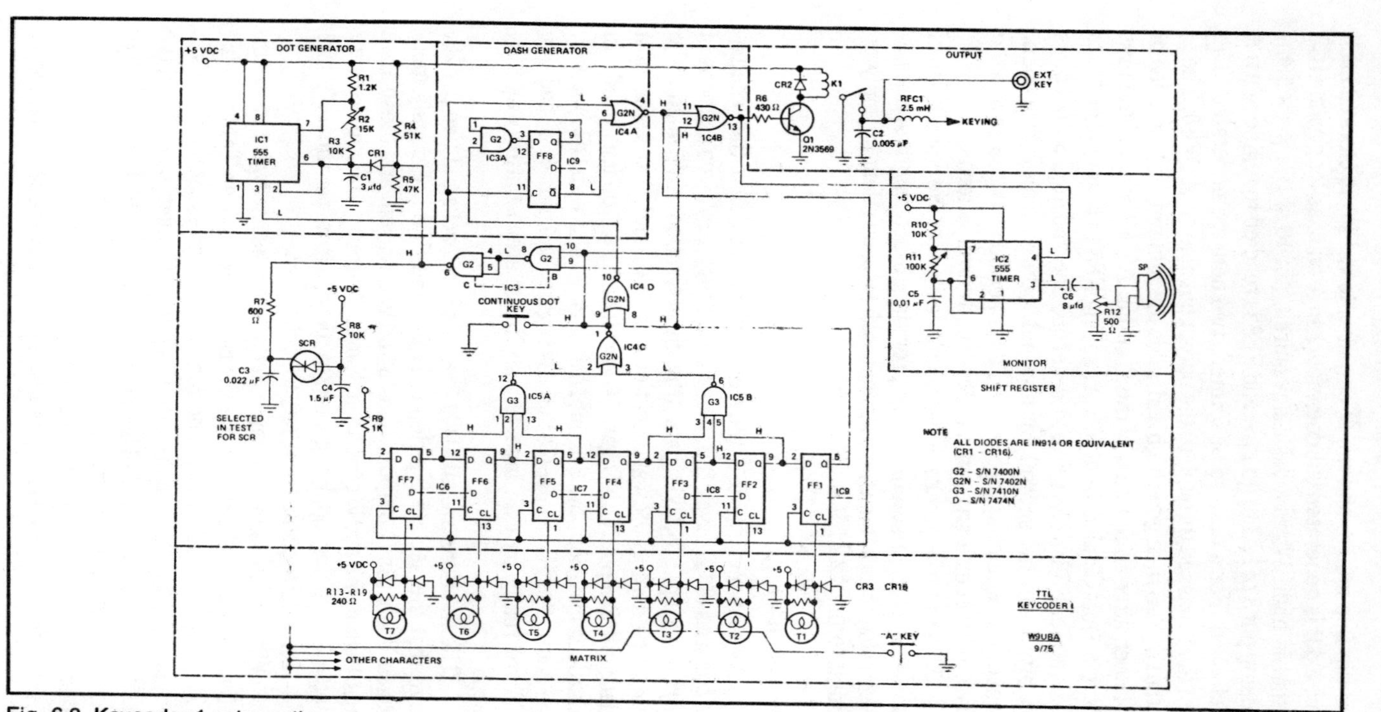

Fig. 6-3. Keycoder 1 schematic.

this flip flop is high and the $\overline{Q}$ output is low. FF8 is clocked each time a dot is generated but does not change state as long as the IC3 output is high. When a dash is required, pin 10 of IC4 goes high, causing pin 3 of IC3 to go low, and FF8 is readied to toggle when it is clocked. For a dash to be formed, two dots are generated. The first dot is passed through to the output but also clocks FF8. The $\overline{Q}$ output of FF8 then goes high and holds the summing gate portion of IC4 open after completion of the first dot. After arrival of the first dot, pin 1 of IC3 also went low, readying FF8 to toggle back to its static state. FF8 is clocked by the second dot, and its $\overline{Q}$ output goes low again, but the second dot now holds the summing gate open. Hence, the summing gate is open for a period of two dots and the space between them, and a 3-dot-length dash is formed.

This same sequence of events takes place at the completion of each character to automatically provide a 3-dot-length space between characters, but this terminal dash is made silent by applying a high to pin 12 of IC4.

Shift Register

As shown in Fig. 6-3, the shift register portion of the schematic consists of "D" type flip flops FF1 through FF7, two sections of a triple 3-input NAND gate (the third section isn't used), the remaining two NOR gate sections of IC4, and two gates of a quadruple 2-input NAND gate IC (the fourth section isn't used). In this mechanization, the 3-input NAND gates serve an OR function, one section of IC4 is used as an AND gate, and pins 4, 5, and 6 of IC3 are really only an inverter, so things aren't entirely as they seem.

The flip flops are all "D" types, and are packaged two per IC. Note that the second flip flop of IC9 is the FF8 previously mentioned. Only the Q outputs (pins 5 and 9) are used, and these are a logic high in the static condition. For our purpose, the flip flops are "set" (Q = "0") by applying a low to their "Clear" terminals (pins 1 and 13). Shifts occur from left-to-right only.

With the foregoing in mind, let's leave the hardware mechanization for a bit and refer to the Character Generation Table (Table 6-1). Here, the "X"s in the horizontal rows represent the dashes in a character, the "X" furthest to the left is a "silent dash," and the blank spaces to the right of the silent dash represent dots. Although data is shifted left-to-right, we "read" Morse characters in this table from right-to-left. For example, in the row for the letter "P", starting from the extreme right we see a blank space which repre-

Table 6-1. Character Generation Table. Xs Indicate the Shift Register Flip Flops That Must be Set to Form a Given Character (Also the Toroidal Core Transformers That the Wire for That Character Must Pass Through).

	Flip Flops						
Character	FF7	FF6	FF5	FF4	FF3	FF2	FF1
A					X	X	
B			X				X
C			X		X		X
D				X			X
E						X	
F			X		X		
G				X		X	X
H			X				
I					X		
J			X	X	X	X	
K				X	X		X
L			X			X	
M					X	X	X
N					X		X
O				X	X	X	X
P			X		X	X	
Q			X	X		X	X
R				X		X	
S				X			
T						X	X
U				X	X		
V			X	X			
W				X	X	X	
X			X	X			X
Y			X	X	X		X
Z			X			X	X
1		X	X	X	X	X	
2		X	X	X	X		
3		X	X	X			
4		X	X				
5		X					
6		X					X
7		X				X	X
8		X			X	X	X
9		X		X	X	X	X
0		X	X	X	X	X	X
.	X	X		X		X	
,	X	X	X			X	X
?	X			X	X		
/		X		X			X
-		X	X				X
$\overline{AR}$		X		X		X	
$\overline{SK}$	X	X		X			
$\overline{AS}$		X				X	

sents a dot. Moving to the left we find two "X"s, which we know are dashes, followed by another blank space for a dot, and finally another "X" (which is the silent dash to separate this character from the next). In terms of the hardware, the "X"s represent the flip flops that are simultaneously set when a keyboard key is struck and also the toroidal cores that must be strung to generate that character.

The character generation scheme, which was originally proposed by James B. Ricks W9T0, has three rules:

1. If only one flip flop is set, a number of dots will be generated, the number being one less than the number of the flip flop set. For instance, in the row for "S", only one "X" appears, in the column headed FF4; therefore 3 dots will be generated.

2. If FF1 *and* any other flip flop are set, a dash will be generated.

3. If only FF1 is set, a silent dash is generated.

As each dot or dash is completed, all entries in the appropriate row of the table are shifted one column to the right, and another of the three above rules is applied. For the letter "B", initially the conditions for rule 2 are met, and accordingly a dash is generated; after one shift to the right, rule 1 applies, and three dots are generated; these three dots produce three more shifts, after which rule 3 comes into effect and the character is completed, followed by a 3-dot-length space.

Returning to Fig. 6-3 observe that the static state for the flip flops is for their Q outputs to be high. This results in the other gates in the register being forced to the logic states shown, ending up with pin 6 of IC3 being high and the dot generator disabled. Through a process to be described later, one or more of the flip flops is set each time a key is struck. The resulting lows on the flip flops cause the outputs of one or both sections of IC5 to change state and, in turn, pin 1 of IC4 and pins 8 and 6 of IC3, thus starting the dot generator. If pin 5 of IC9 goes low and the Q output of any other flip flop also goes low (rule 2), pin 10 of IC4 will go high, and a dash will be generated; otherwise a dot will be formed. The dot or dash causes pin 4 of IC4 to go from a high to a low. This pin is connected to the clock inputs (pins 3 and 11) of the shift register flip-flops, but a high-to-low transition does not affect a "D" type flip flop. However, when the dot or dash is completed, pin 4 of IC4 again goes high, and the low-to-high transition does clock all of the flip flops causing a right shift. That is, the Q output of each flip flop assumes the logic state that its left-hand neighbor had before the clock pulse. The resulting new set of Q outputs now initiates another dot or dash which, in turn, produces another clock pulse to shift the register to the right again, and so on until only the output of FF1 is low. This is the condition (according to rule 3) to generate a silent dash. What actually happens here is that FF1 turns on the dot generator, but since the other flip flop outputs are all high, pin 1 of IC4 is high and provides the output inhibit signal to pin 12 of IC4. Note that the silent dash creates another clock pulse that shifts the "1" output of FF2 into FF1, thus ending the character cycle and restoring the static condition.

In addition to starting and stopping the dot generator, pin 6 of IC3 also activates the gate of the SCR that provides current drive to the keyboard encoder circuitry. Therefore, the keyboard is locked out during character generation, but is automatically reenabled as soon as the character is completed. This feature also provides automatic character repeat: A character is repeated, with 3-dot-length spaces between, as long as a key is held down. This is a fine operating feature for sending words with double "o"s and double "l"s, etc., but is not so hot for sending double "e"s at high speed. If desired, automatic repeat can be disabled by reducing the drive on the SCR.

Note the "continuous dot" key connected between pin 1 of IC4 and ground (this is the switch labeled "DOTS" in Fig. 6-2). This is depressed and held down while adjusting the SPEED control to initially set the keyer to the desired code speed. The continuous string of dots resulting from holding down this switch represents the maximum sending speed for a given SPEED control setting. Holding down the "E" key also produces a string of dots, but with 3-dot-length spaces between each one.

Keyboard Encoding Matrix

Toroidal core transformers, driven by an SCR, are used for encoding keyboard contact closures to set the shift register flip flops in accordance with Table 6-1. These transformers are made by winding 10 turns of #30 enameled wire around half inch diameter powdered ferrite cores. The windings are equally spaced around the circumference of each core, but the spacing is not at all critical (see Fig. 6-4). Only cores with high permeability, such as those in Table 6-2, should be used. The "one turn primary" of the transformer is simply a length of the same type of wire passed through the center of the cores appropriate to the character to be formed (per Table 6-1). One end of this primary is soldered to the keyboard contact of a given character, and the other end to the cathode of the SCR. There is, of course, one such primary for each of the 44 CW characters in the table, and several pass through each core. With #30 wire the cores don't get too crowded, and there is no effect on the cores by wires that pass around the cores but don't go through them (see close-up photo of the logic board, Fig. 6-5).

One end of the secondary winding of each transformer is connected directly to Keycoder's +5 V dc regulated power supply, and the other end to the "Clear" terminal of each shift register flip

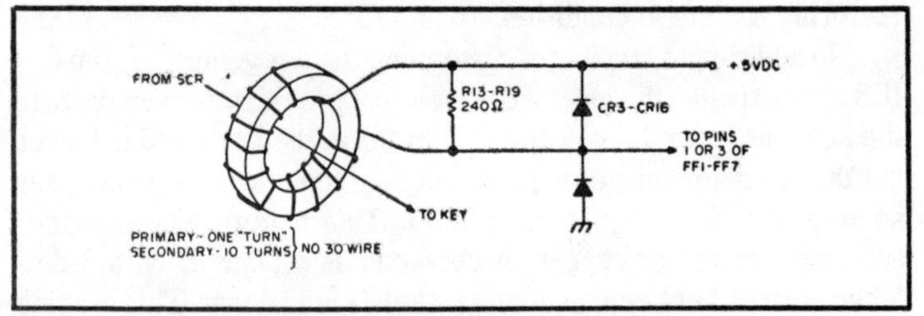

Fig. 6-4. Toroidal core transformers used for encoding.

flop. When the SCR fires, a high current pulse passes through the one turn primary connected to the key pressed. This pulse induces a very short duration high voltage (up to 15 V) pulse in the secondaries of the cores it passes through. These pulses are polarized in opposition to the +5 V applied to the secondaries and drive them to ground potential, thus setting the flip flops to which they are connected. The resistor across each secondary is for loading to suppress transients, and probably wouldn't really be needed (but is cheap insurance). Its value isn't critical. The diode in parallel with the resistor clamps the voltage to +5 V to prevent possible damage to the flip flop from too high a voltage spike. This probably isn't actually required either, but W9UBA and WA9VGS are cautious fellows. The diode in series with the clamping diode to ground prevents the flip flop "Clear" from being driven below ground (negative voltages should never be applied to TTL ICs).

The SCR circuit works like this: C4 is charged up to the power supply voltage through R8. The gate of the SCR is high, but current cannot pass because the keyboard key is open, breaking the cathode connection to ground. When a key is struck, the circuit is complete and C4 discharges almost instantaneously (the actual discharge time depends upon the time constant of C4-R8). This discharge produces the high current pulse to fire the core secondaries and set the flip flops. The instant a flip flop is set, pin 6 of IC3 goes low, closing the SCR gate, and C4 begins to recharge for the next character. The closed gate ensures that the keyboard is locked out until the character being generated is completed and IC3 pin 6 goes high again. C3 is simply a filter capacitor to prevent transients from triggering the SCR.

Monitor

IC2 is also a 555 IC timer connected as a triggered multivibrator. The bias components used in the dot generator are not necessary here since the IC is oscillating at audio frequencies.

None of the R or C values are critical and can be chosen to suit your preferences or what you have available. With those shown, the monitor produces tones from about 50 Hz to several thousand Hz, adjustable by R11. Of course R11 can be a fixed resistor if you'll settle for a single tone. C6 and R12 form the simple but adequate volume control. When set at its low end, R12 completely silences the monitor, and at the high end the IC will drive a 2"-4" speaker to a volume adequate for a medium-sized room. In the dot generator, the "Reset" terminal (pin 4) was tied to +5 V dc and the IC was triggered on and off via pin 2. In the monitor, however, we switch IC2 on and off via pin 4, since our driving signal (pin 13 of IC4) is low in the off state and high in the on state.

Table 6-2. Parts List for Keycoder.

Parts List

IC1-IC2	555 Integrated Circuit Timer
IC3	Quad 2-input NAND gate (SN 7400)
IC4	Quad 2-input NOR gate (SN 7402)
IC5	Triple 3-input NAND gate (SN 7410)
IC6-IC9	Dual "D" Flip Flop (SN 7474)
SCR	2N889, or equivalent
Q1	2N3569, HEP 728, or equiv. NPN audio transistor
Q2	2N5416 PNP high voltage transistor, or equiv. (see accompanying table)
CR1-CR16	1N914 switching diode, or equivalent
C1	3 uF, 15 V electrolytic capacitor
C2	0.005 uF, 200 V capacitor
C3	0.022 uF, 15 V capacitor
C4	1.5 uF, 15 V electrolytic capacitor
C5	0.01 uF, 15 V capacitor
C6	8 uF, 15 V electrolytic capacitor
R1	1.2k, ¼ W resistor
R2	15k miniature potentiometer, audio taper
R3	10k, ¼ W resistor
R4	51k, ¼ W resistor
R5	47k, ¼ W resistor
R6	430 Ohm, ¼ W resistor
R7	600 Ohm, ¼ W resistor
R8	10k, ¼ W resistor
R9	1k, ¼ W resistor
R10	10k, ¼ W resistor
R11	100k miniature potentiometer, audio taper
R12	500 Ohm miniature potentiometer, audio taper
R13-R19	240 Ohm, ¼ W resistor
RFC1	2.5 mH rf choke
Reed Relay	Meshna Electronics #SP-37A, SPST ½ A contacts, 3-6 V coil
Toroids (7 required)	Order from Amidon Associates, 12033 Otsego St., No. Hollywood CA 91607

High Voltage PNP Transistors for Keying Service

Number	Vcbo
2N398A*	105
2N1275	100
2N1476	100
2N1654	100
2N1655	125
2N1656	125
2N2042A*	105
2N2043A*	105
2N2551	150
2N2590	100
2N2591	100
2N2598	125
2N2599A	125
2N2600A	125
2N3062	90
2N3063	90
2N3064	110
2N3065	110
2N3224	100
2N3413	150
2N3495	120
2N3497	120
2N3841	100
2N3842	120
2N3930	180
2N4028	100
2N4357	240
2N4888	150
2N4889	150
2N5400	130
2N5401	160
2N5415*	200
2N5416*	350
2SA305	125
2SA429	150
2SA510	110
2SA511	90
2SA516A	120
2SA637	150
2SA639	180
2SA685	150

*Germanium transistors, all others silicon.

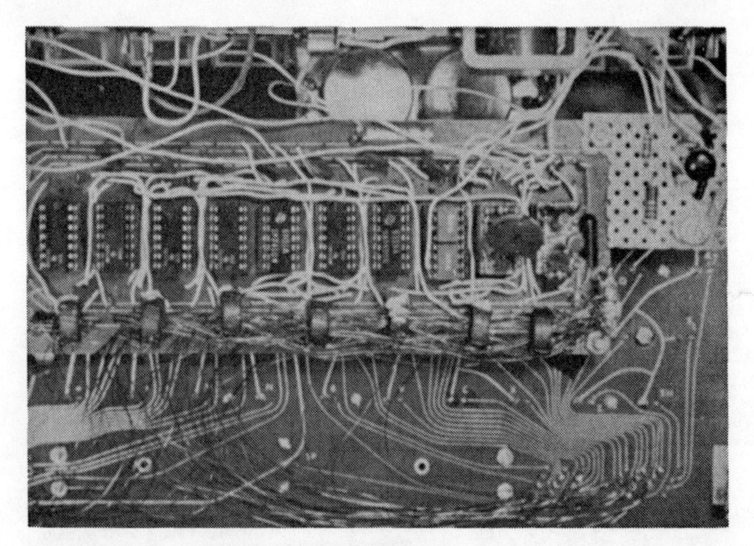

Fig. 6-5. Close-up view of the logic board.

If Keycoder I is to be used exclusively with rigs that have a sidetone, obviously the monitor could be eliminated completely, but for the small number of components involved it's nice to have the unit completely self-contained for code practice without connection to a rig.

Output Circuit

The TTL output of Keycoder I can be used to drive either a keying relay, as shown in Fig. 6-3, or a PNP power transistor, if you can find one with ratings adequate for the keying circuit of your rig.

In the relay circuit shown in Fig. 6-3, Q1 can be any small NPN audio transistor (I used a HEP 728 with R6 = 2200 ohms). The value of R6 can be changed to optimize the base drive if you substitute for Q1. For K1 I used a Griggsby-Barton 831C-2, which is packaged in a 14-pin DIP package and includes a protective diode (CR2). This relay is both small and completely silent, but is relatively expensive and its contacts will only handle about 125 mW. However, this was adequate to key my Kenwood TS-250 transceiver (keying current 10 mA at 12 V). A better choice would be a surplus reed relay, such as listed in Table 6-2. These are quite inexpensive (2/$1), and will handle 500 mA or better. W9UBA has used one of these relays to key his Drake T4X transmitter, Hallicrafters SR160, and several Heathkit rigs, without any difficulty.

C2 and RFC1 are for filtering any rf out of your rig's keying circuit line, and may or may not be needed depending upon how "clean" your rig is. Without either filtering component in the circuit I "froze" the contacts of one IC relay after several hours of troublefree sending, but have piled up several dozen operating hours without any relay problems after adding C2 (I still have no choke in the keying line).

Figure 6-6 shows the circuit for transistor keying. Again, Q1 can be any NPN audio transistor. W9UBA has both a reed relay and the transistor shown in Fig. 6-6 connected in parallel in the output of his keyer, and switches back and forth between the two. Most of his on-the-air contacts have reported no discernible difference between the relay and transistor keying.

Construction

The keyboard unit is enclosed in a rugged and handsome case that is spacious enough for all of Keycoder's component parts. It was obtained from Meshna Electronics. It is built on a heavy cast iron frame, and is trimmed in leatherette and imitation walnut grain sheet metal. The keys are magnetic reed type, very smooth, with about 3/16" travel. An added bonus in this assembly is the 10 rectangular switch positions (as seen in Fig. 6-1). Five of these are dummies—the plastic switch caps are in place but there is no switch beneath them—and I removed them and used the space for mounting 3 control pots (Fig. 6-2). Four of the other 5 switches are momentary action DPDT switches, and the fifth (labeled "PWR" in Fig. 6-2) is a DPDT Push-On/Push-Off switch. The two switches on the extreme right have screw-base "grain-of-wheat" light bulbs installed in sockets inside the translucent caps. I wired one of these, through a 20 ohm, 1 watt dropping resistor and the second set of "PWR" switch contacts to the power supply, to serve as an on-off indicator. W9UBA accomplished the same purpose by installing a red LED in a hole drilled in one of the switch caps (Fig. 6-1).

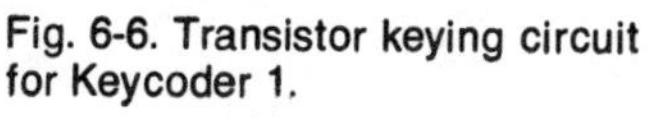

Fig. 6-6. Transistor keying circuit for Keycoder 1.

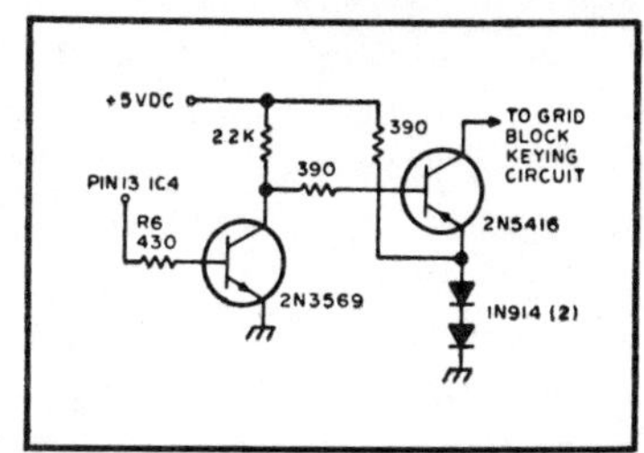

Fig. 6-7. Interior of W9UBA's Keycoder 1.

There are also plenty of unused key positions on the keyboard (SPST), so you should be able to mechanize any control setup you like without additional switches.

There is a diode matrix circuit board just beneath the bottom cover of the keyboard case. This board contains about 160 diodes and several dozen 33k resistors that can be added to your junk box (although the leads are awfully short). This board is attached to the case with a couple of screws and two multiple-lead ribbon cables that are easily removed. Beneath this discarded diode board is the key contact board, with all of the key contacts nicely labeled. Half of the foil patterns from the key contacts go to solder terminals grouped together at the bottom left-hand side of the board, and half lead to similar terminals at the bottom right. I soldered the core primary wires directly to the key contacts wherever they were on the board, while W9UBA made these connections to the bottom sets of terminals, resulting in shorter wire runs and a neater appearance (see Fig. 6-7).

The 9 integrated circuits, the SCR, and their associated discretes were all mounted in DIP sockets on a Radio Shack Universal Display Board (P/N 277-108). This board is predrilled for 10 14-pin DIPs, and etched with solder pads for the DIP connections and 8 separate buses running the length of the board, 4 on each side of the DIP patterns. These buses were used for +5 V dc, ground, SCR common, and the flip flop chock line. We both used #22 hookup wire for all connections; W9UBA ran his wires on the top side of the board while I made most of my connections on the foil side. Either way, this board saves a lot of building time compared with vectorboard construction or laying out and etching a PC board. The board measures about 2½″ × 6¾″.

Since the 555 ICs have only 8 pins, both of them can be mounted in one 16-pin socket; the leftover sockets can then be used for mounting the SCR and 555 discretes. The toroidal core transformers and their diodes and resistors are also mounted on the board, beneath the row of ICs (as seen in Fig. 6-5). The cores are wound and soldered into the circuit, but the leads are left about ¾″ long. Any one of the primary wires connected to the SCR is then strung through each core before it is mounted and the wire is grounded; if nothing happens, the core polarity is incorrect and the long leads permit it to be turned 180°. When a core is operating correctly, it is glued to the board with a drop of Duco cement. Stringing the primary wires through the cores will be easier if the cores are mounted so that their openings are in a reasonably straight line.

The values of R7-C3 and R8-C4 required some juggling to get reliable SCR operation. The most severe test of SCR and core operation is generating the figure "0", since 6 flip flops have to be set simultaneously to produce this character. If you have difficulty generating "0"s, or with repeat operation, vary the values of the above components, and/or try connecting R8 to the unregulated output of the power supply to get a higher voltage and hence more current.

All the integrated circuits were bypassed with .01 μF disc capacitors from +5 V dc to ground, at the buses right opposite each IC. These capacitors are visible at the top of the board in Fig. 6-5. In addition, each side of the power supply transformer primary is bypassed to ground with .001, 400 volt capacitors. The all-metal, completely enclosed keyboard case provides good shielding and also contributes, to the lack of rf interference.

Figure 6-3 does not show the power supply, since the keyer's power requirements are reasonably uncritical and the builder will probably wish to fabricate his supply with whatever parts he has available. The Keycoder draws about 100-125 mA and the +5 V should, of course, be regulated. We decided to build our supply based on the Radio Shack 5 Volt Regulated Power Supply PC board (P/N 277-102, price $1.49). This board measures only 1¾″ × 3¾″, and can accommodate whatever parts you have or whatever circuit variations you wish to make. Figure 6-8 is a schematic of the power supply with the parts values as specified by Radio Shack. In this configuration, the supply is rated at 1 A. The power supply is mounted in the upper left-hand corner of the case in W9UBA's unit, as shown in Fig. 6-7.

Referring again to Fig. 6-7, the output transistor and its driver and associated resistors are mounted on a scrap of vectorboard just to the right of the logic board, and the reed relay is just above the vectorboard, mounted to a partition added to the enclosure. The monitor speaker is mounted on the other side of this partition (beneath and to the left of the relay).

The volume, tone, and speed control potentiometers have to be miniature types if they are to fit in the area shown in Fig. 6-2. The pots are mounted to a small piece of thin aluminum sheet, which is held firmly in place when fitted between the keyboard assembly and the outer case. W9UBA used full-sized pots mounted under the row of keyboard switches, as shown in Fig. 6-1. Also note in this photo the sub-miniature thumbwheel type pot mounted in a rectangular hole cut in the switch cap (extreme left-hand one).

The speed control range can be increased at either the low speed end or the high speed end, or both (by altering the values of R1-R3 and C1), but resolution will become poor if too large a speed range is attempted, and it will be difficult to accurately repeat a given speed setting. Two switch-selected pots could probably be used to obtain a greater range of speeds without sacrificing resolution. Audio taper pots should be used for speed tone, and volume control.

Operation

Both W9UBA and I have had several compliments on having a "gud fist" before informing our contacts that we were using a keyboard keyer. More important, we estimate that our CW speed

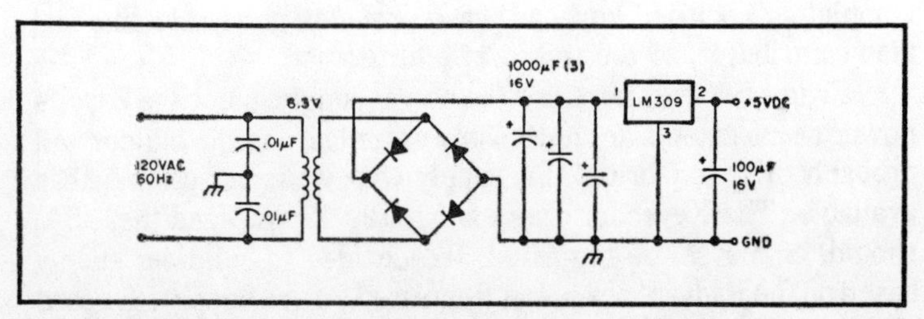

Fig. 6-8. Schematic of +5 V dc regulated power supply.

increased 5-10 wpm within 3 weeks after having our Keycoders on the air. We believe this is due, in part, to hearing perfect code as we are sending, and in part to the natural tendency to send faster, since it's so easy (thus unintentionally inviting a higher speed QSO).

It takes very little practice to get the rhythm of typing on a keyboard keyer, and you don't have to be a good typist. Once you know where the characters are on the keyboard, 20 wpm seems incredibly slow, even for a "hunt-and-peck" typist. A good typist must learn to allow enough time for long letters, such as "q" and "y", to be completed before hitting the next key, and a shorter time for letters like "e" and "i", but with just a little solo practice you'll be ready to go on the air. It was originally thought that a 2 or 3 character memory, and/or a space bar mechanization, would be added to Keycoder I, but operation without these features has been so good that it has been decided to forego this additional circuitry.

A VEST POCKET QRP RIG

There is nothing complicated about this project. See Figs. 6-9 & 6-10. A 2N2222, available for as little as a dime, is used in a conventional crystal oscillator circuit. The oscillator is linkcoupled to a low pass filter and a well-matched dipole, and that's it. The usual rules apply: Keep leads reasonably short, don't use a big soldering iron on the 2N2222, and listen to your signal to be sure you aren't chirping. Output tuning is a bit broad, so adjust C1 for best keying, even if it means giving up a few milliwatts.

Early mornings and daytime seem best for QRP work on 40 meters. Add two or three turns to the coil if you want to work both 80 and 40. Or add a switch and a 75 pF capacitor across C1. At 12

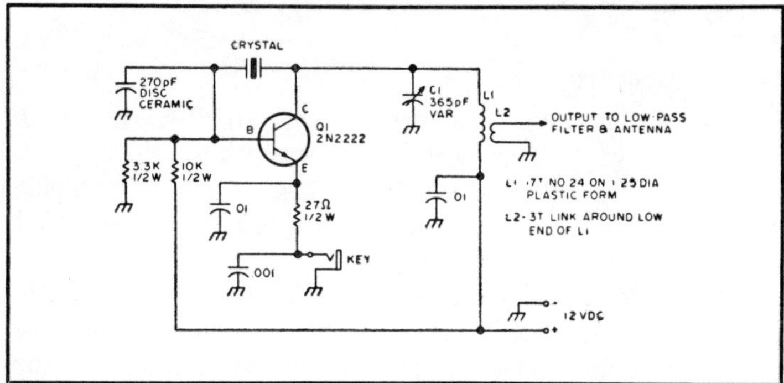

Fig. 6-9. Schematic. L1—17 turns #24 on 1¼" dia. plastic form. L2—3 turn link around low end of L1. C1 — 365 pF variable capacitor.

Fig. 6-10. The completed QRP rig.

volts, my version draws about 45 milliamperes. The 2N2222 will get warm if the key is held down for extended periods. Warm is okay, but hot is not.

A MINIATURE TRANSCEIVER

The entire project occupies 2 PC boards of identical size. Both boards are 3.7″ × 3″ and can be stacked if desired. The receiver portion is a simplified version of the Minicom MK V and covers 3.5 to 4.0 MHz. The transmitting exciter section output is also 3.5 to 4.0 MHz. By adding a suitable amplifier, the rig could be used as is on 75 meters. For use on other bands, additional mixers and a crystal oscillator would be used to heterodyne up to the desired frequencies.

Circuit Description

The Collins filters from my junk box are all housed in the Y-style case which is cylindrical and for which the PC layout is designed. Of the 2 filters from the collection which were suitable for this application, one had a bandwidth of 2 kHz and the other 3.1 kHz. Both worked well in this circuit.

A complement of 6 integrated circuits and 8 transistors provide all the needed functions. One i-f stage, the vfo, and the bfo are common to both receive and transmit modes. The input to the mechanical filter and the output from the common i-f stage are transferred from receive to transmit by means of diode switches.

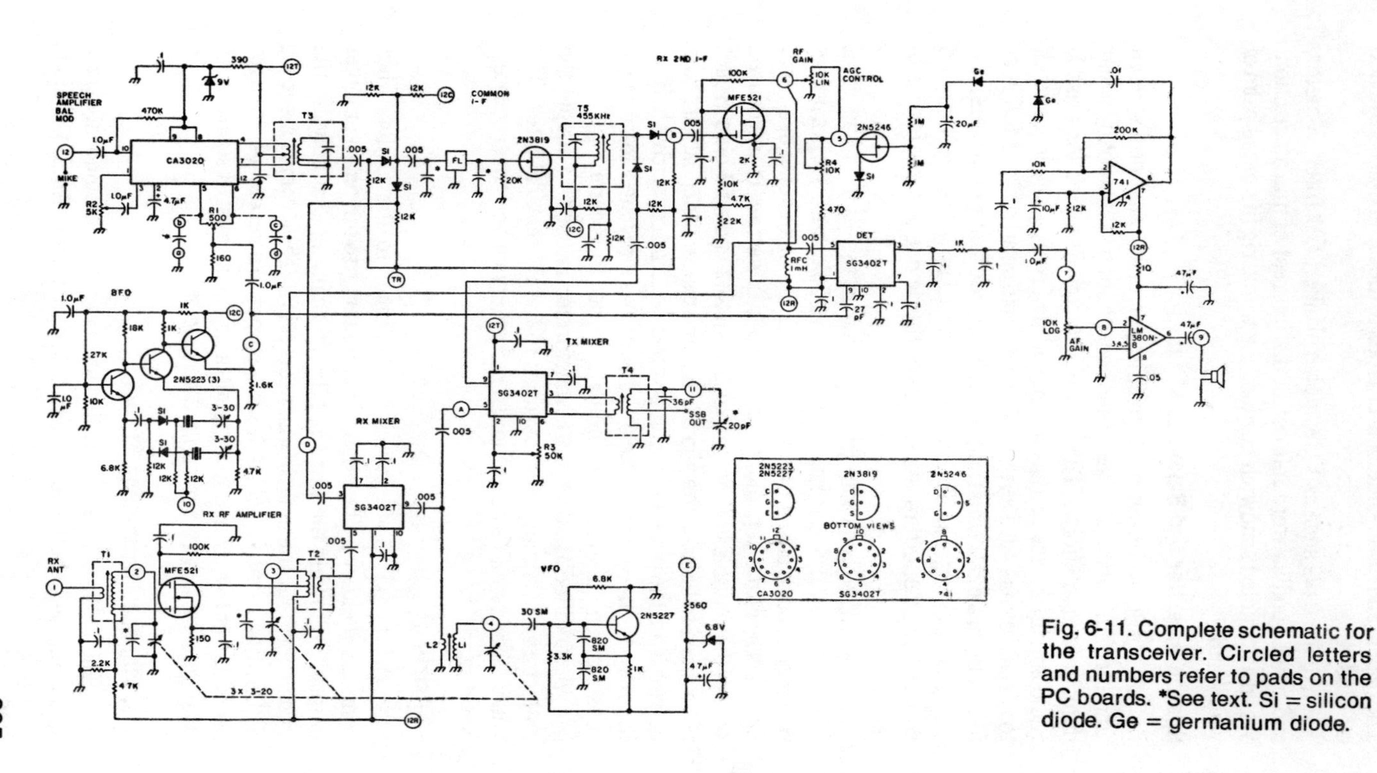

Fig. 6-11. Complete schematic for the transceiver. Circled letters and numbers refer to pads on the PC boards. *See text. Si = silicon diode. Ge = germanium diode.

The upper- and lower-sideband crystals in the bfo are also switched by diodes.

All the frills such as VOX, noise blanker, CW filter, S-meter, and other goodies were left out in the interest of miniaturization. When assembled sandwich fashion, the whole rig fits in the palm of your hand.

Speech Amplifier and Balanced Mixer

A CA3020 integrated circuit performs as both speech amplifier and balanced mixer, thus contributing substantially to our miniaturization efforts. This device, though designed for class B audio amplifier service, has a bandwidth of 8 MHz and lends itself to rf applications such as this. The chip houses an emitter-follower input stage, a differential amplifier, 2 emitter-follower drivers, and 2 output transistors. The output transistors have uncommitted collectors and emitters, which makes the device suitable for use in our circuit.

The mike feeds into pin 10 which is the base of the input emitter-follower. Input impedance is over 50k ohms. R2 is a preset trimmer used to adjust the audio gain for the mike being used and, once set, can be left alone. The 455-kHz carrier is introduced across the emitters of the output transistors and a 50-ohm preset pot is used to balance the signal. In some cases, cancellation of the carrier from the output may not be complete and capacitive balancing may be necessary. The PC board has provision for a small capacitor on either side of the balance pot to ground.

A modified transistor i-f transformer is used to couple the collectors of the output transistors to the mechanical filter. The winding data for this and other transformers will be covered later. See Fig. 6-11.

I-f Amplifier

The SPDT diode switch at the input to the Collins filter connects it either to the transmitter mixer or to the receiver mixer output. Following the filter is a JFET amplifier that stays in the circuit for both transmitting and receiving. Then comes a second SPDT diode switch that routes the output to a second i-f stage for receiving or to a mixer for transmitting. The second i-f stage uses a dual gate FET with agc applied to gate 2.

Bfo, Vfo, and Transmitting Mixer

The bfo is crystal controlled and operates continuously since it is common to both receive and transmit modes. An SPDT diode switch selects the desired crystal for either upper- or lower-sideband operation.

The vfo tunes from 3.045 to 3.545 MHz. It is also a common circuit and operates continuously. It is actually a part of the receiver section. The common vfo and bfo ensure that both transmitter and receiver will be on the same frequency. Since very little energy is required by the mixers, no buffers are used after the vfo. Output is taken from a 1-turn link wound over the tank coil.

The 455-kHz signal and the output from the vfo are mixed in an SG3402T IC balanced mixer. A 3.5-to-4.0-MHz output is produced and the vfo signal is nulled by means of the 50k preset trimmer, R3. The output transformer, T4, is fabricated from a standard transistor i-f transformer. A PC pad is provided for connecting a small variable capacitor of 20 to 30 pF across the coil for front-panel peaking at any frequency. A tiny solid dielectric type from a transistor FM radio is ideal for this purpose.

Receiver

For those not familiar with the DMOS transistors, such as the Motorola MFE521 used in the rf and i-f stages, they require a fixed bias of around 4 volts on gate 1. This is provided by the 2.2k and 4.7k resistors which form a divider across the 12-volt supply. Gate 2 is controlled by agc voltage but, unlike the regular MOSFETs, need not go negative to attenuate the signal. The DMOS transistor works with all positive bias which simplifies the agc. R4 is adjusted so that the no-signal level on the agc line is 6 volts. The transistors will be working at close to maximum gain at this value of bias and a strong signal will drive this voltage down to 3.5 to 4 volts. A manual rf gain control is provided and would normally be mounted on the front panel.

The mixer is an SG3402T IC which provides substantial gain and, as mentioned before, places a very light load on the vfo. Output load from the mixer is routed to the filter via the diode switch.

A third SG3402T is used as a product detector and feeds the LM380N-8 audio IC directly. Some of the audio output from the detector is fed to a 741 op amp where it is amplified and rectified to provide a positive dc voltage which charges the 20-μF tantalum capacitor across its output. Part of this charge is bled off by means of the two 1-meg resistors and applied to the gate of the 2N5246 agc control transistor. As the dc level increases due to strong signals, the transistor conducts more heavily and causes the voltage at the drain to drop. Since the drain is connected directly to the agc line, receiver gain is decreased and the purpose accomplished.

Filling in the Details

There are some numbered pads on the boards and some with letter designations. These notations also appear on the schematic and should aid in keeping the external wiring straight. The pads marked capital A, B, C, D, and E on both boards have to be connected together in pairs, that is, A to A, B to B, etc. Mating pads are directly above one another with the boards properly oriented and component side up. This allows stacking if desired. Pad A is vfo output. B is output from the common i-f, and C is bfo output. Pad D is input to the filter, while E connects 12 volts to the vfo from the other board's constant 12-volt supply connected to pads marked 12C.

Pads to accommodate a balancing capacitor if it should be needed as mentioned earlier are marked with small letters. One pair is a and b, the other c and d.

Pads marked 12R are for 12 volts applied during receive mode, and 12T designates a connection for 12 volts applied during transmit. Only one pad is needed; others are spares. There is a separate pad (12C) for the bfo, however, and it should be connected to a constant source of 12 volts along with one of the other 12C pads.

The pad marked TR is used to control the diode switches in the i-f strip. A 12-volt level is used for transmit and a ground level for receive. A TR relay would normally control this line as well as the 12R and 12T connections. The bfo crystals are similarly controlled by application of either 12 volts or ground to pad 10.

Of the other numbered pads, none needs individual explanation since the schematic clearly indicates their locations and the function becomes obvious.

Coils and Transformers

All the transformers are fabricated from regular 10-mm 455-kHz transistor i-f transformers. T1, T2, and T4 require the bare parts only, which means removing the tuning capacitor and all wire from the bobbin. Carefully salvage all the wire as it will be used to wind the new transformers.

The large winding goes on first and the link is then wound over the top. T1, T2, and T4 should be wound as in Fig. 6-12. Make sure the pinout is correct. Views are looking at the pin end or bottom of the assembly.

For T3, carefully break off the secondary leads where they enter the bobbin. This is the side with 2 pins. Unsolder the remaining wire from the pins and clean off excess solder. Since the new winding is center-tapped, you'll have to steal a pin from a spare assembly and push it into the existing hole in the base. The new link is bifilar wound over the top of the existing winding and connected as in Fig. 6-12. Identify the winding in some way so you won't have trouble orienting the transformer when mounting it on the board. The new winding goes to the CA3020 output.

The vfo tank-coil, L1, is pie-wound on a slug-tuned PC coil form. I used a Gowanda series 7 Velvetork form which is .209″ in diameter by .625″ long. A carbonyl E (red) core was used for the slug. Impregnate the winding with hot coil wax and put a single turn link over the top of the pie for L2.

Construction

I doubt that there will be any mad rush to duplicate this rig exactly as it stands, but parts of it may be of interest to some readers. I even debated the need to supply PC layouts, but decided they might prove helpful in some way. Just in case you do wish to copy the layout as much as possible, the following information may be useful. See Figs. 6-13 through 6-17.

All resistors, diodes, and rf chokes were mounted hairpin fashion to conserve space. Miniature low-voltage ceramic capacitors were used for coupling and bypass applications. Polarized capacitors are dipped tantalum and resistors are ¼ watt.

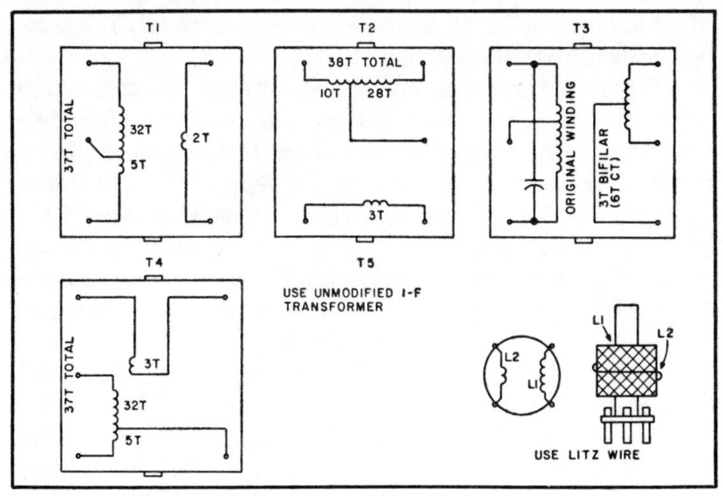

Fig. 6-12. Coil winding table. All bottom views.

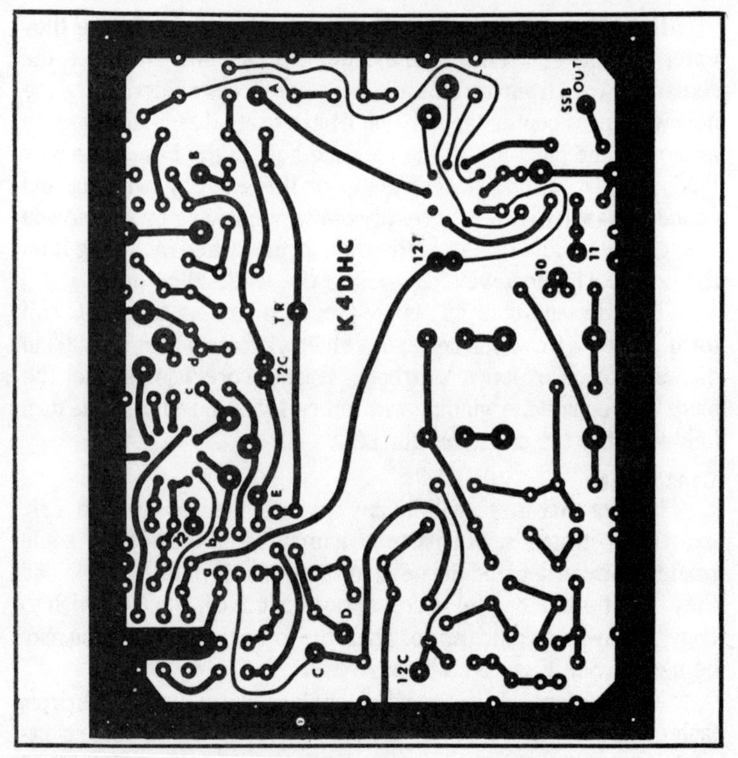

Fig. 6-13. PC board for transmitter.

Before mounting the SG3402T used as the receiver mixer, cut off pin 6. Likewise cut off pin 4 of the product detector. The same goes for pin 11 on the CA3020.

The two trimmer capacitors in series with the bfo crystals are subminiature (5-mm) units in case you were wondering how they squeezed into the space allotted to them. Incidentally, the exact operating frequency for the crystals will depend on the particular filter used, but this circuit will pull quite a bit and allow appreciable leeway in crystal accuracy.

Note that there are two silver mica capacitors across the ends of the mechanical filter whose purpose it is to tune the two transducer coils. Values will depend on the type of filter used. For the F455 Q2 that I used, I had to install a 91-pF capacitor at one end and a 110-pF capacitor at the other end.

The 3-gang tuning capacitor is the same one used in the Minicom receivers and has a range of approximately 3 to 20 pF per section. An additional padding capacitor is required across each of the first 2 gangs. Use 20- to 22-pF silver micas and solder directly to the frame before installation. When mounting this item, use #4-40 screws ¼" long. Put a toothed washer under the head to allow good contact with the copper. Place 2 flat washers over each screw on the component side before mounting the capacitor. This will leave enough space to clear the rivets that hold the compression trimmers for the first 2 gangs.

I used MPN3401 (Motorola) diodes to do the switching since they were in my junk box and are made especially for this use. Regular silicon diodes should work okay, so don't panic.

Tune-Up and Checkout

A DPDT toggle switch can be used for TR switching by using one section to apply 0 or 12 volts to the TR line and the other half to transfer 12 volts between the receive and transmit circuits. An SPDT toggle can be used switch the bfo crystals. A 10k audio gain control with log taper and a 10k rf gain control with linear taper are the only other items needed at this time. With the boards lying on the bench, make the jumpers between boards long enough so that each board can be handled freely while making adjustments. The

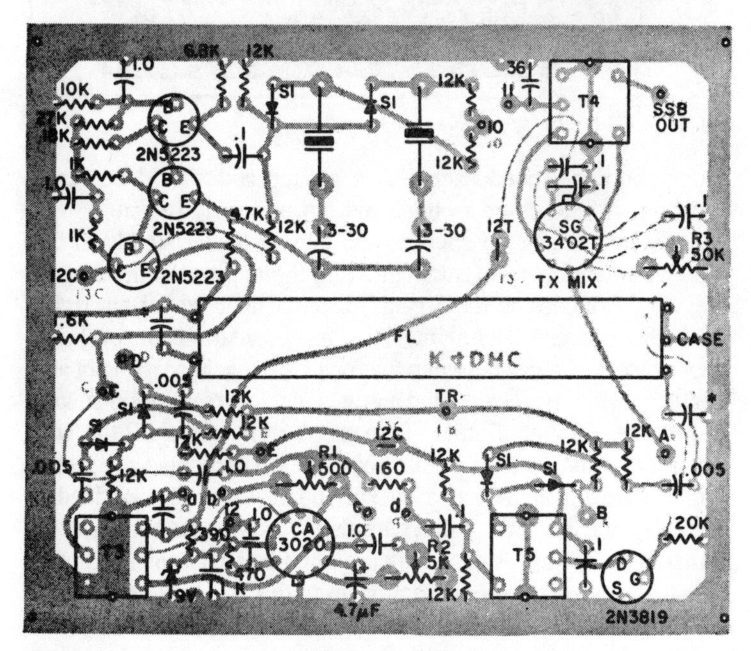

Fig. 6-14. Component layout for transmitter board.

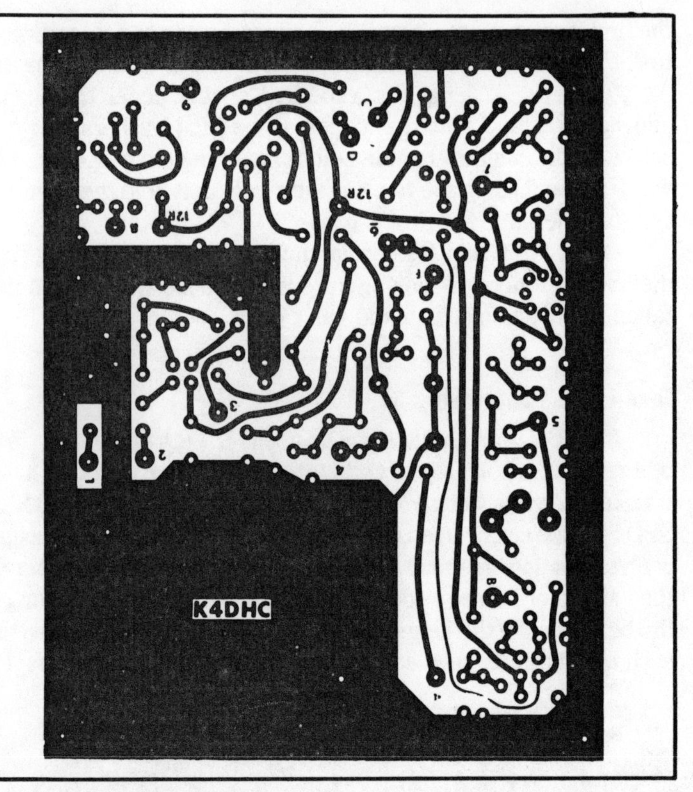

Fig. 6-15. PC board for receiver.

copper border around each PC is ground and should be made common with all other grounds and power-supply return.

Connect an 8-ohm speaker and 12-volt power supply to the rig. Set all trimmers to midpoint. Run the screws on the 2 compression trimmers on the 3-gang capacitor up snug but not tight. Turn the variable to full mesh. Connect a dc scope or high-impedance voltmeter between the top of the rf gain control pot and ground. Switch to receive and apply power. Current drain should be between 55 and 65 mils. Turn the rf gain control all the way down and adjust R4 for a reading of 6 volts. Disconnect the meter and turn the rf gain to maximum. Turn up the audio gain and feed in some signal at 3.5 MHz. Adjust the slug in the vfo tank coil until the signal is picked up. Adjust the cores in T1 and T2 for maximum output. Run the variable capacitor up to the high end and feed in signal at 4.0 MHz. Peak the response by means of the 2 compression trimmers on gangs 1 and 2. Repeat these 2 procedures as many times as needed to achieve good tracking. T5 can be peaked on the noise present. If the band is active, you can connect an antenna and listen on the air to verify proper operation of the receiver. Also make sure both bfo crystals are working. This concludes receiver checkout.

An audio generator should be connected to the mike input during tune-up of the transmitter. Turn R2 to minimum and switch to transmit mode. Current drain should be 40 to 50 mils. Connect a scope to the output side of T3. A piece of discarded pigtail can be temporarily soldered to the pad if necessary to make the scope connections. With no audio input, adjust R1 to null out the carrier. If there is still some trace of rf, tack a mica trimmer across pads a and b and then c and d to see where improvement can be made. Measure the value of the trimmer at the best setting and substitute a silver mica fixed capacitor.

Once the carrier has been nulled, turn up R2 and feed in some audio signal at around 1500 Hz. Adjust the level to a point just below where distortion occurs as observed on the scope. Transfer

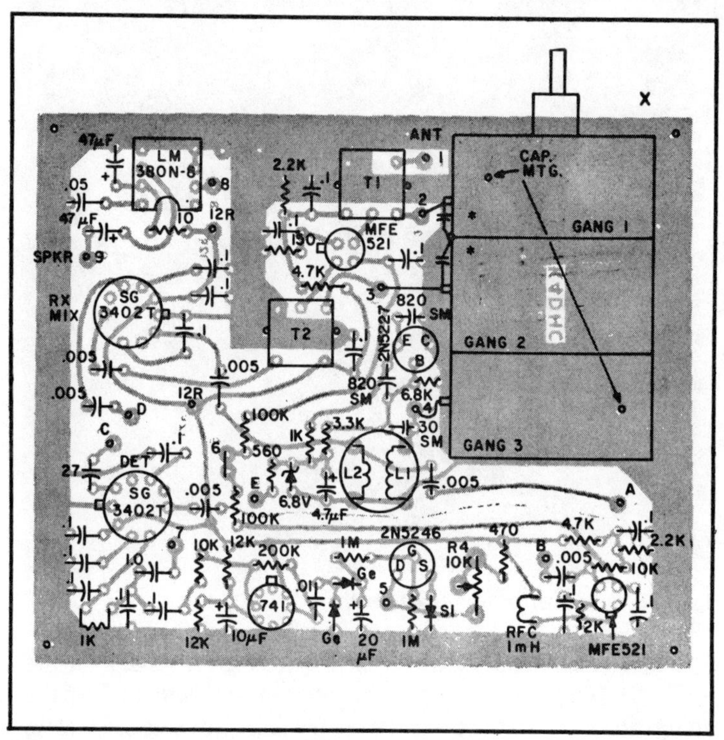

Fig. 6-16. Component layout for receiver board.

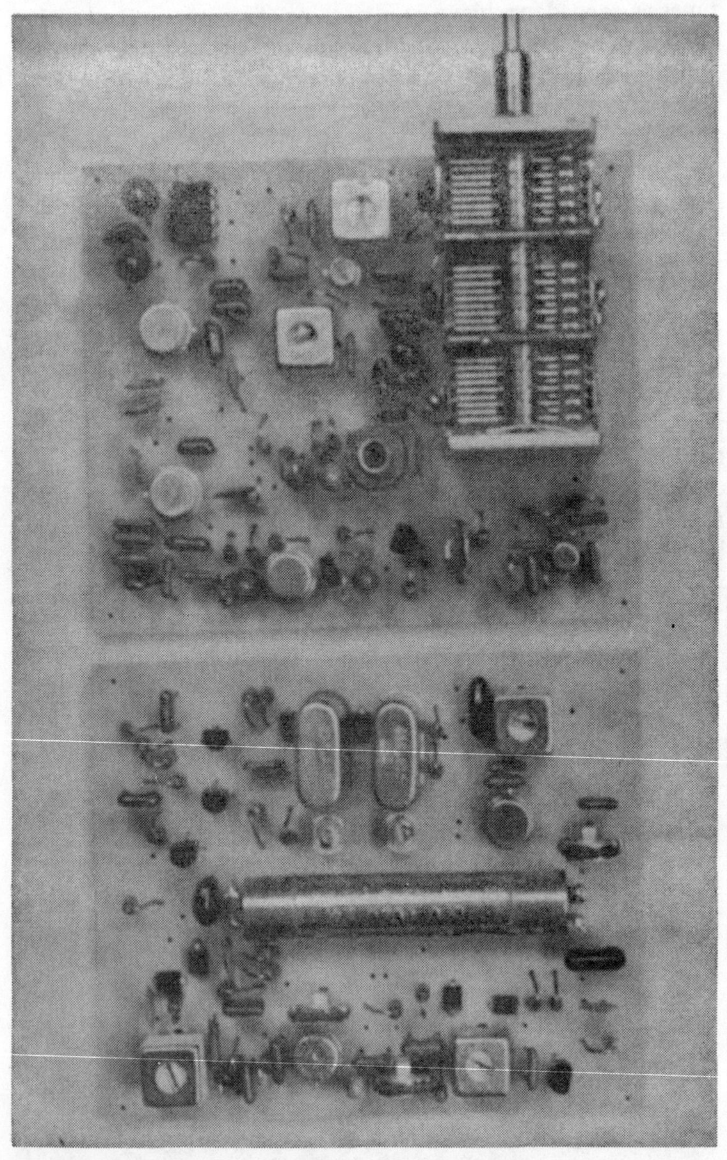

Fig. 6-17. Both boards fully assembled.

the scope to the output side of T5. Determine the values for the capacitors that tune the filter and install them at this time. Turn the tuning capacitor up to the high end of the band just below 4.0 MHz and transfer the scope to the transmitter output at T4. Shut off the signal generator and null the vfo signal by adjusting R3. Turn on the generator and adjust the core in T4 for maximum output. Later on, a tuning capacitor can be mounted on the front panel to allow peaking of the output as frequency is changed. That concludes tune-up of the transmitter.

EASY QRP RIG

The purpose of this project was to build a simple transmitter that could be duplicated easily by any amateur.

My main interest as an amateur lies in designing and building my own equipment. During numerous on-the-air conversations, I discovered that home-brewing is not a forgotten art and that many hams are still interested in building at least some piece of equipment for use in the shack. However, it seems that a good portion of newcomers (and not-so-newcomers) are frustrated when trying to find a project that is simple enough to understand, is cheap, and will produce a useful item which is not time-consuming to build, debug, and get operational.

This 5-watt, 80/40-meter, CW transmitter is all of this and more. All of the parts can be purchased at your local Radio Shack, assembly time is less than an hour, and tune-up time is zero. Using a PC board practically guarantees that the transmitter will work the first time the key is closed. These features should make this a project that both the Novice and old-timer can enjoy.

My original design called for vfo control of the transmitter, but that required five rather than three transistors. Additionally and more importantly, the components needed to construct a stable vfo cannot be purchased at most Radio Shack outlets. To overcome

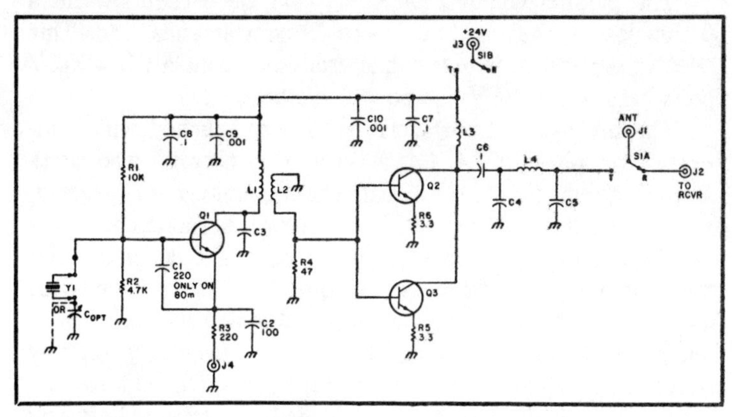

Fig. 6-18. Transmitter schematic.

these problems, crystal control was decided upon. At first thought, crystals conjure up an image of hours of operating without a contact as you wait for someone to happen upon your frequency. That simply isn't the case, as will be shown later. Also, since Novices now can use vfos, there are many crystals lying around in ham shacks everywhere.

The transmitter can be built as a basic unit or with several options, as shown. The basic unit consists of the loaded PC board soldered to an antenna connector and power source. If the transmitter is to be used for serious operation—which it definitely is capable of—then the options, which require only a little more time and money, should be added. Options will allow T-R switching, some frequency variation, two bands in one box, and a package that is more pleasant to look at and show off.

The Circuit

As can be seen from Fig. 6-18, the transmitter meets the design goal of being simple. Only three transistors are used to generate the 5 watts of output power. Resonant circuit inductors are formed using iron-core rf chokes. Common-value ceramic capacitors are used either singly or in parallel to obtain the needed capacitance. See Table 6-3.

Q1 operates as a Pierce oscillator at the crystal frequency. FT-243 crystals, which are inexpensive and plentiful, can be used. Output is taken from Q1 by a five-turn link over L1. Q2 and Q3 comprise the class C final amplifier and are operated in parallel. Parallel operation provides an easy method of obtaining the desired 5-watt output.

The parallel combination of Q2 and Q3 presents about a 60-ohm load to be matched to the 50-ohm antenna load. This collector impedance is determined from the formula $RL = Vcc^2/2po$, where Vcc = 24 V and po = 5 Watts.

The impedance transformation is accomplished with a pi-network composed of L4, C4, and C5. This network also offers harmonic attenuation to the signal. The transmitter, as designed, easily meets the FCC regulations for harmonic radiation.

R5 and R6 are used to equalize current flow in the two transistors. In all of the units built thus far, I have detected no "hogging" of current by either transistor. Nothing special has been done in selecting matched transistors. If they run equally hot, they are matched well enough! Heat sinks are needed on both transistors to dissipate the heat generated. Since the type of heat sink needed is not available at Radio Shack, they must be constructed by hand. Light gauge aluminum can be used by forming a tightly fitting cap over the transistor.

The design goal of using readily-available parts was realized throughout the rig. Radio Shack disc ceramic capacitors are used and have performed well. Unfortunately, there is a limited variety of these parts. To obtain the desired capacitance, the capacitors, where necessary, are soldered in parallel. This allows for the elimination of variable capacitors to tweak the tuned circuits to resonance. In all units assembled, the resonant circuits and matching networks have worked fine with no tweaking necessary.

To construct L1, remove the required number of turns from the Radio Shack choke. Use this removed wire to form the link winding, L2. Wind L2 over the Q1 side of L1. L3 should be made similarly except that no link is needed. The chokes work surprisingly well as resonant circuit inductors at 3.5 and 7 MHz.

Table 6-3. Parts List. (Radio Shack Parts Numbered in Parentheses.)

C1-C10	—Ceramic disc (272-xxx)
C3	—80m: 220 pF; 40m: 47 pF
C4,C5	—80m: 690 pF (220 and 470 pF in parallel);
	—40m: 420 pF (220, 100, and 100 pF in parallel)
Copt	—BC variable, approx. 30-200 pF
J1,J2	—SO-239 (278-201)
J3	—Phono jack (274-386)
J4	—Phono jack (247-252)
L1	—80m: 8.4 uH, 8 turns removed (273-101)
	40m: 10.0 uH, no turns removed (273-101)
L2	—5 turns wound over side of L1
L3	—Approx. 30 uH, 40% of turns removed (273-102)
L4	—80m: 2.4 uH, 16 turns removed (273-101)
	40m: 1.2 uH, 23 turns removed (273-101)
Q1	—RS-2033 (276-2033)
Q2,Q3	—RS-2038 (276-2038)
R5,R6	—Each is 3 10 Ohm, ½-W (271-001) in parallel
S1	—DPDT toggle (275-1546)
Fig. 2.	
C1	—1000 uF, 50 V (272-1047)
R1,R2	—¼-W carbon (R2 can be made 5k variable to provide 3-30-V output)
S1	—SPST (275-324)
T1	—24 V lamp, min. (273-1480 or 273-1512)
Z1	—Full-wave bridge rectifier, 1.4 A, 100 piv (276-1152)

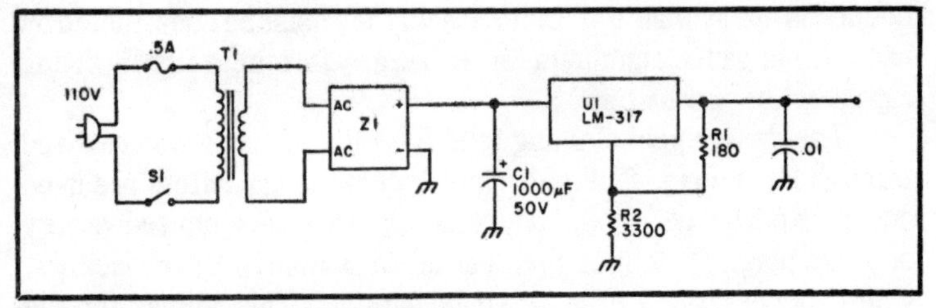

Fig. 6-19. Power supply schematic.

One departure from the norm in this project was the use of +24 V for supply voltage rather than the more common 12 volts. This was done for several reasons. It is much easier to build and get operational a 24-volt supply than it is the additional stages required to realize 5-watts output using a 12-V supply. This also makes the transmitter simpler and cheaper (other design goals).

A schematic for a very simple 24-V supply is shown in Fig. 6-19. This supply can be made variable or fixed. It has performed flawlessly at currents up to 1.5 amps. The regulator contains internal short-circuit protection and is self-contained.

If an ac-operated supply is not desired, four 6-V lantern batteries can be operated in series to provide the needed 24 volts. Many hours of transmitter operation can be achieved from such batteries. Alternatively, and probably cheaper, sixteen D-cells can be soldered in series for the supply voltage. Obviously, the 24-volt supply should not be a deterrent to building the transmitter. It can be used for later projects as well!

Construction

The transmitter is built on a 2¼″ by 3″ PC board. Assembly time is less than one hour due to the small number of parts used. A number of transmitter boards have been constructed and each one has worked fine when power was applied. See Figs. 6-20 and 6-21.

For best operating comfort, the transmitter PC board should be mounted in an enclosure—as mentioned earlier. Any size or type of enclosure will work fine. I used a Radio Shack type, which makes for a nice-looking and compact transmitter.

A crystal socket should be mounted on the front panel to allow for a change of frequency when desired. A variable capacitor can be mounted near the socket to allow for a small amount of frequency excursion from the crystal frequency. On 80 meters, about 1.5 kHz of change has been possible. On 40 meters, this increases to about 3-5 kHz. The amount of frequency excursion will vary, depending mostly upon the crystal used.

Switch S1, a miniature DPDT toggle type, is used to switch the antenna between receive and transmit. All connections between S1, the PC board, and the SO-239 antenna connectors should be made with coax. RG-174 is preferred, but if it is not available, RG-58 will work fine. The only other additions necessary are a phono connector for voltage and a key jack.

If desired, the 40-meter PC board can be mounted in the same box as the 80-meter board to make a two-band transmitter. Another toggle switch will be needed to switch the two boards to the appropriate circuit points. See Figs. 6-22 through 6-24.

Operation

After assembling the PC board and the supporting parts into a cabinet, the transmitter is ready for use. Initially, a dummy load should be connected to the antenna connector. This allows for testing without generating QRM on the air. The dummy load can consist of two 100-ohm, 2-watt resistors in parallel. If a vom (ammeter) is available, it might prove advantageous to hook it in series with the plus side of the 24-V supply. Input power can then be calculated.

After the key is plugged in, the supply turned on, and the crystal installed, switch S1 to transmit and close the key. The vom

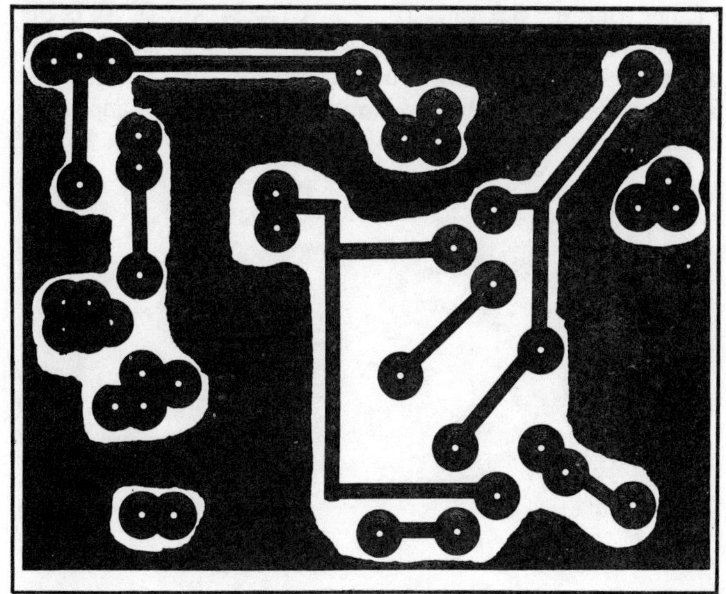

Fig. 6-20. PC board layout, foil side. (Single-sided, fiberglass, copper-clad board.)

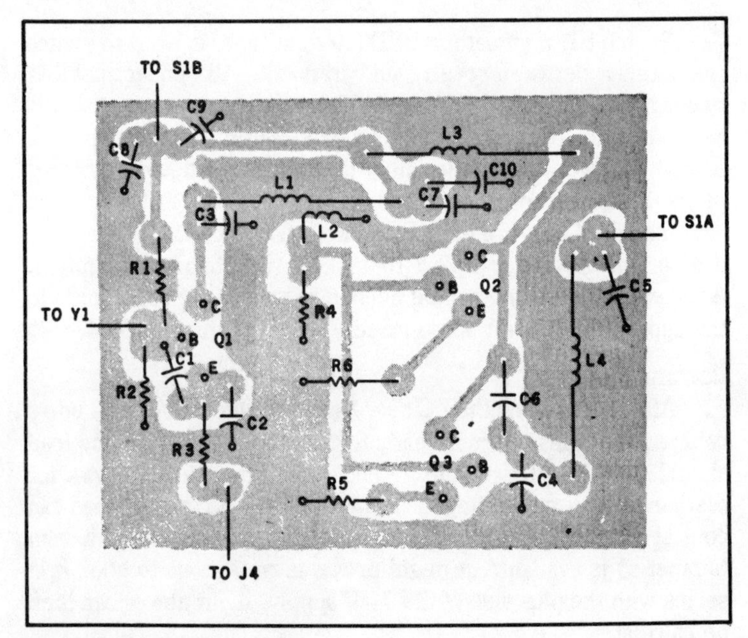

Fig. 6-21. Component locations.

should read about 350 mA of current. This indicates an input power of approximately 8.5 watts (Pi = E × I = 24 × .35). All of the transmitters I have built have had a minimum efficiency (Po/Pi) of around 60%. This indicates an output power of around 5 watts. The input (and output) power of your transmitter might vary depending upon the gain of the transistors used.

Fig. 6-22. Inside view.

Fig. 6-23. Back view.

An antenna now can be hooked to the SO-239 antenna connector. At the same time, a short cable should be run between the receiver-out connector and your receiver.

You are ready now for on-the-air contacts. You probably will be as surprised as I was when you first use your new little powerhouse. Surprisingly, my best success has been in calling CQ. The response ratio has been close to 50%. Using one crystal on 80 meters has resulted in numerous contacts up to 1500 miles away with excellent reports in both strength and quality. The antenna used in conjunction with the transmitter has been a dipole at 20 feet.

Fig. 6-24. Completed transmitter.

Operating QRP is fun. Operating portable is fun. But, operating QRP with the kind of inefficient antenna systems sometimes encountered in portable operation is no fun at all. Sure, with a good matchbox, you can load a barbwire fence, or a set of bedsprings, but those aren't the best radiators in the world, and the puny little 100 mW QRP rigs will have a hard time getting across the street. While facing this dilemma, I decided someone should invent the perfect QRP rig, one that is small enough to be super-portable, low powered enough to run off a lantern battery, and gutsy enough to make some noise across town into the average portable antenna.

Since I am the only inventor I know, I had to call upon myself to come up with this "wonder-rig." And (if I do say so myself) I did a pretty good job. The final outgrowth of the requirements is a muscle-powered mini-watter I call the "Mini-Mite." Now you can build yourself a Mini-Mite for less than twenty dollars. It is the perfect Novice rig with its 7 watts dc input; it is the perfect standby rig for the oldtimers; and darned if it's not the answer to the QRP nut's portable operating dreams. The rig is vfo controlled (why be old-fashioned?), it runs a solid five watts, can be held in one hand, and has a complement of controls you would expect to find only on more expensive rigs. You can build it for any or all the bands of your choice, 80-10 meters. And, finally, if you insist on staying around the one watt level with your QRPing, you can leave out the final for a dandy low-power "low power" transmitter.

Figure 6-25 is a block diagram of the Mini-Mite. The rig is pretty straightforward as far as using conventional circuits. Once you are into the schematic you will see some unique features that save space and money in the rig. The vfo circuit is very stable and

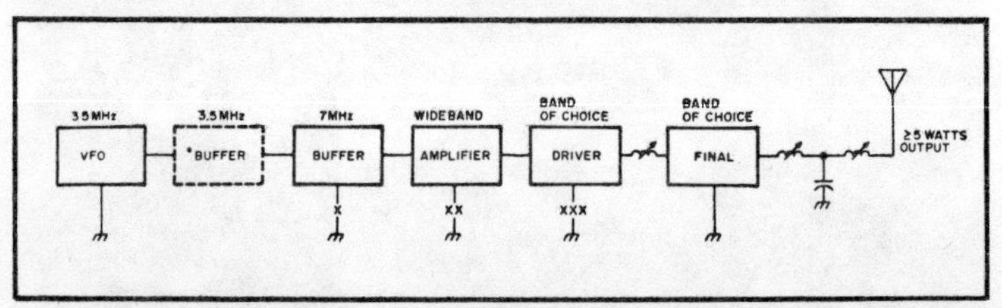

Fig. 6-25. Block Diagram. The Driver and Final stage tuned circuits are selected for the band of your choice. Note they operate straight through on all bands for maximum efficiency. *This buffer is optional; it may be added to reduce vfo pulling—this is especially helpful when the unit is used with break-in keying. The keying points (shown by "X") have various advantages and disadvantages. See the text for a complete discussion.

you have a choice as to using a bipolar vfo or an FET vfo (schematics for both are given). The buffer shown in dotted lines is optional; it provides greater immunity to pulling as the rig is tuned and keyed. This option is a must for the operator who wants "everything" in his rig, or who is interested in using break-in operation. The buffer stage (solid lines) uses a broad-tuned low Q circuit which resonates at 7 MHz, but passes sufficient 80 meter signal to allow the rig to operate multi-band. The buffer operates class A and and runs at fairly low power. The amplifier stage is a wideband outfit which delivers the first real power in the rig; you could use the amplifier output as a transmitter output, except that it is rich in harmonics. The driver serves two purposes; one, of course, is to provide power drive for the final. The other use for the driver is to select the harmonic of the vfo signal that will be used to transmit. The driver tuned circuit alone does not provide adequate harmonic attenuation, so if you decide to build the "QRP" Mini-Mite you may leave out the final stage, but you must keep the final tuned circuit because it provides the necessary harmonic rejection. For normal QRP "plus" operation the driver feeds the final which in turn generates the so-called high-power output signal. The final output circuit is one of those unique circuits I mentioned, it uses slug-tuned forms to provide the tune and load controls. The output will easily load into 50-70 ohms and a little more. Harmonic output at the final is very low and when the rig is properly tuned the output signal is very clean.

Figure 6-26 is the schematic of the complete Mini-Mite. Some of the interesting features are the coils. You may pillage the junkbox to see what you have, or you may buy three sets of the

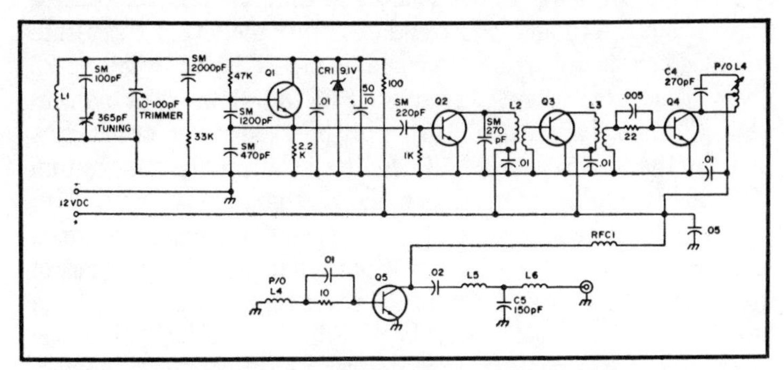

Fig. 6-26. Schematic Diagram. Coils are discussed in the text. All values in pF unless otherwise stated. Resistors are given in Ohms and are all ½ watt carbon types.

Table 6-4. Coil Capacitor Data (Final Values Will Need Trimming).

Coil /Cap	Band 80	40	20	15	10
L1	16t#28E (13t on toroid)	X	X	X	X
L2	X	8t#28E widespaced 5t link	X	X	X
L3	7t#28E 3t link	Same	Same	Same	Same
L4	30t#28E 10t link	15t#28E 5t link	8t#28E 4t link	5t#28E 3t link	4t#18E 2t link
C4	470 pF	270 pF	150pF	100 pF	68 pF
L5	25t#28E	13t#28E	7t#28E	4t#28E	3t#18E
C5	300 pF	150 pF	75 pF	50 pF	37 pF
L6	35t#28E	21t#28E	10t#28E	7t#28E	5t#18E

units given in the parts list (See Table 6-4). For the vfo coil, L1, the buffer coil L2, and the rather unique output transformer L3, you use only the slug from a coil. You do not use the coil form itself. Coils L4, L5 and L6 use the complete slug-tuned form as these three coils are tunable and must have some means of adjustment.

Construction

I am the biggest booster of perfboard construction when only one of anything is built, although Figs. 6-27 and 6-28 show a printed circuit board for building more than one of the units. The coils are available from Radio Shack and are sold two to a package at slightly over a buck.

The vfo uses an Amidon T92 toroidal coil, but other units have been using the slug as described and they offered comparable performance. Although you could construct the vfo in a separate shielded box, it is provided for right on the PC board. Figure 6-29 gives the schematic for the alternate FET vfo. Both vfos tune with a 365 pF tuning capacitor and use a 100 pF fixed capacitor in series to limit the tuning range. Vfo tuning is linear for the bottom 100-200 kHz of the band and then becomes non-linear (but usable) over the remaining portion. This is generally acceptable, since most people will just put the linear portion in their favorite part of the band anyway. The FET vfo will provide more stability than the bipolar unit, but it will run the cost of the rig up a little since the same transistor that is used in the vfo can also be used as the buffer, amplifier and driver stage. Also, you may get slightly less drive to the buffer using the FET.

The buffer stage provides a compromise between actually isolating the oscillator and giving a little power output. That's why there is an alternate buffer, to provide isolation and little or no gain. Figure 6-30 shows the alternate buffer; the circuit uses no tuned circuits. The Mini-Mite buffer is straightforward and uses the slug form previously described. It should resonate around 7 MHz if you want to check your work with a grid dip meter. The windings are wide-spaced along the entire slug, not wound closely together. Hold the windings in place with a little Q dope, or equivalent. The output link is wound on the cold end (Vcc side) of the coil and provides impedance matching for the amplifier stage.

The amplifier runs class B and there are no biasing resistors to worry about. It is very simple: The input goes on the base, the emitter is grounded, and the collector gets +12 through the output transformer L3. L3 is left untuned and serves to couple wideband, harmonic loaded output to the driver stage. The transformer itself is a twist on the toroid. It is fabricated from a high-permeability coil slug that has a hole all the way through it. Either grab one from the junk box, or drill a hole (lengthwise) the rest of the way through the Radio Shack slug after the tuning screw is removed. This stage provides enough power to dimly light a #47 bulb connected across the secondary. The power is to put some kick in the driver stage and if not for all the harmonics, you could use this output for a QRP transmitter all by itself.

Harmonics are taken care of in the driver to a great extent. The higher power drive from the amplifier is used to drive the driver into class C for most efficient operation. When the driver

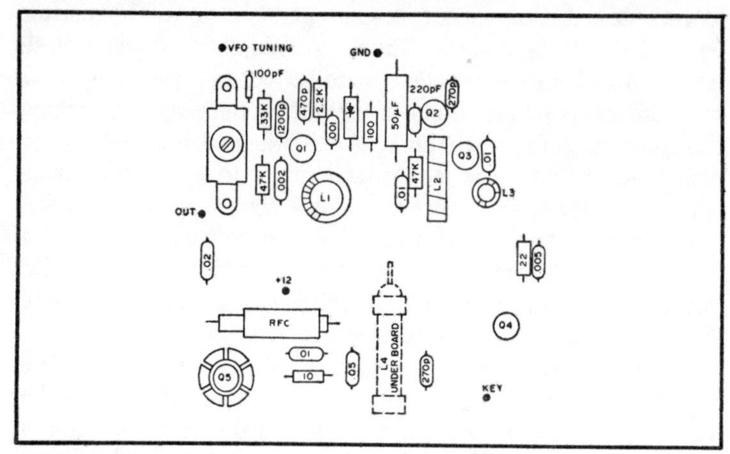

Fig. 6-27. PC board layout, top and bottom views.

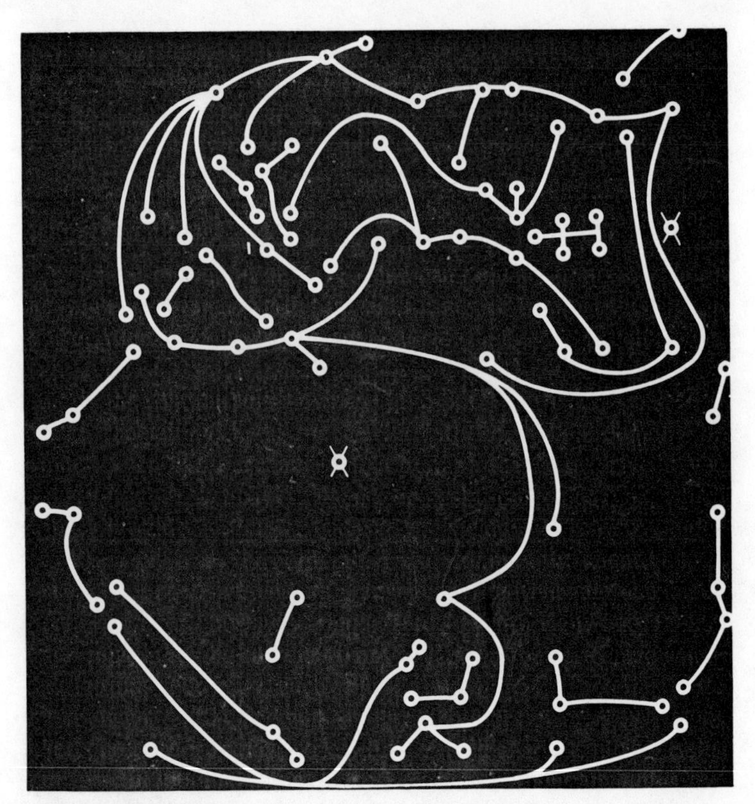

Fig. 6-28. PC Board.

operates, current flows through the 22 ohm resistor, setting up a reverse bias that causes the stage to operate class C. The resistor is bypassed for the ac signal. The tuned circuit in the driver is close to that in the buffer, except that the coil is tapped to yield a higher Q with the higher power output. Harmonics at this stage are nothing like at the amplifier stage, but to use this stage as the QRP rig, you will have to couple the output tank circuit to get an acceptable signal. The tuning of the driver coil is accomplished by a wooden dowel shaft extension. The dowel should be approximately 4½ inches long to reach the front panel from its position under the PC board. As can be seen in Fig. 6-31, the coil mounts on the PC board directly. An angle bracket is satisfactory mounting.

The final stage will run a theoretical 7 watts dc input. The output circuit is a "T" network with slug-tuned coils used as the output tuning and loading controls. The final output transistor is rather expensive, but is hardy and generally works well in these

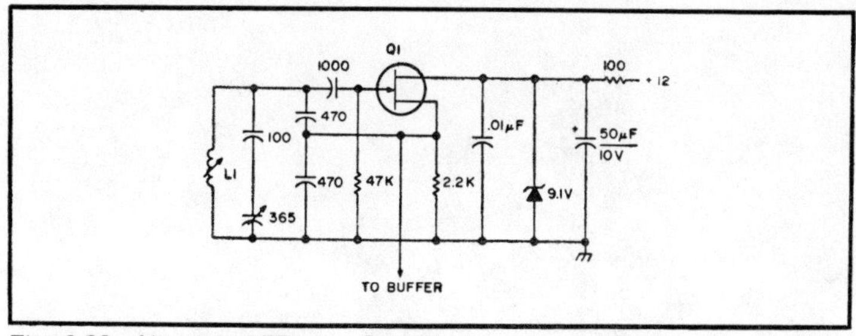

Fig. 6-29. Alternate FET vfo. Capacitors and resistors in pF and Ohms unless otherwise stated.

applications. It will operate at frequencies in excess of the 30 MHz encountered in the 10 meter version of the Mini-Mite, so it is a pretty good choice until research proves the existence of a better all-around transistor. The rf choke used in the final is a high-current jobby. A conventional rfc has too much dc resistance and limits the output power stage can run. Wind forty turns #18 en. close and in 2½ layers directly on one of those slugs.

Table 6-4 gives values for coils and capacitors to be used for multi-band operation. You may wish to switch in the coil/capacitor combinations, or do some experimenting with winding multiple coils on the same form. Some of the problems are that the coil/capacitor combinations are sympathetic to harmonics and you may get false loading. But, as long as the Q is high enough, you could make each band peak at some different setting of the slug in the coil form. This helps eliminate the problem. Also, remember to use shielded solenoid inductance formulas because the overlapping coils act as shields and change the normal calculations. The cost savings of this system would be significant. This could be our answer to the old familiar tapped coil setups in tube-type rigs.

The final tuning coils are mounted on a piece of angle aluminum. Each slug-tuned coil has a wooden dowel extension

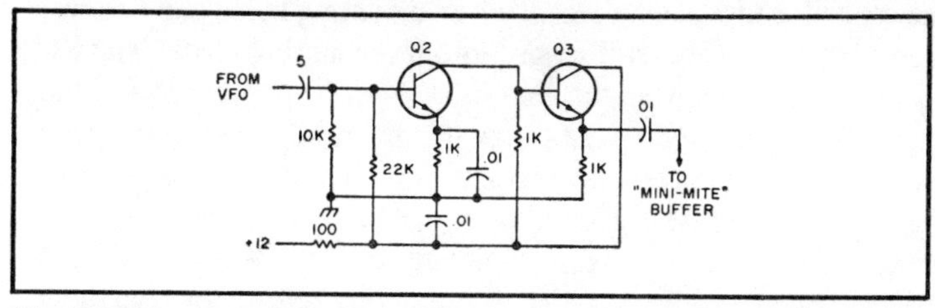

Fig. 6-30. The optional buffer is used in conjunction with buffer in transmitter to provide better isolation for break-in operation or for when some pulling of the vfo cannot be tolerated.

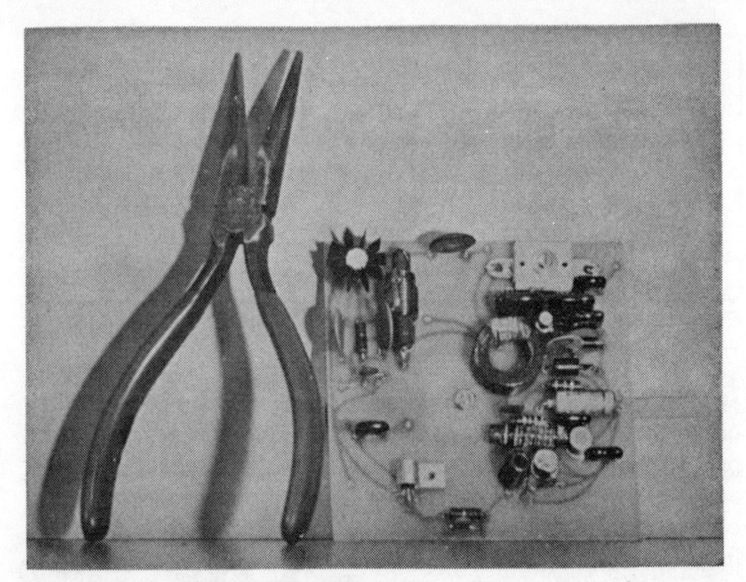

Fig. 6-31. Here is the Mini-Mite PC board.

approximately 3-¾ inches long. The dowels used on all the tuning setups are standard ¼ inch diameter units. They are epoxied to the tuning screws. When the coils are tuned to minimum inductance, the knobs stick out from the front panel a little way. You can build the unit, tune it, then cut off the dowels at the right approximate length to keep the knobs from sticking out too far.

Construction is not too critical. The Mini-Mite is built in a 6×4×2 inch chassis (LMB #642); you may add the bottom plate which is available for this chassis. The circuit board is a very tight fit in the box. In order to get the circuit board in, it will have to be slid into place before the tuning capacitor is mounted. Then push the circuit board back all the way against the back wall of the box. This should leave enough room to insert the tuning capacitor. Hold the capacitor in place by means of its mounting holes and screws, then slide the PC board forward into its mounting place. The PC board is held away from the chassis on ¼ inch spacers. Mount the spacers in place, tighten nuts over the mounting screws, and then the PC board is slipped into position and more nuts are placed over the mounting screws. This allows you to remove the PC board without the hassle of always fumbling to find a way to hold the spacers in place while you stick the mounting screws through them and the PC board. Also, see Table 6-5.

Before you can begin to operate the Mini-Mite, you must

provide for the type of keying scheme which will best serve you. For instance, you may wish to key the oscillator for break-in. Figure 6-25 suggests some keying schemes. The PC board is set up for keying the driver and final simultaneously. To change from this requires some revision of the printed tracks. Keying the driver and final is convenient and easiest, but has one major disadvantage. Dirty key contacts will increase the chance that some voltage will be dropped across the closed contacts. At the higher current levels encountered at this point, you can drop, say, a volt or so, and lose effective input power to the final and driver, and degrade the output signal. Keep the key contacts reasonably clean and you should have no particular problems. Tuning up the Mini-Mite is pretty much like tuning any other rig, except that you must also tune a solid state rig for cleanest output signal. You will have to monitor on a receiver as you tune up. You can use a #47 dummy load, but take care not to leave the key down for periods over a minute or so, as the overvoltage to the bulb will burn it out. Tuning to an antenna can be accomplished with a field strength meter, in-line wattmeter, swr bridge on forward power, or a tuning meter as shown in Fig. 6-32. Tune for maximum all the way around and you should get the cleanest signal. Also, see Figs. 6-33 through 6-37.

Author's Note:

Add this modification (Fig. 6-33) to the Mini-Mite schematic. All it amounts to is adding a 100 Ohm resistor and a 1k resistor in the Buffer Amplifier. The 47k bias resistor is dropped so the stage runs class C.

The best way to run the rig as a 1 watt output QRPer for anyone not interested in going the full 7 watt route is shown in Fig. 8. Leave out the driver transistor and run a 100 pF between the output winding from the amp stage and the collector connection on the driver, then a 100 Ohm across the PA base to ground.

Table 6-5. Parts List.

Q1 2N709 or one of four Radio Shack 276-608
Q2 2N697 or one of above package
Q3 2N697 or one of above package
Q4 GE63 or one of same package
Q5 Motorola HEP53001
Coils (see text) Radio Shack 270-376
Tuning capacitor 365 pF Radio Shack 272-1344 or equiv.
Cabinet LMB 642 (6x4x2 aluminum)

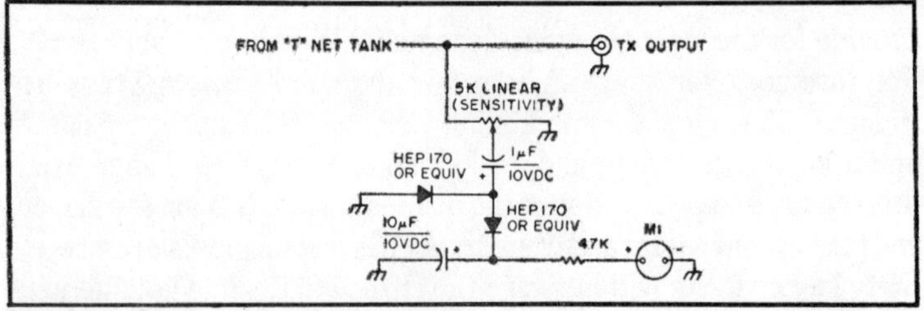

Fig. 6-32. Tuning meter. M1 is a tuning indicator or battery level indicator available from many mail order houses or the XYL's cassette player.

TWO-METER SYNTHESIZER

A low-cost, low-power, compact two meter synthesizer can now be built due to recent developments in COSMOS technology. The synthesizer covers 144-148 MHz and can be adapted to almost any rig simply by programming the output divider and using the proper crystals in the beat oscillator. Any repeater offset can be generated, as the receive and transmit frequencies are independently set in 5-kHz steps. The unit uses a total of six integrated circuits and draws about 60 mA at 12 volts. My unit is interfaced with a Heathkit® HW-202.

See Fig. 6-38. Q1, along with its associated components, forms a vco which has an output in the 22-25-MHz region depending upon what the voltage on VC1 is and if Q9 is turned on or off. Q2 is a buffer amplifier which is connected to one gate of mixer Q5. The other gate of Q5 is connected to Q7, which is the beat oscillator that also oscillates in the 22-25-MHz range depending on which crystal is switched into its base. The output of Q5 contains the sum and difference frequencies of the two signals present at its gates. RFC4 and C21 form a filter which allows the difference frequency to pass on to Q6 which shapes and level-shifts the signal so that it is CMOS compatible.

The first gate of Z4 acts as a buffer to drive Z5, which is a divide-by-N divider. The divider is connected to divide by 800 plus

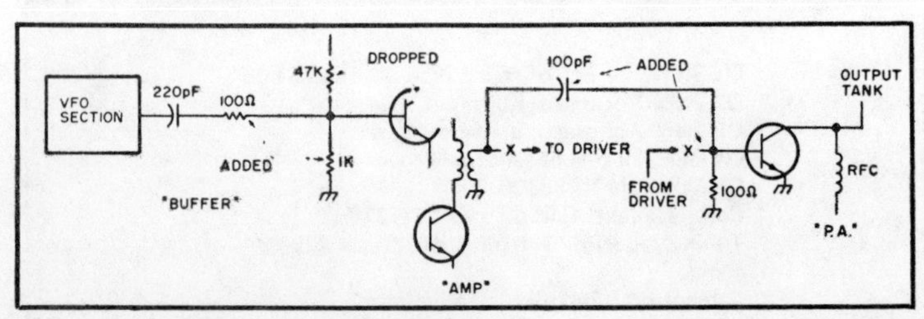

Fig. 6-33. Modifications to the Mini-Mite.

Fig. 6-34. The original bread board view of the Mini-Mite (note the final is missing in this photo) just goes to show that layout is not critical, as this unit worked quite well. Also this prototype used the optional buffer—the two plastic-type just behind the vfo.

twice the switch settings, and then plus one if the 5-kHz switch is on. These switches are labeled as to what decade of the frequency they determine. The output of the divider goes to the input of Z3, which is a phase comparator. The other input of Z3 goes to Z6, an oscillator/binary divider, whose output is 833.333 Hz. This is the reference frequency. The output of Z3 is connected to a low-pass filter whose output goes to varactor VC1.

Let us now trace a complete cycle of the loop (see Fig. 6-39). Suppose we want to transmit on 146.940 MHz, and we set the

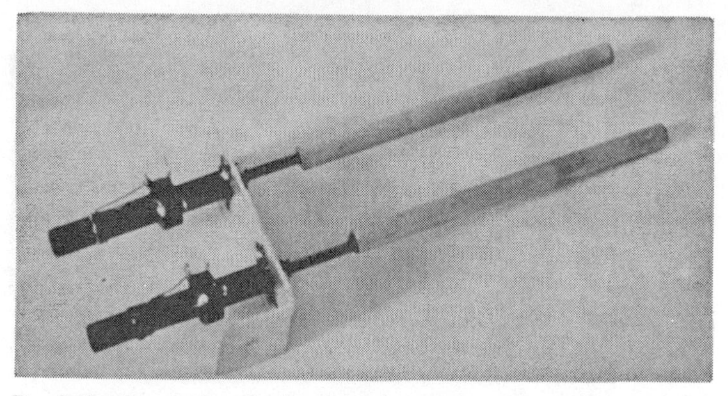

Fig. 6-35. Here is the final tune-load assembly. The angle bracket is mounted to the inside of the cabinet and the wooden dowels serve as tuning shafts to the front panel.

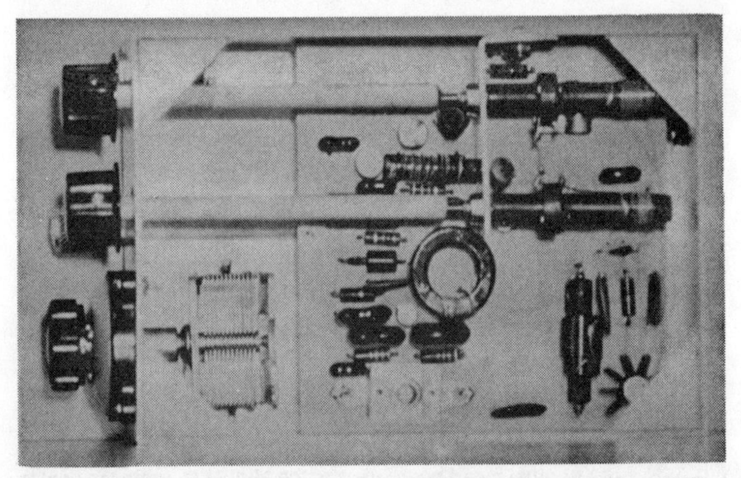

Fig. 6-36. Here is the final version of the Mini-Mite component layout. The output transistor has a finned heat sink. The vfo is in the area shown with the toroidal coil although other coils may be used as discussed in the text.

switches as such. Our divider divides by 800 + 2(294) = 1388. Suppose that the vco is free-running at 24.00 MHz. This mixes with the 23.3333-MHz transmit crystal to give an output of .6667 MHz. This is divided by 1388 to give 480.3 Hz. This is compared to the 833.333-Hz reference, and the 4046 raises the voltage to VC1 to increase the frequency of the vco. When the vco has an output of 146.940 MHz/6 which is 24.490 MHz, the loop will lock since (24.490 − 23.333)/1388 equals 833.333 Hz.

The 4046 will adjust its output voltage so that the two inputs

Fig. 6-37. The Mini-Mite, a "kilowatt" among QRP rigs, runs 7 watts peak power for cutting through the QRM and clearing the frequency for the little QRP rigs.

are identical in phase and frequency. Q3, Q4, LED 1, and associated components form an indicator that lights when the synthesizer is unlocked. This indication is useful when initially tuning the synthesizer, and warns the operator not to transmit if the loop becomes unlocked due to component failure, etc.

The second and third gates of Z4 generate two signals: T and $\overline{T}$ (pronounced not T). T is high in transmit and low in receiver, and $\overline{T}$ is its complement. These signals switch between Y1 and Y2, select which set of frequency switches is connected to the 4059, and turn Q9 on and off, which places C6 in parallel with the vco tank to lower its frequency range in the receive mode. Q8 is a buffer stage which isolates the vco from the output circuitry.

Before the signal from Q8 goes to the receiver, it is passed through the lowpass filter composed of C14, C15, C16, and RFC2. This passes the 24-MHz rf, but keeps the VHF rf from the transceiver from getting into the synthesizer. Q8 is also connected to Z2, which is a quad flip-flop. By connecting the pins of Z2. The chip can divide the 24-MHz signal by 2, 3, or 4, giving a 12-, 8-, or 6-MHz output. The transmitter's signal also goes through a lowpass filter.

Z1 is a five-volt regulator which supplies power to most of the circuit. Some parts of the circuit require 12 volts, and this is obtained at C46. In the HW-202, I take the supply voltage off the 11-volt regulated line within the radio.

Parts layout is fairly critical, and it is recommended that the PC board layout shown in Figs. 6-40 and 6-41 be used. Keep all leads as short as possible and mount Y3 and Z1 flush to the board. The use of IC sockets is encouraged. Resistors R28-37 and diodes D4-15, D17, and D18 are not mounted on the board but directly on the switches concerned. C50 and C51 are .001-μF feedthroughs mounted directly to the metal cabinet enclosing the synthesizer. RFC6 is not mounted on the board but is connected directly to feedthrough capacitor C50. RCA-type jacks are used for the receiver and transmitter output connectors. RG-174/U, 50-ohm miniature coax is used to connect the receiver and transmitter output from the boards to their respective low-pass filters and jacks. The lowpass filters are assembled around the jacks. There are several jumpers that are connected to the bottom of the board (Table 6-6). They are noted on the parts placement diagram as J1, J2, etc. For example, a jumper must be connected from one point labeled J1 to another point labeled J1. Some jumpers go to more than one place. For example, there is a J4a, b, and c. This means

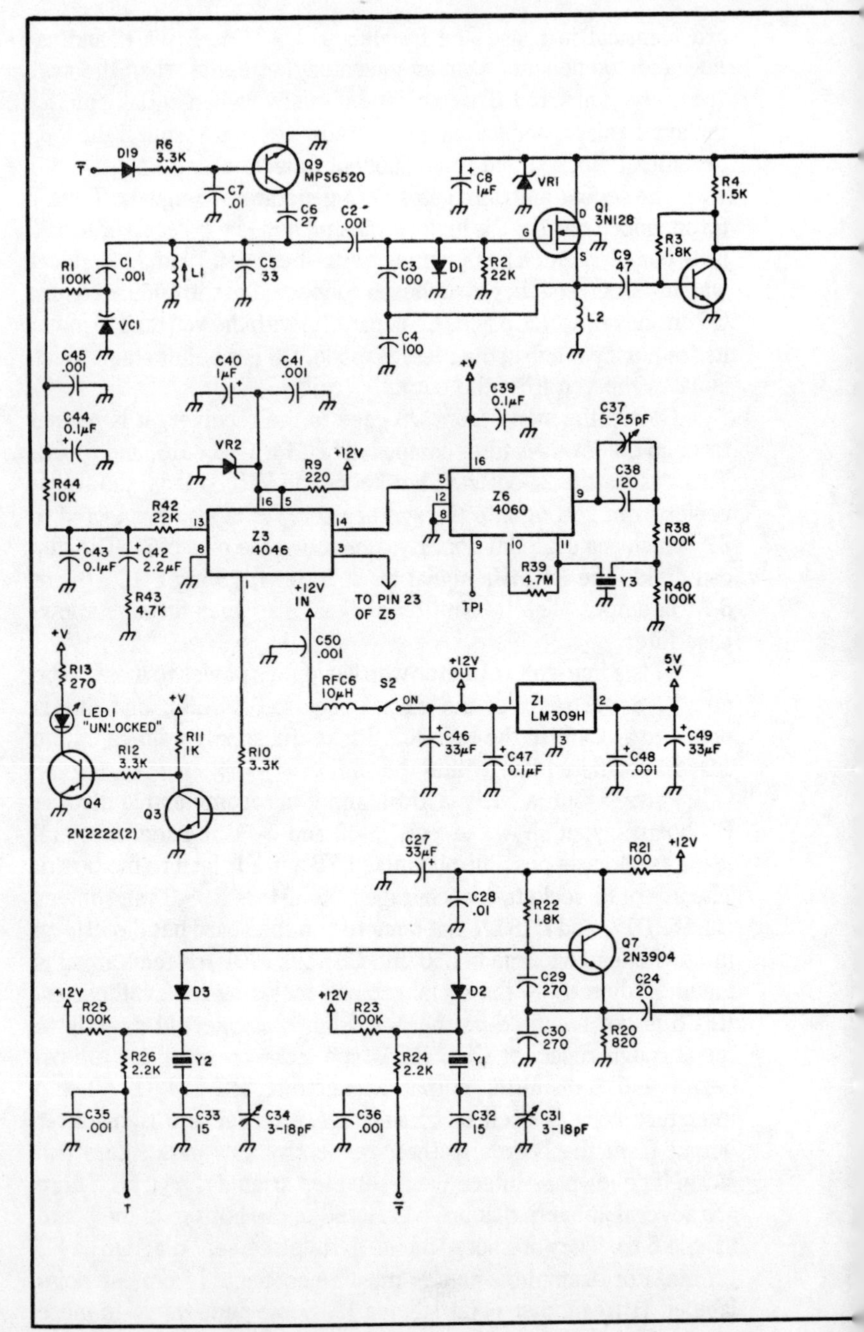

Fig. 6-38. 2m synthesizer schematic diagram.

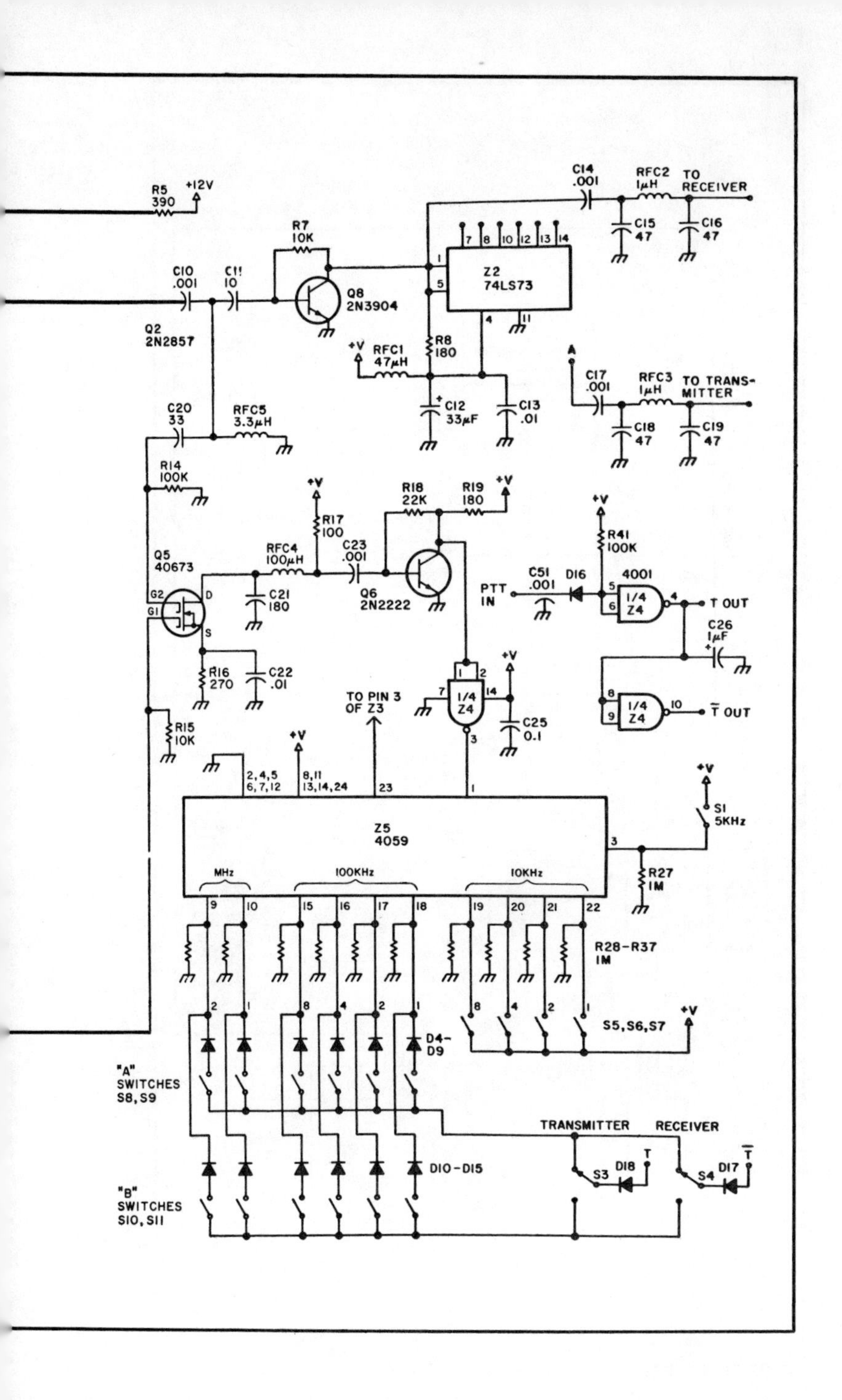

R5
390
+12V
C14
.001
RFC2
1μH
TO
RECEIVER
R7
10K
Z2
74LS73
C15
47
C16
47
C10
.001
C11
10
Q8
2N3904
Q2
2N2857
+V
RFC1
47μH
R8
180
A
C17
.001
RFC3
1μH
TO TRANS-
MITTER
C20
33
RFC5
3.3μH
C12
33μF
C13
.01
C18
47
C19
47
R14
100K
+V
R18
22K
R19
180
+V
R17
100
R41
100K
Q5
40673
RFC4
100μH
C23
.001
Q6
2N2222
C51
.001
D16
4001
PTT
IN
1/4
Z4
T OUT
C21
180
C26
1μF
R16
270
C22
.01
TO PIN 3
OF Z3
1/4
Z4
C25
0.1
1/4
Z4
T̄ OUT
R15
10K
+V
2,4,5
6,7,12
8,11
13,14,24
+V
S1
5KHz
Z5
4059
MHz
100KHz
10KHz
R27
1M
R28-R37
1M
D4-
D9
+V
S5,S6,S7
"A"
SWITCHES
S8,S9
TRANSMITTER
RECEIVER
D10-D15
S3
D18
T
S4
D17
T̄
"B"
SWITCHES
S10,S11

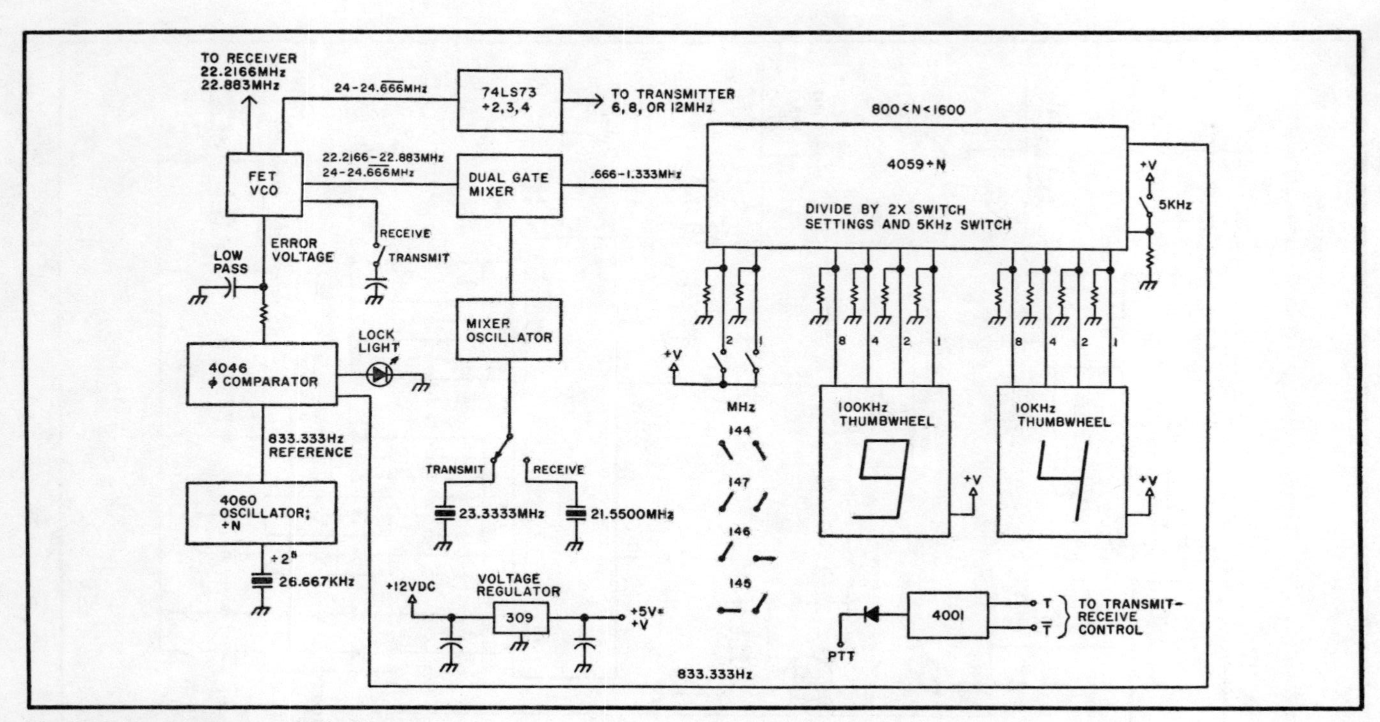

Fig. 6-39. 2m synthesizer block diagram.

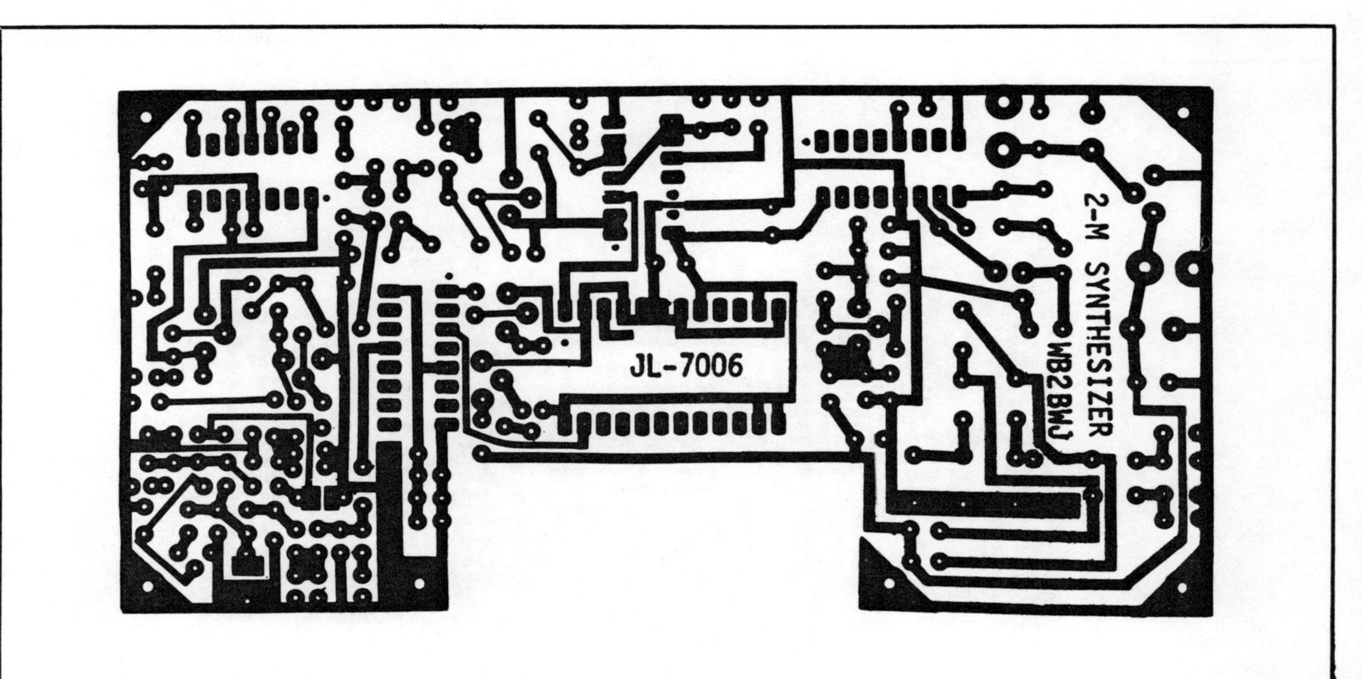

Fig. 6-40. PC board layout.

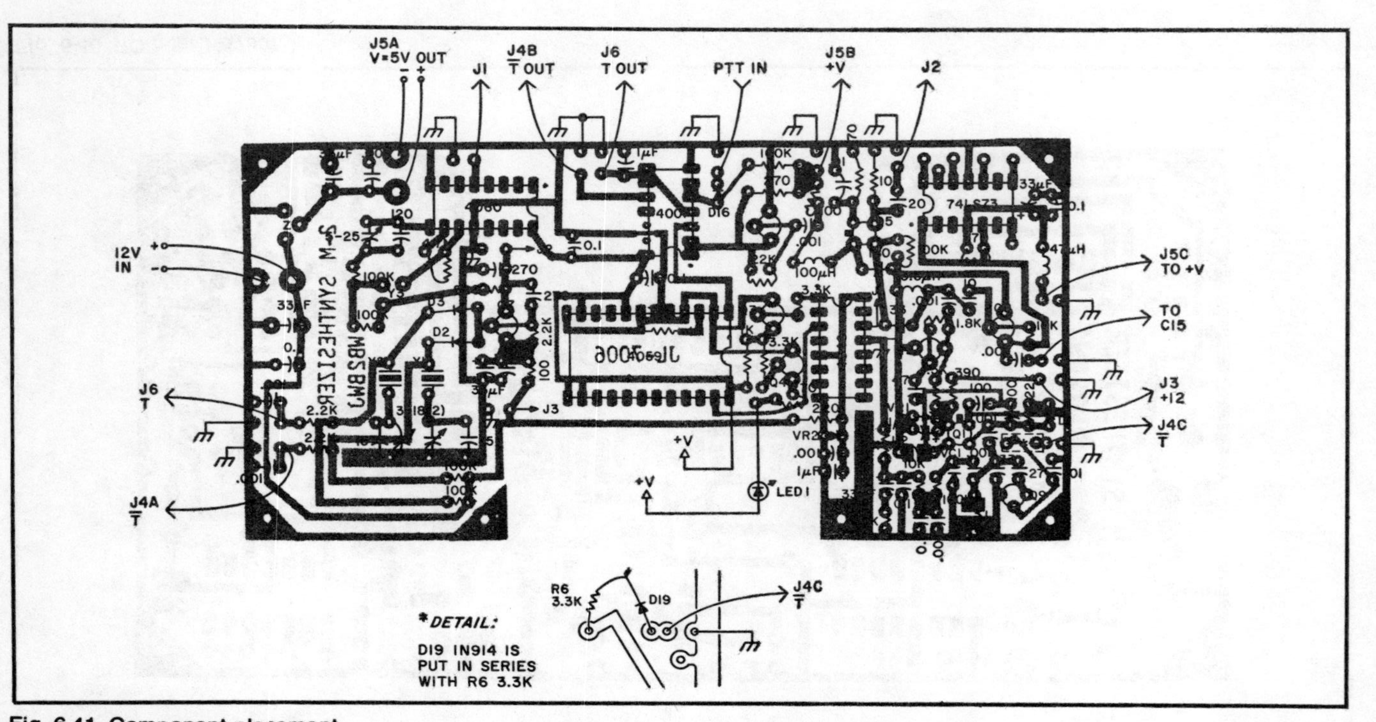

Fig. 6-41. Component placement.

Table 6-6. List of Required Jumpered Connections. Use RG-174/U Coax.

1. Cable J1 from pin 5 of Z6 to pin 14 of Z3; this is the divide-by-N out.
2. Cable J2 from C24 to R20; this is the beat oscillator-to-mixer line.
3. A Cable from J4a to J4b and one from J4b to J4c; this is the $\overline{T}$ line.
4. Cable J3 from R5 to +12 volts at R21; this is the +12-volt line.
5. A cable from J5a to J5b and J5b to J5c; these are +5-volt connections.
6. Cable J6; this is the T line.

that J4a goes to J4b and J4b goes to J4c. All jumpers are RG-174/U coax, and provision is made at each point for the shield to be soldered to ground. Table 6-7 shows the cable lengths.

The switching arrangement I used (Fig. 6-42) was designed to keep the number of thumbwheel switches to a minimum. The arrangement consists of two sets of switches, one labeled A and the other B. S8 and S9 are SPST switches that comprise the MHz selection for the A set. Placing both switches down (turning them off) sets the MHz to 144, S9 up (S8 down) is 145, S8 up (S9 down) is 146, and both up is 147. S10 and S11 work in a similar fashion for the B set. S5 is a thumbwheel switch that selects the 100-kHz step for A; S6 does this for B. S7 is another thumbwheel that sets the 10-kHz step for both A and B. S1 selects whether or not 5-kHz step is used. S3 and S4 select whether or not the A or B setting will be used for transmit or receive.

As an example, if we want to go on 146.34/146.94 with A selecting the transmit frequency and B selecting the receive frequency, we set S8 and S10 to on, S5 to 3, S6 to 9, S7 to 4, S3 to A, and S4 to B. To go on simplex on 146.34, say, to monitor the input, we set S4 to A. To go simplex on 146.94, we set S3 and S4 to B. To go on reverse 146.94/146.34, we set S3 to B and S4 to A. With this

Table 6-7. Cable Lengths (RG-174/U).

Cable	Length (inches)
J1	3½
J2	3¼
J3	4¼
J4ab	3½
J4bc	4½
J5ab	3½
J5bc	2½
J6	3½

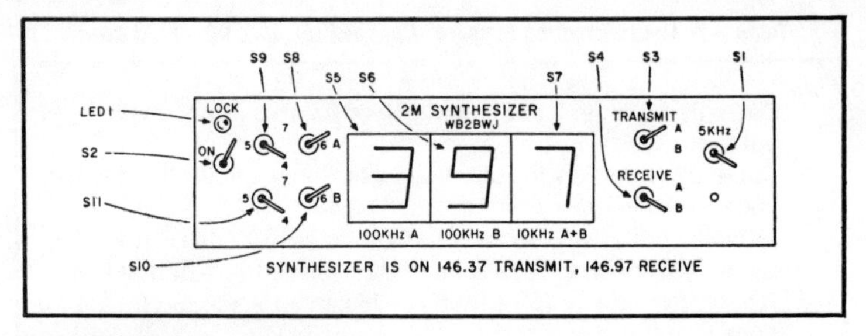

Fig. 6-42. Switch arrangement.

method of switching, most common repeater pairings can be obtained. If a more sophisticated system is desired, an automatic offset could be built in, or a keyboard-type entry system could be used. The important thing to note is that the synthesizer only requires the BCD code of the desired frequency—no look-up table is needed.

Mount all of the parts on the board in the following order: sockets, resistors, capacitors, chokes, transistors and diodes, crystals, and jumpers. Before inserting the IC chips, apply 12 volts to the unit and check for the proper supply voltages at the IC sockets. After turning the supply off, insert all of the IC chips. Turn the unit on again and the unlocked light should come on. The first signal to check on the unit is pin 14 of Z3. There should be a 5-volt peak-to-peak square wave at a frequency of 833.333 Hz. With an accurate frequency counter, preferably set to measure the period, adjust C37 until the frequency, or period, is as stated. This adjustment could also be made by looking at pin 9 of Z6 and setting C37 for a frequency here of 26.6666 kHz. Connect the positive lead of a vtvm or FET vom to the lead of R1 farthest from VC1. The voltage here probably will be either near zero volts or near 10 volts, either of which represents an unlocked condition. With the synthesizer in the receive mode, set the frequency select switches to 147.995 MHz and adjust the tuning slug on L1 until the voltage reads approximately four volts. The unlocked light should now be extinguished. Change the frequency select switches to 144.000 MHz and check to see if the light is still extinguished. Simulate the transmit mode by grounding the PTT line on the synthesizer and check to see that the synthesizer locks over the same frequency range in transmit. The voltages at R1 should be within 0.5 volts of one another for the same frequency on transmit and receive.

Trim L1 until the tuning range is correct. Any constant flickering of the unlocked light indicates an unstable condition, and the

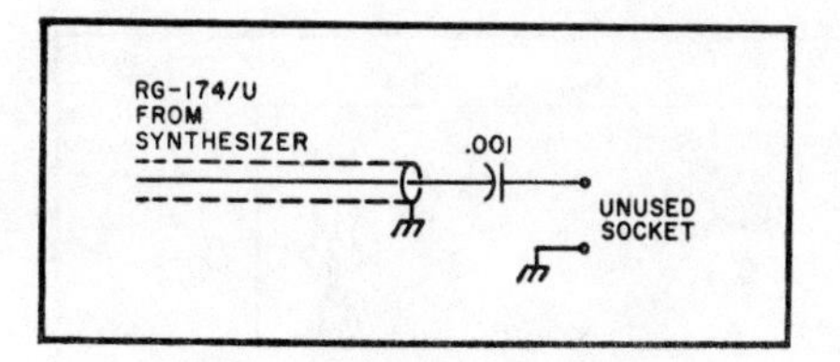

Fig. 6-43. Receiver output.

transceiver should be keyed only if the light is fully extinguished. A brief flash of the light when setting the frequency select switches, or when keying the transceiver, is just an indication that the synthesizer has become unlocked momentarily while changing frequency.

With the PTT line open (receive mode), connect a frequency counter to the receiver output jack and set the frequency select switches to 146.000 MHz. Adjust C34 until the counter reads 22.5500 MHz. Ground the PTT line (transmit mode) and adjust C31 until the counter reads 24.3333 MHz. It should be noted that the reading of 22.5500 MHz in the receive mode assumes that Y2 is 21.5500 MHz, which is the proper crystal for a 10.7-MHz i-f. For any other i-f, the counter should read (24.3333 – i-f/6). This completes the calibration of the synthesizer.

The unit should be built in a metal box with a cover that makes good electrical contact all around its perimeter. This prevents rf from getting into the synthesizer from the transceiver. Connecting leads to the transceiver should be made with RG-174/U. The first step in interfacing the synthesizer to the transceiver is to select Y2. Its value depends on the i-f frequency of your transceiver. For a radio with a 10.7-MHz i-f and receive crystals in the 45-MHz region, Y2 will be (23.3333 – 10.7/6) = 21.5500 MHz. For other i-fs, (23.3333 – i-f/6) will give the value for Y2. Even though the transceiver takes 45-MHz crystals and 22-23 MHz comes from the synthesizer, the receiver's oscillator and multiplication circuits do the proper multiplication. Receivers using 15- or 22-MHz crystals will also work with this scheme.

Connect the receiver output coax to the transceiver at an unused receiver crystal socket. Use a .001-μF capacitor to couple into the socket (Fig. 6-43).

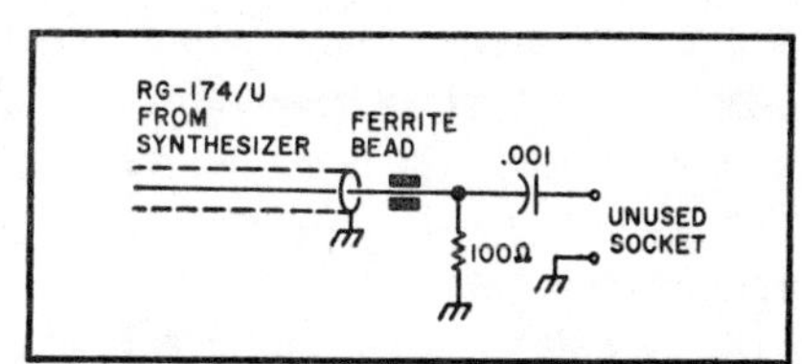

Fig. 6-44. Transmitter output.

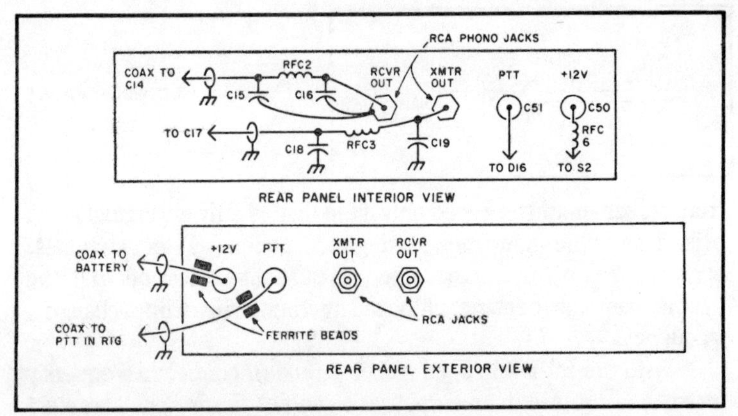

Fig. 6-45. Rear panel connections.

The transmitter output coax from the synthezier also goes to an unused crystal socket, but a 100-ohm resistor and a .001-μF capacitor are connected as shown in Fig. 6-44. The resistor assures smooth operation of Z2. A ferrite bead is placed as shown to act as a choke which keeps the VHF rf from entering the synthesizer.

The PTT input line should go to a line in the transceiver that is open or has at least +5 volts on it during receive and is grounded during transmit. This line will most probably be the PTT line from the microphone to the relay. The +12-volt input line can go to the same place from which the transceiver gets power, but if it is at all possible, connect it to some source of regulated and hash-filtered power within the rig. A ferrite bead should be placed at the ends of the coax on both the PTT and +12-volt connections. See Fig. 6-45.

The most difficult part of interfacing the synthesizer to a particular transceiver is keeping stray rf from the transceiver from getting into the synthesizer. This problem will be noticed when your audio is reported as sounding bassy or distorted. A very severe case of rf leakage will cause the unlocked light to glow on transmit. A less severe case will cause the aforementioned bassy audio.

If you have this problem, listen to yourself on a nearby receiver. Disconnect the receiver's coax at the synthesizer while transmitting, to see if the audio clears up. If it does, then this is the path of leakage. Try more ferrite beads or an additional low-pass filter in series with the other one. To check if the rf is leaking through the power supply line, temporarily run the synthesizer off a 12-volt battery and see if the audio clears up. If it does, try more ferrite beads of a larger value for RFC6. To check if the rf is leaking through the PTT line, disconnect it from the synthesizer and simply short the PTT input to ground. If the audio clears up, try more ferrite beads or another bypass capacitor. If none of these remedies seems to cure the problem, then the transmit coax is probably the path of coupling, and additional ferrite beads or another low-pass filter probably will fix it.

Figure 6-46 is a spectrum analyzer photograph of the 6-MHz output of the synthesizer being sent to my transmitter. The analyzer is set for a 30-Hz resolution, 500 Hz per division, and a 10-Hz video filter in place. The vertical is calibrated at 10 dB per division. The signal is very clean; there is no sign of the 833.333-Hz reference sidebands, and the noise is 60 dB down. Figure 6-47 is the output of my transceiver with the synthesizer set at 146.000 MHz. The analyzer is set for 3-kHz resolution, 100 kHz per division, and 10 dB per division vertically. There are no close-in spurs within 60 dB. Figure 6-48 is also the output of my transceiver, but the resolution is 30 kHz, and every division now represents 5 MHz. The strongest spur is at about 154 MHz, but it is 60 dB down, complying fully with the latest FCC regulations for spurious output.

Table 6-8 gives the parts list and also see Figs. 6-49 and 6-50.

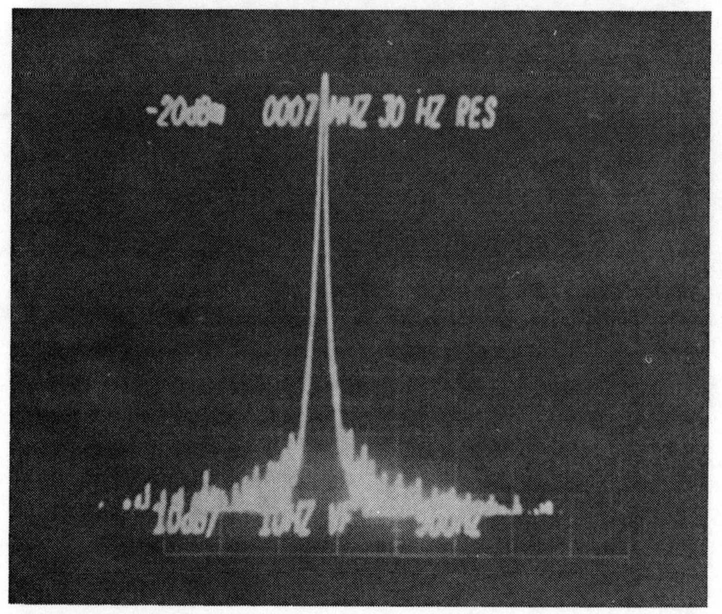

Fig. 6-46. The 6 MHz output.

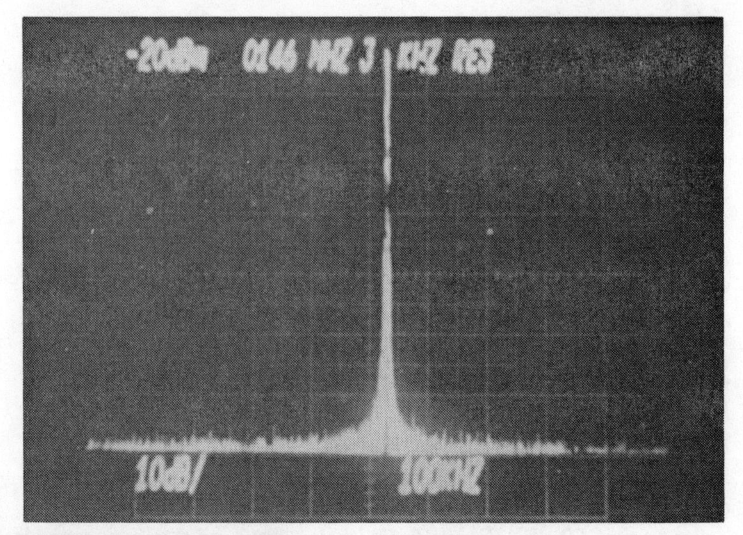

Fig. 6-47. Transceiver output, 3 kHz resolution.

A $10 PHONE PATCH

You don't *have* to spend $40 to $90 for a phone patch to match your new solid-state rig. This simple circuit will give perfectly adequate performance with parts purchased at your local Radio Shack. For a cost of under $10 and one evening's work, you can have a patch that will work with a new solid-state rig or an older

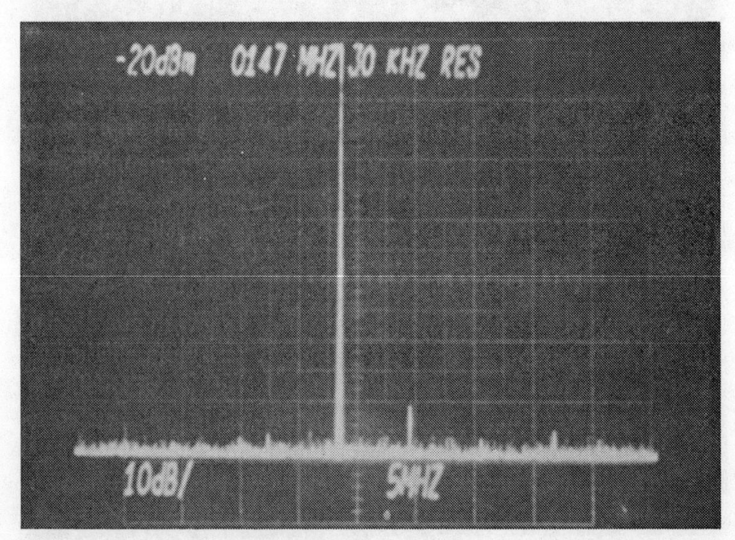

Fig. 6-48. Transceiver output, 30 kHz resolution.

tube-type transceiver. This circuit (or variations of it) has been in use by several hams in this area for a number of years (see Fig. 6-51). With the bypass capacitors shown, no effect from rf feedback has been experienced, even when used with a kilowatt.

S1 is a four-pole, two-position switch (Radio Shack 275-1384 or equivalent) used to switch the patch on and off. The fourth section of this switch (S1d) is used to switch the patch into the circuit in lace of the usual station microphone. If your rig has a phone patch input, you can leave the mike connected to its usual jack and use the alternate connection shown by the dotted line.

S2 is a DPDT toggle or lever switch. It allows you to select XMIT (transmit) or RCV (receive) from the front panel of the patch, rather than fumbling with the microphone PTT switch, telephone handset, and receiver audio gain control all at once.

S2 is wired in such a way as to permit operation with many types of solid-state rigs. These rigs often have the novel little problem that the receiver audio is not cut completely off while in the transmit mode. During normal operation this is not a problem. However, with many of the usual phone patches, an audio oscilla-

Fig. 6-49. Top view of the synthesizer.

Table 6-8. Synthesizer Parts List.

Resistors—all ¼-Watt, five percent	
R1, R14, R23, R25, R38, R40, R41	100k
R2, R18, R42	22k
R22, R3	1.8k
R4	1.5k
R7, R15, R44	10k
R13, R16,	270 Ohms
R43	4.7k
R6, R10, R12	3.3k
R17, R21, and one in radio	100 Ohms
R20	820 Ohms
R24, R26	2.2k
R27-R37	1 megohm
R39	4.7 megohms
R9	220 Ohms
R5	390 Ohms
R8, R19	180 Ohms
R11	1k
Capacitors – all disc ceramic, unless otherwise noted	
C1, C2, C10, C14, C17, C23, C35, C36, C41, C45, C48 and two in radio	0.001 uF
C50, C51	0.001-uF feedthroughs
C3, C4	100-pF silver mica
C9, C15, C16, C18, C19	47 pF
C12, C27, C46, C49	33-uF tantalum
C7, C22, C28	0.01 uF
C42	2.2-uF tantalum
C43, C44	0.1-uF tantalum
C13, C25, C39, C47	0.1 uF
C24	20 pF
C29, C30	270-pF silver mica
C32, C33	15-pF silver mica
C8, C26, C40	1-uF tantalum
C38	120-pF silver mica
C5	33-pF silver mica
C20	33 pF
C21	180 pF
C6	27-pF silver mica
C11	10 pF
C31, C34, C37	5-30-pF subminiature trimmers

RF chokes and coils	
RFC1	47 μH
RFC2, RFC3	1 μH
RFC4	100 μH
RFC5	3.3 μH
RFC6	10 μH
L1	¼-inch slug-tuned form wound with 8 turns of #22 wire
L2	20 turns #30 wire on Amidon #73-801 ferrite bead
Semiconductors	
Z1	LM309H
Z2	74LS73
Z3	4046
Z4	4001
Z5	4059
Z6	4060
Q1	3N128
Q3, Q4, Q6	2N2222
Q5	40673 or HEP F2004
Q7, Q8	2N3904
Q9	MPS6520 or HEP S0009
Q2	2N2857
VC1	HEP R2503 varactor
D1-D19	1N914 diodes
LED1	any type red LED
VR1, VR2	5.1-volt ½-Watt zener diode
Crystals	
Y1	23.3333-MHz, Heath #404-586*
Y2	21.5500-MHz, Heath #404-584*
Y3	26.667-kHz, Statek type SX-1H
Switches	
S5, S6, S7	10-position BCD switches with endplates
S3, S4	SPDT toggle switches
S1, S2, S8, S9, S10, S11	SPST toggle switches

*International crystal cat #435274

Miscellaneous

2 RCA phone plugs and jacks
RG-174/U miniature 50-Ohm coax
Amidon #64-101 ferrite beads

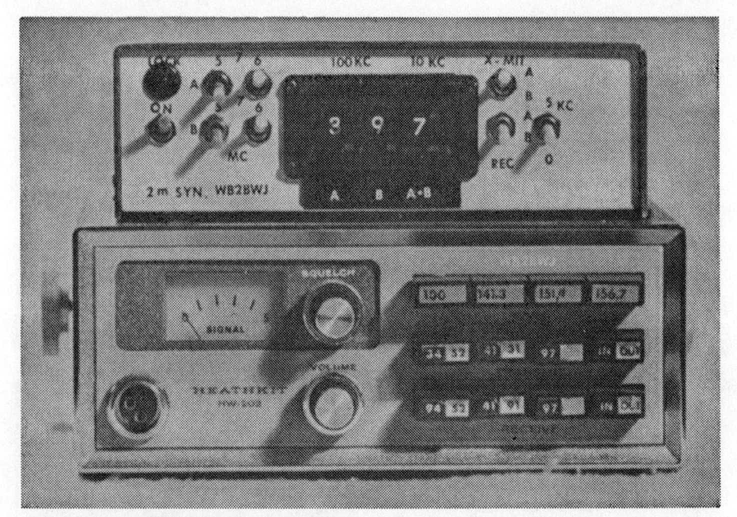
Fig. 6-50. The finished synthesizer.

tion will result, with the phone patch acting as the oscillator coupling element. Switch S2 disconnects the receiver audio from the patch when in the transmit mode, thus eliminating the problem.

The second section of this switch (S2b) grounds the PTT line when you wish to transmit. On a few of the new rigs, this line is called MOX rather than PTT. MOX stands for Manually Operated Xmit, similar to VOX for Voice Operated Xmit.

Impedance matching is provided by T1 and T2, which are identical 8-ohm-to-1000-ohm (center-tapped) audio transformers. Radio Shack lists this item as 275-1384. They are not critical, and any 8-ohm-to-1000-ohm or 8-ohm-to-500-ohm audio transformer will do. Good performance has been obtained even with a pair of 12-volt filament transformers. The 12-volt secondary is connected in place of the 8-ohm winding and the 115-volt primary in place of the 500- or 1000-ohm winding.

The transmit level is set by the 500-ohm pot connected to T2. Bypass capacitors are shown on all input and output leads to prevent rf feedback. A metal enclosure for the patch is recommended.

To use the patch, set the S1 to ON and S2 to RCV and listen on the telephone handset. You can get a clear line (no dial tone) by dialing the first digit of a local exchange. Tune in a station on your receiver and set the audio gain control on the receiver for a comfortable level in the telephone handset. Telephones are quite tolerant and level setting is not critical. If the audio sounds com-

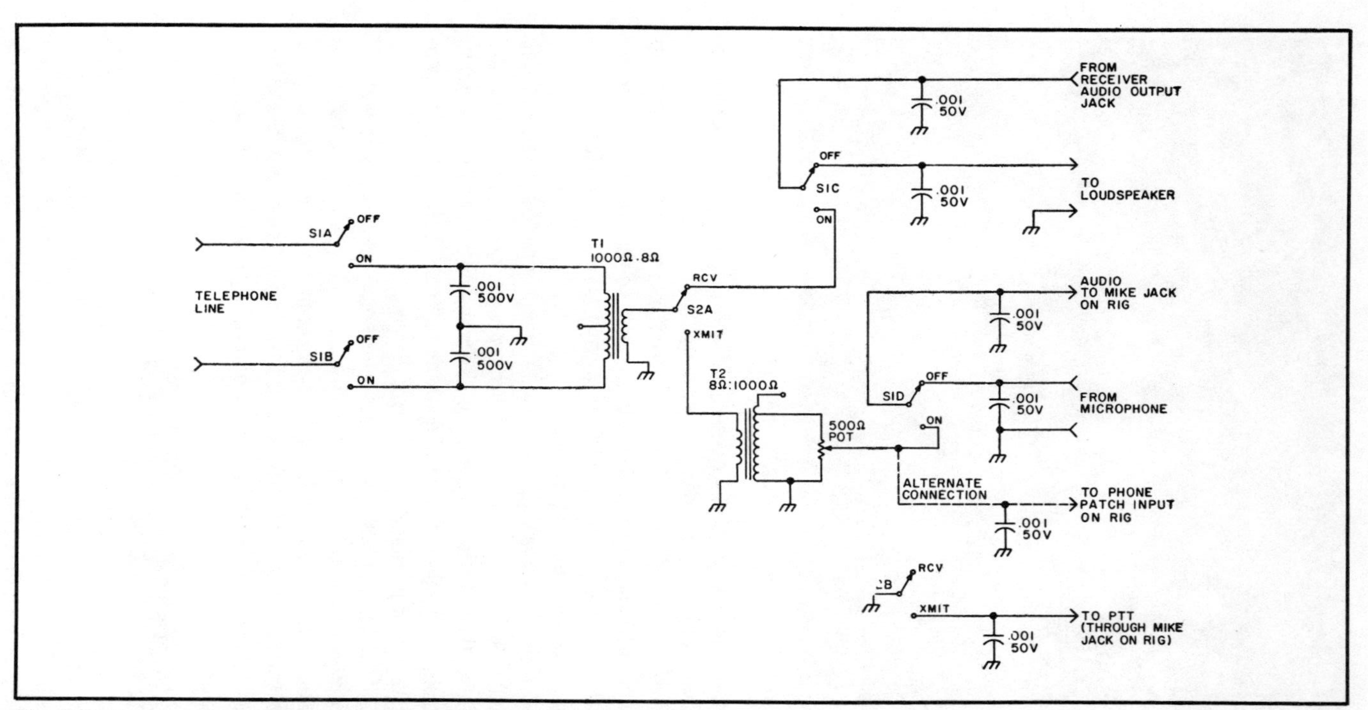

Fig. 6-51. Inexpensive phone patch uses readily available components.

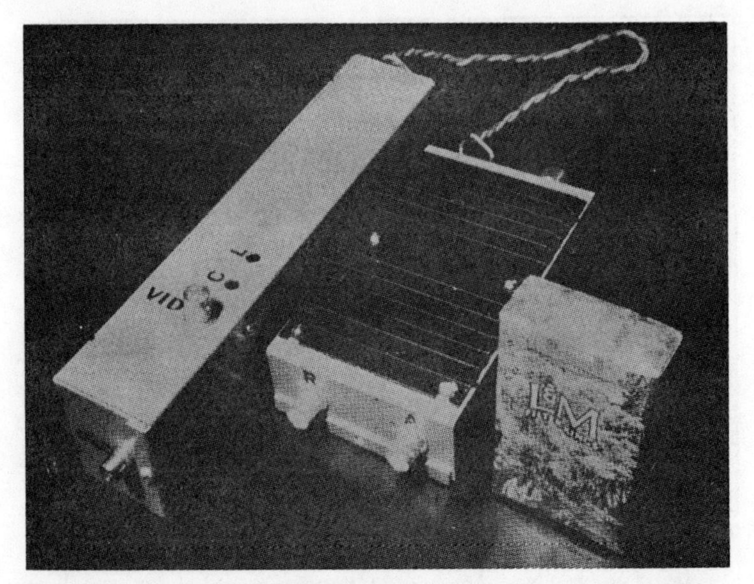

Fig. 6-52. Low cost 10 watt ATV transmitter (exciter left, amplifier right).

fortable in your ear on receive while listening through the telephone handset, it probably is acceptable.

Next, put S2 in the XMIT position and talk normally into the telephone handset. Set the 500-ohm pot so that the meter on the final in your rig swings into its normal area as though you were using the station microphone.

There is no provision for VOX operation. Most hams prefer manual RX/TX since it both prevents an operator's accidental sneeze or cough from turning on the transmitter and allows you to cut off the speaker if he or she attempts to say something inappropriate for transmission over your station.

One tip on phone patch use—for some reason, when you tell someone on the telephone to talk louder, they will do so for a few minutes and then lapse back to their original volume. However, if you turn the audio gain control down so that they hear the other station more softly, they will automatically speak up as though to compensate.

FAST SCAN ATV TRANSMITTER

Here's a compact 10 watt fast scan amateur television (ATV) transmitter with audio on the video carrier and T/R switching that can be built for about $120 (Fig. 6-52). The rig incorporates the video exciter described in the June 1976 issue of *73* to drive a quasilinear 10 watt ¾ meter amplifier. No amplifier tuneup is required since it utilizes the Motorola MHW-710 sealed power module. (For theory of operation of this module in the ATV mode, refer to Nov/Dec 1975, page 37 of *73*.)

Operating at 13.8 V dc, the transmitter draws about 2.7 Amps from an external regulated power source. Linearity and frequency response performance is shown in Fig. 6-53.

As noted above, the construction details for the exciter have already been given; therefore only the amplifier circuit will be described here. Several different mounting arrangements are possible, so you may wish to deviate from the following procedure. Of course, both the amplifier and exciter can be located in the same enclosure; however, experimentalists may prefer the two-box modular approach to effect rapid exciter or amplifier interchange with future designs.

Amplifier Construction Procedure

Refer to Table 6-9 and Fig. 6-54.

1. Drill holes in chassis and heat sink per Figs. 6-55 and 6-56. Make sure that holes in heat sink line up with holes in chassis.
2. Referring to Fig. 6-57, mount all components to PC board. (Foil layout for board is shown in Fig. 6-58).
3. Using two #4-40 screws, lockwashers and nuts, bolt PC board on two "L" brackets as shown in Figs. 6-59 and 6-60.
4. Spread heat sink compound over back of heat sink and Motorola MHW-710 module. Place module on inside of chassis and

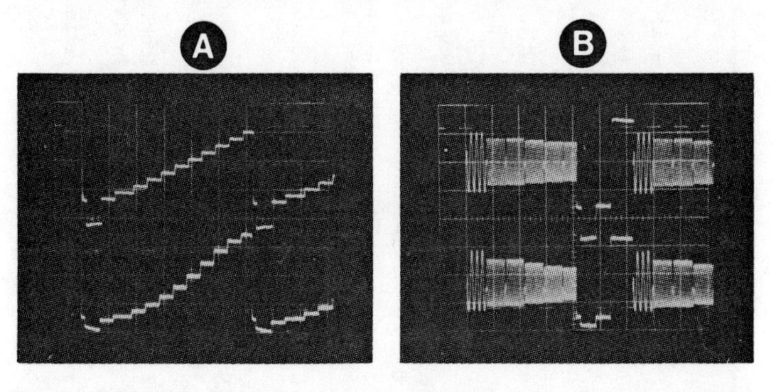

Fig. 6-53. Performance curves, 10 watt ATV transmitter. All vertical scales uncalibrated. Power: 13.8 V dc 2.7 A. (a) Linearity: top scale—video in; bottom scale—detected rf output; 10 usec/div horizontal; 10 watts out (average). (b) Frequency response: top scale—video in; bottom scale—detected rf output; 10 usec/div horizontal; burst order (in MHz)—0.5, 1.5, 2.0, 3.0, 3.58, 4.2.

Table 6-9. ATV 10 Watt Amplifier Parts List.

Part #	Description	Qty	Unit Cost	Total Cost	Source of Supply
1	5½"x3"x1½" Chassis; LMB #139	1		$2.00	Electronic Supply Store
2	UG-1094 BNC Bulkhead Connector	3	$ .85	2.55	Electronic Supply Store
3	.001 uF feedthrough cap; Erie #327-005-X5UO-102M	1		1.62	Electronic Supply Store
4	3/8"x3/4" "L" bracket; Calectro #J4-641 (2 brackets in package)	2		.49	Electronic Supply Store
5	Heat sink, 3"x4.75"x0.46" International Rectifier HE330-C or Wakefield 623-K	1		2.72	Electronic Supply Store
6	Heat sink compound; Archer 276-1372	1		.89	Radio Shack
7	½" #8 screws, nuts and lockwashers (to mount heat sink to chassis; also for gnd lug)	5			Hardware Store
8	½" #6 screws, nuts and lockwashers (to mount MHW-710 with "L" brackets to chassis)	2			Hardware Store
9	#8 hole terminal lug; Waldom #KT-198	3			Hardware Store
10	#8 nut (to secure ground lug soldered to relay)	1			Hardware Store
11	¼" #4 screws, nuts and lockwashers (to attach PC board to "L" bracket)	2			Hardware Store
12	RG-188 cable	18"		3.00	Cable & PC Brd both from Stu Mitchell WAØDYJ, 14761 Dodson, Woodbridge VA 22193
13	Amplifier PC board; cut, etched and drilled	1		Ppd	

14	MHW-710-1 or -2 Power Amplifier Module, Motorola. The 710-1 covers 400-440 MHz; the 710-2 covers 440-480 MHz. Either device will give equivalent performance in the 435-450 portion of the band.	1		42.50	Call local Motorola sales office for source
15	#20 stranded wire, insulated	20"			Electronic Supply Store
16	Stick-on lettering kit				Stationery Store
17	DPDT relay, 12 V dc; Archer #275-206	1		3.99	Radio Shack
18	Copper foil, Circuit-stick #9252	1		1.49	Electronic Supply Store
L1	2½ turn ferrite choke; Ferroxcube VK200-20/4B		.51	1.02	Eastern Components 1407 Bethlehem Pk, Flourtown PA 19031 $10 min. order
C1	500 uF, 35 V dc, Axial #272-1018	1		.89	Radio Shack
C2	.05 disc, #272-134	1		.39	Radio Shack
C3	33 uF, 35 V dc, PC Type, Lead aluminum	1		.30	Lafayette; Elec. Supply
C4-6	4.7 pF (or 5 pF), #272-120	3	.29	.87	Radio Shack
R1	270 Ohms, ¼ Watt, 10%	1		.10	Electronic Supply Store
R2	10 Ohms, ½ Watt, 10%	1		.12	Electronic Supply Store
Z1	15 V zener, 1N4744	1		.40	Electronic Supply Store

A 13.8 V dc power supply with a rating of 4 Amps continuous is required.

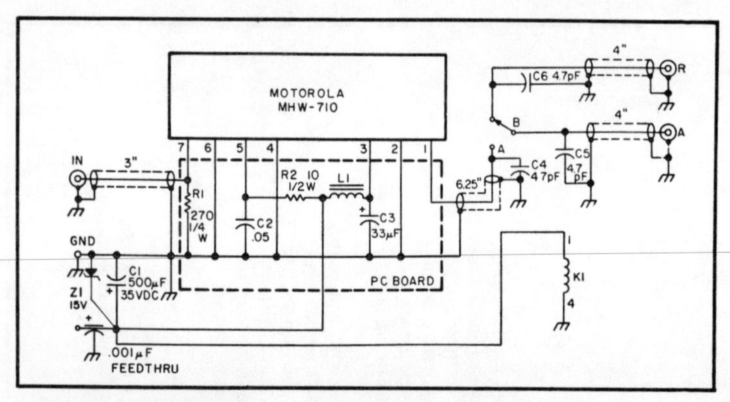

Fig. 6-54. 10 watt amplifier schematic. K1 is Archer (Radio Shack) #275-206 relay. L1 is Ferroxcube. VK200-20/4B.

heat sink on outside. Place PC board mounting brackets on module. Position brackets, module and heat sink so that all holes line up. Bolt all together with two #6-32 screws, lockwashers and nuts.

5. Using four #8 screws, lockwashers and nuts, bolt the corners of the heat sink to the chassis.

6. Solder all seven pins of the module to the PC board (pin numbers shown in Fig. 6-58).

7. Mount 3 BNC connectors, feedthrough capacitor and ground lugs to chassis. As shown in Fig. 6-60, also secure a #8 terminal lug to one of the screws holding the heat sink to the chassis. This is the relay ground lug.

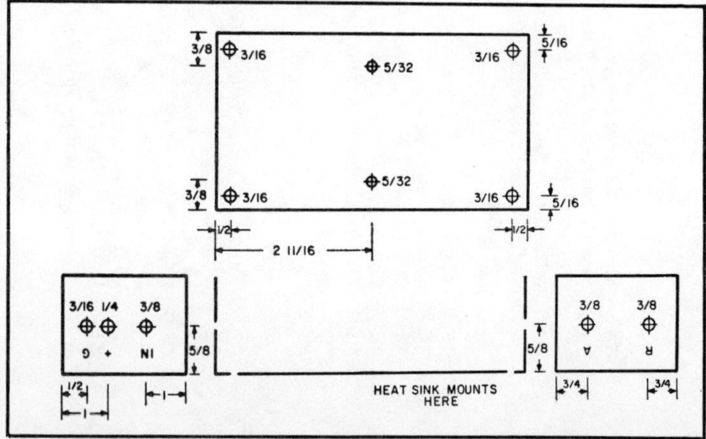

Fig. 6-55. ATV 10 watt amplifier chassis drill guide. Notes: All dimensions are in inches. All measurements are from outside edge of chassis. Chassis is LMB #139. Guide is not drawn to scale.

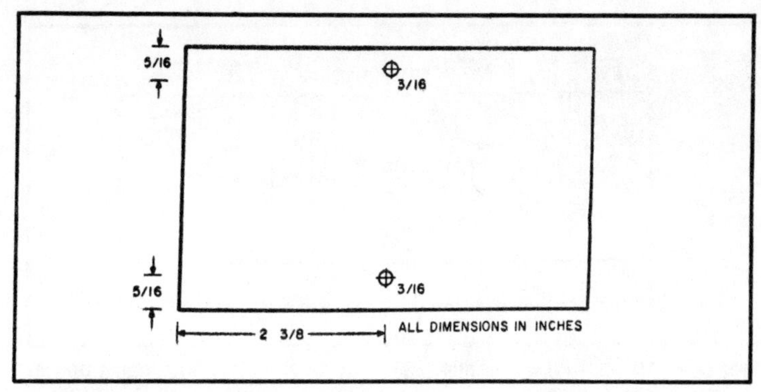

Fig. 6-56. Heat sink drill guide. Heat sink is International Rectifier HE330-C or Wakefield 623-K.

8. Run twisted #20 wires from the feedthrough capacitor and ground lug (next to feedthrough) to + and GND on the PC board respectively. Solder a 500 μF electrolytic capacitor across the feedthrough and ground lug.

9. Completely cover the transparent plastic sides and end of the relay with copper foil. Also run a .2″ strip of foil across the bottom of the relay (where the contacts are). The relay socket is not used. Solder all copper foil pieces together to insure a good shield.

10. Position the relay as shown in Fig. 6-60. Terminals A and B will be up. Solder the #8 terminal to the copper foil at the end of the relay opposite from the contacts. This will partially secure the relay in place while also grounding the relay's shield.

11. Solder wire from the ground terminal to terminal #4 of the relay. Solder a wire from the feedthrough capacitor (+V dc) to terminal #1 of the relay.

12. Prepare four RG-188 cables as shown in Fig. 6-61.

13. Connect the 3″ cable from the "IN" BNC to "IN" on the PC board (module pin 7). Solder the cable shield directly to the

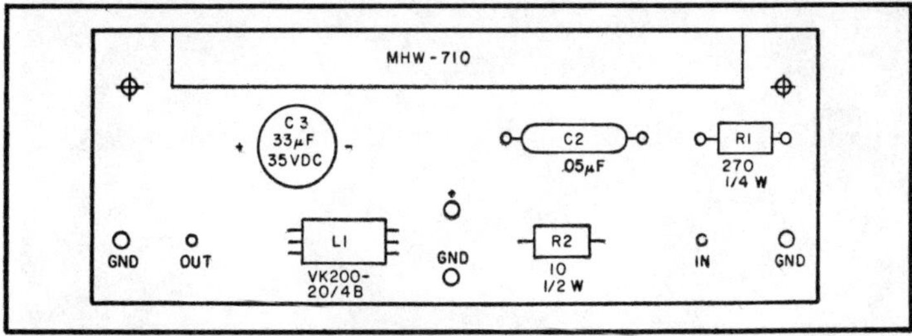

Fig. 6-57. 10 watt ATV amplifier PC board, component side.

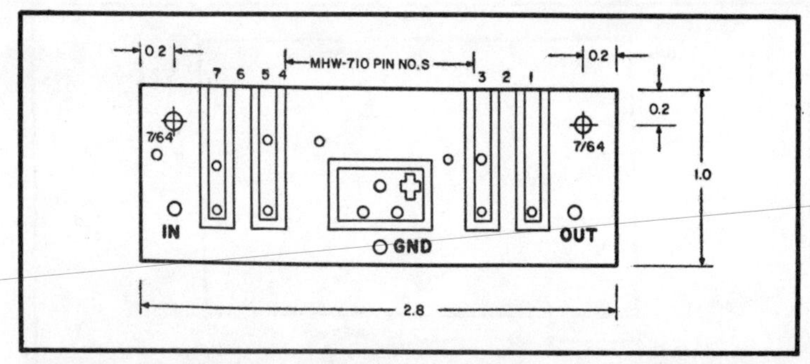

Fig. 6-58. 10 watt ATV amplifier PC board, foil side. All dimensions in inches. Board is glass.

grounded portion on the BNC connector. The other shield is inserted in the hole provided on the PC board and soldered to the foil. All shield lengths must be as short as possible.

14. Solder the ½″ long center conductors and shields of the two 4″ long cables to BNCs "A" and "R." Also solder the ½″ long center conductor and shield of the 6¼″ cable to "OUT" on the PC board (module pin 1). Again keep shield lengths as short as possible.

15. Solder the 1¼″ long center conductor of the cable from the PC board to terminal A of the relay. Solder the shield to the relay's

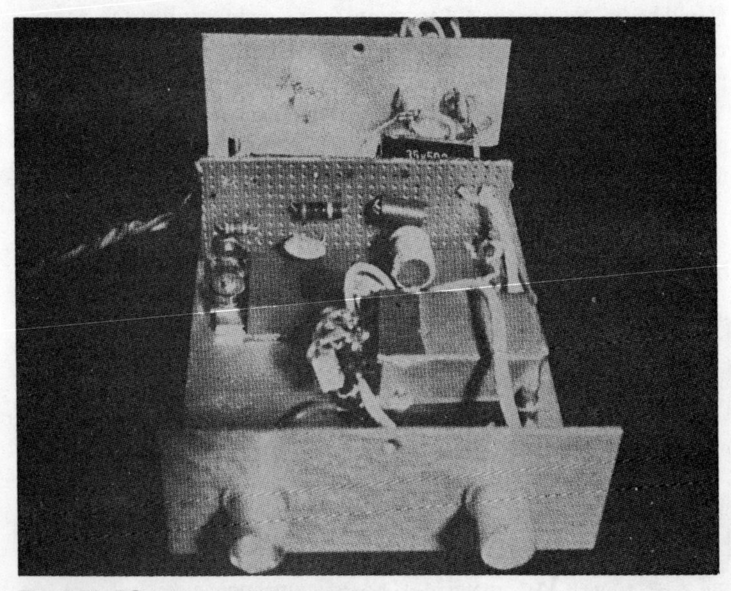

Fig. 6-59. PC board, prototype shown.

Fig. 6-60. Internal (bottom) view.

copper foil near the terminal thus creating a small loop. Solder a 4.7 pF capacitor between the grounded shield and the terminal (see Fig. 6-62). Solder the cable from BNC "A" to relay terminal B in like manner. In the same way, solder the cable from BNC "R" to the relay terminal immediately below terminal A. Again all shield lengths should be as short as possible. Also keep loops as close as possible to their respective capacitors.

16. Screw on bottom cover of chassis and label using stick-on lettering.

17. The amplifier is now complete (see Fig. 6-63).

Tune-Up With QRP Rig

Connect a short length of RG-58 coax from the "OUT" connector on the QRP transmitter to the "IN" connector of the amplifier.

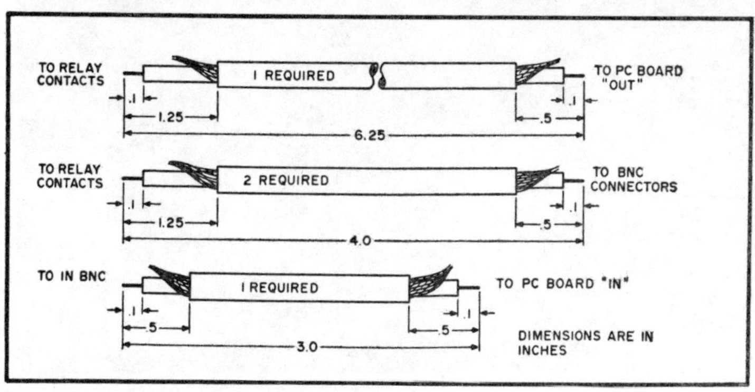

Fig. 6-61. RG-188 cable preparation.

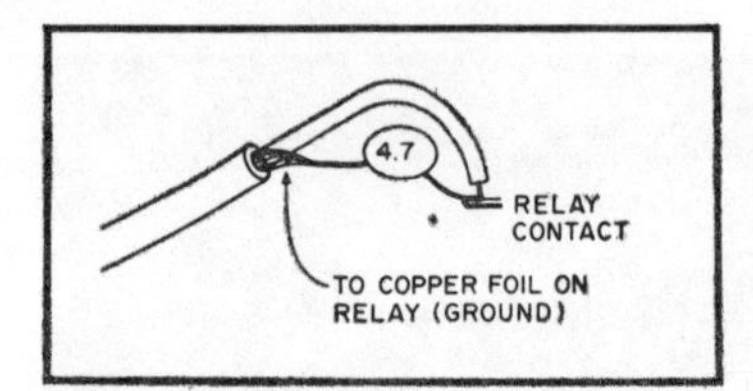

Fig. 6-62. Relay contact connection.

Also connect a through-line wattmeter and dummy load or antenna to the amplifier's output (BNC "A"). Apply 13.8 V dc from a regulated 4 amp continuous supply to both the amplifier and exciter.

Basically follow the tuneup procedure given in the QRP rig article. Of course, now you will be aiming for a good picture at about 10 watts instead of ¾ watts. The following suggestions may be of help:

☐ Remove the core from L6 on the exciter. Set the "L" (level) control fully counterclockwise. These actions will knock down the drive level which should make tune-up easier.

☐ Start with "C" (contrast) control fully clockwise.

☐ Adjust output modulation and power levels using the 4 variable capacitors in the exciter's output circuitry.

Don't rely too heavily on the picture you see on your local TV monitor since the 10 watt transmitter will probably overload it. Try to have an on-the-air station, remotely located, assist you. If you

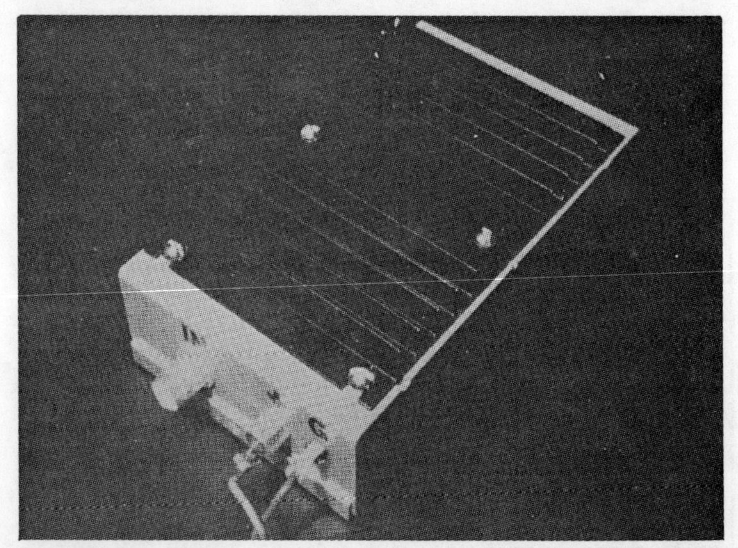

Fig. 6-63. Amplifier, top view.

use a coupler, detector and oscilloscope to tune up the rig, you may note a 15 to 20 MHz oscillation on the signal. It will generally not be observable on your TV monitor. To attenuate this parasitic signal, adjust L1 through L5 for minimum oscillation amplitude.

A complete ATV station is shown in Fig. 6-64. Be sure to use hardline and a good antenna for best performance. As explained in the QRP article, a separate receiver is required to derive audio from a signal using the audio-in-the-carrier format.

Important Design Notes

Amplifier power output is highly dependent upon power supply voltage. A 1 volt difference can result in a 3 watt difference in output power. To achieve a good video signal at 10 watts average power, 13.8 V dc must be used. Current drain will be slightly less than 3 amps. If the amplifier is driven hard into a class C mode (no video), it will be possible to initially obtain about 15 watts. You will note that as the amplifier warms up, the power output will drop. This is natural operation for the Motorola module. Also don't be alarmed if the heat sink and case get very warm. This, too, occurs in normal operation. If you should use the amplifier without video, try not to overdrive it. Use the minimum drive power necessary to achieve full output power (about 300 mW). You will generate fewer spurs while also reducing possible damage to the input of the 710 module. **Warning.** When exciting the amplifier, always make sure that BNC "A" is loaded. I smoke-tested the amplifier with about ¾ watts drive and no load and found that the amplifier self-destructed in 2 minutes. The MHW-710 is rugged and can handle short periods of misuse but don't overdo it.

When procuring the 710, you will note that two models are available: the 710-1 for 400-440 MHz and the 710-2 for 440-480 MHz. I have used both types and found that they perform equally well in the 435-450 MHz portion of the ham band.

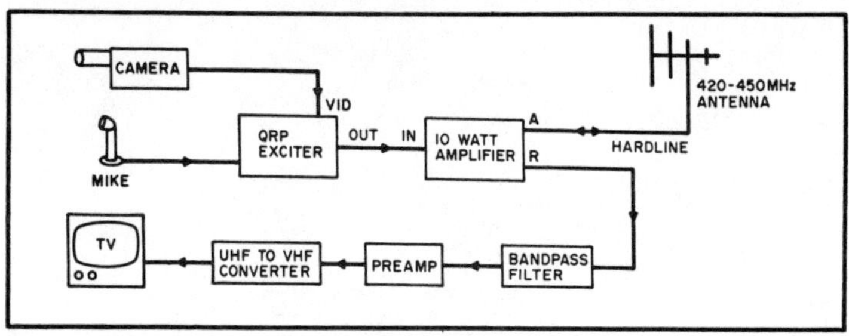

Fig. 6-64. Typical operational configuration.

If you can't get at least 15 watts from your amplifier at a cold start using 13.8 V dc, you may be experiencing high losses in the relay circuitry. To verify this, connect a cable directly from the wattmeter to "OUT" on the PC board. Normally the relay will exhibit a 1 watt loss in the 15 watt range. Relay efficiency is highly dependent upon the length of cable between the relay and PC and "OUT." If you do have a loss problem, experiment using different cable lengths.

The rig is placed in transmit by applying voltage to both the exciter and amplifier power terminals. This arrangement is rather unconventional for normal PTT use, but has been implemented here for simplicity. You may wish to use the spare set of relay contacts and mount additional feedthrough capacitors to achieve a standard switching scheme.

Chapter 7

Antennas, Mounts, and Matchers

Every receiver and transmitter needs some kind of antenna. Strangely enough, antennas are probably the simplest electronic devices to build and yet at the same time the least understood. This chapter starts by explaining the basics of antennas and then goes on to give you some easy-to-build antenna projects.

One of the questions most asked by newly-licensed Novices is, "What should I use for an antenna?"

While no one solution is right for everyone, here is one which should satisfy the requirements of most Novices (and some more experienced hams), even if they live in places where materials are hard to get. This idea won't cost an arm and a leg, either.

The half-wave dipole antenna is probably the most widely-used amateur antenna in the world, because it is cheap, easy to build, easy to erect, easy to feed, and it gives good results. So build your antenna system around the dipole. The "ingredients" you'll need are wire, insulators, feedline, and supports.

Start with the wire. There are several good sources which are right on the beaten track. One is your local electrical contractor. He always has scraps of wire left over that can be bought quite reasonably (sometimes for nothing, if he is sympathetic toward amateur radio). Ask him if he has any scraps of No. 12, 14, or 16 wire. He may have 10- to 30-foot long pieces, but don't worry about that. These pieces can be soldered together with good results. Be sure the joint is mechanically sound and that the wire is heated enough to let the solder flow. Also, be sure you get copper wire and use rosin-core solder (acid-core solder is for guttering). Don't worry about the insulation. It won't harm the antenna's sending and receiving properties. Don't get any insulation in the solder joints, however.

Another source of wire could be your local radio and TV repair shop. Ask the repairman if he has any old burned-out TV transformers he will give you. These can be taken apart and the wire taken from the core for your antennas. Be sure to sand off the varnish used for insulation at any solder joints.

Egg insulators can be bought at an electronics store, but you probably have the makings for some excellent insulators right at home. Many consumer products now come in shatterproof plastic bottles (shampoo, mouthwash, etc.). The bases of these bottles, especially oval-shaped bases, can be cut out and made into very good insulators.

Many hams have the misconception that coaxial cable must be used as feedline for dipoles. Coax if fine, but it isn't the only kind of feedline that works. For years, I have used flat line, simply two parallel wires, with excellent results. So where do you get flat line? Just go to the local hardware or discount store and buy either plastic-covered lamp cord or four-conductor flat TV rotator cable (you have to separate the rotator cable into two-conductor line, but that's easy to do). Either the lamp cord or the divided TV rotator cable has conductors spaced so that the characteristic impedance is in the 50- to 75-ohm range—perfect for feeding dipoles.

Sometimes you may be lucky enough to have trees spaced just right so that you can erect your dipoles between them. Nylon twine purchased at the hardware or discount store is fine for this. But many subdivision lots have neither the space nor the trees to allow convenient dipole placement. However, don't be discouraged. You can make a support for the center of your dipoles using two 10-foot TV masts (see your local discount store) nested together. Use more than two masts if you can support them. Drive a 4′ × ½″ or ¾″ water pipe halfway into the ground, and slip the mast over the pipe. For a 20-foot mast, no guying will be needed if you mount 80 and 40 meter dipole center insulators at the top and swing the elements out in "semi-inverted-V" style at 90 degree angles (or as close as possible) around the top of the mast. Use an eyebolt and nylon twine to support the insulators, and apply plastic electrician's tape about 12″ down from the top of the mast to prevent the antenna from touching the metal. Use nylon twine to tie off the ends of the dipoles (don't tie them too low where people walk!). See Fig. 7-1 for details.

A 40 meter dipole can be used for both 40 and 15 meters quite effectively. A 10 meter "piggyback" antenna can be added to the 80 meter dipole (using the same feedline) with good results without affecting performances on 80 (see Fig. 7-2). However, it is recommended that two different feedlines be used for the 40/15 and 80/10 meter dipoles, due to harmonic problems which can arise when the same feedline is used to feed two antennas, one of which is resonant at the second harmonic of the other.

Now, what about the lengths of the dipoles? Use this formula: Length of ½-wave dipole in feet (or ½ λ) = 468/frequency in MHz. So an 80 meter dipole for 3.775 MHz is about 123′ 11″ long, a 40 meter dipole (7.125 MHz) = 65′ 8″, and a 10 meter dipole (28.150 MHz) = 16′ 8″.

Be sure when you put up this or any antenna system that you stay well clear of any overhead power lines. And be extra careful when climbing trees or roofs. It is far better to have an antenna "not quite as high as you wanted it" than to risk broken bones or worse!

THE "NO ANTENNAS" ANTENNA

Are you one of the unfortunate few who happens to be an apartment dweller ham? And is your landlord or apartment man-

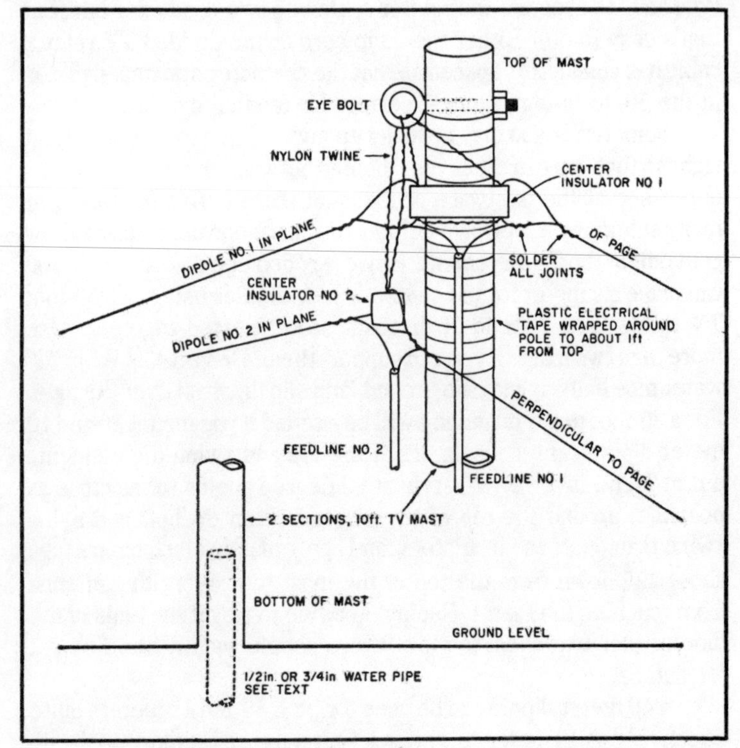

Fig. 7-1. How to connect the dipoles.

ager one who forbids outdoor antennas of any sort? If so, this article may be for you. The problem of erecting an antenna in a "no antennas outdoors" situation is a tricky one, indeed. This limits one to indoor antennas, the logic being that what is not seen will not be noticed.

After experimenting with many different types of indoor antennas, all with disastrous results and much TVI, I finally came upon a coaxial dipole suggested by a fellow ham friend. This antenna appealed to me because of its greatly attenuated harmonics, thus lessening the TVI problem. Unlike a conventional dipole, this antenna is very broad-banded, covering from 500 kHz to 1 MHz, depending upon band used, and with an swr under 2:1 at band edges.

Its broadband characteristics are due, in part, to the feedline being matched to the antenna and the electrical incorporation of its own balun, with the result that no add-on antenna tuner or balun is required. The coaxial dipole has a slight amount of gain over a conventional dipole has a slight amount of gain over a conventional dipole, and since the vinyl jacket covers the entire antenna, it reduces static charge build-up considerably, which causes a popping noise in the receiver when discharged. Thus, the coaxial dipole is a very "quiet" antenna with slightly stronger signal punch than a conventional dipole.

Construction of the antenna is simple. One may use either RG-8/U or RG-58A/U coax, the latter being lighter in weight and easier to work with. Maximum legal power can be used with either choice of coax. For antenna lengths, see Fig. 7-3. The 40-meter antenna will be used as an example.

Begin construction by removing 2.5 cm (1″) of vinyl jacket (½″ each side of center) at the center of the antenna. Cut the shield in the center all the way around the coax. Care must be used so that you do not cut the dielectric or the center conductor. Next, form two leads with the shield as shown in Fig. 7-4. This is the feedpoint of the antenna.

From this center feedpoint, measure out each side of center 5.1 m (16′9″) and cut the coax at that point. Remove approximately 2.5 cm of vinyl jacket from each of the ends and fold back the shield so that the dielectric is exposed. Cut and remove about 2.5 cm of this insulation, being careful not to cut the center conductor. Then twist the shield and center conductor together and solder. This must be done at both ends and forms the 52-ohm matching section and balun.

Next, cut two lengths of coax, each 4.4 m (14′ 9″) long. Then remove 2.5 cm of vinyl jacket from all four ends, fold back the

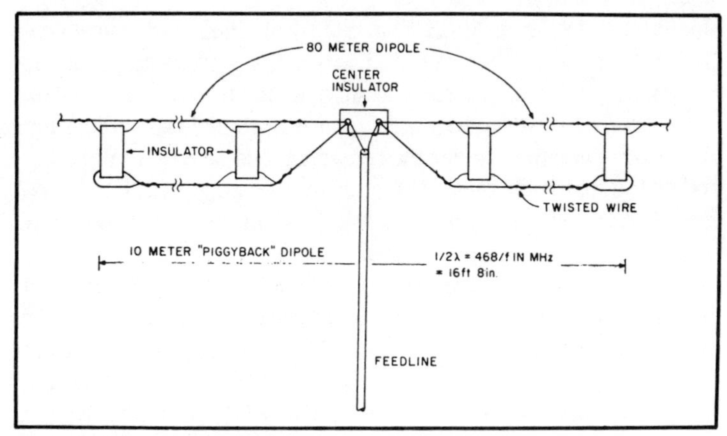

Fig. 7-2. 80 and 10 meter dipoles.

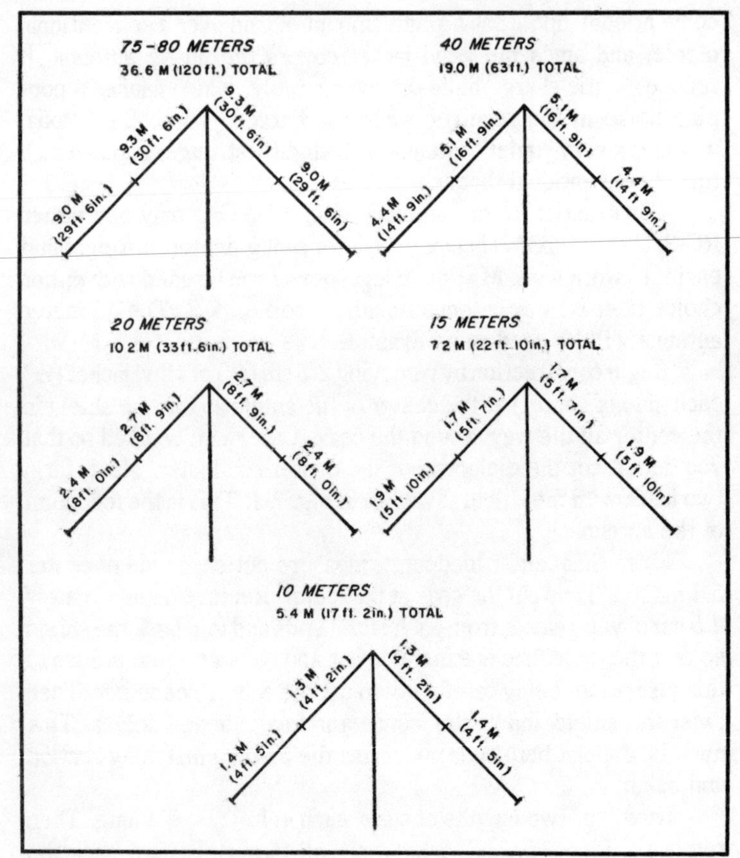

Fig. 7-3. Antenna lengths.

shield, remove center insulation, and twist shield and center conductor together as before. This forms the end sections of the antenna. Attach one of these end sections to one end of the matching section by twisting together the prepared ends and soldering. In the same manner, solder the remaining end section to the other end of the matching section. Waterproof these joints as best you can. Waterproofing of the ends will be done later, for they may need cutting for tuning purposes.

The next step is to attach the feedline. Any random length of coax will do, but it must be of the same type used for construction of the antenna. Remove approximately 2.5 cm of vinyl jacket from the end of feedline, fold back the shield, and remove center insulation. Form two leads with the shield and center conductor. At the feedpoint of the antenna, connect the feedline by soldering the feedline center conductor to one of the feedpoint leads. Then solder the feedline shield to the remaining lead. You may wish to waterproof this area, making sure that the feedpoint leads do not touch each other and short out. Follow this procedure for antennas on other bands.

Erecting the antenna is next. If you have access to an attic or crawl space in the roof of your apartment building, so much the better. Using monofilament fishing line as anchor ties, a series of half hitches along the vinyl jacket ends of antenna will do nicely for anchoring the antenna. The monofilament line will bite into the vinyl as it is pulled taut.

If you are not fortunate enough to have access to an attic, the antenna may be stapled to a living room or bedroom ceiling using plastic cable ties or any other non-conducting material as support. Wrap the cable ties around the antenna at intervals and staple the free end(s) of the ties to the ceiling. Do not staple directly through the antenna itself.

This antenna can be used as a dipole or inverted vee. If used as a dipole, try to erect as much of it as possible in a straight line, keeping it as far away from large metal objects as feasible. The ends may hang down as long as they don't touch any nearby metal objects. More than one antenna may be erected in the same area, providing they are run at angles to each other rather than being parallel. The reason for this is that the inactive antenna could absorb some signal from the active antenna, thereby attenuating the signal output.

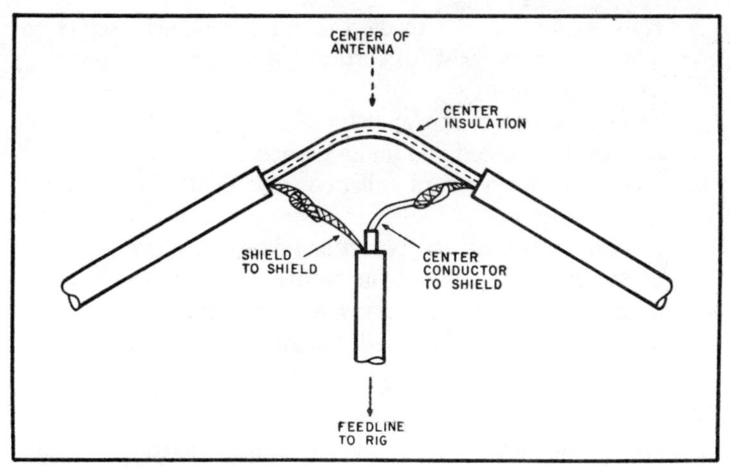

Fig. 7-4. Feedpoint connection.

After erecting the antenna, check swr and trim the ends if needed. Be sure to twist the ends of the antenna as before (shield to center conductor), then recheck swr. The antenna will interact with any hidden wiring in walls, so a considerable amount may have to be trimmed from each end. Once you have gotten the swr down to an acceptable level, solder the ends of the antenna and waterproof them if desired. This completes construction.

Aside from a low-pass filter, no other add-ons are needed, the filter being only a safety precaution. And, since the antenna is basically omnidirectional, orientation can be determined by the space available at your location.

Once you start enjoying the pleasures of operating from your apartment with this antenna, you will be amazed at what you can work and the signal reports you get with it. I have used coaxial dipoles on 10, 15, 20, and 40 meters from an apartment, and all are stapled to a ceiling in "inverted-U" fashion rather than as an inverted vee. Signal reports received vary from S-6 to 60 dB over S-9. TVI is minimal, considering my TV is only a mere ten feet from the antenna and I run 200 watts PEP.

With these coaxial dipoles in use for over two years now, I have gotten Worked All Continents, Worked All States, and DXCC with 121 countries worked to date. So there's no telling what you can do with this antenna and you may be pleasantly surprised at the results. It sure beats non-operating just because you live in an apartment! Happy DXing!

THE BETTER VERTICAL

How would you like to be a proud owner and user of an inexpensive (around $60-$70) vertical DX antenna which—

☐ Is self-supporting.

☐ Is attractive in appearance.

☐ Can be installed in a limited space.

☐ Gives a low vertical radiation angle even when it is one wavelength long.

☐ Can be used on all present and future amateur bands.

☐ Minimizes TVI because its radiation is vertically polarized and because harmonics are radiated at high vertical angles.

☐ Is safe from shock hazards because its base is grounded.

☐ Has a built-in lightning protection system.

Theory

An elevated-feed vertical antenna is not a vertical antenna which is elevated. It is a vertical antenna which is fed at a point which is ⅓ of its height from the ground—see Fig. 7-5A.

I first came across the discussion of this antenna in *Amateur Radio Techniques*. It contains a discussion of how an elevated-feed vertical antenna can be applied to amateur work to obtain " . . . low-angle radiation, without unwanted high-angle lobes, from vertical aerials of appreciable electrical length." It explains how feeding a vertical antenna at the ⅓ point produces a current distribution different from that of a base-fed antenna. This is true only in cases where the antenna element is ¾ λ or 1 λ long. If element length is ½ λ or less, the elevated feed will perform approximately the same as a base-fed vertical antenna of the same height.

The comparisons of the current distributions and approximate vertical-radiation patterns for the base-fed and elevated-feed antennas are shown in Fig. 7-5. To understand how a low vertical radiation angle is achieved in the elevated-feed antenna, one should study the current distribution along an antenna element. The *ARRL Antenna Book* states that current is reversed every ½ λ

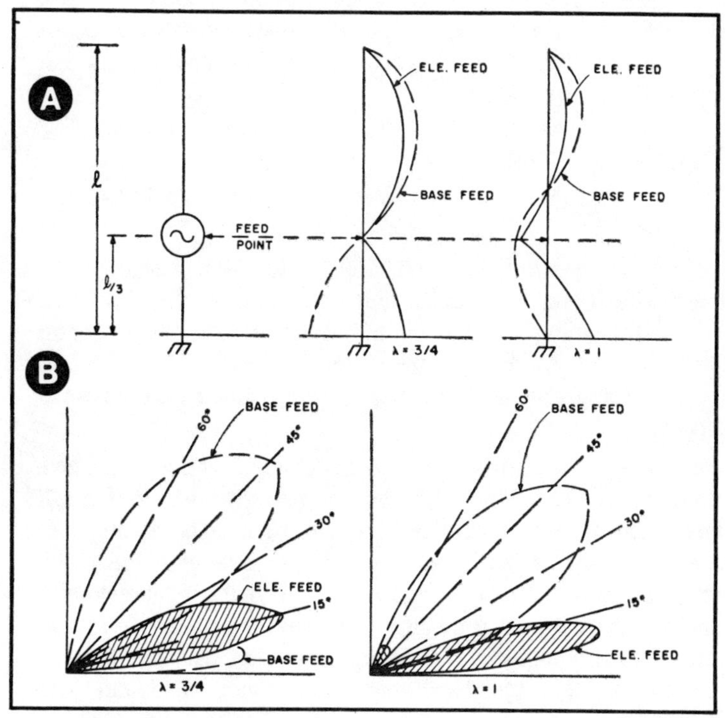

Fig. 7-5. (A) Current distribution and (B) vertical radiation patterns for ¾ λ and 1 λ elevated-feed vertical antennas.

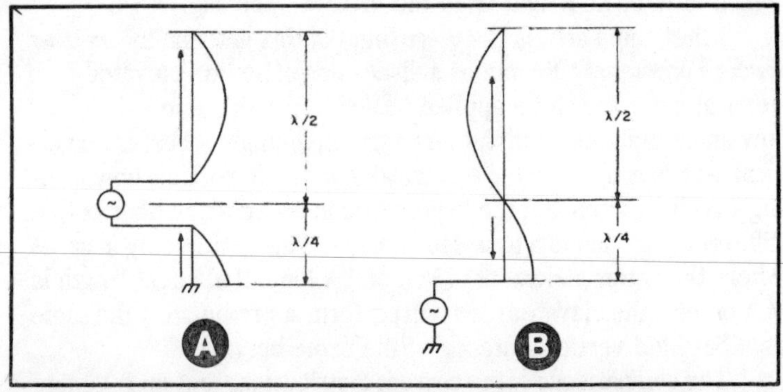

Fig. 7-6. Currents along antenna element for ¾ λ, elevated feed (A) and base feed (B).

along the element. Figure 7-6A shows how this results in an inphase collinear array in the elevated-feed antenna which is ¾ λ long. This inphase current distribution along the antenna element is the reason for its low vertical angle of radiation. If the same antenna were fed at the base, the current distribution would not be in phase—see Fig. 7-6B—and an unwanted high-angle lobe would appear in the vertical plane as shown in Fig. 7-5B.

Design Considerations

In the design, I gave priority to the following considerations:

☐ Limiting the design to a reasonable height.

☐ Incorporating a top hat to dissipate static charge.

☐ Positioning the tuning unit near the ground.

☐ Designing and building a strong yet inexpensive center insulator from readily available materials.

☐ Designing the antenna strong enough to be self-supporting.

I chose an overall antenna length of 33 feet as this would give me a full wave-length—the longest practical length for DX operation—on 10 meters. The 33-foot length meant that the upper section must be 22 feet because the antenna is fed ⅓ of the length from the ground. This would make it ½ λ from the feedpoint on 15 meters, so I detuned it slightly to lower the impedance at that point. The optimum length of the upper section, as determined by 15-meter and 20-meter band impedance curves, was found to be 24.5 feet.

Because aluminum tubing comes in 8-foot sections, and because I would lose 2-feet in the bushings and overlap, the available length from three sections was reduced to 22 feet. To get around this limitation, I decided to enlarge the considered top-hat section to achieve the desired effective length. Four top-hat radials, each 1.3 feet long, were experimentally found to provide the missing link.

The prospect of climbing a stepladder to adjust the tuning unit did not appeal to me. To avoid this, I chose to place the tuning unit near the ground and to use a 12.5-foot section of RG-8 foam coaxial cable to carry the power from the tuning unit to the feedpoint. Theoretical approximation showed that, at the worst, an swr of 7:1 would be present. The additional power loss for a 12.5-foot section of RG-8 foam cable with an swr of 7:1 was found to be 0.25 dB. I preferred this to climbing the ladder.

Power limit at an swr of 7:1 is found by dividing the power rating of the cable at an swr of 1:1 by the swr under operating conditions. For this application, this is 2200/7 = 314.3 watts. The output from a kW linear should be approximately 500-600 watts. For intermittent duty, the average power would be half of this figure, or some 300 watts. This does not give much of a safety

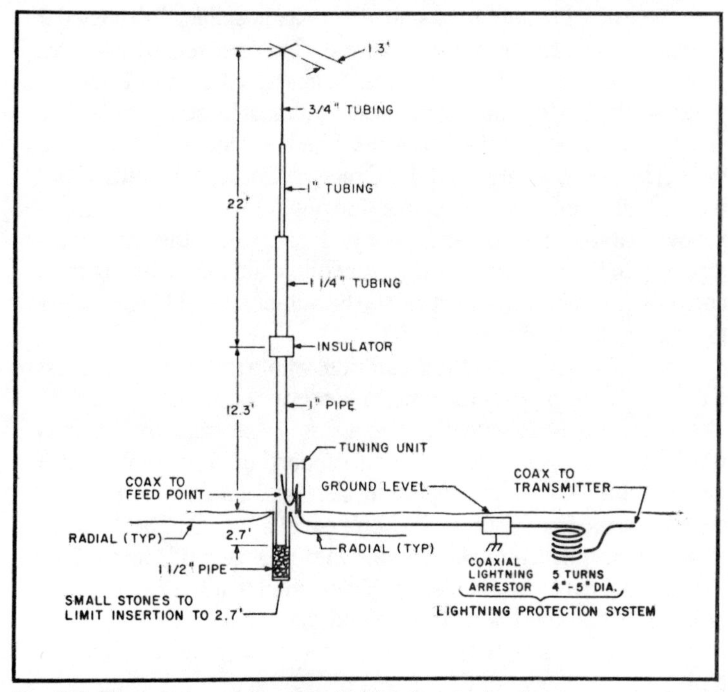

Fig. 7-7. Final design of the elevated-feed vertical antenna.

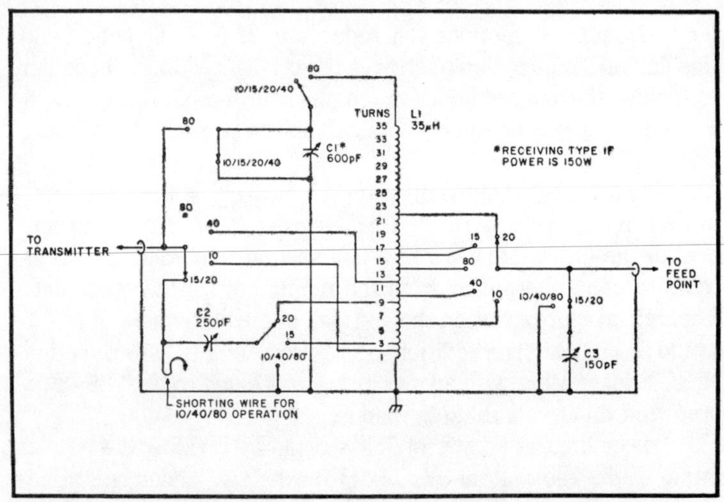

Fig. 7-8. Antenna tuning unit set for 20 meters. All air variables not in use are shorted and grounded.

factor but I decided to go ahead and worry about it when and if I acquired a linear.

To make the antenna as attractive as possible, I designed it strong enough to be self-supporting. This presented no great problem except for the center insulator, which proved to be the greatest challenge of the whole project. It must be strong enough to support the upper ⅔ of the antenna without guying. I finally settled on building the insulator from PVC pipe reinforced with plexiglas™ panels and nylon cord, the whole thing held together with silicone rubber bathroom caulk and epoxy. I calculated the insulator's strength to be much greater than that of the aluminum tubing right above it. So, theoretically at least, the antenna should break at the tubing and not at the insulator.

By calculating stress values for the whole antenna, I found that if I used 1" steel pipe for the bottom ⅓ section and 1-¼", 1", and ¾" aluminum tubing for the 3-piece upper ⅔ section, the antenna would be strong enough to be self-supporting. The weakest link would be the 1-¼" aluminum tubing section. Moral support for this decision came from Capt. P. H. Lee's excellent book, *The Amateur Radio Vertical Antenna Handbook*, where he used this size tubing to construct his Mark II antenna. He claimed that the antenna was flexible; it bent with a high wind and did not break.

The final design of this antenna is shown in Fig. 7-7, and the tuning unit is shown in Fig. 7-8.

Construction Procedure

The construction is started by the assembly of the center insulator. Figure 7-9 shows how one piece is cut and inserted into the other piece. Use PVC pipe cement to bond the two pieces together.

Figure 7-10 shows how an inexpensive jig can be constructed from a screwdriver and a piece of soft wood. This jig will hold the cemented PVC pipe in a vertical position to ease the task of cementing the plexiglas panels. The panels can be epoxied to the pipe first so that they will stay in place when applying the silicone rubber bathroom caulk.

Before cementing the plexiglas panels, insert the steel pipe and aluminum tubing into the PVC pipe to the dimensions shown in Fig. 7-11, i.e., to within ½″ from each other, centered at the center of the insulator. Mark the radial direction on the pipe, aluminum tubing, and insulator. Drill holes 90° apart in the pipe and tubing for the mounting bolts, drilling through the PVC pipe. When drilling in pipe, use a ¼-20 tap drill and enlarge the hole to ¼″ when the pipe is removed. The position of all holes is shown in Fig. 7-12. To avoid weakening the pipe, stagger the tap holes. This procedure will align all the holes and assist in the final assembly.

When the bathroom caulk has cured, wind five bands around the panels as shown in Fig. 7-12. Use nylon or dacron line approximately ⅛″ in diameter and space the bands evenly. Epoxy the line for extra strength and to prevent it from unwinding. Drill the two vent holes between the panels in the center of the insulator. Build a little roof over the vent by using caulk. This will prevent moisture from seeping into the insulator.

The three sections of the aluminum tubing are assembled as shown in Fig. 7-13. The bushing is made by cutting 6″ from the

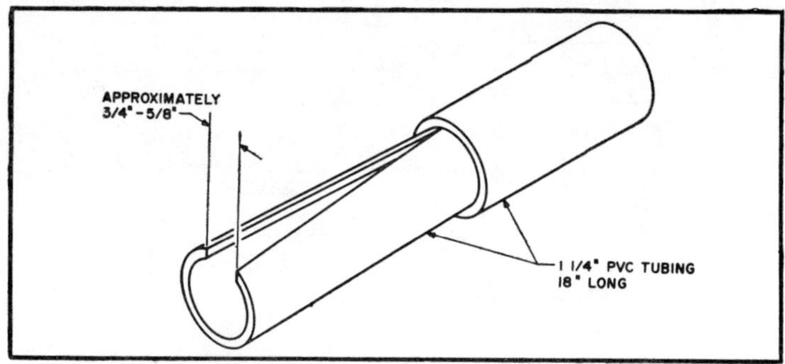

Fig. 7-9. Joining the two 1-¼″ PVC sections.

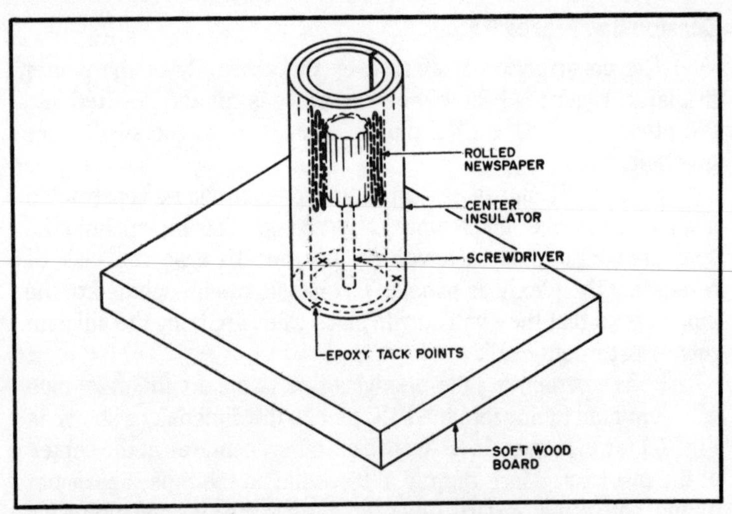

Fig. 7-10. Jig for the construction of the center insulator.

smaller of the two pieces at the junction, splitting it and forcing it over the shortened piece.

The top hat is made by cutting a 3-foot length from aluminum clothesline, bending it in the center, and bolting it in place as shown in Fig. 7-14. After it is bolted in place, bend it until it is perpendicular to the tubing. After bending, cut it to the dimension

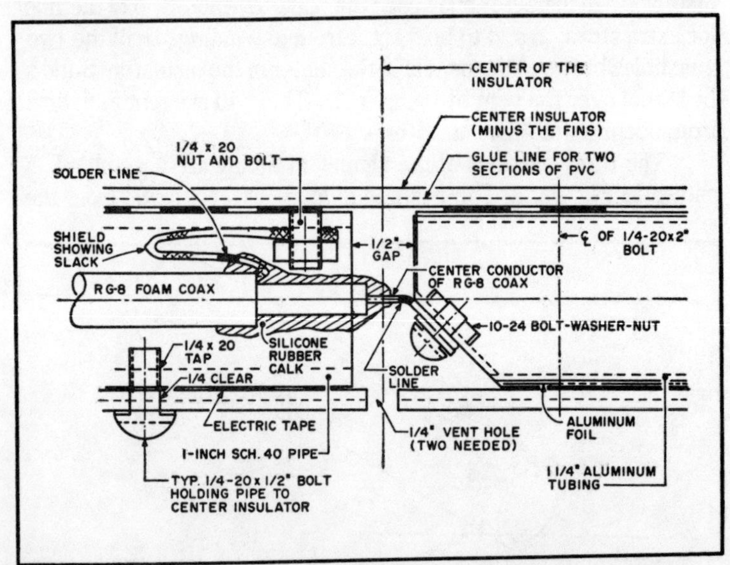

Fig. 7-11. Locations of the pipe and tubing within the center insulator.

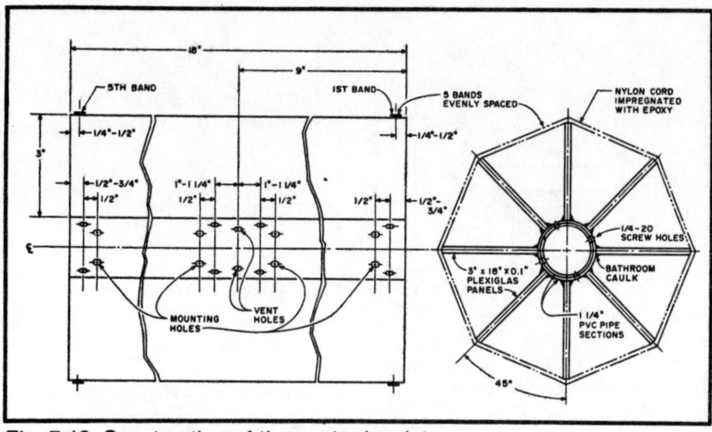

Fig. 7-12. Construction of the center insulator.

shown (1′4″) and spread the two wires until they are 90° apart. Install the top button and seal the whole area with bathroom caulk.

If possible, obtain a piece of 1″ Schedule 40 pipe which is 15 feet long. If this cannot be obtained, use one 10-foot and one 5-foot section. Position the 5-foot section next to the insulator and join the two pieces together by using a 12-inch piece of 1-¼″ Schedule 40 pipe and ¼-20 bolts. Use aluminum sheet between the pipes for a tight fit. Drill and tap the holes at this junction by following the same procedure as outlined previously when drilling holes in the center insulator. Drill one 7/16″ hole approximately 4-5 feet from the bottom end of the pipe. This is the exit hole for the coaxial cable.

Cut a piece of RG-8 foam coaxial cable 15 feet long. Strip one end as shown in Fig. 7-11. Allow sufficient length of shield to

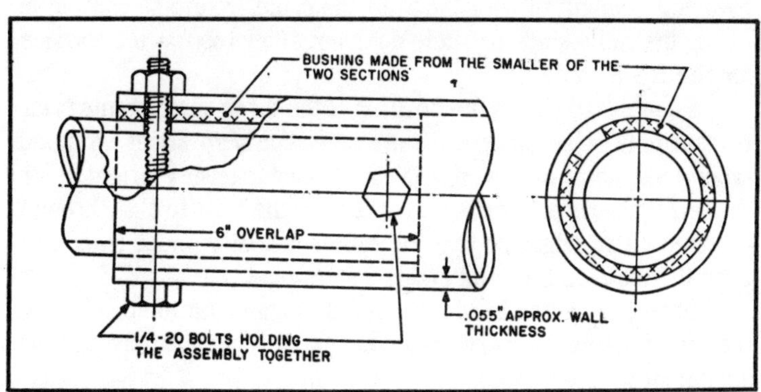

Fig. 7-13. Assembly of the aluminum tubing sections.

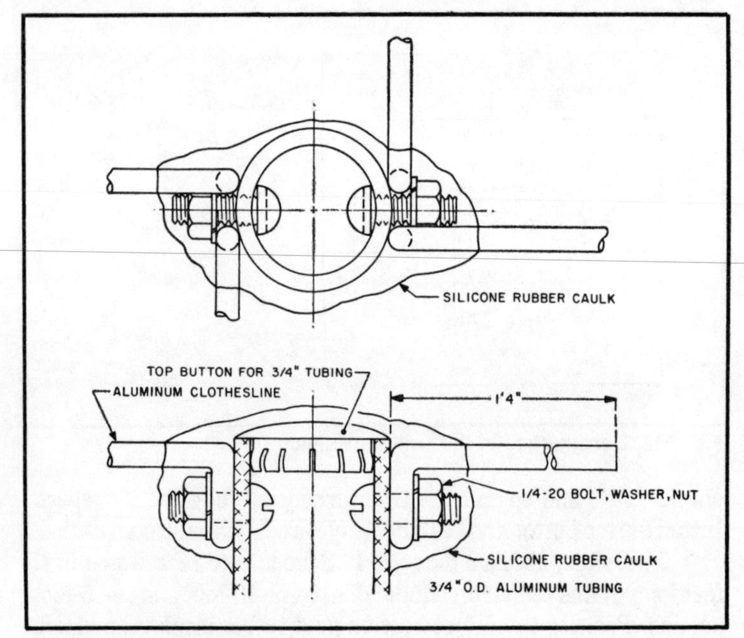

Fig. 7-14. Top-hat assembly.

produce the slack. During assembly, the pipe and aluminum tubing will come together across the ½″ gap forcing the coax down. The slack is needed to prevent bending or damaging the center conductor. Impregnate the center conductor and the shield with solder so that about ¼″ of soldered length will protrude from the silicone rubber caulk when applied. Apply silicone rubber caulk as shown in Fig. 7-11 to seal the cable from moisture.

Thread the cable from the insulator end to the 7/16″ exit hole by using a length of wire taped to the cable. Exercise caution in taping the cable since the hole does not allow too much clearance for the RG-8 cable.

Figure 7-15 shows the position of the three components prior to assembly. Use electrical tape and aluminum sheet wrapped around the tubing and the pipe as necessary to ensure a tight fit for the center insulator. Cut holes in them for the bolts to pass through and smooth all edges so that the center insulator slides smoothly over the aluminum tubing and the steel pipe.

Slide the center insulator over the aluminum tubing. Verify the markings which were made during the drilling to avoid hole alignment problems.

Attach the shield of the coaxial cable to the pipe first. To do it, drill and tap a ¼-20 hole in the pipe about ½″ from the end, as shown in Fig. 7-11. Screw a ¼-20 bolt from the outside of the pipe. Secure the shield to the bolt inside the pipe with a nut. Tighten the nut. Cut the bolt flush with the outside of the pipe wall.

Bend one edge of aluminum tubing and drill a 10-32 clearance hole in the bent section, as shown in Fig. 7-11. Attach the center conductor to the tubing by using 10-32 hardware.

Slide the aluminum tubing until it butts against the pipe. If the slack in the shield is of correct length, the two pieces should butt without any problem. If they do not butt properly, more slack in the shield will be required.

With the two sections butted, slide the whole antenna until it rests against a wall or other stationary object. Slide the center insulator over the pipe until the mounting holes are in alignment. Secure the insulator to the pipe by using ¼-20 × ½ bolts. Gently slide the aluminum tubing out of the insulator until the mounting holes are in alignment. Secure the insulator to the tubing using ¼-20 × 2 bolts and nuts.

Install the antenna in a 1-½″ pipe, 5 feet long, which is driven into the ground to a depth of 4-½ feet. Small stones are dropped into the pipe to limit the depth of insertion. Aluminum or hardware shims are used to hold the antenna in place.

A ground radial system is needed for optimum performance, especially on the 80- and 40-meter bands. I have five radials, each 33 feet long, and I plan to install eight more. As with every vertical antenna installation, a low ground resistance is necessary for good performance. A high ground resistance (few or no radials) results in high power losses because the ground resistance is in series with the radiation resistance of the antenna.

For this installation, I attached the ground radials to the 1/½″ buried pipe. I grounded the antenna to the pipe by using a ½″ × ⅛″ aluminum grounding strap.

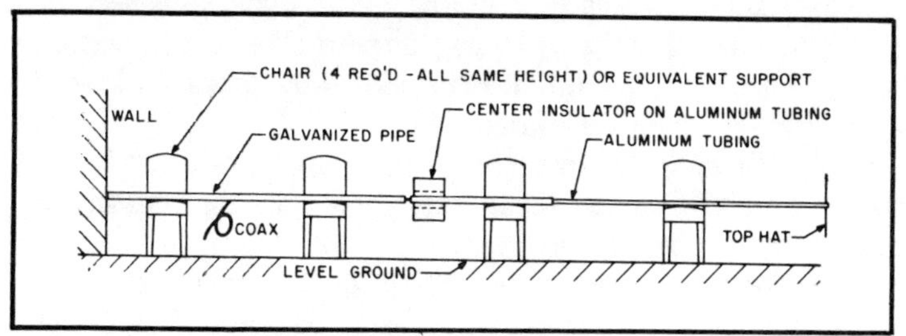

Fig. 7-15. Assembly of the elevated-feed antenna.

Tuning Unit Construction

The schematic of the tuning unit is shown in Fig. 7-8. The unit is installed next to the antenna, but not grounded to it. It is grounded only to the shield of the coaxial cable.

I constructed my tuning unit on a piece of plexiglas 7-½"×16-½" and mounted it inside a watertight cabinet. Since I had enough air-variable capacitors in my junk box, I decided to be extravagant and use separate C2 and C3 air variables for the 15-meter and 20-meter bands.

One word of encouragement: The construction of this unit is not complicated. The cost to build it need not be high. I obtained all the parts and the cabinet for about six to seven dollars at two hamfests held in my local area. The real bargain find was an old Army surplus tuning unit which was priced at $5.00. This unit yielded two air variables, the coil, and the enclosure. To those of you reading this article who have not been to a hamfest, my advice is to go to one! It is lots of fun plus being a place for some real bargains.

Once the tuning unit is built, connect it to the coax feeding the antenna and to the transceiver placed next to the unit. Follow the procedure below to obtain tap points for your coil.

Tuning Procedure Using SWR Meter

(1) Connect the swr meter in the line between the tranceiver and a dummy load.

(2) Tune the transceiver as usual for maximum output on the 80-meter band. Adjust the swr meter sensitivity for a full-scale forward power indication.

(3) Do not change any of the transceiver or swr meter settings. Switch the swr meter to read reflected power.

(4) Disconnect the dummy load and connect the tuning unit in its place.

(5) Using the turns ratio in Fig. 7-8 as a guide, connect the appropriate wires to the coil using alligator clips or equivalent.

(6) Position all tuning unit switches to 80 meters and adjust all air variable to *minimum* capacitance.

(7) Watching the swr meter, place the transceiver in the transmit mode. The swr meter may show anything from an off-scale reading to an swr of 1:1.

(8) Not changing any of the settings on the transceiver or the swr meter, note the swr reading. Place the transceiver in the standby mode.

(9) If the swr was high, adjust the taps on the coil and repeat steps 7 and 8. If the swr was low (swr meter deflection is ½-⅔ scale, equivalent to an swr of about 3:1 to 5:1), leave the taps alone and adjust the air variable for an swr of 1.3:1 or lower.

(10) Repeat steps 7 to 9 until the swr of 1.3:1 or lower is obtained. Record all settings for future reference.

(11) Repeat this procedure for the other bands.

The procedure is designed to obtain the best possible match by adjusting the turns on the coil first. Once this is accomplished, air variables are used to reduce the swr still further. Always adjust one component at a time and fight the temptation to tinker with knobs. It took me two days to learn this lesson.

Connecting the Antenna to the Shack

After installation and tuning, connect the antenna to the shack by using buried RG-8 coaxial cable. Install the lightning protection system as shown in Fig. 7-7. It consists of a coaxial lightning arrestor grounded to a 5′-6′ ground rod, followed by the turns in the coax. Tape the arrestor well with electrical tape to prevent moisture damage.

Performance

The theoretical performance calculations were hammered out with N9CR during various coffee breaks. He has a newly installed three-element tribander atop a 60′ tower. We chose to compare the relative merits of the elevated-feed vertical antenna to those of the beam 60 feet in the air.

Theoretical data for this comparison came from *The ARRL Antenna Book* and P. H. Lee's book, *The Amateur Radio Vertical Antenna Handbook*. The summary is presented below. We chose the 20-meter band for this comparison.

A three-element beam 1 λ above ground has a vertical pattern consisting of two lobes. Only the lower lobe is good for DX. It has a horizontal beamwidth of about 60° (−3-dB points) and a vertical beam width of about 15° in the lower of the two lobes. Judging by the published patterns, we assumed that the power going into the antenna is divided equally between the two vertical lobes.

The beamwidth of the elevated-feed vertical antenna on 20 meters is approximately 20° in the vertical plane. Since it is non-directional, the horizontal beamwidth is 360°.

For DX operation, the spherical area illuminated by the beam is 60°×15°=900 "square degrees." The spherical area illuminated

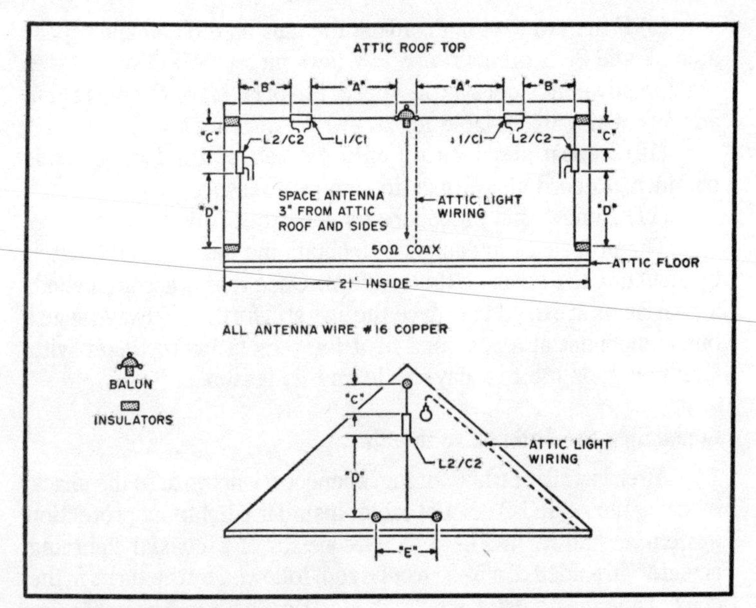

Fig. 7-16. Antenna layout.

by the elevated-feed antenna is 20°× 36°° = 7200 "square degrees." The power gain of the beam relative to that of the elevated-feed antenna can be calculated theoretically as:

$$\text{Gain (dB)} = 10 \log P_1/P_2$$
$$10 \log 7200/900$$
$$9.03 \text{ dB over elevated feed}$$

Because only half of the power (3dB down) goes into the "useful lobe," the actual gain that the beam realizes over the elevated-feed vertical antenna is 9.03 dB − 3.0 dB=6.03 dB, or 1 S-unit.

On the air, the antenna performed beautifully for DX on 28 MHz and 21 MHz where the radiation is at low vertical angles. On 14 MHz, the antenna performed very well over the United States and Canada, and fairly well for DX. On 7 MHz and 3.5 MHz, the antenna lays down a strong ground wave; I had very good signal reports from stations 30 and 40 miles away. Many fine 80- and 40-meter QSOs were also had with stations as far as 800 miles away.

A TRIBANDER FOR THE ATTIC

Figure 7-16 illustrates the layout for a small 21′-long attic. If you are fortunate enough to have a longer attic, by all means install segments C, D, and E horizontally in the same plane as A and B for a slight, but measurable, gain of approximately 1 dB on 40 meters. Test equipment required to optimize this antenna on your favorite frequencies on each band consists of a grid-dip oscillator and swr bridge. If you do not have a GDO, just follow directions, as both the first and second traps are high inductance/low capacitance units with resultant wide bandwidths.

The antenna segments with traps are resonant as follows: Segment A is resonant on 10 meters, segment A+B+C on 20 meters, and segment A+B+C+D+E on 40 meters.

Trap L1/C1 is parallel resonant at 28.7 MHz, offering a high impedance and thus isolating the rest of the antenna at 10 meters and providing a loading inductance for shortening the 20- and 40-meter segments. Trap L2/C2 is parallel resonant at 14.2 MHz and presents a high impedance, thus isolating the rest of the antenna at 20 meters and providing a loading inductance for shortening the 40-meter segments, D and E.

Construction Detail

L1 is a 3-inch length of ⅞" diameter broomstick using 5 feet of no. 16 double cotton-covered (DCC) copper wire space-wound with 19½ turns as shown in Fig. 7-17. No. 16 DCC copper wire should be used if available, but ordinary bare bus wire may be used if you carefully wind and space the turns on L1 and L2 to ensure that there are no shorts. C1 is a 10½-inch length of no. 18 zip cord. Grid-dip L1/C1 and adjust to 28.7 MHz. L2 is a 6-inch length of ⅞"

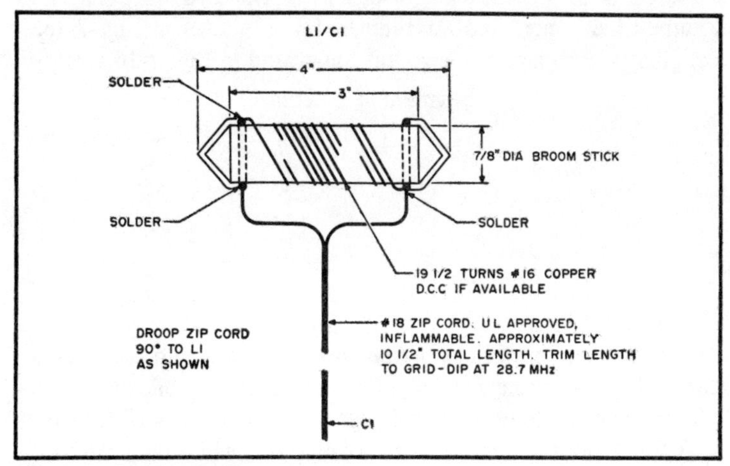

Fig. 7-17. Coil L1 details.

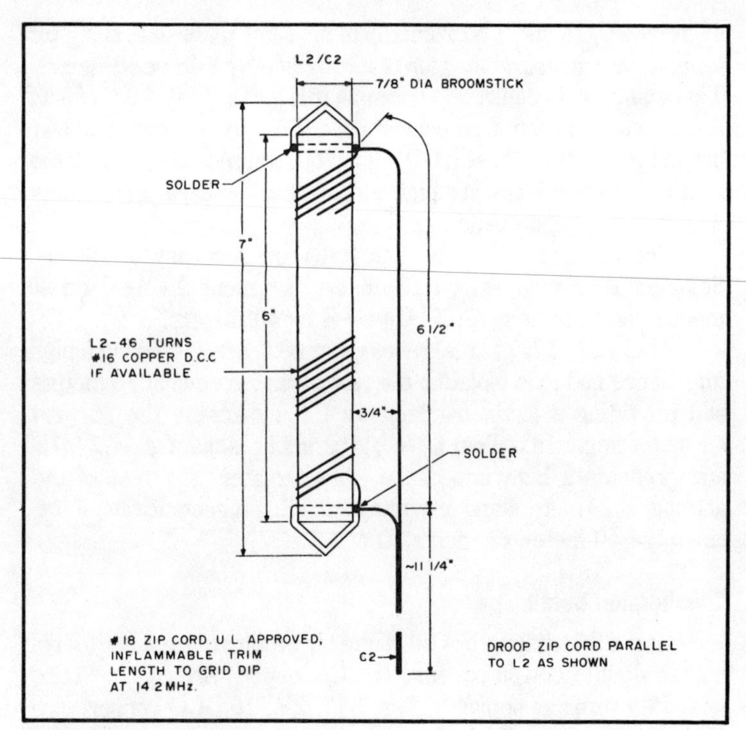

Fig. 7-18. Coil L2 details.

diameter broomstick using 12 feet of no. 16 DCC copper wire, slightly closer than space-wound with 46 turns, as shown in Fig. 7-18. C2 is a 17¾-inch length of no. 18 zip cord with one end trimmed 6½″ short and attached to L2 as shown in Fig. 7-18. Lengths of each antenna segment illustrated in Fig. 7-16 are:

Segment	Length
A	96″
B	24″
C	22″
D	46″
E	20″

Tuning

As the lengths of all loaded segments interact with each other, this multi-band antenna should be tuned exactly as follows or you will surely come to grief! Using your GDO, tune traps L1/C1 and L2/C2 individually, unconnected to anything, while they are balanced on a glass mayonnaise jar (empty) at least 8″ above your wood workbench or desk. This is to avoid obtaining misleading GDO readings due to stray capacitance. Start with both C1 and C2 an inch longer than specified and trim off ¼″ between GDO readings until the GDO null (max dip) is exactly at the desired frequency. Also, use your station receiver to check your GDO frequency reading, as all GDO readouts are only approximate and may be as much as 1 or 2 MHz off actual frequency when coupled to the trap under test.

After the traps are tuned with the GDO, leave them alone. Install the entire antenna system as illustrated in Fig. 7-16. Using *very low power* from the station transmitter, with the swr bridge in the coax line, check swr at 28.5 MHz, 28.7 MHz, and 29.0 MHz. Swr should be less than 2:1 if the antenna is installed correctly. There must be *no* electrical power/lighting wiring parallel to or close to any of the antenna segments if you wish top performance. An overhead attic light is OK if installed close to the balun at the center of the antenna and the light's wiring is run down the inside of the roof, 90 degrees to the plane of the antenna, as shown in Fig. 7-16.

If you wish to change the center frequency on 10 meters, add or subtract approximately ⅜″ per 100 kHz. After 10 meters is satisfactory, check the 20-meter swr at 14.0 and 14.3 MHz, and, if necessary, shorten or lengthen segment C for minimum swr at the desired frequency. After 20 meters, adjust the width of segment E for minimum swr on your favorite 40-meter frequency. A few inches either way will make a considerable difference as segment E is, in effect, a capacity hat for the 40-meter dipole.

Harmonics

Being a multi-band antenna, this system is an extremely efficient harmonic radiator. If there is any question in your mind about your transmitter's harmonic output, you would do well to include a coax antenna tuner between the transmitter and coax for your own sake, your fellow amateurs, your neighbors, and the FCC.

15-Meter Option

Although I do not operate 15 meters, the second method described here was satisfactorily developed for a young friend who does operate on that band. There are two obvious ways to include 15-meter coverage in this antenna system, if desired. The first method uses a separate 10-meter trap with ½ turns of L1 and double the length of zip cord C1. The 15-meter trap, placed about 14″ out from the new L1, is tuned by additional capacity across the old L1 (now the 15-meter trap). Both segments B and C are shortened accordingly. The second method, and surely the simplest, is another dipole from the balun in parallel with the original antenna system. It should be slanted so that the outside ends are dropped 2′ below segment B, and, most importantly, not less than 12″ away from segments C and D to avoid disastrous interaction/detuning of the original system. Ordinary plastic clothesline can be used to support the dropped 15-meter dipole at point X. See Fig. 7-19 for details.

TRIBAND DUAL DELTA

The formula for the length of the driven element of a delta loop is $1005/f_{MHz}$. Since I was bound to need some extra wire at the ends, I cut the antenna to 71 feet overall in length for a 20-meter loop. I mounted the antenna to the rafters inside the attic, using screw hooks and ceramic insulators to hold the wire at three corners.

I used #14 solid, Formvar-insulated wire simply because it was cheap and available; you could use smaller wire if cost considerations were important, however. Try to select a section of the attic that doesn't have a lot of metal ductwork or plumbing lines that could detract from the performance of the delta loop. Also, make sure that the wire isn't touching any wood or metal inside the attic.

I initially fed the antenna with a random length of 50-ohm coaxial cable and a 1:1 balun. However I found out that the swr was substantially higher than I wished. I measured the antenna with a noise bridge and found out that an impedance-matching device would be necessary to use the antenna on 20 meters. Rather than changing the length of the antenna, using a length of 72-ohm coaxial

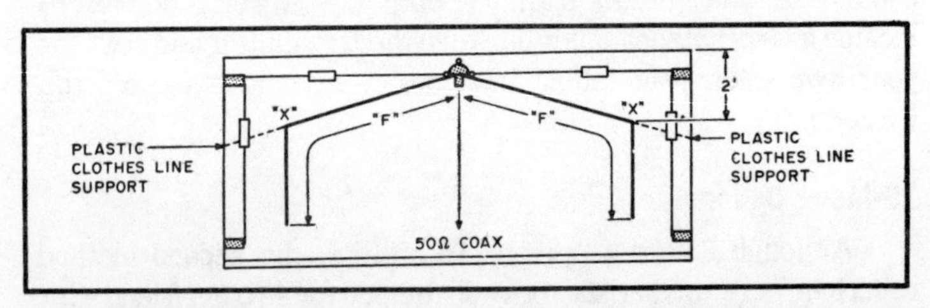

Fig. 7-19. 15-meter dipole option. Start with each segment "F" at 11′6″. Trim one inch at a time for minimum swr at your favorite 15m operating frequency. Do not allow drooped ends of "F" closer than 12″ to either L2/C2 or segment D, to avoid detuning other bands.

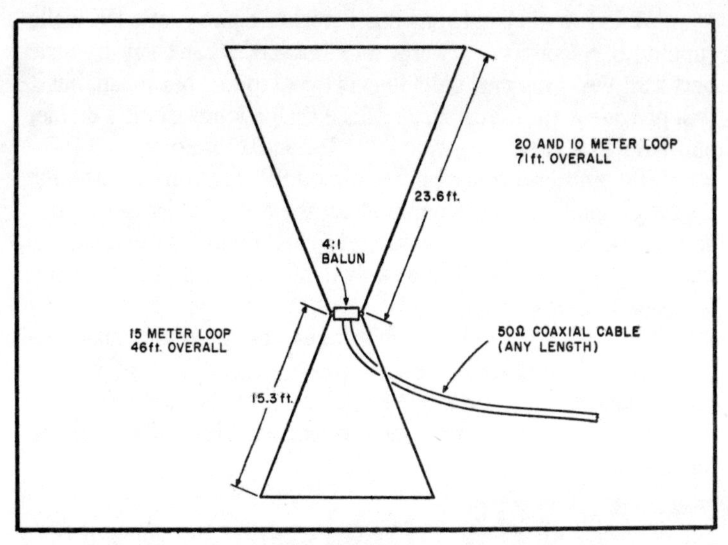

Fig. 7-20. Top view of the dual delta loop.

cable to match the impedance, or using a gamma match, I chose to substitute a 4:1 balun for the 1:1 balun already on the antenna. I was rewarded with an swr of 1.3 to 1 across almost all of the 20-meter band. The broad-band characteristics of the antenna were helpful with respect to swr, although overall efficiency suffered due to the relatively low Q. As a bonus, the antenna worked very well on the entire 10-meter band also, with an swr of 1.8 to 1 on the entire band.

For a total investment of $25, I had a 20- and 10-meter antenna that had a theoretical gain of 2 dB over a dipole and, best of all, it was completely invisible to the neighbors.

Not happy with missing out on the action on 15 meters, however, I added a second delta loop inside the first loop. The second loop was cut to 46 feet and also was laid out in the form of an equilateral triangle. The 15-meter loop was soldered to the 4:1 balun at the same point that the 20- and 10-meter loop was soldered.

Needless to say, adding the 15-meter loop increased the swr of the 20- and 10-meter loop to over 2.5 to 1 on both bands. Since I wasn't ready to give up yet, I took the 15-meter loop and rotated it 180 degrees so that the balun was now at the apex of two delta loops (see Fig. 7-20). This variation was a winner, with the swr on 20 and 10 returning to its original values and the 15-meter loop giving a 2.1 to 1 swr across the entire band.

While I would like to say that I worked some exotic DX while running barefoot with 3 watts on sideband, I can't say that the performance of my dual delta loop is equal to that of a beam, but it will provide a bit better performance than a longwire or a dipole, and with an acceptable swr on 20, 15, and 10 meters.

The antenna appears to be omnidirectional, although ductwork inside the attic might affect the radiation pattern somewhat. Needless to say, mounting the dual delta loop outside and much higher than the 10-foot height of mine will provide some increase in performance.

The low price ($30) and unobtrusive nature of the dual delta loop make it attractive to hams who are faced with restrictive covenants regarding towers and antennas. Try the dual delta loop; you can't beat it for performance, price, and simplicity. It really beats a dipole!

A TRAPPED DIPOLE

Often the need arises for a permanent low cost antenna. A dipole or inverted vee is a good choice. They are easy to install and cheap to build. One of the disadvantages of such antennas is that they are only usable on a single band, unless they are fed with an open feedline and an antenna tuner.

Most traps used in amateur radio multiband antennas are made of a lumped inductance and capacitance in parallel. I tried to overcome this.

By placing a trap 32 feet 6 inches from the feedpoint, a current maximum will occur at 7200 kHz. With the correct wire length on the outside end of the trap, the antenna can also show current maximum at the feedpoint for 3900 kHz. In both cases, the dipole functions as a half-wave dipole.

Why not add another antenna under the existing 80 and 40 meter wire, fed at the same feedpoint, with another trap tuned for 21300 kHz? An outside wire of the correct length will give current maximum on 80, 40, 20 and 15 meters, all functioning as a halfwave dipole.

The dimensions given here are resonant at 3.9 MHz, 7.2 MHz, 14.3 MHz and 21.3 MHz. For 40 meters it's 160 turns, for 15 meters, 55 turns. Number 12 magnet wire is wound on a ½ inch rod, close wound. The coil is removed from the ½ inch rod and placed inside the ½ inch PVC pipe. See Figs. 7-21 and 7-22 and Table 7-1.

The PVC pipe is cut to 18 inches for 40 meters, 10 inches for

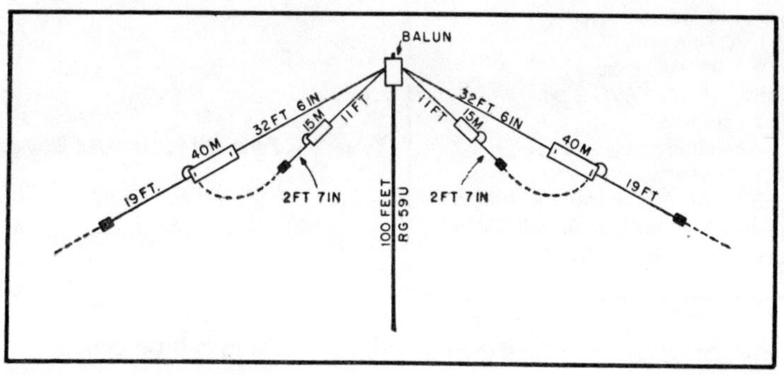

Fig. 7-21. Trapped dipole.

15 meters. The PVC is then placed inside the ⅞ inch ID, 1 inch OD aluminum tube. The aluminum is cut to 16-½ inches for 40 meters, 8-½ inches for 15 meters.

Drill a hole in the center (ends) of eight ½ inch PVC caps, and mount stainless steel eye bolts on them. (Cut off the eye bolts as short as possible, so they will not go into the PVC tube.) Now drill a hole to fit the #12 magnet wire below the eye bolt in each end cap. See Fig. 7-23.

Cement one end cap onto the PVC tube after bringing the end of the coil wire through the small hole. Secure a tin solder lug on one end of the aluminum tube, as shown in Fig. 7-24, with a pop rivet or small screw. Do not use aluminum or copper for the solder lug. Slide the aluminum tube over the PVC with the solder lug end first, and solder a jumper from the lug to the coil wire as close to the PVC cap as possible.

You are now ready to tune the traps. The traps were adjusted to frequency through the use of a grid-dip meter (checking on a receiver for accuracy). The coil can be changed quite easily if an extra turn or two is put on for adjusting purposes. The coil can also be wound with spacing and compressed or extended to get the traps exactly on frequency. Tune to 7.2 on 40 meters. Tune to 21.3 on 15 meters.

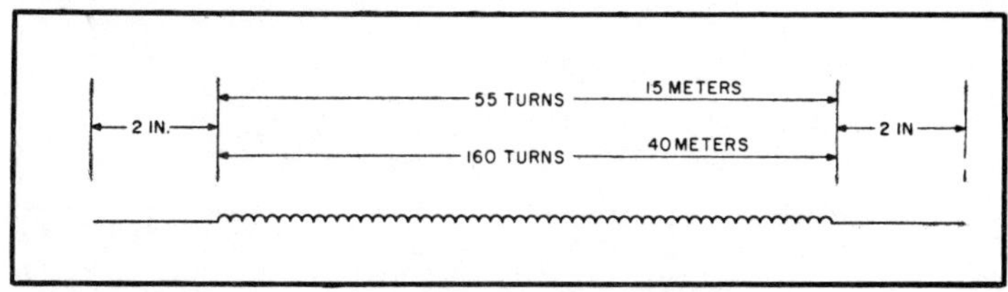

Fig. 7-22. Don't forget to leave 2" one each end of each coil.

Table 7-1. Parts List for Trapped Dipole.
PVC cement
8 1/2" PVC caps
56" of 1/2" PVC pipe
1 balun, 1:1
4 ceramic insulators
135' of antenna wire
50" of 1" aluminum tubing (a discarded lawn chair will do)
80' of #12 magnet wire

After the tuning is completed, the end cap can be cemented on. The two wires sticking out of the end caps are to be soldered to the antenna wires.

My antenna is supported in the center about 32 feet high and 10 feet at the ends. I show an swr of 1.2 to 1 on 3.9, 1.3 to 1 on 7.2, 1.3 to 1 on 14.3 and 1.2 to 1 on 21.3. The CW bands can be worked with the swr less than 2 to 1 on all CW bands.

The overall length is 106 feet, and it can be installed as an inverted vee in a lot less than 90 feet.

SUPER LOOP ANTENNA

To the urban amateur or apartment dweller, operation on the 80 and 40 meter bands often is out of the question. The main problem is where to put the antenna. Various schemes, some of them ingenious, have been tried with varying degrees of success. Each has had its drawbacks. In preparing a book on amateur antennas, I felt that there was a need for a simple indoor antenna that the average Novice could build economically, that would produce the results needed to encourage the neophyte further into the hobby. The worst problem exists on 80 meters, and it was there that this experiment centered.

After trying numerous configurations, it was decided that a closed loop offered the best hope. Loop antennas are often treated as a specialized class. Little has been written on them, in compari-

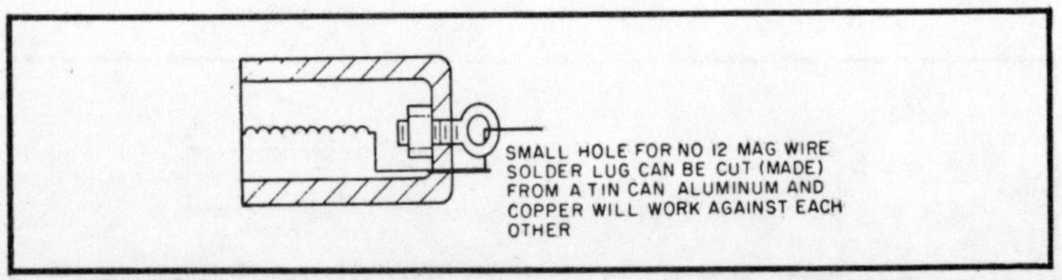

Fig. 7-23. The coil will expand to make a nice fit inside the PVC tube. The aluminum fits snugly over the PVC, and the cap rims help hold the aluminum tube in place. It all makes a very nice looking assembly.

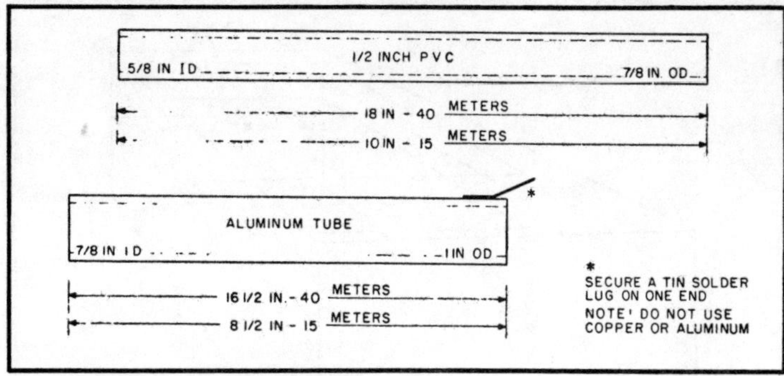

Fig. 7-24. Securing the solder lug.

son to the volumes written on other configurations, so little is known of them by the average amateur. Except that they can be a bit tricky to tune, there is no rational reason—it's just a class of antennas that has never been fully explored. See Figs. 7-25 and 7-26.

One thing that is known, however, is that a loop can be a very efficient radiator *if* it is properly matched to the transmitter. There is where the big hole is; few loops are matched to their associated equipment, and the consequent poor results quickly discourage the user.

In spite of its apparently large size, the loop described here is in the class known as small loops. A small loop is one in which the total length of the wire used is small compared to a wavelength. Current in a small loop is all in one direction, and is fairly uniform in magnitude. This loop uses a full 130 feet of wire—just ½ wavelength. It is consequently about as big as you can get and still have a small loop.

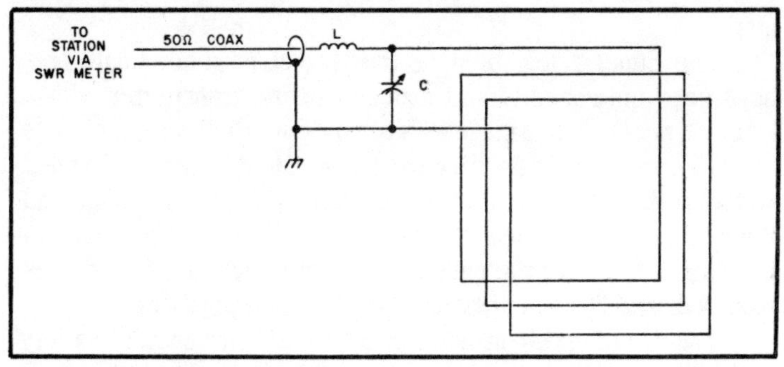

Fig. 7-25. 130′ wound in 3 turns 2″ apart. L—7 μH, 24 turns, ¾″ diameter, 1-⅛″ long. C—365 μF.

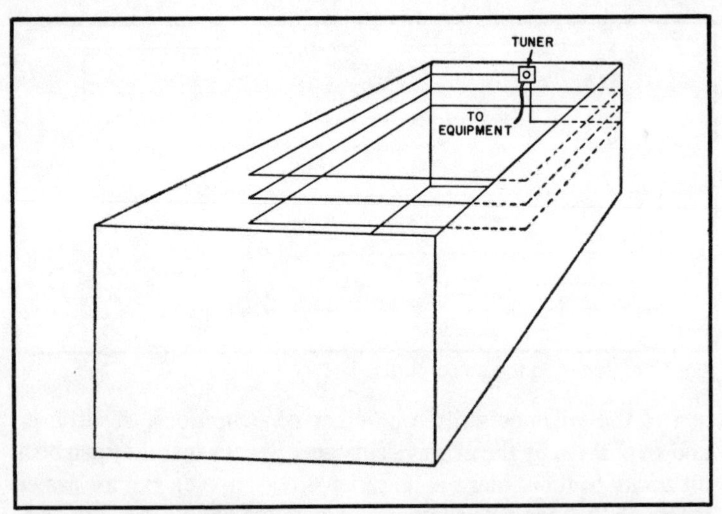

Fig. 7-26. The loop is wound three times around the room near the ceiling. It's nice if it just fills the room, but it doesn't have to. Also there's no reason why it cannot be mounted on the roof.

Small loops behave as large inductors. They can be tuned to any frequency at which they're still small loops with the appropriate capacitor. Their radiation is polarized perpendicular to the plane of the loop, and nearly omnidirectional in the plane of the loop, with virtually no radiation in the directions perpendicular to the plane of the loop. This may seem a contradiction to the next statement because of the polarization. A small loop is considered in engineering circles to be a magnetic dipole. That is, it does with the magnetic component of the wave what a dipole does to the electric component. Being primarily magnetic in its behavior, it is relatively insensitive, when receiving to lightning and man-made static.

The Super Loop, then, is the largest possible small loop, positioned horizontally, and matched to the transmitter with a simple L network. It is made with common doorbell wire, since it is used indoors. Doorbell wire comes in standard lengths of 65 feet, so two rolls make a nice half-wave antenna. It was wound around the wall of the room near the ceiling, the runs spaced about two inches apart. Where the ends came together, the L network was mounted, and 50 ohm coax fed down to the equipment.

The L network consisted of a 365 μF "broadcast" variety variable capacitor, one of the few parts still easily obtained, and a 7 microhenry inductor. The inductor was made by winding 24 turns of #18 wire on a scrap of ½ inch PVC pipe, also easy to obtain (½ inch pipe has an *outside* diameter of ¾ inch). When close wound with #18 wire, the length of the winding is 1-⅛ inches. If other wire size is used; it should be space wound to fill the specified length.

The initial tune-up or Super Loop can be a stinker, and a reflected power meter is recommended. Set the antenna tuning capacitor to about ⅔ of its full capacity. With your transmitter set for reduced drive, tune it up as you would a conventional installation. Then set the reduced power meter for swr and tune the antenna capacity for a dip in swr. Finally, repeak the plate tuning capacitor of the transmitter. Because of the relatively small impedance transformation, the output tuning capacitor (sometimes called the load capacitor) of the transmitter can interact considerably with the antenna tuning capacitor. It is here that things can get sticky. The trick is to find that point where the two seem to produce minimum swr. Once that point is found, the loop can be relatively easy to use.

The first experiments with the loop brought almost fantastic results until it was discovered that the loop was proximity-coupling into my outside wire. When the outside wire was taken down, results seemed more rational. Good signal reports were obtained with satisfactory QSOs as far as 1000 miles or more. The worst results came from just over the horizon, which is to be expected, since the loop is a low radiation angle device because of its horizontal position.

Swr averaged around 1.4 throughout the band, thanks to the L network. I have good reason to suspect that the dimensions of the loop have some tolerance, since it has the variable capacitor across it. There being no "ends" in the wire, the so-called end effect does not apply, and the loop dimensions are governed by the wavelength in space. As long as there is a fairly small side-to-end ratio, the form factor isn't too important. Simply adjust it so that the half wavelength of wire makes three full turns.

I have good reason to suspect that the behavior of the loop will vary somewhat from one location to another, since there is no way of predicting what length of conductor will be within its field to affect it. Nonetheless, based on the reports I've had, I think I might be on to something. The only truly valid test is its use in many different locations, and while I have several amateurs working on loops, it will be quite a while before the results can be fully evaluated.

A FORTIFIED TWO-METER WHIP

Are you in need of a good 2 meter mobile antenna? How about an antenna for your base station? Or, do you just plain have the feeling that your station has become too commercialized and that a portion of your setup should be home-brewed? Then, why not try this antenna project? In actual checks it was found to compare favorably with the commercially-made antennas tested. Whether you decide to use the antenna for mobile or base station operation, you'll be pleasantly surprised with its performance. The antenna is easy to construct and tune, and best of all, it's inexpensive.

Design

One unique feature of this antenna is the construction of the whip. It consists of a ¼"-diameter fiberglass rod with a shield of

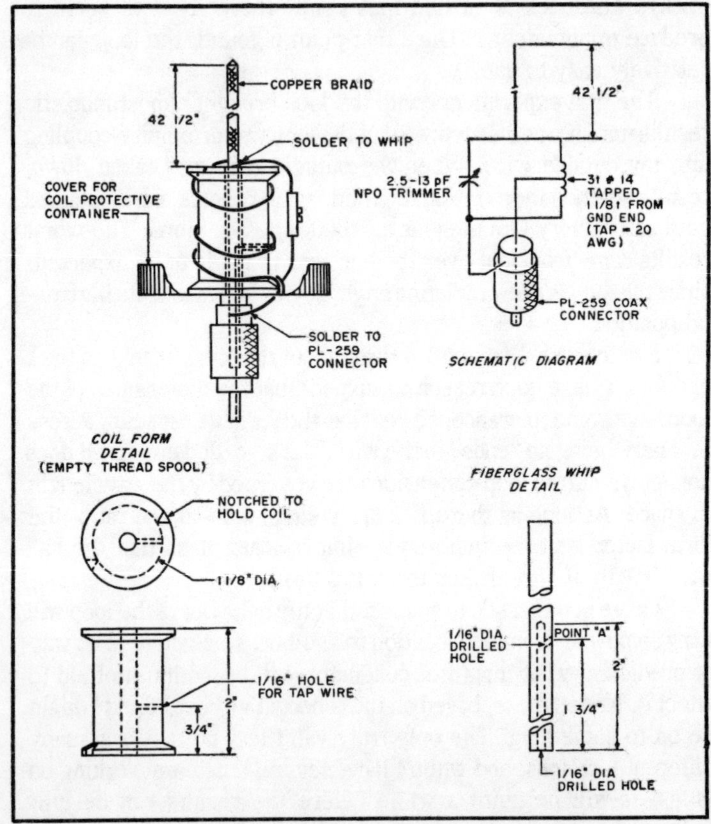

Fig. 7-27. A rigid 2 meter antenna for base or mobile operation.

Fig. 7-28. The coil protective cover has been removed in this picture to show details of the base coil.

copper braid. This design was selected because it provided rigidity to minimize deflection during high winds or mobile operation. Research has shown that deflection of the flimsy-type whip caused degradation of the vertically polarized signal. In some instances, the efficiency of a ⅝-wave antenna actually becomes less effective than a ¼-wave antenna.

The fiberglass rod is from the pennant-topped-type whip that is made for mounting on bicycles. Many retail stores have given away these whips as promotional items. They also are readily available from department stores and bicycle shops for approximately $1.25.

The impedance matching coil is 3 turns of No.14 tinned copper wire wound on a wood thread spool from your XYL's sewing basket. It is tapped 1-⅛ turns from the ground end. A small ceramic trimmer capacitor across the coil provides a precise match in conjunction with the base coil tap. The impedance matching circuit is protected from the weather by enclosing it in an empty plastic container. Fish food had come in the container we used.

Construction and Assembly

Since no tricky construction or special tools are needed, no problems should be encountered. The fiberglass rod is prepared by drilling the 1/16″-diameter hole from the bottom as indicated in the diagram. Another 1/16″-diameter hole is drilled on the side of the rod at point A. This should be drilled at a slight angle towards the bottom to make the routing of the coil tap wire easier. The depth of this hole is only to the extent of meeting with the hole previously drilled from the bottom. When drilling these holes in the fiberglass rod, it is important to use a sharp drill and not allow the drill to heat up. It is best to cut the whip to proper length after the coil form is secured in place.

Prepare the coil form and other parts as indicated. Check that the hole in the spool is of the proper diameter to permit the spool to slide on the rod. The notches filed into the coil form prevent the coil from slipping. The hole for the coil tap is displaced ⅛ of a turn from the alignment of the bottom notch.

After the coil form is prepared, feed the 20-AWG tap wire through the tap-wire hole and out the bottom of the spool. Slide the form over the fiberglass rod and carefully route the tap wire through the drilled hole at point A and downward through the rod, out the bottom. Allow sufficient length for the tap wire to be soldered later in the PL-259 connector. Apply epoxy glue to the appropriate rod area, slide the coil form into its proper position on the rod, and take up any slack in the top wire. The final position of the coil form should be such that the tap-wire holes in the spool and the rod line up with each other and the rod extends sufficiently below the bottom of the spool to accept the UG-176 adapter. To hold the tap wire securely in position, apply a small amount of epoxy to the tap-wire opening in the spool and at the bottom of the whip.

Two holes must be drilled in the cover of the coil protector. A ¼" hole in the center will permit it to be slipped over the bottom of the whip. With the cover in position on the whip, use the notch of the coil form for determining the position of the second hole. This is a ⅛" hole and should be drilled in the correct position to allow the ground end of the coil wire to pass through the cover and be soldered to the PL-259 connector. After both holes are drilled in the cover, epoxy the cover (threads towards the coil form) to the bottom of the wood spool. Position the feedthrough hole in the cover so that the tap occurs at 1-⅛ of a turn when the coil is added.

With epoxy applied to the bottom of the fiberglass rod, slide the UG-176 reducer onto the rod and up against the container cover. Check that none of the other parts has slipped from its proper position. At this point of construction, it is best to allow the epoxy to harden before proceeding.

After the epoxy hardens, the UG-176 reducer can be screwed into the PL-259 and the tap wire soldered in the center pin. Measure 42½ inches from the top of the coil form and cut the whip to length.

The next step is to slide the copper braid shield over the fiberglass rod. Tinned braid is recommended. However, if this is not readily available, the shield from RG-8/U coax cable will work fine. If the braided shield is too snug to readily slip over the rod, the diameter of the shield can be enlarged by squashing the braid together a little bit at a time. If the diameter of the shield has to be enlarged to any extent, be sure to allow for the shrinkage in length that will occur. The braided shield is slipped over the full length of the rod down to the coil form. The shield can be snugged to the rod by running your hand tightly along the braid.

One end of the coil is secured by routing a 24" piece of 14-AWG wire through the hole in the cover and soldering it to the side of the PL-259 connector. With the one end secured, the 3-turn coil can then be easily wound on the form and soldered to the whip shield. Solder the tap wire 1-⅛ turns from the coil bottom (ground) and the trimmer capacitor across the entire coil. Cut off the braid shield so that it extends ⅜" above the top of the rod; twist and solder. To protect the whip from the weather, shrink tubing, plastic electrical tape, or a protective spray can be used.

Mounting

For base station operation, I used a simple L-shaped aluminum bracket with an SO-239 connector RG-58 coax, and four

19¼″ ground radials. This arrangement is secured with U-bolts to the mast above a triband beam. For mobile operation, the bracket design is dependent on the type of car and individual desires. For my mobile operation, I mounted a simple bracket and connector arrangement directly to the luggage rack.

Tuning

The easiest method to tune the antenna is with a field-strength meter at a distance of approximately 2 to 3 feet. With the antenna connected and the transmitter keyed on an unused simplex channel, adjust the trimmer capacitor with a non-metallic screwdriver for a peak field-strength indication. The vswr will be minimum at this point. Numerous antennas have been built, and, on all occasions, vswrs of less than 1.2 to 1 were obtained. The antenna, of course, should be situated away from all objects and as high off the ground as practical during tuning procedures.

With a 5/16″-diameter hole drilled in the bottom of the plastic container to accommodate the whip, slide the container over the whip and screw it into its cover. If the container affects the tuning of the antenna, drill a hole in its side and retune the antenna with the container in position. With RTV, seal the top opening of the container, but not the bottom. The hole in the cover will help prevent any moisture from accumulating.

Do you want to generate conversation? Just mention on your local repeater the fact that you're using a home-brew antenna.

THE POTTED J

Some antenna is better than no antenna at all. In many circumstances where adverse weather conditions are encountered, a primary consideration should be to have an antenna that will function dependably even though it might not provide the last dB of gain. This antenna is designed primarily for 2 meter repeater usage. The antenna has only a modest amount of gain over a simple ground-plane antenna. But, it has far better structural integrity. The materials used to construct the antenna might vary a bit depending on what is available locally, but the design permits construction by anyone using only simple hand tools.

The antenna form is a commonly used variation of the old-fashioned J antenna shown in Fig. 7-29. The variation, as shown in B, simply uses a closed metal cylinder for the lower ¼ λ section instead of the open-style construction of the original J antenna. This type of construction has a number of advantages, such as easy mounting on any type of mast, relatively inconspicuous appearance, and an all-metal, dc grounded structure. Since the antenna's central element and the metal cylinder are electrically shorted at the base of the antenna, that point will be a low impedance point. At the other end of the ¼ λ cylinder, there will be a high impedance point where consideration has to be given to proper insulation between the metal elements of the antenna.

The mechanical dimensions of the antenna are given in Fig. 7-30. Depending on the materials available locally, the diameters of the cylinder and the radiating element can vary a bit from those shown. But the lengths should be closely maintained and dimensioned using the formulas shown in Fig. 7-29 for the particular segment of the 2 meter band of interest. The mechanical construction can vary a bit from that shown as long as the dimensions are maintained, the central element is securely grounded to the bottom of the ¼ λ cylinder, and the coaxial feedline is connected properly.

In the method of construction shown, the end of the central element is flattened out at the bottom and bent and bolted to the bottom side of the ¼ λ cylinder. Additionally, a small piece of the same material as the central element is flattened out and used as a brace to the other bottom side of the cylinder. The coaxial cable shield is connected via a ground lug to the inside of the cylinder and the inner conductor is either soldered or connected to the central element at the point indicated via a ground lug. The connection to the central element can be made before that element is inserted in the cylinder. The connection of the cable shield must be made after insertion. This requires a bit of dexterity working in the narrow cylinder, but with good lighting and patience it can be done. One

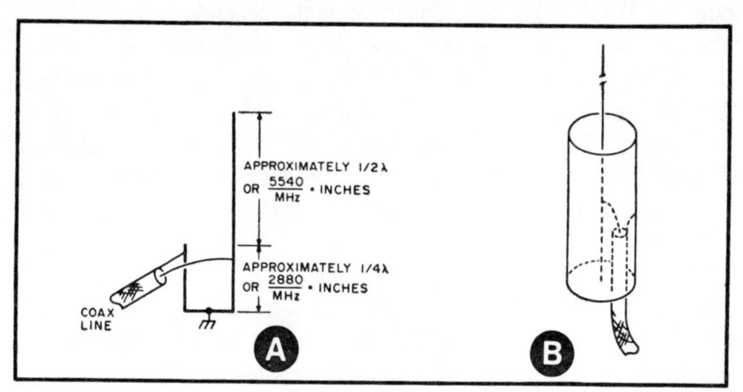

Fig. 7-29. Basic J antenna (A) and variation where lower ¼ λ section is in the form of a cylinder (B).

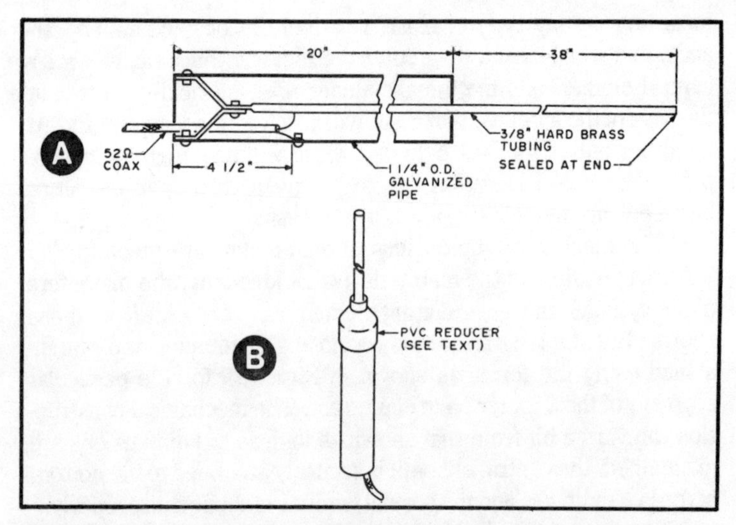

Fig. 7-30. (A) A cross-sectional view of the antenna giving dimensions centered in the 2 meter band. (B) A PVC pipe reducer used to insulate and center the vertical radiator at the top of the ¼ λ cylinder.

handy way to help things along is to temporarily glue the nut for the bolt holding the ground lug on a thin piece of wood. Once the nut is started on the bolt, the wood piece can be broken away.

Before the final assembly step, check the electrical performance of the antenna for swr. It should check out with a very low swr if proper dimensions have been maintained. If one wants to optimize the swr, the feedpoint on the central element can be varied slightly up or down. Usually this can be done without having to change the connection point for the ground shield of the coaxial feedline. This does mean going through the procedure of having to take out and reinsert the central element, but it is not at all that tedious after one does it once or twice. During this test, the central element can be held centered in the upper end of the cylinder by a PVC reducer fitting as shown in Fig. 7-3. These reducer fittings can be found wherever PVC piping fittings are available and one can be found which will fit exactly over a pipe having a 1¼″ outside diameter.

The final and most important step in assembly is to fill the cylinder in completely with an insulating compound. This will give the antenna its final mechanical rigidity and, more importantly, completely exclude moisture leakage or condensation in the cylinder. Many potting or sealing compounds can be used as long as they contain no form of metal filler. The plastic resin body fillers sold in automotive stores, with or without fiberglass reinforcement, are readily available. However, to make the filler flow readily in the cylinder, the filler should be heated so that it is fairly liquid. Don't use an open flame to make the filler liquid, but rather insert the container (a tin can will do) containing the filler into a bath of very hot water. Temporarily plug the top end of the PVC fitting where the central element protrudes and, with the cylinder initially held at an angle, pour in the filler from the coaxial cable end.

The use of a coaxial connector on the antenna was deliberately avoided. Simple coaxial connectors, when used in a harsh outdoor environment, will almost always eventually become the source of a problem. This is the same reason why screw-in elements, etc., are avoided. Of course, this all means that the antenna becomes a throw-away unit in case something should really damage it. But the cost of materials involved to build another antenna is relatively low. As long as it does last, one can use the antenna with the confidence that none of the electrical or mechanical connections inside the antenna are likely to become corroded.

The ¼ λ cylinder need not be insulated from a metal mast as long as it is fastened to the mast at the *bottom* of the cylinder with the usual metal U-type mast clamps.

One can add parasitic or phased elements around the basic antenna if it is desired to obtain some directivity. Parasitic elements, of course, require no cable connection to the main element, but then one has to go through a careful process of seeing that the feedpoint of the main element is altered to compensate for the presence of the parasitic element. In spite of the increased cable cost, etc., if one really wants to keep the weather-ruggedness for a directive antenna installation paramount, it would be better to have two spaced antennas of the type described with separate feedlines. Then, the necessary phasing, matching, and switching can be done from the protected environment inside the shack.

BREW UP A BEAM FOR TWO

Here is a 2 meter beam you have been looking for. It has the gain, directivity, and simplicity that put it in a class by itself. No cut and try, no tuning, it has a 300-ohm feedpoint, or you can use 50-ohm cable with the balun at feedpoint with the same low swr. The pattern is such that there is one lobe in front with no other side lobes.

It is rugged enough to stand 3 inches of ice with no ill effects during last winter's ice storm. There are many of these in use in

Fig. 7-31. Two-meter beam.

this area. I have two pairs of these up at 50 feet, the top pair vertical, and 40 inches below, another pair horizontal. With my IC-22 I have worked a great many DX stations on 2 meters in the last five years. In this area, there are many repeaters and many on the same frequency, so it is important to be able to work the one you want and not bother the others.

I recently made checks with the antenna you see in Fig. 7-31, taken in Sarina, Ontario, 35 miles away at the VE3SMQTH. It is up 38 feet, and the front lobe read 20 over 9; the ends of the elements on me were S1 while the back was S4, so that would make the gain about 30 dB. Stan is using an IC-22S with 10 watts out. See Figs. 7-32 and 7-33.

Ordinary hand tools are all you need to make it up. Use plated bolts and nuts, and aluminum for the brackets. The U-bolts are from Radio Shack, and the booms and elements are from old TV antennas. The cross boom is ¾-inch aluminum conduit. The insulators are ¼-inch clear plastic. I use Belden feedline UHF no. 9085. It has low loss even when wet, and I put the balun at the transceiver. It is quite broadband with the center at 146.52 and works either lower or higher with very little difference in swr.

I have a similar pair of 4-element beams that I take camping with me, with four sections of masting that go together with thumbscrews. It all goes in the trunk of my car. They, too, give me excellent coverage on the 2 meter band. So spend a few hours and get one together; you'll like it.

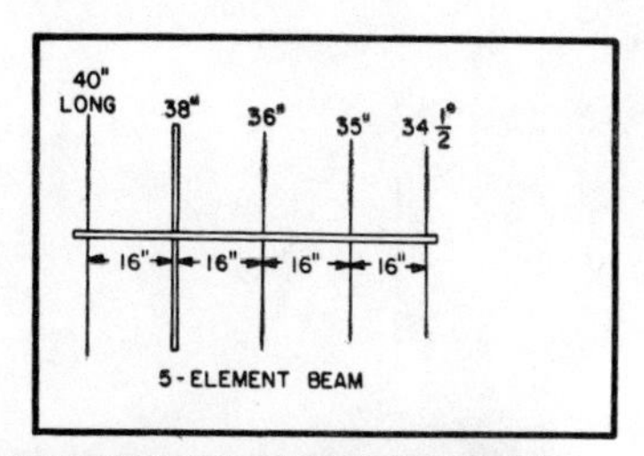

Fig. 7-32. The 5-element beam (2 required) with about 13 dB gain.

THE COLLINEAR BEAM

Ever since the ready availability of aluminum tubing after the second World War, yagi beams have been popular with hams. More recently, quads have joined as popular antennas for the HF bands. These parasitic array antennas have a number of attractive features, including high gain, unidirectional radiation, and rotatability. Realization of their full potential, however, requires relatively expensive support equipment, including tower and rotor.

I have enjoyed using collinear wire beams, which give reasonably narrow bidirectional radiation patterns. These arrays are composed of elements arranged in a straight line. Each element is connected to the next by a phasing stub so that all elements operate in phase. The collinear antenna is often used at VHF and UHF in vertical orientation, but at HF it is only practical to erect horizontal collinear antennas. Collinear antennas have long been used in military applications, or in any situation where need exists for a rugged, directive HF antenna which can quickly be erected and which can deliver good performance. Especially for the Novice or for the ham on a budget, the collinear should be considered because it is easy to construct and it is inexpensive.

Many antenna texts discuss collinear antennas, but few provide complete design information. Particularly omitted is information on impedances. Collinear antennas may use elements of varying electrical length, but most common is the one-half wavelength element. Collinears may have as many as 6 elements. Because of mutual coupling and increasingly uneven current distribution obtained as one adds elements, most hams have built collinears with 2 or 3 elements. Highest gain is obtained if the antenna elements are spaced with 0.4 or 0.5 wavelengths of space between the end of the elements, but this introduces construction problems. Less gain but much simpler design results from separating the elements only by use of an insulator.

I give design criteria below for 3-element 20 meter and 5-element 15 meter collinears, each of which requires less space to

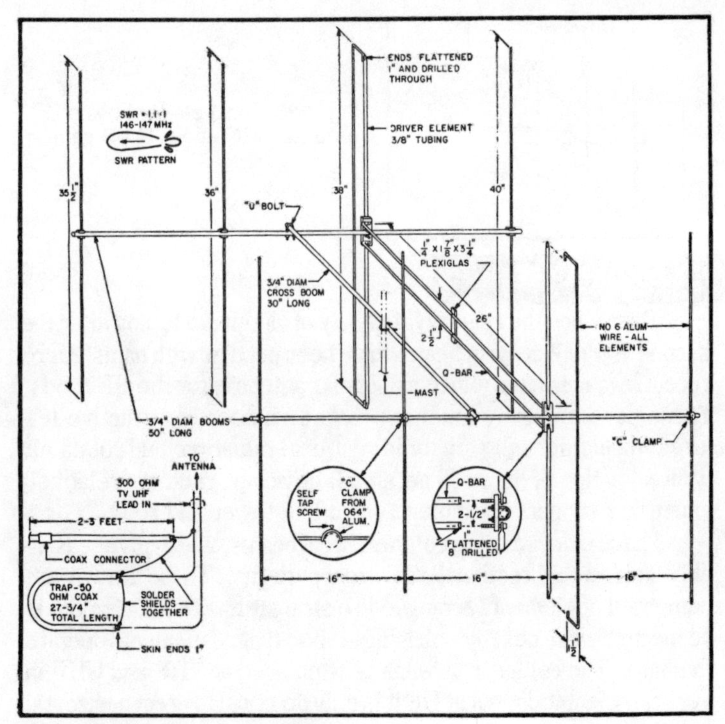

Fig. 7-33. The 4-element model.

erect than an 80 meter dipole. I also give data on impedance characteristics at and near resonance to aid in selecting a suitable feedline. Many collinears are fed between the ends of the elements, where a very high impedance exists. In the antennas described below, feed occurs in the center of an element at a relatively low impedance, high current point. This permits a good match to 300- or 600-ohm balanced lines. If one prefers, a 4:1, 6:1 or 9:1 ratio balun may be inserted so that the antennas may be fed directly with coax. My collinears are fed with balanced line, and a balun is introduced just outside the shack.

I was not able to find information on the impedances met when feeding collinears at current points, so I performed measurements on 3- and 5-element collinears suspended ¾ to 1 wavelength above ground. The impedances were measured on my Boonton 250-A "RX" meter. The impedance at resonance of the 3-element centerfed antenna was found to be 372 ohms resistive; the 5-element antenna showed an impedance at resonance of 600 ohms. Each of these antennas operates across an entire ham band with vswr of less than 1.5:1 (±1% of design frequency), using no matching device except a properly chosen balun. The transmission line is chosen to be an integral number of half wavelengths. Of course, a balanced line tuner may be employed and the balun omitted. One can also tap up and down the transmission line for impedance matching purposes if this method is desired.

It is worth pointing out that the velocity of propagation of electromagnetic waves along transmission line varies with the dielectric material used in constructing the line or cable. This velocity factor (V) must be included in calculating electrical wavelength in various types of lines. For example, open-wire line can be assumed to have a velocity factor of nearly 1.0 (or perhaps more exactly, 0.95 to 0.975). Thus, a wavelength in free space is nearly the same as one measured along open-wire line. The coax cable usually used by hams has a V of 0.66, and most 300-ohm polyethylene balanced line ("twinlead") has a V of 0.82. This means, for example, that a wave of 15 meters length in free space has a physical length in coax of about (0.66×15m)=10 meters and a physical length in 300-ohm twinlead of about (0.82×15m)=12.3 meters. In the designs below, I use 300-ohm twinlead for the ¼-wave matching sections.

The 3-element collinear achieves a gain of about 3.3 decibels over a simple half-wave dipole and has a beamwidth to the half-power points (where the field strength voltage drops to 0.707 of its maximum value) in the horizontal plane of about ±18°. This is for a horizontally oriented antenna, of course, and the radiation is greatest at right angles to the axis of the array. The 5-element antenna achieves a gain of about 5.3 decibels and a beamwidth of about ±10° or so. Each antenna's horizontal radiation pattern is bidirectional with minor side lobes. I have designated the centerfed element halves as "A" in Fig. 7-34 and in Table 7-2, the pairs

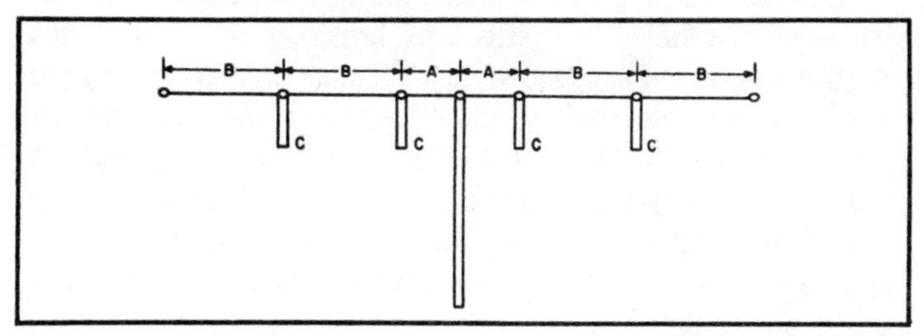

Fig. 7-34. Centerfed multi-element collinear wire beam. A = ¼ wavelength (ft.) = 246 (0.95)/f_{MHz}; B = ½ wavelength (ft.) = 492 (0.95)/f_{MHz}; C = wavelength (ft.) = 246 (0.82)/f_{MHz}.

Table 7-2. Dimensions as Measured for 15 and 20 Meter Collinears. A = Halves of Center Element; B = Outer Elements; C = Matching Stubs.

Frequency	Dimensions			½-wavelength in feedline	
				coax V = 0.66	twinlead V = 0.82
	A	B	C		
21.2 MHz	11' 4"	22' 7½"	9' 6"	15' 4"	19' 0"
14.15 MHz	16' 6½"	33' 1"	14' 3"	22' 11½"	28' 6"

of outer elements as "B", and the matching sections as "C", for the case of the 3- or 5-element centerfed collinear.

All half-wave elements should be of equal length in a given antenna. One can erect the centerfed element and adjust it to exact resonance, if desired, then add the stubs and outer elements. Velocity factor V was found to be 0.95 for the 3-element collinear. The 5-element collinear resonated slightly higher than the calculated frequency; evidently a V of 0.95 overcorrects for end effects. I found V here to equal 0.975 and I give the actual determined dimensions for each antenna in Table 7-2.

I used home-brew nylon insulators which are 2 to 3 inches long and cut out of scrap. They are light and tough. One could also use Plexiglas™ or some other material or commercially-made ceramic insulators.

The phasing sections could be of open-wire line (remember, then, V would be about 0.95), but I chose 300-ohm twinlead because the stubs tend to blow around in high winds and ladder line might twist up and short. (Also, I had a few scrap pieces of 300-ohm twinlead in the shack.)

The antenna elements themselves I made of odds and ends of #16 and #14 hard-drawn copper wire, but almost anything will do here.

Care should be taken to fasten the phasing stubs to the insulators so that they do not fatigue and break off. Solder the stubs closed at the ends. I suspended the collinears high in some maple trees using clothesline pulleys and ¼-inch nylon line. One has been up for two winters and has survived numerous ice storms and high winds. You might wish to silicone or wax the twinlead to minimize swr changes during rainstorms. I'm not fussy about these antennas, except that I make sure that I never erect an antenna near, over, or under a power line.

In Table 7-2, I have given actual dimensions for center frequencies of 14.15 MHz for the 3-element and 21.2 MHz for the

5-element antenna. I use these collinears for working into Europe with my Triton IV. The 15 meter collinear would make a dandy antenna for Novice DXing. If one has the space for an 80 meter dipole, then 2 or 3 hours invested in construction and erection of a multi-element collinear will probably result in surprise and pleasure that such a simple antenna brings such good results for only a few dollars.

THE MONSTER QUAD

For years I had used a four-element monobander, and after the loss of two towers, I decided to try the quad antenna. My first try was with two-elements on an eight-foot boom, but it did not compare with my four-element beam. Next, I used a four-element quad on a 20-foot boom. However, my beam still worked better. I was plagued with a low front-to-back ratio, high swr, and interaction between bands. So out came the books for many hours of research. The results were a quad with high forward gain, high front-to-back ratio, no interaction, and low swr with a wide bandwidth. (The following specifications as to gain are approximate but can be considered accurate by amateur standards.)

Four-Element Triband Quad Specifications

Boom length—30 feet; element spacing—10 feet, all equal; gain—13 dB; front-to-back ratio—30 dB; wire size—#14 enameled copper; five-percent difference factor between elements; design frequency—14.250, 21.300, and 28.600 MHz.

Directors 1 and 2 are the same size. I used the formula $975/f_{MHz}$. The frequency and wire lengths are 14.250 MHz—68′4″, 21.300 MHz—45′8″, and 28.600 MHz—34′1″.

For the driven elements, I used $1005/f_{MHz}$. The frequency and wire lengths are 14.250 MHz—70′5″, 21.300 MHz—07½″, and 28.600 MHz—35′1″.

For the reflectors, I used $1030/f_{MHz}$ to obtain wire lengths of 14.250 MHz—72′3″, 21.300 MHz—48′4″, and 28.600 MHz—36′0″.

Spreaders

I used one-piece fiberglass spreaders 13-feet long and screwed eyes through the arms to run the wire (see Fig.7-35). This lets the arms move in the wind and not break the wire, and also lets the wire draw and sag with temperature changes and not bow the arms. A note of interest: Bamboo can be used but should be wrapped with two-inch wide duct tape and then sprayed with krylon® or varnish.

Placement of the screw eyes is done by taking the wire length in feet for each band, dividing the result by four, and inserting that number into the formula $A=C/\sqrt{2}$, where A is the distance along the spreader from the center of the boom to the drill point and C is the length of the element divided by four.

Example: Find drill point for 20-meter driven-element wire:

14.250 MHz=70′5″
70.5 divided by 4=17.625=C
Using $C=\sqrt{2}$, $A=17.625\ \sqrt{2}$, $=17.625/1.414$,
=12.46′ or 12′5″ from center.

Below are the drill points for each element:

Directors 1 and 2:
14.250—12′1″
21.300—8′1″
28.600—6′0″

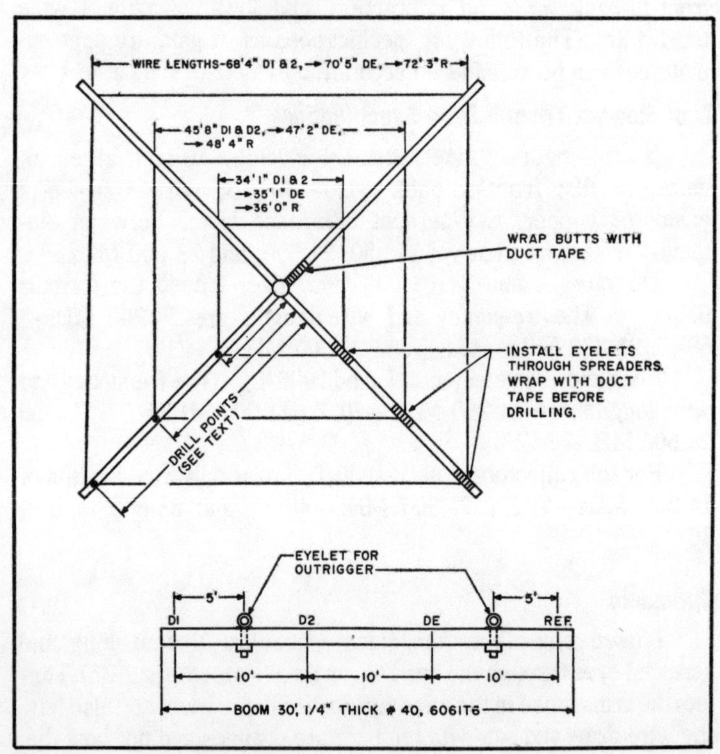

Fig. 7-35. The monster quad.

Driven element:
14.250—12′5″
21.300—8′3″
28.600—6′2″

Reflectors:
14.250—12′8″
21.300—8′6″
28.600—6′5″

These figures are to be used if you measure from the center of the boom out. To measure from butt of the arms, add 1⅜″ to each figure. This way the arms may be drilled before attachment to the boom spreaders. Each hole should be wrapped with duct tape after drilling, then a small nail can be used to punch a hole in the tape. Each spreader should be sprayed with krylon® or other type of coating to increase its life and prevent the eyelets from rusting. I also wrapped the butt ends with duct tape for added strength.

Feeding the Quad

I decided to use ¼-wave stubs after burning up a one-kw ring transformer. It's no fun waiting two weeks for a new transformer before you can operate! I used 72-ohm coax, but kW-rated twinlead can also be used.

Below are the lists of lengths for both coax and twinlead using the formula $L = 246(VF)/f_{MHz}$, where VF is the velocity factor of the transmission line used.

Stubs: RG-11A/U coax, Z=72 ohms, VF=0.66. Length to match driven elements: 14.250—11′4″, 21.300—7′6″, and 28.600—5′6″. For 1-kW twinlead, Z=72 ohms and VF=0.71, 14.250—12′3″, 21.300—8′2″, and 28.600—6′1″.

The stubs should be cut as close to the lengths shown as possible, a PL-259 and barrel connectors installed on one end, attached to 52-ohm coax to the shack. I tuned each 52-ohm feedline to the shack using my noise bridge and R-4C so I would have little swr on my feedlines.

One problem many hams have is how to string the spreaders. I drove a 2″-diameter, 4′-long pipe into the ground and attached the arm supports to this pipe. I then drove 2 wooden 3′ stakes into the ground for each arm to keep them straight. By using this type of jig, each element can be wired, removed, and then placed on the boom. I covered all nuts with General Electric clear silicone rubber, and then I sprayed them with krylon®.

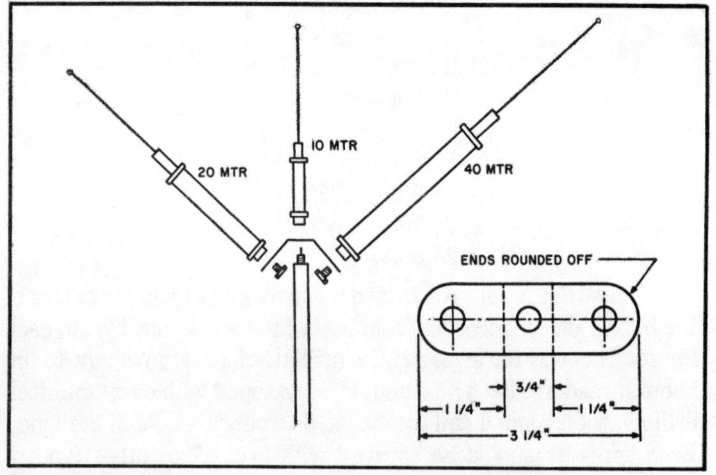

Fig. 7-36. Four-band mobile antenna construction details.

Conclusion

After the antenna was installed, measurements were made. The swr was 1.6:1 at its highest point on any band, with very flat response across each band. I can operate either the CW or phone portions with the swr never going above 1.6:1. I have been using the antenna for about two years and have yet not to make it through the pileups. The work involved is well worth the time, considering the results obtained.

FOUR-BAND MOBILE ANTENNA

Do you ever find yourself cruising down the highway working forty meters and wishing you could switch to twenty, fifteen, or ten without having to stop to change resonators? You can! I experimented with this contraption in 1960 and have been using it ever since. There is even a commercial version that came out a couple of years ago.

Any set of three resonators may be used but I prefer the 40-20-10 combination since you also get a 15-meter fallout from it. When I first started using it with an old Galaxy V, I installed a remotely-operated super-tuner gizmo in the trunk, but later found that with patient stinger adjustments on the three resonators, the tuner was not really needed. I am presently using an Atlas 210 for mobile, and the broadbanded rigs are supersensitive to swr over 1.5:1. I have worked many foreign countries with this rig with good signal reports. See Fig. 7-36.

The strap that holds the resonators must be of sufficient strength to prevent the angles of the forward and aft resonators from changing, Changing this angle affects the resonant frequency. The strap I happened to start out with was about one inch wide and 3¼ inches long. A hole large enough to accommodate the threaded extension on the mast is drilled in the center. (I happened to have had all Hustler equipment so that is what I have used since.) The other holes on each end are large enough to hold non-corrosive bolts that will screw into the bottoms of the other resonators. The strap is bent as shown with each end dropped at forty-five degrees. The assembly is attached to the mast and held in place by the center resonator, and the time-consuming tuning is started.

Begin with the lowest frequency resonator, adjusting to the lowest swr, then proceed to the next higher and then the last. The resonators interact, and this procedure must be repeated several times until the swr no longer can be improved. I have 1.1 at 7.260 and 1.35 at 7.225 and 7.295. On twenty meters, the swr is less than 1.35 across the band and even better on ten. The swr does peak up to 1.5 on fifteen meters. I use the ten-meter resonator to hold the strap to the mast with the twenty-meter resonator in front and the forty-meter aft. This streamlines the assembly in the direction of travel and reduces wind resistance. Also, with the larger resonator aft, it tends to stabilize the assembly at normal highway speeds.

I have found that the majority of the noise associated with mobile reception can be eliminated by using the copper braid out of

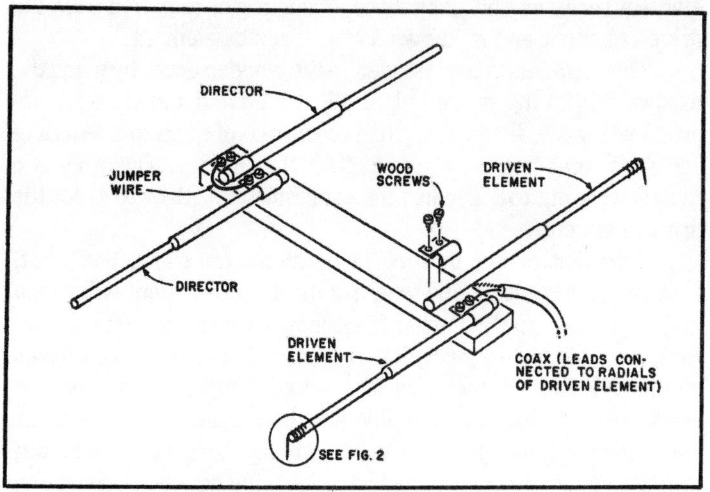

Fig. 7-37. Beam construction.

RG-8 coax as grounding the car hood and trunk lid to the frame of the car. In addition, the exhaust should be grounded in at least two places, one in front of the muffler and one aft. Be sure to scrape the rust off to bare metal when attaching the grounding straps. The braid can be cut to the desired length, the ends shaped to hold a bolt, and then heavily soldered to make a good connection.

INEXPENSIVE BEAM

Looking for something fast, easy, inexpensive, and effective on 10 meters? Well, this could be right down your alley.

A good majority of amateurs in the ham world started out as CBers. What these ex-CBers don't realize is that they probably have the materials for a good two-element 10 meter beam around the house just sitting there and collecting dust. How about that old CB ground plane you used to use? If this is one of your household treasures, half the battle is won. Here's what I did.

I took my half-wave ground plane, (the particular one I used was a Super Mag), and disassembled the four radials and hardware that fastened the radials to the base. I then went to the local lumber store and got a five-foot piece of 2×2. Jerry Swank W8HXR was the supplier of the technical information, such as length of the driven and director elements, the spacing, etc.

What I was going to attempt to build was a 10-meter beam incorporating 2 elements. In this beam, one element is fed directly from the feedline and is a half wavelength of the operating frequency (Fig. 7-3). This is called the driven element. The other element receives power by either induction or radiation from the driven element and is known as the director element.

The elements were fastened to the wooden boom by using the hardware from my ground plane. Since the four radials were approximately 8½′ long each, I had to cut two of them to a length of 7′7″ for a combined length of 15′2″ for the director. The other two radials were cut to 8′2″ each, for a combined length of 16′4″ for the driven element.

If you cut one or both of the elements too short, don't fret, because all is not lost. The length of the driven element will vary a little according to the design frequency. I cut my driven element too short and coiled a piece of bare wire on both sides of the driven element. I then let the wire drape down until I got a good swr reading (Fig. 7-38). Coaxial cable was then connected between the two radials of the driven element, being sure the shield was insulated from the center conductor. Spacing between driven element and director was around 4′3″.

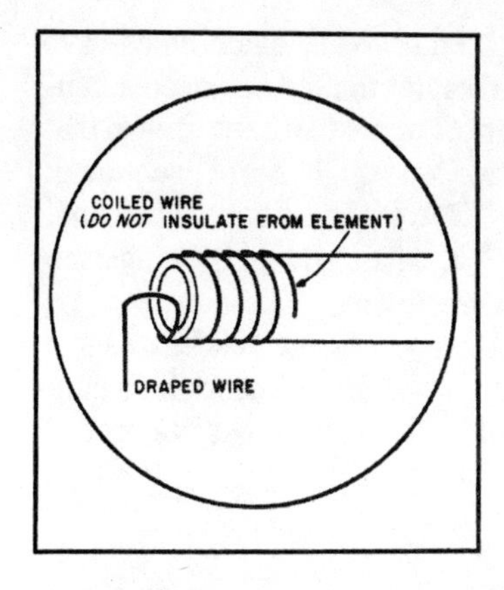

Fig. 7-38. Wire coiled on end of element.

The gain of this antenna was figured to be somewhere around 5 dB, but don't let this small figure scare you. Using the Kenwood TS-520S, this 80-cent cheapie (the cost of the 2×2) was found to make a difference in receiving of 2 to 4 S-units higher than my 4-BTV vertical, and 2 S-units when transmitting. (Polarization of the other station will, of course, play an important role.)

So, as you can see, for a very small amount of space you can have a pretty effective DX antenna for a fraction of the cost of a commercially-made beam. However, if you really feel ambitious and have the extra aluminum around the house or can obtain it for a reasonable price, another element can be added for an additional 3 dB of gain. A balun is not essential, but it could prove to be quite helpful in preventing your coax from radiating.

If you're thinking that it takes an expert craftsman and years of experience to build your own beam, you're wrong. When I built this antenna (about two years ago), I was 15 years old and had just a Novice class license; I know that if I can build my own antenna, anybody can!

MULTIBAND GROUND PLANE

The idea of feeding a vertical antenna—more specifically, a ground-plane antenna—with balanced transmission line is not new. Balanced, open-wire line lends itself well to multiband antenna applications because of its low loss in comparison with prefabricated coaxial lines. The transmission line used with the antenna described here is designed to have the least possible attenuation within reasonable limits.

How can balanced line possibly work well with an unbalanced antenna? Won't there be radiation from the line, and won't this cause horrible TVI and other problems? Well, since hams have been using unbalanced line (coaxial cable) to feed balanced antennas (dipoles and beams) for a long time and it has been shown that this is a perfectly satisfactory practice, one might be inclined to ask whether the situation should be any different the other way around. In the case of the antenna at W1GV/4, balanced line is being used with success to feed a ground-plane antenna.

Let's take a look at some of the theoretical considerations of balanced transmission lines and antennas before I describe a multiband ground-plane antenna without coils or traps that has proved very effective.

Parallel-Wire Line

In the old days of radio, the type of transmission line most often used was of the open-wire variety. The reason this kind of line works is that at every point along the line the currents in the two wires are always equal in magnitude and opposite in direction. Since the two wires are very close together with respect to the wavelength, they may be considered to occupy the same space. The field produced by one wire therefore cancels that produced by the other wire, and no radiation takes place.

For the currents in the two wires to be exactly equal and opposite, the antenna must have certain characteristics. If one side of the antenna presents a different impedance than the other, the currents in the two feedline wires will not be exactly equal or exactly out of phase. This may occur because one side of the antenna is longer than the other, or because one side of the antenna presents a different capacitance with respect to ground. See Fig. 7-39.

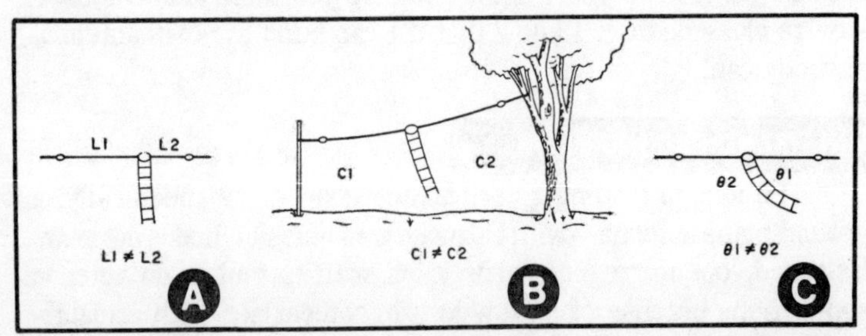

Fig. 7-39. Causes of feedline radiation. At A, antenna not fed at center; at B, one side of antenna closer to ground or obstructions; at C, feedline brought away from antenna is a non-symmetrical position. These factors can cause radiation from coaxial as well as parallel-wire lines. A perfectly balanced system is, fortunately, not usually necessary for satisfactory antenna performance.

There is a third reason for radiation from a parallel-wire line: antenna currents. If one side of the antenna is closer to the line than the other side, the electromagnetic fields from the two halves of the antenna will not cancel each other in the vicinity of the line. This will induce a current in the line equal in magnitude in both wires but in the same direction (Fig. 7-39C). Consequently, the line radiates because this current produces its own electromagnetic field. These antenna currents cause the net current flow in the two wires not to be equal and opposite, and this can result in trouble with an all-too-familiar gremlin for hams: rf in the shack!

Usually, a small amount of radiation from a transmission line is not a great handicap. This is fortunate because it is almost unavoidable. We certainly would want to minimize line radiation if we were using a highly directional antenna where front-to-back ratio is important. But with a simple antenna such as a dipole or ground plane, we need not worry about some deviation from theoretical perfection.

Coaxial Line

Many hams got it into their heads that coaxial line is shielded and therefore cannot radiate. This is not the case! Coaxial lines are just as susceptible to radiation-causing factors as parallel-wire lines. Antenna currents, induced in the outer conductor of coax, will produce electromagnetic fields and rf in the shack. These currents can be caused by any of the three situations shown in Fig. 7-39.

Coaxial line has the advantage of being easy to install. It can be run close to or directly over metal objects such as gutters and pipes, and its attenuation characteristics and impedance will not be affected. This is not true with parallel-wire line. Metal objects very close to the latter type of line will cause "impedance bumps" and possible imbalance.

The main disadvantage of coaxial line is that it has relatively high attenuation. The swr becomes important at high frequencies or with long runs of line. Generally, the antenna must, impedance-wise, be fairly well matched to the line if coax is to be used with maximum success. Any time the swr is 2:1 or better, the line will function at essentially full efficiency. But an swr of, say, 20:1, will almost always cause significant signal loss. Furthermore, such a severe mismatch can cause conductor or dielectric breakdown because of extreme currents and voltages at nodes along the line.

A heavy-duty, balanced transmission line such as is used at W1GV/4 can be operated at amateur power levels with utter disregard for the swr. Thus, all the matching can be done conveniently at the operating position by means of a transmatch.

The Ground Plane

A full-size ground-plane antenna consists of a quarter-wave vertical radiator and several quarter-wave radials (usually three or four), and the base is at least a quarter wavelength above the ground. Such an antenna exhibits excellent low-angle radiation characteristics and consequently is good for DX work. It is less effective for local communication where the angle of radiation must usually be nearly 90 degrees with respect to the horizon. However, the low-angle reputation of vertical antennas has been somewhat overemphasized. Even at a radiation angle of 45 degrees the field strength is nearly as great as it is parallel to the horizon—see Fig. 7-4.

The ground plane is an uncomplicated and versatile antenna.

As the base is lowered to heights of less than ¼ wavelength above ground, losses begin to occur because the *ground currents*, which should be confined to the radial system, will begin to flow in the lossy earth. Three radials comprise a nearly perfect ground if the base is sufficiently elevated. At a height of ⅛ wavelength, some of the ground currents will flow in the earth unless radials are added. The closer the base gets to the ground, the more radials are necessary.

Suppose we tune a full-size, 20-meter ground plane by means of loading coils so that its resonant frequency becomes 7 MHz. If the antenna was ¼ wavelength above the ground on 20 meters, it will be ⅛ wavelength above the ground on 40 meters. The radial system will not be as good at the lower frequency because of the lower height. Also, the radials are not ¼-wavelength long on 40. Although they are an electrical quarter wave in length, their physical length is just ⅛ wavelength. This fact, too, will cause more of the ground currents to flow in the soil. But suppose it is impractical to make the radials any longer; how are we to improve the efficiency without raising the antenna?

The answer is, of course, to add radials. More radials will be required if their physical length is ⅛ wavelength than would be necessary at the same height with ¼-wavelength radials. I did not try to mathematically figure out how many radials I would need. I just decided, arbitrarily, that eight would be a good number. But

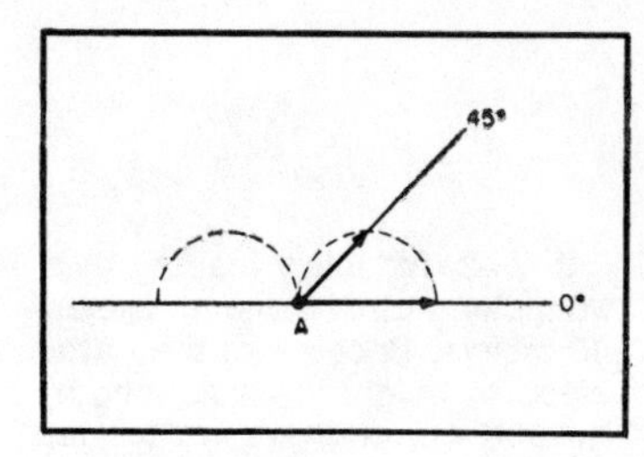

Fig. 7-40. Vertical-plane radiation pattern for a quarter-wave ground-plane antenna. Although it is generally thought that such an antenna radiates only at very low angles, we can see that it radiates quite a lot of energy at high angles. Only near the zenith is the radiation level very low.

maybe I'm getting ahead of myself. The loading system I used did not require any coil winding; this would have restricted me to 40 meters, anyhow. I wanted a 40-through-10 system without any traps or coils or stubs or multiple radiators. The mission was accomplished simply be feeding a 20-meter ground plane with open-wire line.

The Feed System

The actual installation is roughly illustrated in Fig. 7-41. The antenna acts as a full-size, quarter-wave ground plane on 20 me-

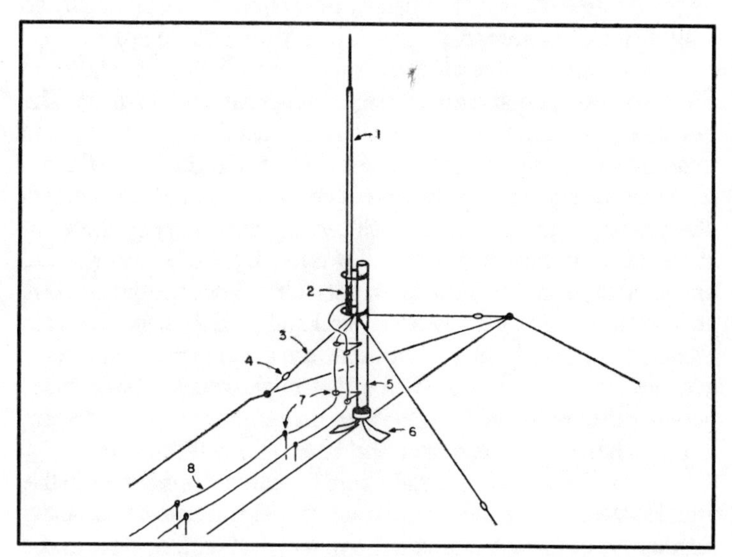

Fig. 7-41. The system at W1GV/4. Only three radials are shown, but there are actually eight. A feeder-wire spacing of 5 inches is maintained. (The viewing angle may give the impression that they are unevenly spaced.) Illustrated by number: 1—16-foot vertical radiator; 2—hose clamps attaching feeder to vertical radiator; 3—radial wire, 16 feet long, no. 8 solid aluminum ground wire; 4—strain insulator; 5—mast, 5 feet long; 6—TV base plate; 7—TV standoff insulator; 8—feeder wire, #8 solid aluminum ground wire.

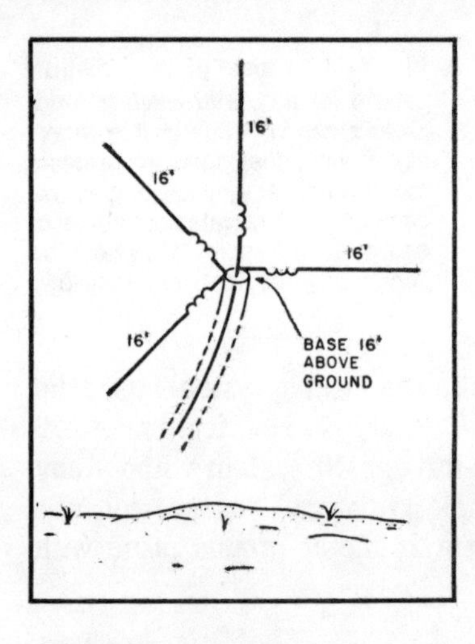

Fig. 7-42. Inductive loading of a 20-meter ground plane for use on 40 meters. Because of the lower electrical height and radial length, ground currents are no longer confined to the radial system. To restore a good image plane, we must either raise the antenna or add radials.

ters. On 15 and 10 meters there is some gain at low radiation angles and correspondingly less radiation at higher angles. This works out very nicely since only low-angle radiation will be returned to Earth by the ionosphere on these bands; the skip distance is usually so long that high-angle radiation will pass through into space.

On 40 meters, the antenna is the equivalent of that shown in Fig. 7-42, except that there are eight radials rather than three. The radiation resistance at the feedpoint is very low on 40, so it is important that the connections between the feedline and the antenna be excellent. The feedline is constructed of #8 soft-drawn aluminum TV ground wire, which can be found in most hardware stores. The spacing used at W1GV/4 is five inches, but any spacing between three and six inches is satisfactory. Since the dielectric is air and the wire is very heavy, this line has just about the least attenuation possible. All connections are aluminum-to-aluminum, avoiding any corrosion problems that might result from contact between dissimilar metals. (Perhaps immersing the entire system in liquid helium would reduce the attenuation still further!)

The antenna itself is unbalanced; the current is different in the radial system than in the vertical radiator. This is true on all bands. Furthermore, the antenna is non-symmetrical with respect to the transmission line, and some feeder radiation is thus inevitable. It should be noted, though, that coaxial line also will suffer from the non-symmetry of the ground plane.

This non-immunity of coax to radiation-causing effects has already been discussed. How much difference, if any, is there in practice? This was determined by means of a field-strength meter.

Originally the system was a coax-fed 20-meter monoband antenna. Tests were conducted at various locations in the vicinity of the feedline on 20 meters both before and after the changeover to balanced line. There is a little bit more radiation from the open wire, but the difference is hardly noticeable. Radiation levels in the shack are the same with both types of line (at the same power level, of course).

On 40 meters, the antenna is "tuned" by the feed system rather than by coils as in Fig. 7-42. (The ⅛-wavelength section of line closest to the antenna has replaced the inductors.) On 15 meters, the feedline also "tunes" the antenna. On 10 meters, the feedline has no actual tuning effect; the antenna is theoretically voltage-fed on this band.

We can be pretty sure that the swr is very high on all bands. But the line has such low attenuation that the swr is of no practical concern. We haven't even paid any attention to the characteristic impedance (Z_0) of the line! It is probably about 500 ohms. Assuming the feedpoint impedance is 40 ohms resistive on 20 meters, the swr is about 12:1 on this band. It is no doubt quite high on 15 and 40 meters also, because of the reactance at the feedpoint on these bands. On 10 meters, the feedpoint presents a pure resistance, but its value is difficult to predict. There is a possibility that the line is nearly flat on 10. But it doesn't really matter.

Construction Details

In order to minimize losses, every effort was made to ensure that there are no electrically "weak" points in this antenna system. The feedline spacing is five inches; long TV stand-off insulators are used to support each wire individually at 10-foot intervals. The wire is wrapped with electrical tape until it fits tightly in the large opening in the plastic part of the stand-off insulator. No splices should be made in the line; soft-drawn no. 8 aluminum wire is usually available in lengths that are any multiples of 50 feet.

The wires should be positioned so that they do not come within five or six inches of metal objects. If the wires must be run parallel and close to a metal pipe or downspout, both wires should be kept at the same distance from the object. If the feedline must cross over such an obstruction, it should cross at a right angle. These precautions minimize chances of imbalance.

The connections at the antenna are made in such a way that the primary electrical contact is aluminum-to-aluminum. One feeder wire is clamped directly onto the vertical radiator at the base, using three hose clamps spaced one inch apart. The other wire is connected to the aluminum base mount, using the nut on one side of the lower U-bolt holding the base mount to the mast. (The base mount at W1GV/4 comes from a 14AVQ that has been multilated from experimentation.)

The eight radials, each 16-feet long, also are made of #8 aluminum ground wire. The radials double as guy wires; strain insulators are used to obtain the correct lengths for radial purposes. The mast is five feet tall. This, in addition to the height of the house, puts the base about 16 feet above the ground.

It should not be necessary to go into much more construction detail. The builder can put the antenna together to suit particular needs and passions. The radials should be the same length as the vertical radiator, but 16 feet is not a magic length. Actually, the longer the better. However, any length over about 18 feet will raise the radiation angle on 10 meters; lengths greater than 25 feet also will raise the radiation angle on 15 meters. This may or may not matter, depending on band preferences.

The length can also be less than 16 feet. However, as the length is shortened from this value, the radiation resistance will decrease markedly on 40 meters and will rapidly become so small that losses will occur no matter how hard we try to prevent them. I can't tell you an exact minimum length because there is no real cutoff. You'll probably be able to make some contacts on 40 even if the length is four feet, but if you opt for this size antenna, you can rest assured that it will not be very efficient.

Performance

As of this writing, the antenna described here has been in use at W1GV/4 for about four weeks. Many contacts have been made, both long and short haul. Several thick DX pileups were cracked on the first or second attempt. Europe and Japan have been worked on 40 meters, where the antenna is half-size. Since it is a vertical antenna, we should expect that the low-angle radiation will be good even on 40 meters, so this is not too surprising. The antenna seems to work exceptionally well on 10 meters; including the image, it acts as a vertical 2-element collinear on this band.

Mostly out of curiosity, I decided to try tuning the antenna on 80 meters and found that the transmatch did provide about 1.5:1 match at 3.5 MHz. The turning was quite sharp, and I was indeed surprised to work a midwestern station and get a report of 589! Several other midwestern and northeastern stations have been worked on 80 with good reports. (Even so, I really can't believe that this antenna is very efficient at that frequency.) The antenna was designed with 40 through 10 in mind. Performance has been eminently satisfactory considering the unobtrusiveness, small expense, and simplicity of the open-wire-fed ground plane.

THE MAGNETIC MOUNT

The magnets I used are 1 inch by ⅝ inch by ⅛ inch and cost 10 for a dollar at Radio Shack, but any similar square or rectangular magnets would work. I used 8 magnets stuck together in 4 piles of 2 (Fig. 7-43). The body of the mount is the cover from a can of Krylon spray paint (in the color of your choice). Another brand of cover will work, provided that it has the inner cup, as in Fig. 7-44. Almost any good socket can be used, but I chose a phone socket, since it holds the plug firmly, is simple to mount, and is all I had in the junk box. Avoid RCA phono sockets; the antenna may fall out. If you use an SO-239, solder up the coax before setting it in epoxy. Mount the socket as shown in Fig. 7-44, and fill the inner cup with epoxy to the level necessary to hold the socket firmly without covering the electrical and mechanical parts.

Next, put the magnets together as shown, and paint them with epoxy. Before letting it dry, unravel some cable shielding and lay the wires across the top of the magnets evenly. Place a piece of waxed paper over the wire, and hold it in place with a weight until it's dry. Twist the excess wire together, and solder it to the shield connection of the plug. Then put a small hole in the side of the body of the mount, pass the coax through it and once around the small inner cup, and solder it to the socket.

Fill the larger cup of the cover about ¼ full with epoxy. Place the magnets down on a piece of waxed paper, quickly turn over the cup, and place it over the magnets so that the edges of the cup are tight against the paper. The glue will flow down and cover the magnets, sealing them in place. Put a weight on top of the unit

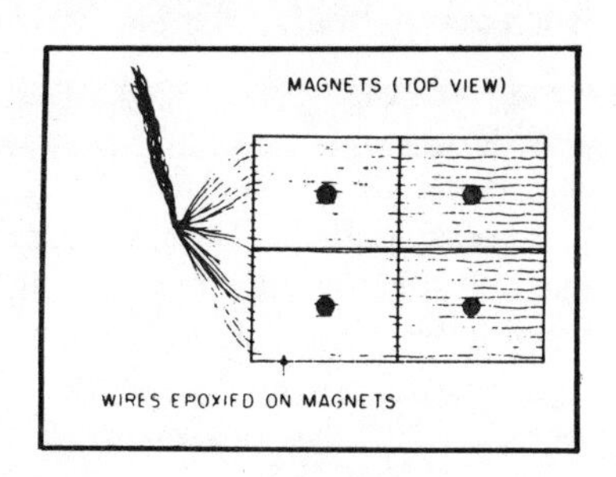

Fig. 7-43. Base of the magnetic mount.

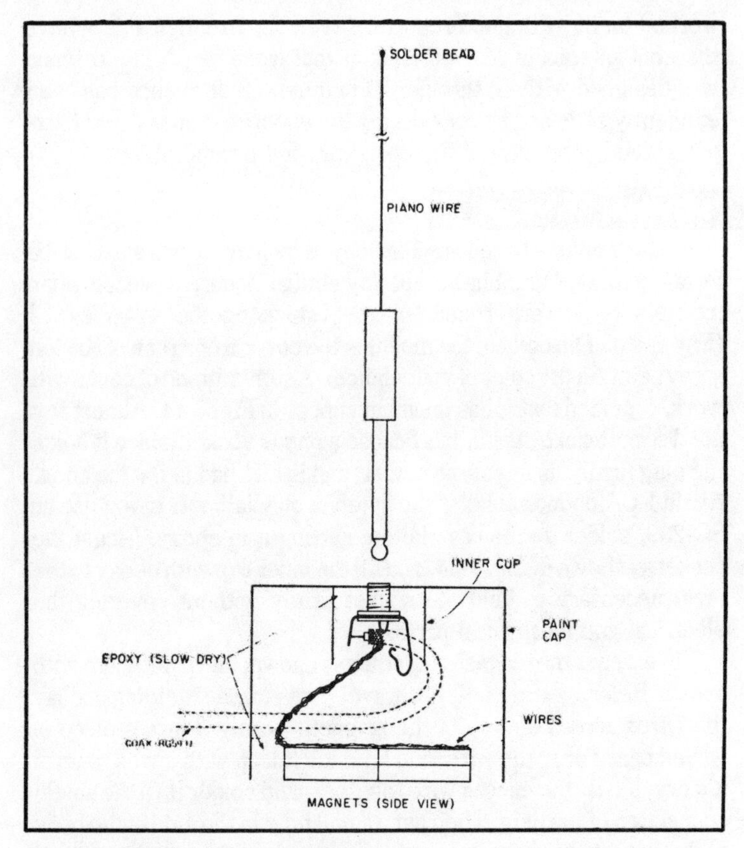

Fig. 7-44. Details of the magnetic mount.

while the epoxy sets to minimize leakage. Nineteen inches of piano wire or clothes-hanger wire soldered to the plug completes the antenna.

On-the-air tests have been excellent, and the magnet mount stays put under all road and wind conditions. And, even better, when you get out of the car, it's very easy to hide the antenna.

MAGNET MOUNT ANTENNA FOR TWO-METERS

Here is a truly versatile magnetic mount two meter antenna that is easy to build. The heart of this antenna is the magnet, which must be able to retain its grip at high driving speeds and when low hanging branches hit the antenna. After some searching in my basement for a suitable magnet, I found a burned-out twelve inch loudspeaker. The speaker had originally been used in a rock and roll guitar amplifier and had a large 2½ lb. ceramic magnet. Such a speaker could be picked up gratis from any musical repair shop.

Assembling The Antenna

Before handling the magnet, remove your watch. Treat the magnet gently, since the ceramic is brittle like glass. Carefully remove the magnet assembly from the loudspeaker frame with a small cold chisel. Now remove the metal pole pieces that sandwich the magnet. These are usually lightly glued to the magnet and can be removed with the aid of a rubber hammer. The large pole piece is turned over and glued to the top of the magnet with epoxy (see Fig. 7-45).

On the top of the magnet assembly, a loading coil is placed. I used an old high voltage insulator, but almost any insulating material such as phenolic, plastic rod or pipe will do. If copper pipe caps are epoxied to each end, this would form a suitable coil form. Another excellent coil form would be a large blown-out cartridge fuse which your local power company may be able to provide. The dimensions used are not too critical and can be anything from ¾ to 1¼ inches in diameter and about 8 inches high.

Fig. 7-45. Close-up view of base and loading coil.

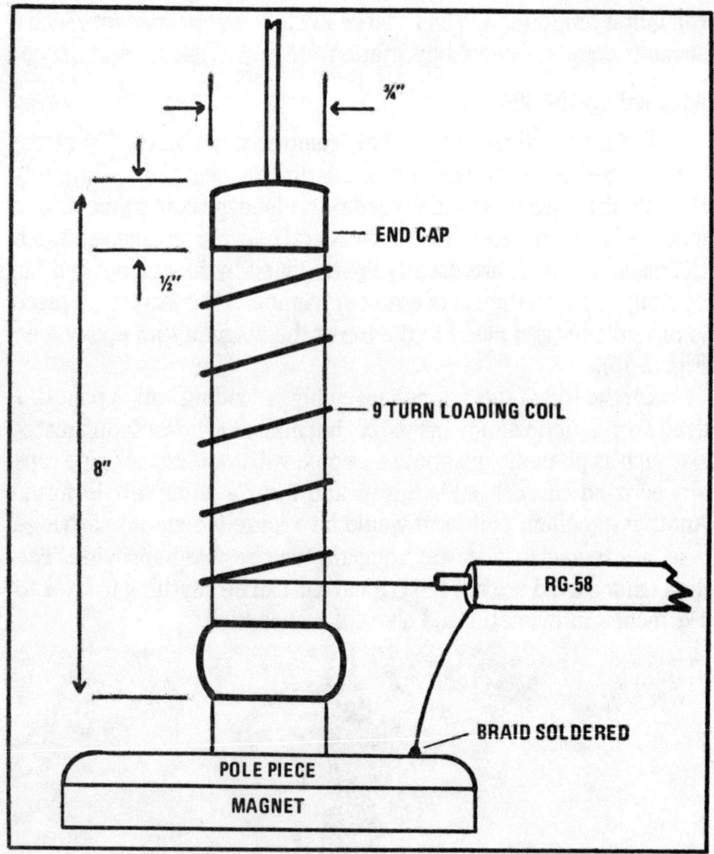

Fig. 7-46. Loading coil and base.

On this form 9 turns of #14 gauge copper wire is wound. The type of wire is not terribly critical and thinner wire could be used. As a radiating element, a 39 inch piece of coathanger or other stiff wire is used. For my antenna, I straightened out a heavy coathanger in a vise and soldered it to the top cap. The coil assembly is now either soldered or epoxied to the metal base.

Finally a 14 foot length of RG-58 coaxial cable is connected to the antenna, with the inner conductor soldered to the bottom end of the loading coil. The outer braid is soldered to the pole piece with a large soldering iron. Alternatively, a hole could be drilled and tapped in the pole piece. The braid is then secured by means of a screw and lock-washer. As a final step, the exposed braid and inner conductor are sealed with silicone RTV to prevent water from seeping into the coax.

Tune-up

The completed antenna is now placed on a car roof or other large piece of sheet metal and the radiating element trimmed for lowest vswr. If a vswr meter is not available, cut the antenna to 39 inches in length, and as long as reasonably low power levels are used, the antenna should perform well. A good rule of thumb when testing any home brew antenna is to make sure that it receives correctly. Check it with an ohmmeter before firing rf into it.

Conclusion

My antenna, powered by the one watt from a TR-22 and placed atop a 1960 Chevy, produced truly spectacular results. The WR1ABV machine in Boston could be hit solidly while driving through southern New Hampshire, at an airline distance of over 40 miles. Operating stationary mobile from Easton, New Hampshire, it was possible to hit the WR1AEA machine in northern Vermont with good results, a total air distance of 60 miles. I hope this antenna works as well for you as it has for me. Also, see Figs. 7-46 through 7-48.

A DELUXE QRP TRANSMATCH

I have found that QRP has many advantages. High voltage components are unnecessary, bringing the cost of components down to a reasonable range.

Because of my low budget and the low cost of the Argonaut 509, it was love at first sight. The Argonaut has many advantages. At present I have no plans to use a linear, so I have no worries about an antenna switch or a tuner.

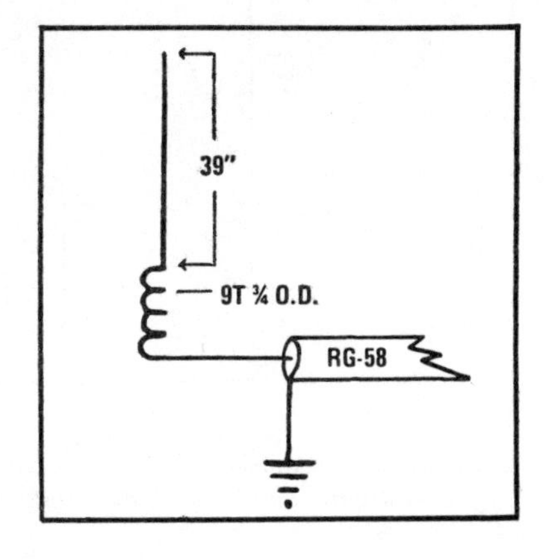

Fig. 7-47. Schematic diagram.

Fig. 7-48. Antenna mounted on top of author's 1960 Chevrolet.

Figure 7-49 shows a complete schematic of the circuit. The tuner can be switched in or out of the circuit. At position 11 of the switch, I have provided a 50 ohm dummy load. This can be left out, if not needed. If used, it should be capable of handling your transmitter's output. Position 12 is left unused on my setup, to provide the option of a 100 kHz calibrator.

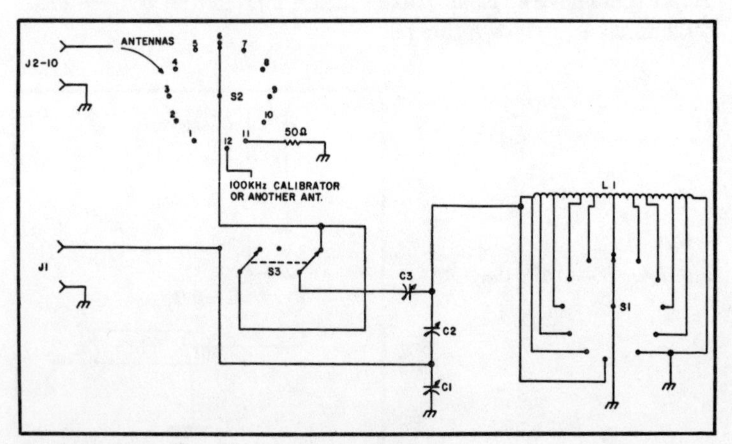

Fig. 7-49. Note: Tap L1 every several turns.

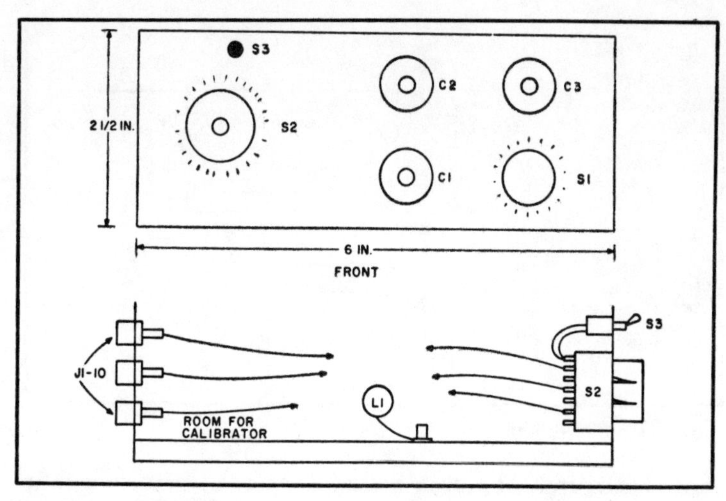

Fig. 7-50. Circuit layout.

The tuner circuit was taken from the *Radio Amateur's Handbook.*

L1 could be a roller inductor, but that would drive up the cost of the circuit. The tuner works well on 80-10 meters.

Figure 7-50 shows the layout of my circuit. Extra room was provided for the calibrator option.

For compactness, I used miniature 365 pF variables. C1-C2 could be a gang tuned capacitor, but that would cost more and take up more space.

The phono plugs should be the shielded type, to prevent unwanted radiation. See Table 7-3.

The cost of the unit should be 16 or 17 dollars, depending on the cost of the case and what options are used.

NO-WIRE ANTENNA SWITCH

Originally I came up with this idea to feed a tri-band quad. Now, the quad is a fine antenna and the fact that it can easily handle

Table 7-3. Parts List.

C1-C3	365 pF variable capacitors (Radio Shack 272-1341 or equivalent)
L1	B & W 3012
S1, S2	1 pole, 12 position (Radio Shack 275-1385)
S3	DPDT (Radio Shack 275-1546)
J1-J11	Single hole phono jack (Radio Shack 274-346)

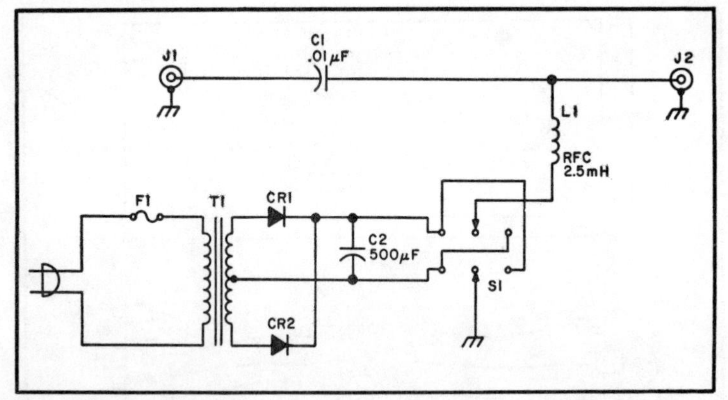

Fig. 7-51. Control unit.

three bands makes it even better. However, there is a little problem in feeding three driven elements on one antenna. Neither three feedlines nor a matching system appealed to me, and neither did using three hundred feet of RG/8U. The result was a remote switch mounted on the boom of the quad with one feedline to the shack and separate lines from the boom to each element. This saved quite a bit of coax but added a control line for the remote switch. There had to be a better way.

After some experimentation, a remote switch was perfected that did not need a separate control line, used inexpensive relays for switching, was small enough to mount on the boom of the antenna, and did not cause a noticeable increase in swr. In order to eliminate the control line, the antenna coax was used to carry dc to the relays while also carrying rf to the antenna. This dual use of a feedline is a very common practice in TV antenna preamps where the preamp is mounted on the antenna. The feedline carries power to the preamp and signal back to the set.

To accomplish the necessary switching, two relays are required. With both relaxed, the 20 meter element is connected to the main feedline. With one relay energized, 15 meters is connected, and when the other relay energizes, 10 meters is connected. To make independent selection of the relays possible, steering diodes are used in series with each relay. Then by simply reversing the polarity of the control voltage, either relay may be energized.

Remote control is handled at the operating position by the control unit, which houses a simple power supply. Referring to Fig. 7-51, rf is fed into the control unit at J1, passes through C1, and comes out at J2. RFC1 is used to prevent rf from entering the power supply while allowing the dc voltage to be applied to the rf line. Capacitor C1 blocks the dc from getting back into the transceiver or linear. Switch S1 is a center-off slide switch and is used to provide either a positive or negative voltage to the rf line.

At the remote switch end of the system, the rf passes through C2, through the contacts of relays K1 and K2, and out to the appropriate antenna. If no control-voltage is applied to the line, rf will be fed out to J4. If a positive voltage is applied at the control unit, relay K1 will energize and rf energy will be fed out to J6. If a negative voltage is applied, relay K2 will energize and the rf will be switched to J5. See Fig. 7-52.

In constructing this project, it is necessary to use high quality relays for K1 and K2. Since they will probably be mounted up on the tower or even on the boom of an antenna, they must be reliable. The relays used here were surplus types of the hermetically sealed variety and had 24 volt coils. Keep in mind that, if you are using long runs of coax, there will be some loss of voltage at the relays due to the dc resistance of the cable and connectors. For this reason I do not recommend using relays with less than 12 volt coils. For example, the 24 volt units I used draw very little current and thus cause very little voltage drop to occur in the line. Also, the

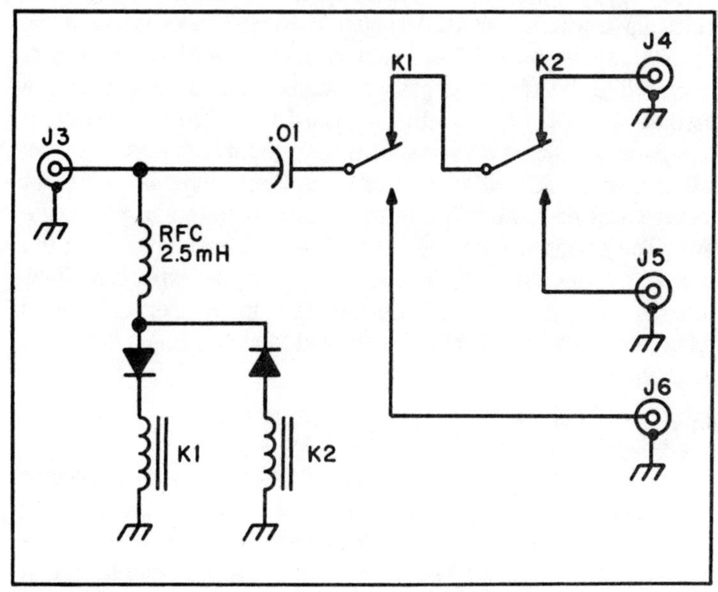

Fig. 7-52. Remote switch.

capacitors used at C1 and C2 should be capable of handling the rf power you plan to use. At the 200 watt PEP level, I have found high quality disc ceramics satisfactory, but at higher power levels I recommend using transmitting type capacitors.

The power supply components are not specified because they will depend upon the relays used. Typically, a filament transformer and 1 A diodes should be satisfactory. Use enough capacitance in the power supply to keep the relays from chattering.

The remote end of the system should be built into a watertight container having appropriate mounting hardware to attach it to your tower or wherever you plan to locate the unit. Use coaxial cable to connect between the jacks and the relays, and anywhere the rf path is more than 1½ inches. This should result in a system with almost no reactance to cause high swr. Although I could not measure insertion loss on my unit, it appears to be very low. Below 30 MHz, I cannot tell any difference on my swr indicator when using the switch or connecting directly to my dummy load. At 2 meters it results in about 1.5:1 indicated swr. Since I only use it on HF, the swr is no problem.

This system of remote control has many possibilities other than those mentioned here. For example, if you only need it for two antennas, use only one relay. Perhaps with a little work this system could be used to switch the pattern on a set of phased verticals or wire antennas. All you have to remember in expanding upon this system is that the capacitors block dc and look like a short to rf, while the rf chokes pass dc and block rf. I have used this method in conjunction with a standard rf switch to work up a five-band antenna switch using three feedlines. Separate lines were used for 80 and 40 meters since these antennas were not located near each other, while 20, 15 and 10 meters used a single line. The saving in coax has been more than enough to offset the cost of building the switch, and not having the extra coax lines running into the shack has helped to clean up my operating area. I think you too will find this a useful and money-saving addition to your shack.

HOME-BREWING A PARABOLIC REFLECTOR

Some months ago, I began searching for a parabolic reflector to augment some gear for the reception of GOES weather satellite transmissions on 1691 MHz. A reflector on the order of 6 to 8 feet in diameter seemed like a good compromise, so the search went on . . . and on. After looking over some "finds" that several more

Fig. 7-53. 8-ft. parabolic reflector. Elements are ⅝″ aluminum tubing spaced 3″. Reflective surface is reinforced with ¼″ hardware cloth. The cylindrical horn is a 2-lb. coffee can with probe.

fortunate hams had made, I began to realize that if you did locate a dish this size, you'd better have a fat wallet and a strong back.

Luckily, I happend upon an article by Norm Foot WA9HUV, in a May, 1975 issue of *Ham Radio* on open-grid parabolic reflectors. This excellent article, along with some others and suggestions and formulae from Roy Cawthon who has had numerous articles in 73 regarding weather satellite pictures during the past several years, started me on this project. I'm grateful to both sources. You won't find any theory here—no proof of performance curves, etc. This is more in the nature of a nuts and bolts kind of article to encourage you along the same route. I can assure you that with a generous amount of dexterity (and little over $100 at this date), you'll have a good parabolic reflector. See Fig. 7-53.

A word about materials: This reflector is 2.3m (7½′) in diameter. 8-foot lengths of ⅝″ o.d. aluminum tubing (.049″ wall thickness) were used. The tubing is the type found in hardware store do-it-yourself displays, a type that is easily bent and remains so after being bent. More rigid material, ⅝" electrical conduit, was used for the two main supporting members to which the elements are attached.

An F/D (focal length to diameter) ratio of 0.4 had been recommended as optimum to facilitate illumination of the parabolic reflector. Following this, I chose a focal length of 36" and a diameter of 90". Element spacing is 3" on center (close to the 0.439 wave figure for 1-dB loss). To get at the business of element cutting (and use tubing cutters, by all means!), Fig. 7-54, Table 7-4 and the relations shown in Table 7-5 were used.

Element lengths, together with number required, are shown in Table 7-6.

Reflector Construction

A template will be needed to check bending and forming of the aluminum elements to fit the parabolic curve. Assigning 36 inches for focal length and solving $x = y^2/144$ produces the results shown in Fig. 7-54 and Table 7-4. Nails or screws at the x and y points will define the curve. The principal tool in the project is a tubing bending jig—that is homemade, too. I used a 4' × 4' sheet of ½-inch plywood and mounted a 5-inch pulley and hardwood block according to the scheme shown in Fig. 7-55.

Mark the center of each of the elements. Slip the end in between the block and pulley, advancing the tubing in small bites and gentle bends. The pulley will aid in keeping the tubing parallel to the plywood surface. After just a few bends (even on the very first element), you'll get the feel of it.

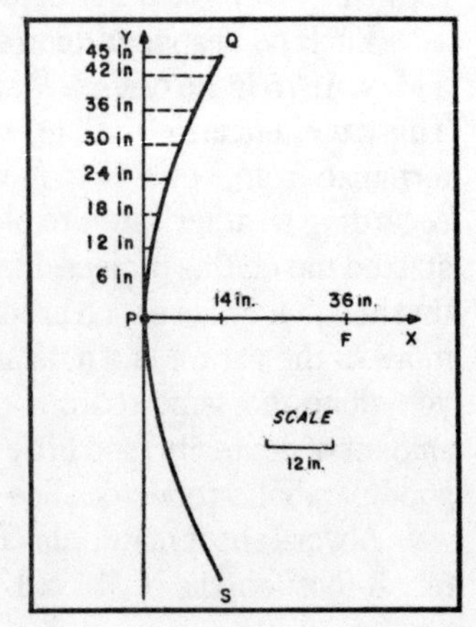

Fig. 7-54. Parabolic curve generated using $x = y^2/144$. Results are shown in Table 1.

x	y
1/4	6
1	12
2 1/4	18
4	24
6 1/4	30
9	36
12 1/4	42
14	45

Table 7-4. Curve Coordinate Dimensions (inches).

I began with one of the longer (95″) elements, fashioning it to the parabolic curve, then using it for a pattern for the remaining elements. After all the aluminum elements are formed, proceed to the two conduit supporting members. These will require a little more "tug," but you'll be a pro by that time.

The two supporting pieces (conduit) now are ready to be marked for drilling. An easy way to mark the center line is to make a scribe, using a wooden block and finishing nail. Drill a pilot hole in the block at half the tubing o.d., then insert the finishing nail. Place the conduit member on a level surface (the bending jig) and scribe the center line. Starting at the center of the conduit piece and working toward each end, mark off 3-inch intervals. Punch and drill these intersections with the center line, using a 7/64″ bit. Drill through just the front wall. A 9/64″ bit will be needed for the aluminum elements.

Number 6, 1-inch zinc chromate tapping screws (pan head) are used to fasten the aluminum elements to the conduit supports. Begin with the 32-inch end elements. Measure in 4 inches from each end of the 32-inch elements and using the 9/64″ bit, drill through the elements. Attach the elements to the conduit supports, squaring up the assembly. From now on it's just a matter of centering each element on the frame, marking and drilling the elements, and attaching them with tapping screws. (A pair of sawhorses can be used to hold the frame, making the job easier on

Table 7-5. Math Box.

1) $r^2 = x^2 + y^2$—for sides and hypotenuse of right triangle.
2) $y^2 = 4Fx$—parabola at origin.
3) $\text{Arc QPS} = \sqrt{4x^2 + y^2} + \frac{y^2}{2x} (\log_e) \frac{2x + \sqrt{4x^2 + y^2}}{y}$ — length of arc of parabola.

Table 7-6. Element Lengths and Number Required.

2 each (lengths rounded to nearest inch)	
32	84
46	87
55	90
63	92
69	93
78	95
80	95

1 central section, extending to 96 inches, to be made of two 46-inch lengths installed on either side of a 1-inch floor flange centered on the boom. (27 lengths of tubing should do it.)

the back.) After all elements are in place, the conduit supports are prevented from sagging by inserting small "S" hooks in each end and connecting these with supporting lines of number 18 galvanized wire.

Boom Construction

A 26-inch length of 2 × 4 (treated with several coats of redwood preservative stain) is used for the top member of the boom. The assembled reflector is centered on the boom and attached with U-brackets and brass screws, using two brackets for each conduit support. See Fig. 7-56.

Install a 1-inch floor flange at the center of the boom to accept the cylindrical horn assembly. The horn support is made of 1-inch

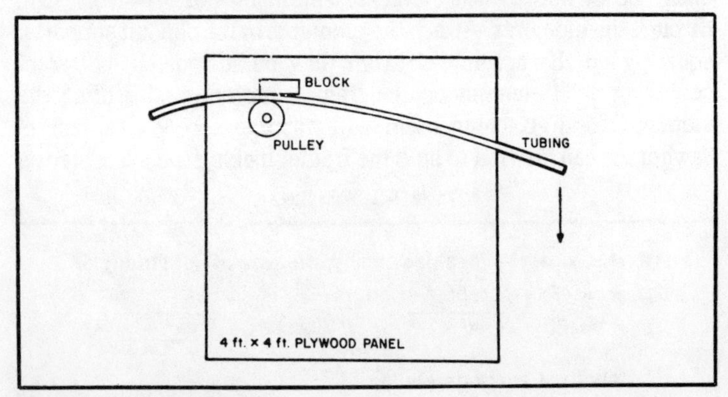

Fig. 7-55. Tube bending jig.

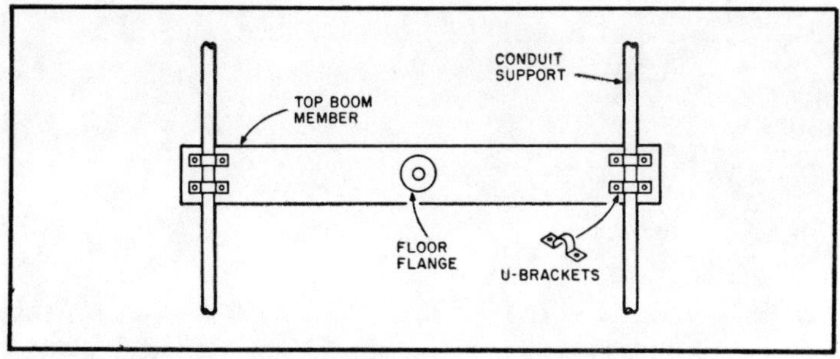

Fig. 7-56. Boom construction details.

PVC (plastic) pipe. Cut a 24-inch piece of pipe and install a 1-inch male adapter at one end, using PVC cement. Cut a 12-inch length of 1-inch dowel rod, give it several coats of redwood stain, and install it in the open end of the PVC pipe section so that about 8 inches of dowel extends from the pipe. Attach this assembly to the floor flange.

Elevation adjustment is provided by two pieces of 2-inch aluminum angle, 2 inches long, coupled by a 6-inch length of ¼-inch threaded rod. See Fig. 7-57. The lower boom section is a 26-inch length of 2 × 6 (treat with preservative stain). Another 1-inch floor flange is fastened to the bottom center of this member (use brass screws). The two boom sections are joined by 3-inch brass butt hinges. Attach the elevation control assembly as shown.

The completed parabolic (open-grid) reflector should weigh about 40 pounds. In order to thread it onto the mast, I inserted a ¼-inch lag screw about 4 inches long into the center of the lower

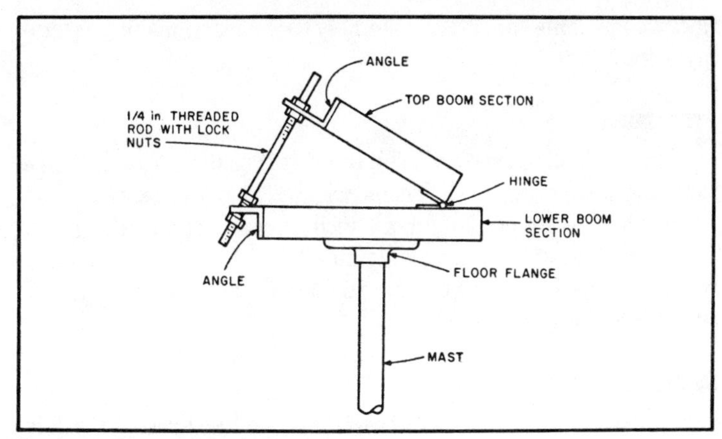

Fig. 7-57. Elevation adjustment details.

Fig. 7-58. Pedestal mount.

floor flange. This "pins" the assembly to the mast, making assembly a lot easier.

Pedestal

Posthole diggers were used to a depth of about 2 feet, then a form for the pedestal 18 inches square by 6 inches deep was constructed. A 10-foot length of 1-inch galvanized pipe, along with a section of 1-inch conduit (containing RG-8/U and 24 V ac lines) was set in the form. The block was poured, using several bags of ready mix. See Fig. 7-58.

Cylindrical Horn

The horn was constructed using ⅜-inch Plexiglas™, a disk is cut to fit the open end of the 2-lb. coffee can (about 5.9 inches).

Using an expansion bit, cut a hole in the center of the disk large enough to accept a 1-inch male PVC adapter. Cut a 1-inch PVC union into two parts. Install a 1-inch male adapter at one end of a 10-inch section of 1-inch PVC pipe. Insert this into the disk and snug it up with the piece of union just cut. Install the disk in the open end of the coffee can, drilling and tapping the Plexiglas disk for six 4-40 screws.

Downconverter Housing

A weatherproof box (11″ × 12″ × 15″) of ½-inch outdoor plywood houses a MicroComm RX1691 downconverter. The box was treated with preservative stain and caulked with a generous amount of clear silicone seal. Actually, it's a box within a box, the converter being further enclosed by sections of 1-inch thick styrofoam. A regulated 12-volt dc supply shares the housing with the downconverter. See Fig. 7-59.

Fig. 7-59. Downconverter housing mounted on mast.

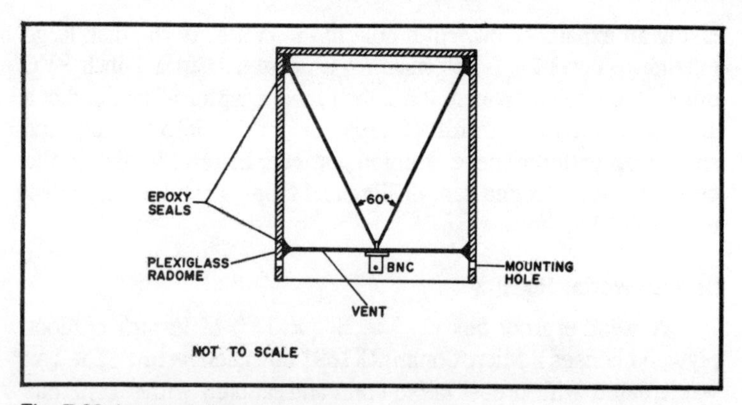

Fig. 7-60. Inverted discone antenna.

I used the open-grid reflector for several months, getting good results (usually full quieting) on the two GOES satellites I monitored: GOES Central and GOES East. Recently, I covered the reflector, installing sections of ¼-inch hardware cloth (24-inch width) to the inside of the reflector with loops of number 18 galvanized wire: I'll admit that there has been some slight improvement in performance. I will certainly say that the open-grid performance was well within what one could expect.

The project looks much more formidable than it really is and the method easily could be extended to the fabrication of larger dishes, say, for TVRO satellite use.

Fig. 7-61. Antenna cone.

Ask most amateurs why they do not try the microwave bands, and you will receive four standard answers:

- ☐ Nobody to talk to.
- ☐ No commercial equipment available.
- ☐ Communication is limited to line of sight.
- ☐ Construction requires lathe and mill precision work.

The first reply is self-serving—if you get on the band, then there is another operator to talk to. The second reply is sadly true, with the exception of the Microwave Associates "Gunnplexer." The third reply reminds us of the response to two meter FM before repeaters came on the scene. The fourth is a myth—many of the pioneers in microwaves started with a soldering iron and tin snips. Good examples of what can be done with simple tools are the many fine construction articles by Bill Hoisington K1CLL.

This project shows how to construct simple and efficient broadband antennas for 1296 MHz and up. The construction is not difficult and can be done with simple hand tools. The antennas can be classified as inverted discones. You can find the theory of operation elsewhere; this will be concerned with the practical construction.

Figure 7-60 shows a cross section of the inverted discone. The conic portion has a 60-degree apex angle. The cone's base and sides are all made three-eighths wave long at the lowest frequency of interest. Most texts specify this length to be one-quarter wave, but we have found that three-eighths wave gives a significantly lower vswr. The disc diameter is also three-eighths wave. The cone is made by cutting a semicircle from copper or brass sheeting with a radius equal to the desired length and rolling it into a cone. The edges are sweat soldered. The BNC connector is sweat soldered to

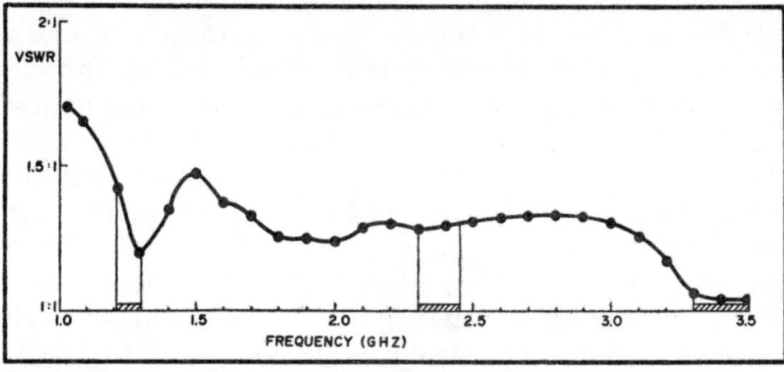

Fig. 7-62. Vswr of 100 mm inverted discone antenna.

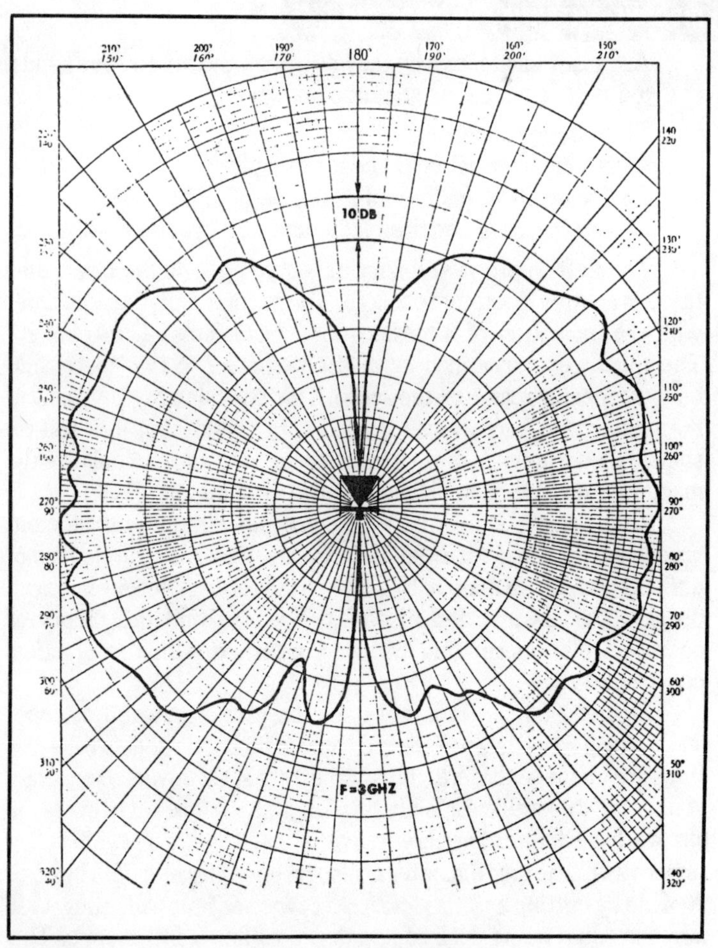

Fig. 7-63. Polar plot of 100 mm inverted discone.

the disc, and then the cone peak is soldered to the connector's center pin. The entire assembly is slid into a Plexiglas™ radome which serves to support the cone and weatherproof the antenna (see Fig. 7-61). If the randome is allowed to extend below the disc, it can serve as a convenient mounting ring. The antenna in the photograph has a base diameter of 4 inches (100 mm). The vswr from 1 GHz to 3.5 GHz is shown in Fig. 7-62. A typical radiation pattern is shown in Fig. 7-63. The antenna is vertically polarized. The horizontal radiation pattern is a circle. The useful bandwidth for an inverted discone is 5 to 1; a 1GHz design is useful to 5 GHz. A smaller unit would be usable to still higher frequencies.

Chapter 8

Batteries And Battery Chargers

Batteries are necessary for portable equipment and very handy to have around in case of an emergency. Nicads are probably the most popular type of battery used by the electronics experimenter because they can be recharged over and over again.

Virtually all problems involving batteries come with complaints like: "Battery life too short"; "Won't hold a charge"; or perhaps, "Battery too weak." Sometimes there is a real problem—and sometimes the battery is not getting what it must to do a good job.

Here are some practical tips:

☐ Fully charge the battery. Some chargers have a NORMAL-TRICKLE switch. In the TRICKLE position, it would take 24 to 60 hours to fully charge a dead battery. On NORMAL, it would take 12 to 14 hours. Nicad batteries can be charged continuously at the NORMAL rate with absolutely no damage to the batteries whatsoever. Leaving the radio on while charging will cause the charging rate to be longer.

☐ Don't over-discharge the batteries. Turn OFF the radio when the batteries become low (the SQUELCH control usually won't silence the radio).

☐ Never insert batteries backwards. This will almost certainly ruin something.

☐ Inspect your batteries occasionally for any indication of rust or corrosion. A white, powdery deposit around the rubber seal at the positive end of the cell or an oily discoloration on the label may be the first sign of an upcoming failure.

☐ If your batteries have a short life, check the battery-charging system. Two simple checks will be enough to find the problem. First, check to see if the charger is putting out enough current. Second, check to see if the radio draws too much current. If the charger and radio are OK and you are allowing enough time on charge, then the battery is probably at fault.

What is a Nicad Battery?

The nicad battery is two or more nicad cells connected together. The nicad cell is called a secondary (storage) cell and is used to store electrical energy until needed. It may be recharged many times during its life. The cell may be described electrically by its voltage and capacity.

Cell voltage is determined solely by the materials from which the cell is made. Nickel and cadmium in a potassium-hydroxide electrolyte produce a cell with a nominal voltage of 1.2 volts. There is only a relatively small change in cell voltage from fully-charged to discharged conditions. Refer to the section on battery-testing (following) for cell voltage-measuring techniques. Cell voltage varies from 1.4 volts when just charged to 1.0 volts, at which point it is considered discharged. Nominal cell voltage is 1.2 volts since the cell is very near 1.2 volts for most of the time it is in use. (Of course, if you have a 10-cell battery, the battery voltage is nominally 12 volts.)

Cell capacity is defined as the maximum current the cell will deliver continuously for one hour. This capacity is given by the battery manufacturer in milliampere-hours (mAh) for small cells, and ampere-hours (Ah) for large cells. Capacity is determined by the size of the cell. For example, an AA-size cell is rated around 350 to 500 mAh and a D-size cell is rated at 2.0 to 4.09 Ah. A very important figure associated with cell capacity is the one-hour discharge rate (C) which is numerically equal to the capacity. For example, for a quantity, C, we can discuss the charge and discharge of nicad cells conveniently without concern for actual cell capacity.

Temperature

Battery operation should be at temperatures between minus 20 and plus 40 degrees C. They may, however, be stored indefinitely at temperatures between minus 60 and plus 60 degrees C. Most batteries will self-discharge at rates dependent upon the storage temperature involved. At 0° C, discharge amounts to 90% in 60 days. At 20° C, it is 50% in about 55 days, and at 50° C, it is 50% in about 20 days.

Life

Generally, batteries may be expected to last several years under normal use. A minimum of 300 cycles of complete charge and discharge is to be expected. If only a partial (say, 203½) discharge is used, the life may extend to 5000 cycles. However, if the battery is partially discharged continuously, it should be periodically deep discharged to realize its full capacity.

Charge and Discharge

Most batteries are normally discharged (in-circuit) at rates less than C and charged at a rate of 0.1 C. If a trickle charge option is available, the charge rate is 0.01 to 0.05 C. Most batteries may be left on NORMAL (0.1 C) charge for indefinite periods without damage. At the normal rate, a completely discharged battery will recharge in 12-14 hours. Less time is required for partially discharged batteries. Charge rates above 0.1 tend to overheat the cell

Table 8-1. Cell Voltages at 20°C.

Condition	Voltage
Fully discharged, open-circuit	*1.2 V
Fully charged, open-circuit	*1.27 V
Fully charged, charging at 0.1 C	1.45 V
Freshly charged, begin discharging at C	1.4 V
Fully discharged, discharging at C	1.0 V

*These voltages are reached slowly as the cell is allowed to stand for a time.

and cause damage. Special "Rapid-Charge" cells are required for fast-charging applications.

Table 8-1 (showing cell voltages) may be of help in understanding battery function during charge and discharge.

Testing

The battery, charger, and radio constitute a small system which is one end of a communication link. When this system fails, testing each element is necessary to determine the proper correction. Based on experience, the charger is the most likely to fail, followed by the battery and then the radio. However, due to ease of testing, test the charger and radio first.

For the 12-volt, hand-held radio chargers, connect a milliammeter using a D'Arsonval movement (such as: Simpson 260 or Triplett 630), capable of measuring 55 mA, in series with a 240-ohm, 1-watt resistor. Connect the meter-resistor combination across each and every set of charging contacts for a 12-volt battery. Observe correct polarity. The charger current should be 45-55 milliamperes.

Consult the appropriate data sheet for the radio under test. Measure all applicable maximum current drain on: full squelch receive, full volume receive, and transmit. Readings should not exceed spec maximums.

A quick battery check would be: Charge at normal (0.1 C) rate for 15-30 minutes. Measure battery or cell voltage. Less than 1.2 volts per cell (12.0 volts for a 10-cell battery) indicates possible defective cells.

For a more complete battery test for a hand-held radio battery with 10 AA cells, fully charge the battery for 12-14 hours at the normal (0.1 C) rate. Connect a 27-ohm, 10-watt resistor across the battery and monitor the time required to discharge the battery to 1.1 volts per cell. The time should be close to 60 minutes.

This test will vary according to ambient temperatures. The time will run short if the ambient temperature is much over 25 degrees C, or if started with the battery more than slightly warm to the touch.

Conclusions

The nickel-cadmium batteries will perform excellently if used within their limitations. Poor performance usually results when the limits are exceeded.

THE NICAD CONDITIONER

Quite a few nicad battery-charger circuits have appeared in print lately. However, as none of the circuits would fill my requirements, considerable research and planning was done before starting on this project. Most of the chargers described in articles were for the preferred constant-current method. Others used the constant-potential system. Still others used a combination of both. Each of the articles detailed the advantages of its method, neglected to mention the disadvantages. Let us look briefly into each of the systems mentioned.

Constant-current charging is usually done at the 10-hour or .1 C rate (C is the rated capacity of the battery). This is fine, except that it takes from 14 to 16 hours for a full charge—an awful long time to be without your HT! Some of the newer chargers use constant-current charging at the .3C rate so that full charge can be obtained in as little as four hours. Remember that we must replace between 130% to 150% of the battery capacity due to inefficiencies in charging. This is a step in the right direction, but we must be careful about overheating as this can ruin the cells.

Constant-potential charging is definitely not recommended. With this method, the charge would start out at quite a high level, resulting in some heat being generated within the cells. As we approach the overcharge condition, additional heat will start to build up. As the battery heats up, the cell voltages will decrease somewhat, leading to even more overcharge current and greater heat buildup. This is called thermal runaway, which will eventually destroy the cells. Because of this problem, constant-potential charging is generally not recommended. The voltage cannot be set low enough to prevent thermal runaway and still fully charge the cells.

Combinations of constant-current and constant-potential systems have also been described. These systems charge the battery at perhaps the .3C or an even higher rate. A voltage sensor is provided so that when the battery reaches some predetermined level, the constant-potential method will take over. The voltage here can be set to hold additional charging current to a suitable value. One drawback to this system, however, is that cell voltages can vary with repetitive charges. It can also vary with ambient temperature. Then, of course, suppose the battery develops a shorted cell. All of these problems mentioned will prevent the battery from reaching the sensor cutoff, so that high-rate charging can continue, eventually damaging the battery.

At this point, let me digress. Some experimenters and suppliers of chargers believe that the charge should be terminated after replacing 130% to 150% of the battery capacity. This is to prevent losing electrolyte within the cells. Others prefer the battery to remain on a trickle charge which may be as low as .01C of the battery capacity. General Electric Company's *Nickel Cadmium Battery Handbook* recommends that nicad batteries be charged at a fairly high rate, but that at completion of the normal charging period they be kept on a .1C-rate topping charge. As all cells in the battery may not have identical characteristics, the topping charge will permit the weaker cells to get a full charge without harming the other cells. G.E. goes on to say that most nicads may be left on the .1C charge rate for extended periods of time without harm.

Another item worth mentioning here is the memory effect of nicads, but since we are all probably aware of this condition by now, no discussion should be necessary.

Very few chargers described in magazine articles or that are available from the HT manufacturers take into consideration the state of charge remaining in the nicad battery. If the battery is not depleted when put into the charger, this can lead toward developing the memory effect just mentioned. To prevent this from happening, the conditioner shown in Fig. 8-1 was developed. This can be built as a stand-along unit or may be incorporated into a complete charger system. Before the battery is placed on charge, it is put into the conditioner. It will immediately go into discharge at the 1C rate. When the battery voltage drops to the 1-volt-per-cell cutoff point, the relay will drop out, thereby terminating any further discharge. This procedure will not harm the battery and it will erase any memory effects.

Another advantage gained by this conditioning is more operating time per charge. In Fig. 8-1, the relay used should operate on a low current and have a coil rating somewhat lower than the battery voltage. Resistor R is chosen so that the total current drain on the battery, including the pilot light and relay, should total the ampere-hour rating of your battery.

To describe the operation of the conditioner: When a battery which is not fully discharged and has greater than 1 volt per cell is connected, current will flow through the zener diode. This will place a positive bias on the base of the transistor, causing it to conduct. The relay is pulled in and the discharge cycle starts. When the battery voltage drops to the 1-volt-per-cell level, the zener diode stops conducting, cutting off the transistor. This causes the relay to drop out, terminating any further discharge.

The next consideration in the design of the charger was the method to be used. I wanted a constant current to charge at a fairly high rate. As it is necessary to replace about 130% of the battery capacity for a full charge, I decided to charge at the .5C rate for approximately two hours to replace 100% of the capacity. At the end of this time the charger should switch automatically to the 1C rate to complete the charge, or for battery-topping.

Consideration was given to the control device. Some rapid-charge batteries have a thermistor built in. As the battery approaches full charge, the resistance of the thermistor will change, triggering the control device to reduce the charge rate. I found that the thermistor resistance would sometimes change. Then again, different manufacturers would use different thermistors. Sometimes they were different from battery to battery. This method, then, was discarded.

What was finally decided upon was a timer method of control. The Motorola 4058 Programmable Timer proved ideal, as it can be set for any time period desired. The circuit of the timer is shown in Fig. 8-2. Once the timer is triggered, the countdown begins. At the end of two hours, my timer operates, pulling in the relay for a low

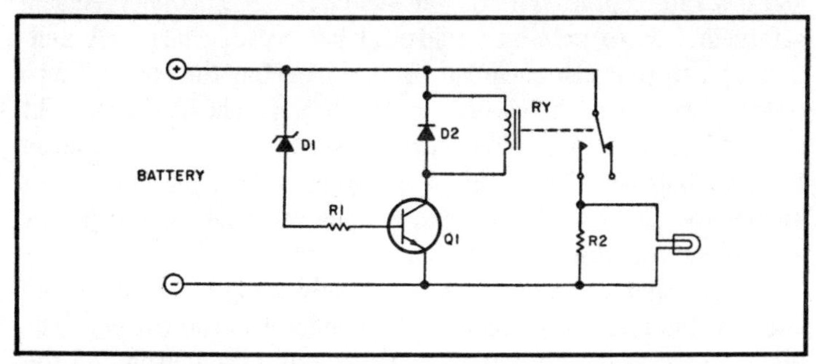

Fig. 8-1. Conditioner circuit.

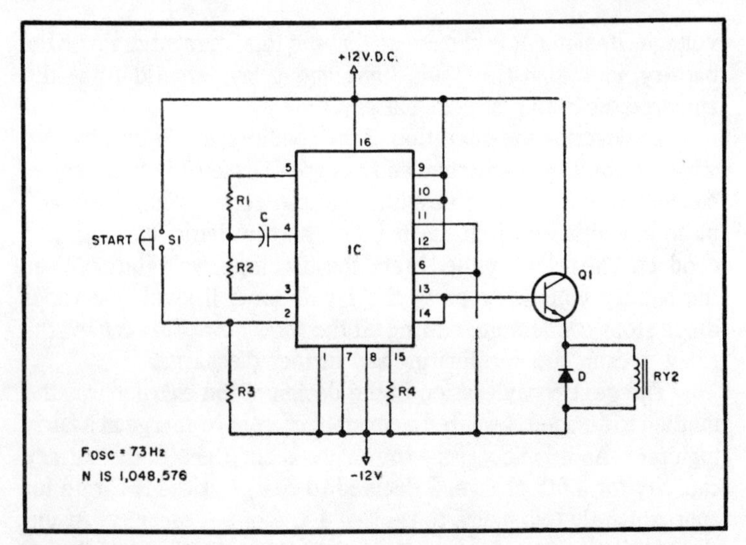

Fig. 8-2. Timer circuit.

charge rate. This system has proven itself most reliable, with none of the defects listed for other charge-control methods. A copy of the 4058 application bulletin may be obtained by writing to Motorola.

Figure 8-3 shows the circuit of the complete conditioner/ charger. Although the parts listed apply for use with my Motorola HT 220 Slimline using a 225-mAh rapid-charge battery, the same circuit can be applied to any other battery. Only a few resistors may have to be changed to vary the charging rates. The length of the high-rate charge time can be changed very easily by changing two or three jumper wires. See Tables 8-2 through 8-4.

Several refinements were added, such as pilot lights to indicate the charger/conditioner status, discharge, high-rate charge, and low-rate charge. The power switch has been connected as a safety switch to prevent accidental battery discharge. A self-testing feature is also included. Before inserting the battery, turn on the power switch and press the start button. The discharge light and the high-rate charge indicator lamps should come on. Release the start button and both lamps will go out. If you press the start switch and the low-charge lamp comes on, shut off the power switch and then turn it on again.

Operation of the unit is extremely simple. Once you've checked the charger as described, connect it to the battery. The high-rate lamp should come on. Press the start button and the discharge lamp should come on. When the battery is down to 1 volt per cell, the high-rate charge lamp should come on again. This will now indicate that the battery is receiving its charge.

As soon as the battery goes into the charge mode, a pulse is sent to the timer to initiate the countdown. At the end of the scheduled period, the timer will pull in relay 2, shifting the charge to the low rate. High-rate charging was purposely limited to 100% return rather than 130% to prevent overheating of the cells.

This charger has now been in use for over a year, and it performs flawlessly. A fully automatic system at last!

LOW COST TRICKLE CHARGER

You can build a trickle charger for just about zero cost. This one was for a 12-V lead-acid storage battery in a recreational vehicle that sits idle for considerable periods. It belongs to one of my sons, and one day he came up with, "Hey, Dad, do you remember that trickle charger you built for the Mercury back in

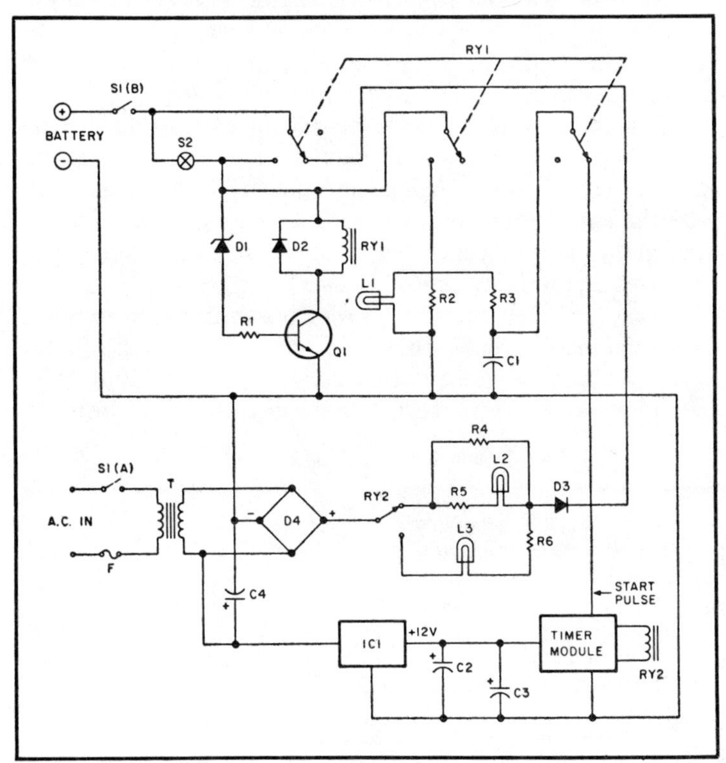

Fig. 8-3. Circuit for complete conditioner/charger.

Table 8-2. Parts List for Fig. 8-1.

D1—Zener diode (1 V per cells in battery)
D2—Silicon diode
R1—1000 Ohms, ½ Watt
R2—75 Ohms, 10 Watts
RY—Coil rated at less than battery voltage
Q1—Motorola HEP S0038
Lamp is type 387 in suitable holder.

1947? Could you build me one like it?" What a memory that boy has!

Yes, I remembered it . . . and yes, I could build one like it. The original was a zero-cost project, and the reproduction would fall in the same category.

A bit of pawing through what remains of my once extensive junk collection revealed a transformer with three 5-volt windings. There's nothing sacred about the total voltage needed from the transformer. Anything above about 15 volts and below about 35 volts can be used. The hyper-simple method of controlling the charge rate can cope with a wide range of ac voltages.

The three windings were hooked in series-aiding. Yes, there's "polarity" to ac . . . use a voltmeter to ensure series-aiding and not series-bucking! A bit more digging in the junk heaps produced an ancient base-mount socket for a light bulb. Sniffing through my diode collection nosed out one with a 3-A rating; one with a lesser capability would have served quite well.

What could be more simple! All that remains is to ascertain the size of the light bulb you'll need for the desired charge rate. Here my multimeter came into use. Starting with the 10-A range and with a 25-W bulb in the socket, the charger was hooked across a

Table 8-3. Parts List for Fig. 8-2.

IC—HEP C4058P programmable timer
Q1—HEP S0038 transistor
D—Silicon diode
RY—12-volt relay with dc coil resistance greater than 250 Ohms
R1—2700 Ohms, ¼ W
R2—6200 Ohms, ¼ W
R3—1 megohm, ¼ W
R4—10k, ¼ W
S1—Momentary contact push-button switch
C—2.2 uF. 35 V

12-V battery. The meter barely moved. A milliampere meter was substituted. It showed about 35 mA, a bit on the light side. Other light bulbs were tried, revealing that the charge rate could be varied from a few milliamperes to over an ampere just by swapping bulbs. The size bulb you might need will depend upon what voltage your transformer produces plus what charging rate you desire. A few moments of trial-and-error will show just what you need. Be sure you start with a low-wattage bulb, or you might end up with a popped diode or a bent needle!

Other than that small precaution, it's a foolproof project, one providing a useful product at a very minimum cost. See Fig. 8-4.

HOME-BREW AN HT CHARGER

Having a set of dead nicads in your HT is no fun—especially when you really need them! If you find yourself in this situation and haven't bought your handie-talkie complete with charging base at the outset, chances are good that you will want to purchase one in the very near future. The Wilson HT's charging posts, which are recessed into the battery slide tray on the underside of the unit,

Table 8-4. Parts List for Fig. 8-3.

T—Stancor P 6469, 25.2 V@1 A
F—1-A fuse
D1—Zener diode (1 V per cells in battery)
D2, 3—Silicon diodes
D4—1-A bridge rectifier
R1—1000 Ohms, ½ W
R2—75 Ohms, 10 W
R3—1000 Ohms, ½ W
R4—250 Ohms, 10 W
R5—430 Ohms, ½ W
R6—620 Ohms, ½ W
RY1—4PDT 12-volt coil (Allied Control TF 154C-C) (P and B KHU 17D11)
RY2—SPST 12-volt coil (greater than 250 Ohms)
S1—DPST toggle switch
S2—Momentary contact push-button-type switch
L1, 2, 3—Type 387 lamps in appropriate holders
C1—10 uF, 25 V
C2—4.7 uF, 35 V
C3—.47 uF, 35 V
C4—50 uF, 50 V
Q1, 2—Motorola HEP S0038
IC1—7812 regulator

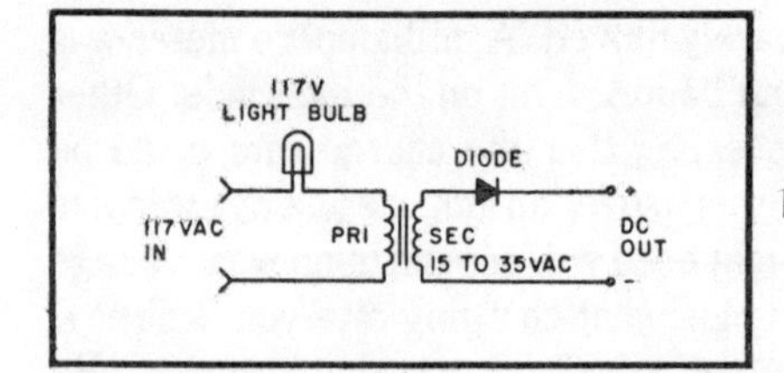

Fig. 8-4. Low cost charger.

make it very difficult to charge the unit without one. Custom charger bases, accessories designed to service a specific manufacturer's HT, are expensive. There are ways to charge the Wilson 1402SM and other similar units having tray-mounted charging terminals, though doing so takes a bit of ingenuity.

I've experimented with a number of alternatives to charging my unit, including removing the battery tray from the case and charging it separately. Most of the approaches I tried were rather cumbersome. I've found the simplest approach to recharging the 12-volt pack to involve no more than an inexpensive CB-type hand-held recharger and a junked commercial charging base. Here's how it's done:

Purchase a Radio Shack #21-516, or similiar nicad charger, of the type used to charge their line of 12-volt CB hand-helds. The battery packs of these units are electrically similar to the Wilson's, though they are mechanically very different. Both use 10 series-connected AA-size nicads, each having an open-circuit voltage of 1.2 to 1.25 volts, to produce a supply voltage of 12 to 12.5 volts. This charger, which resembles an ordinary transistor radio ac adapter or battery eliminator, has a stated output voltage of 17.4 V dc as indicated on its case. Don't worry about this. It won't harm the nicads, as the voltage tapers off as the cells charge, and the charging rate is limited to 40 or 50 mA.

As supplied, the charger comes with a cord which is terminated in a standard "inverted" dc power plug. This is a standard charging and power plug, and it is intended to plug directly into any of the Radio Shack CB hand-helds' charging jacks. To use the charger with the Wilson or similar HT is more of a mechanical problem than an electrical one. You need to scout out a junked commercial charging base for any kind of radio pager, monitor, pocket scanner, or small HT. We're not looking for a good *charger*, just the *base* that can be modified to physically support the Wilson and provide a means of feeding it the charger's output. I located a defunct "Page-ette" base (used to recharge a small pocket-radio pager) at a local hamfest for all of $2.50, purchased on an as-is basis. Other possible sources for these units, other than the flea-market circuit, are local fire and police departments, two-way radio repair shops, and other commercial users. They may have unrepairable or obsolete bases lying around either free for the asking or which can be obtained for a very small sum. Use your scrounging skills here! One word of caution, however: If you should find a charger base that appears to be intact electrically and you want to use it *directly* to charge your unit, be *sure* to check out the charging voltage and current. Many of these units are low-voltage, high-current types, which would not be suitable for charging the Wilson, even if it might fit snugly. Using such a base might even damage your HT or battery pack, as well.

In any case, the charger base you locate will almost certainly not accommodate the Wilson without modification to the base. You will find that the Wilson is a physically large HT (at least the 1402SM is), and is much larger than most pagers and commercial units. So, the base must be modified to accept it. To do this, first remove the plastic sleeve generally used to hold the pager or HT securely in place atop the charger base. It can be discarded, as it won't fit the Wilson. Next, carefully file the rectangular opening in the base until the Wilson can be freely inserted without undue binding. Be sure, however, not to make the hole too large, which would allow excessive slippage when the unit is inserted (it won't have the support of the sleeve, which has been discarded).

Next, install a chassis-mount "inverted" dc power jack on the rear of the base to accept the plug on the Radio Shack charger. These jacks, despite their strange name, are standard items and can be obtained from Radio Shack, Olson Electronics, Burstein-Applebee, and most other mail-order stores. Remove or disconnect the existing charger wiring—you won't be needing it. Route the leads from the jack to the base's charging clips, being especially careful to observe correct polarity. The tip on the charging plug is positive and the shell is negative. Wiring is shown in Fig. 8-5. Note that on the Wilson 1402SM HT, when looking at the unit from the *front*, the *positive* terminal is on the *left*, while the *negative* terminal is on the *right*. Failure to observe correct polarity is guaranteed to be an expensive mistake!

You're almost finished, but getting a good, solid connection to the recessed charging terminals on the Wilson can be a bit tricky. Usually, a very short (¼" or less) 6-32 screw, soldered upside down to the top of the base's charger clip, will do the trick and make good contact with the Wilson's terminals. In some cases, a "pointed" solder blob on each clip will be what is needed to make good continuous contact when the HT is inserted and standing vertically in the charger (Fig. 8-6). You will, of course, have to inspect your charger base to determine what kind of clips it uses and what must be done to accept the Wilson or whatever HT you want to use with it. Usually, modifying and adjusting the clips is no problem, but it takes a little patience to get the HT to seat just right.

The Radio Shack charger I used will fully charge the nicad pack in about 12 to 14 hours. Although it can be left connected to the HT's terminals, it's best to disconnect it to prevent any possibility of overcharging the cells and to eliminate any possibility of fire from overheating or failure of the charger unit. This can be done very simply by unplugging the charger from the ac outlet, removing the Wilson, and reinserting it so that it's no longer resting directly on the charger clips, but, rather, is reclining on the charger base at a 45-degree angle so that charging current cannot

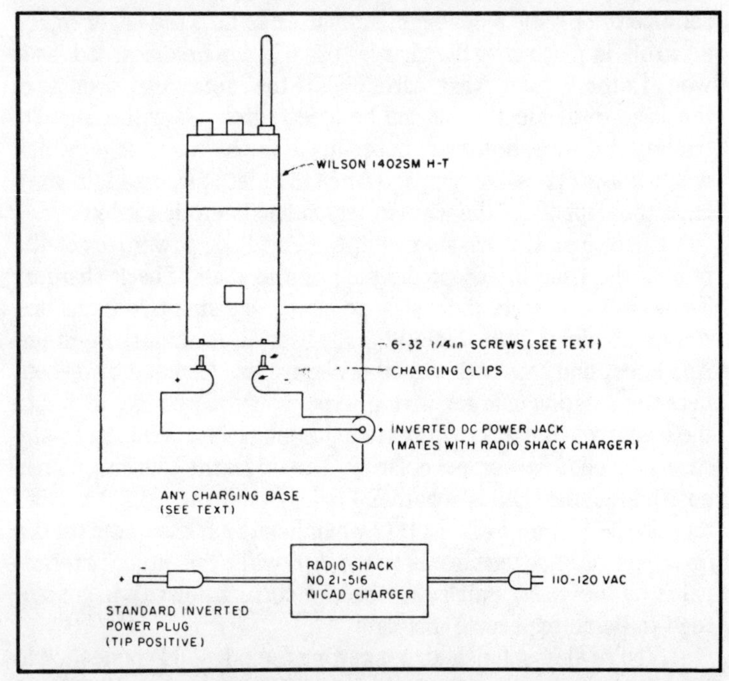

Fig. 8-5. Charging system shown above provides an inexpensive means of charging the Wilson 1402SM and similar handie-talkies that are charged by means of terminals recessed in the battery pack. An old, defunct charger base is modified to accept the Wilson and provides mechanical support while charging.

Fig. 8-6. Wilson 1402SM HT is shown in the charge position, ready to be charged up from the small CB-type battery charger.

flow to the terminals. This position also makes for a dandy fixed-base operating position for the HT (Fig. 8-7).

While the best charging philosophy to follow is a highly-charged subject (pun intended), I believe that you can help ensure maximum battery life by recycling the nicads three or four times a year. This means letting the batteries discharge completely before recharging them. Doing this tends to inhibit "plating" which reduces the batteries' efficiency. By exercising a little care in charging (not over charging and not charging in too many short spurts), you should get well over a thousand charge cycles from a set. Be sure, when charging, that *each* of the ten nicads in your battery tray is a good one, and that each is seated properly so that the charging current flows through it in the right direction.

REGULATED NICAD CHARGER

More and more nicad cells are becoming available and many hams are utilizing them in portable rigs and test equipment. To avoid damaging these batteries though, a few precautions are necessary:

1. Always utilize the full capacity of the cell. Nicads have a sort of "memory" action and a unit that is habitually required to provide only ½ of its rated capacity will go dead at that half way level when the whole bit is needed.

2. Don't reverse charge a nicad. Keep the charge condition on all cells in a series string at the same percentage rate. Substitution of a partially charged cell into a series string of fully charged units may ruin the weaker cell through reverse charging.

3. When charging standard nicads (other than "Quick-Charge" units), limit the charge current to about 1/10 the rated ampere-hour capacity. Excessive charge current causes overheating, which may result in seal rupture and venting of excess pressure. Once the seal is broken, the cell will rapidly dry out and become useless. Figure 8-8 is representative of a "universal" type nicad charger circuit. The transformer, rectifier, and filter capacitor are conventional design. The transformer itself is an 18 volt doorbell unit which gives a rectified dc output of 25 volts.

The current regulator is somewhat less conventional, as most

Fig. 8-7. HT is shown in the non-charging reclining position. Once charging is complete, the unit can be disconnected from the charger and is ready to operate portable or fixed-base as shown in the photo.

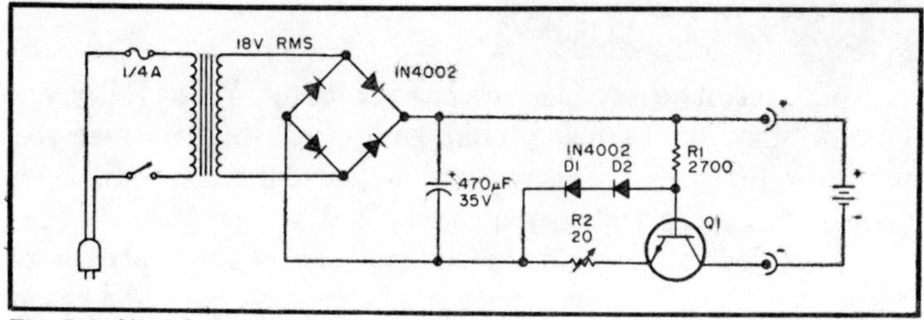

Fig. 8-8. Nicad charger.

hams are familiar with the emitter follower circuit in Fig. 8-9. Placing the load in the collector circuit as in Fig. 8-8 allows a measure of gain and results in better current limiting action.

In Fig. 8-8, resistor R1 is used to provide forward bias to the base of Q1, bringing that transistor into conduction. With no collector load (batteries) in the circuit, the emitter current is very low. Thus the resulting voltage drop across the base-emitter junction and R2 is not adequate to forward bias the two diodes, D1 and D2. This leaves the transistor in a full-on state with the whole supply voltage present at the output terminals.

Now, if we put a heavy load (0 ohms) across the output terminals, the current will increase(!), but how much? Watch what happens. As the current increases, the voltage drop across R2 also increases. When the base-emitter drop plus the R2 drop reaches approximately 1.2 volts, the two diodes go into conduction and limit any further increase in base potential. Thus the current is limited to that point where the emitter circuit voltage drops equal the series turn-on potential of D1 and D2.

For silicon diodes, the turn-on potential is about 0.6 volts. This also holds true for the base-emitter junction of silicon transistors. This means that the required value for R2 is about 0.6 volts divided by the current limit desired.

Varying the load (using 1 to 18 nicad cells) reveals that the current limiting action will hold within 1 to 2 mA from 0 to 24 volts. In other words, you can charge any random number of cells from 1 to 18 without adjusting the charger.

Transistor Q1 should be chosen for a reasonably good hfe and a power capability of twice the total supply voltage times the current limit value. Since my primary interest is in 450 mAh penlight cells, my charge current is set at 45 mA. This means that my transistor must dissipate 25 volts times 0.045 amperes, or 1.125 watts. Double that for safety and a 2 watt transistor is about right.

A JUNK-BOX HT CHARGER

Wouldn't it be nice to have a charger for the Wilson HT on the bedside stand, in the living room, garage, or wherever else you might like to monitor? This would allow you to listen and still keep your batteries charged so you could pick it up and go.

I decided that I would like one, so I checked the spare parts department (junk box) for necessaries and home brewed a cheap charger that works as well or better than the factory model.

Its features include:

☐ Constant charge rate in both high and low mode;

☐ Low charge rate adjustable so batteries will stay charged while monitoring;

☐ Use of voltmeters and milliammeters if desired and available, but they are not necessary (more on this later).

Construction

Any small box will hold the parts; I prefer the 5¼" × 6" × 3" box from Radio Shack. The transformer should deliver about 25 volts at the secondary. I found one with 48 volts center-tapped and used one side for 24 volts. A bridge rectifier is used. The capacitor value is not critical; in fact, you can even leave the capacitor out and the charger will work. I measured the current drain of my Wilson (in standby) and found it to be 25 mA instead of the 14 mA stated in the specs, so I adjusted resistance to give 30 mA on low charge and 55 mA on high charge. This is about the correct rate (50-60 mA) for slow-charging AA nicads. You could fastcharge at 150-200 mA without any problems, but I think it is easier on the cells to use the slow rate of about one-tenth the amp-hour capacity.

A 0-15-volt dc meter is a helpful option across the output terminals to determine the condition of the cells. At full charge, the meter will show approximately 14 volts with the HT in the charger. Also the condition of the cells is determined by the voltage drop observed by transmitting with the Wilson in the charger. If the cells are good, a 2½-watt HT will cause a voltage drop of ½ to 1 volt. If the voltage drop is much greater, it is probably caused by a

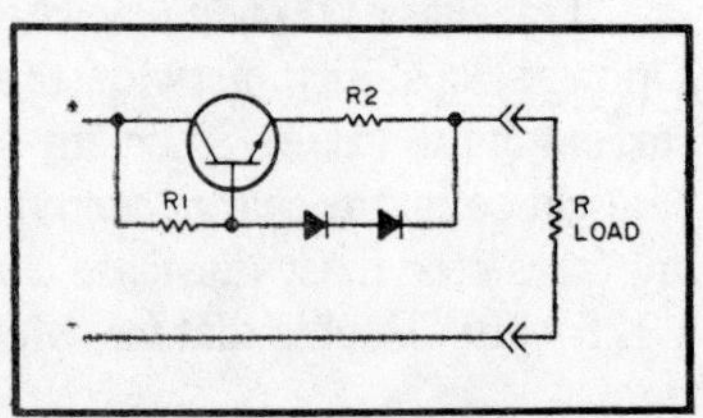

Fig. 8-9. Current regulator.

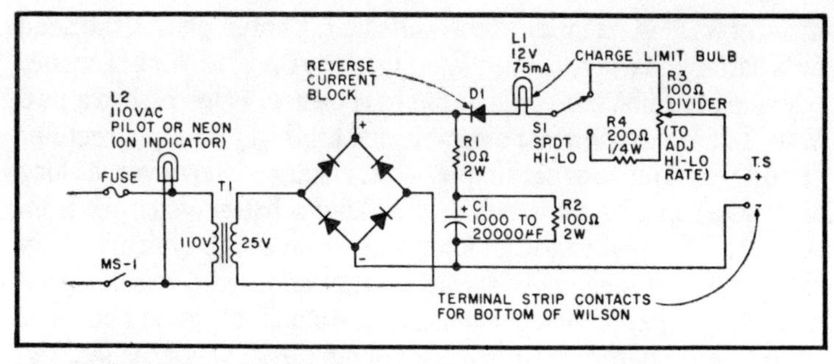

Fig. 8-10. Junk-box HT charger.

weak or dead cell. A milliammeter is an option, but one should be used to adjust the values of resistance to set the proper charge rate when constructing the charger.

The problem of contacts for the charge terminals on the bottom of the Wilson is solved by using a barrier terminal strip and spade lugs bent to 90 degrees with a short length of #12 solid copper wire soldered into the spade lugs. The hole in the top of the case is cut with tin snips and the edges are smoothed and covered with rubber molding or tape. Pop rivets are handy for mounting the barrier strip and transformer, and a hot-melt glue gun can be used to hold some small parts. The 100-ohm voltage divider makes adjustment of high-low charge rates simpler. Be careful to handle the 110-volt primary side of the circuit with care. I advise grounded plug and chassis, a fuse at one amp, and the use of a microswitch to turn the primary on/off when the charger is in or out of use.

It is also a good idea to insulate or cover all 110-volt connections inside the case. Don't forget the rubber grommet to protect the power cord and to provide some sort of strain relief (a knot will do).

Summary

The fourth charger I built was completed in about two hours from mostly junk parts, and it works like a charm. So get busy and have some fun building one or more. They are not critical; just watch the milliamp charge rate. Considering that the commercial version is about $40, these are very nice at about $5, depending on what you find in your junk box. I found the box, which I bought, to be the most costly item at about $4. See Fig. 8-10.

A CHARGER WITH AUTOMATIC SHUTOFF

While operating mobile on occasion, I have found it necessary to recharge a run-down battery. This is particularly true when the ham rig was left on inadvertently all night. Contemplating the need for a battery charger, but not wanting to pay the price for a new one, I decided the junk box could be put to good use. I already had a 12.6 V ac, 10 amp transformer available and I thought a simple rectifier circuit would make a real simple battery charger. However, as long as I was going to the trouble of making a battery charger from scratch, I decided to add another dimension to the typical battery charger circuit design. An automatic turn-off feature was designed into the charger so a maximum charge voltage could be obtained. After the predetermined battery voltage level was reached, the charger circuit would automatically turn itself off. Thus, overvoltage protection would be inherent in this design. Since I had already thought of other applications for the hefty dc power source in the ham shack, I put in an override circuit to the over-voltage sense circuit. By using a Variac at the input of the transformer, an adjustable dc voltage was possible at the output when the manual override was enabled. Figure 8-11 shows the basic block diagram of the power supply with the automatic cut-off feature. The trip voltage for the cut-off is adjustable with a variable resistor. Under normal circumstances, the trip point is pre-set with a good, fully-charged battery connected across the output terminals.

Figure 8-12 shows the schematic diagram of the battery charger. With the output voltage below the trip point level, transistor Q2 is turned on and the relay RL1 is closed, applying full output voltage to the charger's terminal. As the battery is charging, its voltage increases. When the battery voltage reaches the pre-set level, zener diode D1 conducts, turning on transistor Q2 and the relay. The charger's dc power source is then disconnected from the battery, preventing overcharging. If the battery voltage goes

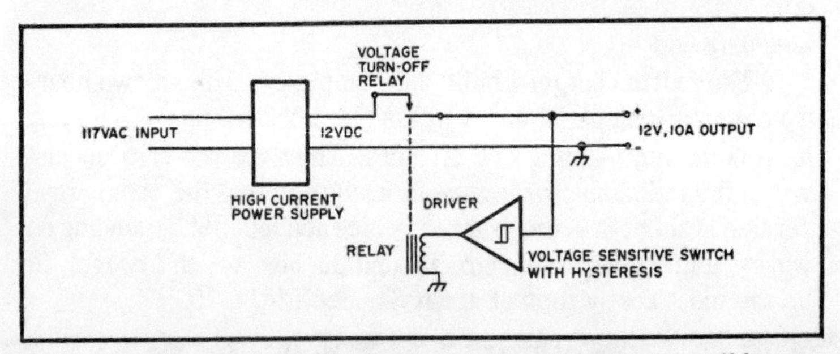

Fig. 8-11. Block diagram of the battery charger with auto-turn-off feature. Power supply can stand alone for general use. High current power supply is a standard design. Relay interrupts output when the battery voltage reaches the pre-set threshold.

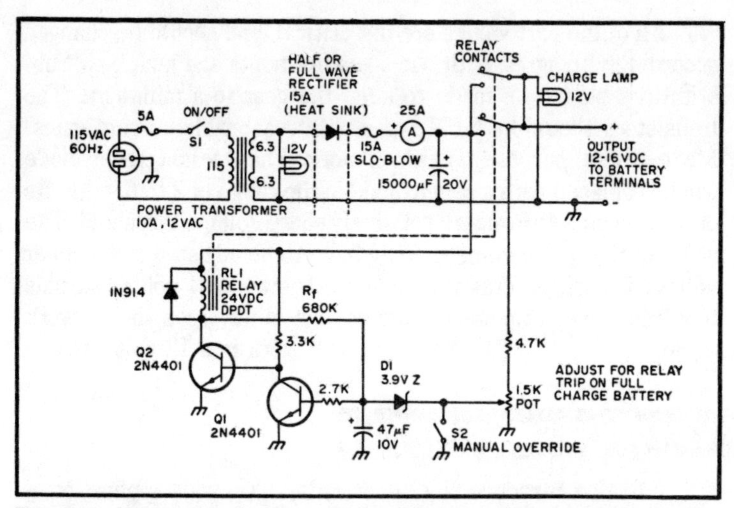

Fig. 8-12. Schematic diagram of the battery charger and high current power supply with the automatic cut-off feature. A manual override permits use of the power supply by itself for general bench use.

below the threshold voltage by a small amount, the relay will automatically connect across the battery, again charging it. Thus, the charger may be left connected to the battery without fear of overcharging. Resistor R_f is used to provide some hysteresis so the relay does not chatter when the threshold voltage is reached. R_f is chosen for about 0.5 volt hysteresis. A 12 volt lamp is used to show when the charger is charging the battery.

To furnish the power source at all times across the charger's terminals, switch S2 grounds the threshold potentiometer so the output voltage does not have any effect on the operation of the relay.

If the ripple is too great, a full wave bridge rectifier may be substituted for the 15 Amp diode. The meter may be left out if it is not necessary to monitor the charging current from the charger.

This charging technique may also be used to charge nicad batteries. A resistor should be connected in series with the batteries however, to prevent excessive currents during the initial charge process. The resistor should have a value to correspond to the maximum charging current permitted. As the nicad batteries come up to voltage, the charging current will drop accordingly. When the predetermined battery voltage is reached, the current will go to zero as the circuit is automatically disconnected. The batteries will be thus protected from excess currents and destruction.

All of the part values are not critical, and should be changed accordingly to agree with one's requirements and junk box. Substitutions should be made to keep the cost to a minimum. The transistors should be NPN types with reasonable current gains. Most general purpose types will work. The 3.9 volt zener diode can be replaced with a zener diode in the range of 2 to 6 volts. Its only purpose is to furnish a relatively sharp voltage threshold. The value of the potentiometer may have to be adjusted if the zener voltage is changed drastically. The relay was a 24 volt dc surplus type capable of 15 amps of current, but other types should work equally well. Most 24 volt relays will work in a 12 volt circuit.

A BATTERY VOLTAGE MONITOR

A device introduced by Litronix, Inc., has wide application as a voltage monitor in all types of battery-operated equipment. The RCL-400 Battery Status Indicator is a current-controlled LED which has a voltage sensing integrated circuit incorporated into a small LED package.

The only additional circuit component necessary to build a voltage monitor is a suitable zener diode, or string of forward biased diodes, to bring the device into its normal operating range. The RCL-400 is designed to turn *on* at 3 V and *off* at 2 V; thus normal operation can be provided by selecting Vz = Vcc-3 V (See Fig. 8-13). When Vcc drops to Vz + 2 V, the LED is switched off by the internal IC voltage sensing circuit to give a low voltage indication. Since the device has a relatively constant current demand in the *on* region (~ 10 mA), the zener power rating need only be ¼ W for most battery-powered equipment. One precaution is necessary: You must be sure that the voltage across the LED does not exceed 5 V (its maximum rating).

For low voltage IC circuits using a nominal 4.5 V battery pack, the required value of Vz is only 1.5 V. It is easy to obtain this value by simply substituting a pair of silicon diodes in series with the LED See Fig. 8-13.

STORAGE BATTERIES

Storage batteries have been with us a long time. The Babylonians used them, but Allesandro Volta rediscovered the galvanic battery in 1800, and Gaston Plante came up with the first lead acid storage battery in 1859.

Until the advent of the solid state age, amateur use of storage batteries was usually limited to trying to re-charge the family car battery, or a once a year cumbersome field day exercise. Inexpensive surplus storage batteries have made really portable operation not only within the means of the average ham, but a real pleasure.

Two types of storage batteries are generally encountered on the surplus market: Lead acid, similar to your car battery, and alkaline storage batteries. The most common alkaline batteries are the nickel-cadmium and the nickel-iron (Edison) cells.

When buying a surplus battery carefully inspect the case for cracks. Check the terminal posts for looseness or signs of internal damage. Remember, there may be a good reason for Uncle Sam to sell the battery! Buy the most recent battery possible (if they are dated). The negative plates of dry charged lead-acid batteries tend to oxidize rapidly on exposure to moist air, so look for the ones that still have the seals over the filler plugs.

To get the best performance and life out of storage batteries requires careful maintenance and charging techniques. Dirt, corrosion and mechanical damage all shorten the life of a battery.

Terminal corrosion is best removed with a stiff fiber brush. After the corrosion is removed, wash the battery with water containing a mild detergent, and rinse with plain water. After replacing the connectors, coat the connectors and terminals with a light coat of cup grease or petroleum jelly to prevent further corrosion.

Connectors should be the proper size to provide maximum contact area with the battery terminals. Remember, lead has about 12 times the resistivity of copper! Always loosen and spread the connector for easy fitting. Never drive connectors on with a hammer or mallet as this may cause internal battery damage if it doesn't crack the case.

It is necessary to use completely separate sets of hydro-

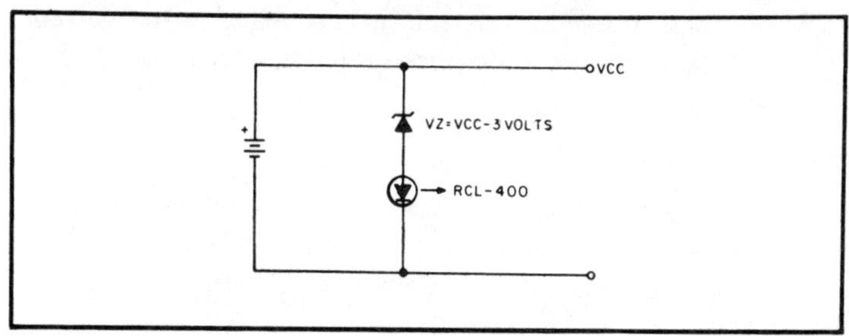

Fig. 8-13. Battery voltage monitor schematic.

meters and syringes for lead-acid and alkaline batteries. The alkaline cells are readily damaged by impurities, especially acid. The Army even insists that completely separate sets of tools be used.

The electrolyte should be kept at the recommended level. If electrolyte is lost by spilling, replace it with electrolyte. If the loss is due to evaporation or electrolytic decomposition bring the level up with distilled water. Don't use tap water as it may contain mineral impurities harmful to the battery.

Lead Acid Batteries

These are by far the most common surplus batteries. The small 6 volt Sniper Scope units, and the cased 4 volt Marine Corps walkie-talkie units are most desirable. Most of these batteries come dry charged. Dry charged batteries are activated by addition of sulphuric acid electrolyte. This is usually available at local battery shops. The most common specific gravity is 1.280. The electrolyte should be at least 60 degrees F before filling. Fill slowly until the electrolyte fills about half the battery. Allow the battery to stand for 15 minutes. Tap it gently several times during this period to release the trapped gas. Continue to fill with electrolyte to the operating level. The battery should then be placed on a constant voltage charge (2.5 V/cell) until it is fully charged.

Battery life is largely determined by the charge-discharge cycle, charging at too high a rate will cause the battery to heat up rapidly and gas excessively. This can damage the separators and buckle the plates destroying the battery. A temperature corrected hydrometer is desirable for checking electrolyte specific gravity during operation and charging. An approximate state of charge can be found from the specific gravity. For most high discharge rate batteries, a specific gravity of 1.280 represents full charge, and 1.110 completely discharged.

For constant voltage charging, the voltage should not exceed 2.75 volts per cell. For constant current charging the current should be approximately 10% of the rated battery capacity. See Fig. 8-14.

Alkaline Batteries

Nickel-cadmium and nickel-iron (Edison) are the two alkaline cells commonly found on the surplus market. They have the same positive plate material and use the same electrolyte (Potassium Hydroxide).

Unfortunately, all of the Edison cells I have found in surplus

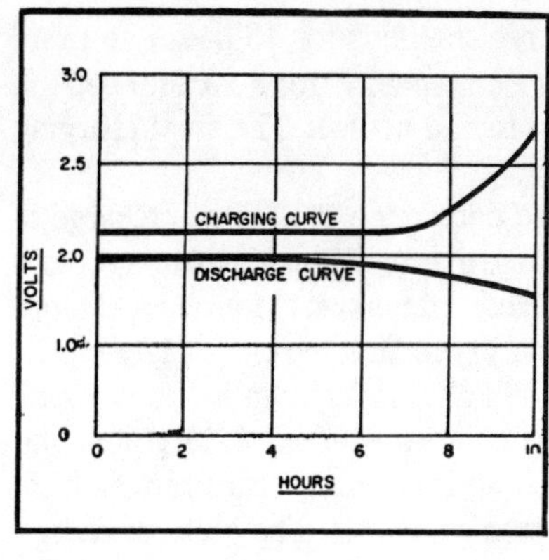

Fig. 8-14. Charge and discharge curves for lead-acid cells.

stores have been badly damaged. Thanks to design changes in certain anti-aircraft missile systems there is a good supply of small nickel-cadmium cells. These cells are of two types, the hermetrically sealed type containing electrolyte, and the dry charged type. If you get the dry charged type you will have to obtain vent plugs to replace the shipping plugs or you will rupture the cases on charging. Vent plugs are packed in the carton with the cells, but they have a habit of getting lost in surplus stores.

The potassium hydroxide electrolyte can be obtained from three sources: 1, Surplus. The electrolyte is packed in cartons ready for use. 2, Chemical supply houses. It can be ordered as a powder or in tablet form. Order potassium hydroxide, reagent grade and specify that it is for batteries. You will need distilled water to dilute it to the proper specific gravity. *WARNING:* Always add the potassium hydroxide to the water, never the water to the potassium hydroxide! Proceed slowly. The chemical reaction liberates heat which can cause the solution to splatter if you aren't careful. Both the chemical and the solution are highly caustic. They will do a fine job of dissolving skin. It is quite poisonous so store it carefully out of the kids reach. Mix according to the directions, or approximately 9 ounces of the powder to a quart of distilled water. The specific gravity of the solution should be 1.32. Check it with a hydrometer, *but* not the one you use for lead acid batteries! 3. Your local druggist can make up small quantities for you.

Fill the cells to the level mark, or just above the top of the plates using an electrolyte syringe or a large eye dropper. Tap and squeeze the cells to release the trapped air. Allow the cells to stand for at least 15 minutes before charging them.

A rule of thumb is to charge the cells for 15 hours at a rate equal to 10% of their rated service capacity (for a 10 amp cell the charging rate would be 1 amp for 15 hours). The final charging voltage will be 1.45-1.50 volts.

Since it is very difficult to determine the state of charge of these cells it is advisable to charge used cells in this manner before use. After charging the cells let them stand two to four hours before adding distilled water (if necessary) to adjust the electrolyte level.

The cells are rated at 1.25 volts under nominal load. Open circuit voltage of fully charged cells runs about 1.33 volts. The voltage when loaded to the ten hour discharge rate ranges from 1.2-1.35 volts depending on state of charge. The state of charge cannot be determined by measuring the specific gravity of the electrolyte. One method is to measure the cell voltage under load. The standard ten hour discharge rate is normally used. Figure 8-15 shows the approximate relation of cell charge to cell voltage. As you can see from Fig. 8-15 this is a very flat discharge curve. A more accurate way is to measure the current flowing into the cell when it is connected to a constant 1.50 volt source. If the current is less than 5% of the 10 hour rating you can assume that the cell is fully charged.

Paralleling of nickel-cadmium cells to increase the current capacity of the battery is not recommended. The internal resistance varies sufficiently so that it is difficult to prevent cells from over charging, failing to charge, and from reversing polarity.

Charging

Two methods of charging are commonly used for either lead-acid or alkaline batteries. They are the constant voltage and constant current methods.

The constant voltage method is the easiest to use providing the charger has adequate capacity. Initial charging rates for a fully

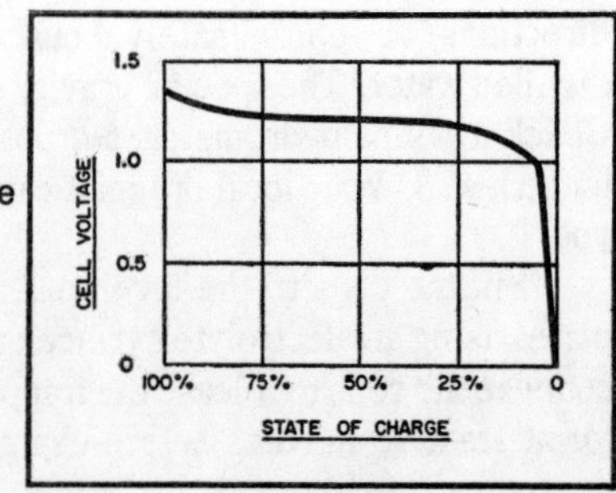

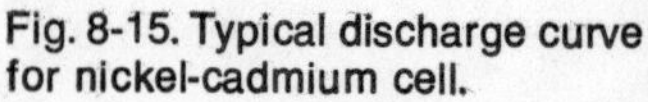
Fig. 8-15. Typical discharge curve for nickel-cadmium cell.

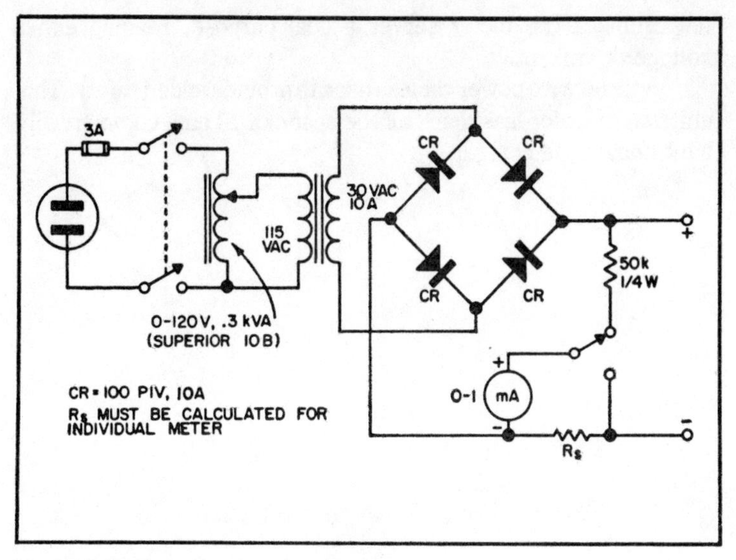

Fig. 8-16. Battery charger schematic.

discharged battery are high, running from 3-5 times the 10 hour rate for nickel-cadmium batteries, and 6 times the finishing rate for lead-acid batteries. The advantage of constant voltage method is that it is difficult to over charge the battery.

When using the constant voltage method on lead acid batteries you can take advantage of the "quick charge" technique used by service stations. A high percentage of the charge can be put back into the battery in a short period of time—if you are careful. The rule is to keep the charging rate (in amperes) less than the number of ampere-hours out of the battery. For example, if you have used the battery for two hours at a five amp rate you can charge it at the 10 amp rate.

When using the constant current method care must be taken to prevent over charging. For lead acid cells this can be done by checking specific gravity of the electrolyte. For nickel-cadmium cells charge at the rated current until the cell voltage reaches 1.45-1.50 volts.

Figure 8-16 is the schematic diagram of the battery charged. It is a simple bridge rectifier device built into a case made from a BC-375-E tuning unit. A Variac provides continuous voltage adjustment, while the transformer provides isolation and voltage step down. All components except the rectifiers are mounted to the front panel. The rectifiers, on insulated heat sinks, are mounted to the cabinet. The meter serves a dual purpose, reading either voltage or current.

A three wire power cable is used to provide added safety. This unit was built for less than half the cost of a 10 amp commercially built home type charger.

Chapter 9
Various Electronic Devices and Gadgets

Here's a real variety of electronics projects! From an automatic thermostat to a junk box anemometer and even special projects to aid the visually and hearing impaired, this chapter contains quite an assortment. Want to make your own crystal filters or build a counter? Well, these projects and more are just begging to be built!

AN AUTOMATIC THERMOSTAT

It is common knowledge that heating bills can be reduced by setting the heat lower at night. As a matter of fact, most heating engineers will tell you that a savings of up to 16% can be realized by lowering the nighttime temperature in a home by 10 degrees Fahrenheit for at least 8 hours every day during the winter heating season. While it is very easy to use your index finger to push the thermostat back 10 degrees every night, this manual method does not turn out to be very reliable. Either someone forgets or the attitude of "I'm not getting up in the cold to turn up the heat" seems to prevail. Thus, the use of an automatic "day/night" type of thermostat is a better way to cut down on heat bills. This type of thermostat, however, is usually rather expensive ($50-75) and must be used in place of each thermostat in the house. In a dwelling with multiple zones and multiple thermostats, this can be quite expensive.

Here we describe a "day/night" heating control which will automatically turn the heat down at night and back up again in the morning. While this device sounds like a "day/night" thermostat, it is different in that it connects to the 24 V ac electrical system of the furnace and will control all of the thermostats in a typical house. It is easy and economical to build and can be installed quite simply on either a hot water or hot air heating system. It has been designed so that an inexperienced experimenter can build, install, and maintain it.

Principles of Operation

A typical heating system has a circuit diagram similar to the diagrams in Figs. 9-1 and 9-2. Note that, in these diagrams, all thermostats have one side connected to a common point or common wire. The entire heating system (except for the control for hot water reservoir) is disabled if the connection between the 24 V ac transformer and this common point is broken. The "day/night" heating control has a 117 V ac relay connected to a cyclic timer. The "normally on" contacts of this relay are used to provide the connection between the transformer and the common point. At night, when it is time to set the heat back, the timer turns on the 117 V ac breaking the connection between the thermostats and the transformer disabling the heating system.

When I was building my first unit, I asked a neighbor who was a heating engineer if there was any reason why I should not let the house get as cold as possible during the night. His answer was a definite, "Yes, there's a very good reason." It turns out that a very complex relationship exists between the outside temperature, the day/night setback temperature differential, and the amount of fuel needed to bring the house up to temperature the next day. Because of this relationship, there is an optimum "setback" temperature, which for the average house is about 10 degrees for a maximum savings of 16%. If you exceed this setback temperature by very much, your savings will decline and it is possible to reach the point where you are ultimately using extra fuel rather than saving fuel. For this reason, an additional thermostat was included in the system to provide an overall nighttime temperature for the house. This additional thermostat uses the 24 V ac transformer on the furnace for power and is connected to a 24 V ac relay. This thermostat is set at a temperature which is 10 degrees less than the daytime temperature. When the house temperature falls below this nighttime setting, the thermostat closes, turning the 24 V ac relay on. Since the power leads for the 117 V ac relay go through the normally closed contacts of the 24 V ac relay, the 117 V ac relay will be inactivated and the system will go back to normal. At this point, all of the daytime thermostats take control and call for heat as required. When the temperature rises above the nighttime temperature, the heating system is once again disabled. Note that the nightime thermostat does not turn on all zones automatically but merely puts the heating system back to daytime or normal. This was done to allow a given maximum temperature in a particular zone and to permit a zone to be turned off. If a zone is turned off during the day, it will not be activated at night by this system. I wanted this feature for my own house where the basement zone is off except for rare occasions. If I had used the nighttime thermostat to turn on all zones, I would be wasting heat in the basement.

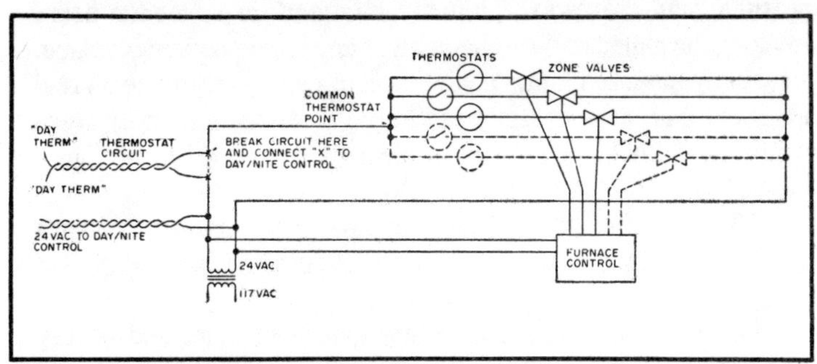

Fig. 9-1. Typical hot water heating system.

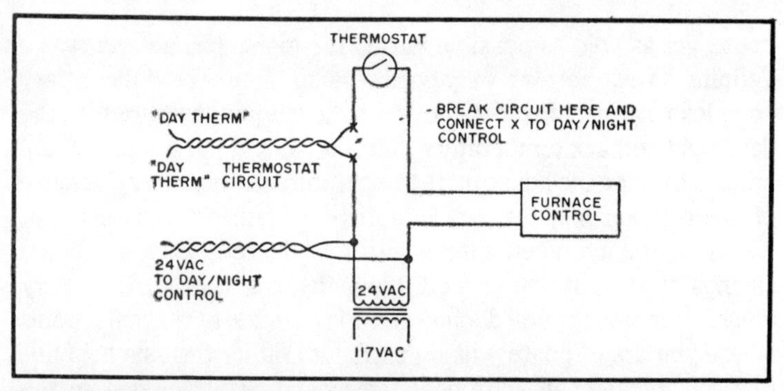

Fig. 9-2. Typical hot air heating system.

Construction

Simple, quick construction was one of the design goals of the system since many neighbors expressed interest in this device and contemplated building it for their homes. Reliability was another goal, since few of the neighbors had an electronics background and thus could not troubleshoot the system should a malfunction occur. Consequently, parts were chosen on the basis of ease of construction and reliability, rather than on a minimum cost basis. Those readers with a good junk box and electronics knowledge can improvise and substitute, possibly saving a few dollars here and there.

I might mention that an all solid state version of this unit was built and tested; however, since it was not totally reliable and was expensive to build, it was abandoned. It is difficult to beat the relay in regard to reliability and cost.

A small metal box with dimensions of 3″ × 4″ × 5″ or larger is used to hold the two relays. Top mounting sockets are used for the relays so that it will not be necessary to punch large holes in the box. A small quarter inch electric drill and a screwdriver are the only tools needed for the project. After assembling the unit, both the timer and the control unit are mounted on a wooden board which can be nailed to the cellar wall or any other convenient place. I originally mounted lights on the box, but since they have no real value and add unnecessarily to the cost of the unit, they were omitted in all subsequent units. Hence the circuit has no lights shown.

All connections made inside the box are brought out through a ½″ grommet in a ½″ hole. Connections should be labeled as shown in Fig. 9-3, using masking or adhesive tape.

The timer used is a heavy duty industrial timer and is very inexpensive for its quality.

Installation

Install the nighttime thermostat in a central part of the house. Do not mount the thermostat on an outside wall or a wall that gets cold. Use twisted pair, solid thermostat wire or equivalent for the run from the thermostat to the control.

After the control box and timer are mounted on the board, connect the timer and control together as shown. (If you use a timer other than the Tork timer, then connect the 117 V ac timer leads to the terminals marked "load" on the timer.) Plug the timer into an outlet and test the unit as follows: Throw the lever in the timer to "on" and the 117 V ac relay should pull in. Throw the lever to "off" and the relay should drop out. Unplug the timer, cut the wire going from the "common thermostat point" to the transformer, and connect the two ends to the control box leads marked "day therms." Turn on one of the thermostats in your house and the heat should go on. Leave the thermostat on, plug the timer in and turn the lever to "on." The heat should go off. Connect the leads marked "24 V ac" to the 24 V ac transformer and connect the "night therm" leads to your night thermostat. Turn the timer on so that the 117 V ac relay turns on. Turn the nighttime thermostat to ten degrees or so below the actual house temperature. The heat should stay off. Click the night thermostat on. The relays should click and the heat will go on. At this point the system is installed and ready for operation.

Note that all connections should be made with the appropriate circuits de-energized. Never work on a "live" circuit. When making connections to the 24 V ac transformer, be sure that the circuit to the furnace is turned off. In some cases, it may take two to three minutes for the heat to come on due to inherent delays in the controls for the furnace and the zone valves.

Operation and Use

In our household, the timer is set so that the heat is turned off ½ hour before bedtime and turned on ½ hour before rising. With

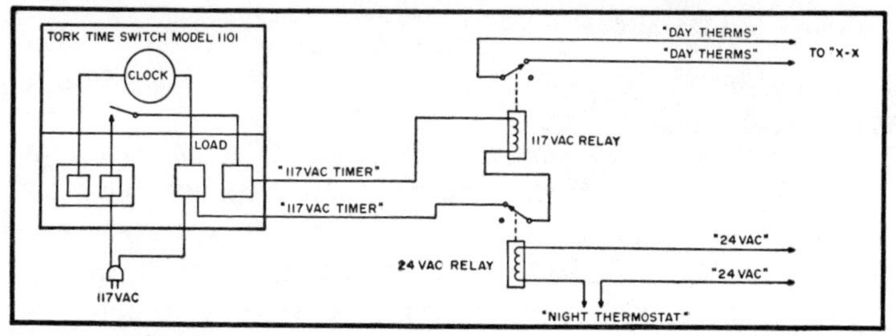

Fig. 9-3. Day/night furnace control.

these settings, our house remains comfortable before going to bed and warms up before rising. In cases where we will be staying up late beyond normal bedtime, we merely turn the nighttime thermostat way up until it "clicks in." This puts the system back to normal without having to change the timer. In this case we turn the thermostat down before retiring and the nighttime system takes control.

It should be noted that when you set the trippers on your timer, the trippers are set to turn "the control on" and "the control off." This corresponds to disabling the thermostats and enabling the thermostats. In cases where no one is at home because of school or work, additional trippers may be added to turn the control on at 8:30 am and off at 4 pm or any other time. By doing this, the heating system would be in "nighttime" mode from 8:30 to 4. In this example, an additional fuel savings would occur for 7½ daytime hours. Additional trippers cannot be added to the inexpensive timer noted in the parts list (see Table 9-1).

This system has been working for several years now and has saved us a real 10% in fuel utilization. We are quite pleased with the system; however, I will admit that it did take a while to get used to it. First of all, the bedroom does get cold at night, which means that we had to learn to sleep with three blankets and a quilt. Second, visiting the "head" in the middle of the night can be a chilling experience. Third, the temperature throughout the house at night is not as uniform as when the system is in normal. This is of little consequence since you should be asleep at this time.

Very obviously, the advantage of this system is fuel savings for the individual homeowner. It would not be unreasonable to save twice the cost of the unit during the first winter. From the savings that our family is seeing right now, we may do considerably better. Aside from the savings experienced in the individual household, this unit could have an impact on nationwide fuel economy if it were used on a national basis. It could help us conserve our scarce national fuel resources.

JUNK-BOX ANEMOMETER

While most hams are worrying about their antennas, my concern is my newly hewn wind generator propeller. Since my garage doesn't come close to a machine shop, and my frugal nature abhors expenditure for an air brake or complex feathering device, the natural thing to do is crank down the tower or immobilize the prop for the high winds. But, when you can't spend all of your time

Table 9-1. Recommended Parts List for Thermostat.

Quantity	Description	Supplier
1	120 V ac relay Line Electric Co. Model MKH2A	Hatry's
	or	
	HI-G, 115 V ac Model 4SLRP-215	See notes
1	24 V ac relay Line Electric Co. Model MKH1A	Hatry
	or	
	HI-G, 26.5 V ac Model 4SLRP-126	See notes
	Sockets for Line Electric Co. relays – Potter and Brumfield #27E122	Hatry
	or	
	HI-G #4SLRP	See notes
1	Tork Time Switch Model 1101	Graybar
	or	
	Sears #34H5870	Sears
	or	
	Sears #34H6442	Sears
1	Thermostat Sears #42H9235	Sears
	Thermostat wire as required Sears #42H9151	Sears

in the yard watching the weather or stay awake all night listening, the first step is a wind indicator. Then, if you have a counter and an alarm circuit, you've got the system down pat. Well, the counter alarm comes next, but here's my answer to the indicator. This is an adaptation of one by Hank Olson W6GXN. Almost any small signal NPN transistors will do, and Olson's Electronics (Akron, Ohio) has the Fairchild μL914 (part TR 297). The RTL circuit is a little antiquated but very effective. It is easily mounted on the small printed circuit board from Radio Shack that accommodates the 914 and most of the smaller parts nicely. In case you are thinking of adapting to TTL, the circuit for the μL914 is in Fig. 9-4.

The wind spinner is made from three small kitchen funnels (49¢ each) attached using a pop-rivet tool. The spinner plate may be cut from 1/16″ aluminum plate. A layout drawing (NTS) is in Fig. 9-5.

The mast can be any type you choose, but aluminum is best since it will not affect the field of the small magnets that pulse the reed switch. This was a small worry for me, however.

The reed switch was mounted in a hole through an expired felt-tip pen. It is offset mounted from the spinner pipe shaft by a bracket to be under and ¼″ below the magnets.

Silicone compound from the hardware store filled the holes in the funnels (tips cut off) and weatherproofed the reed switch in its pen mounting. Silicone was found to be better than epoxy for holding the magnets to the aluminum spinner. Epoxy kept weathering loose no matter how clean the aluminum spinner was. An overall mounting detail is provided in Fig. 9-6.

Try several NPN transistors from your junk box, since some are more responsive than others. The wire connecting the reed switch is your own choice, though coax is probably best. I used speaker wire, expecting to replace it every year or so. Rf induction did not appear to be a problem with the choke inside a 4″ × 4″ × 5″ minibox along with the circuit.

Parts and Assembly

The single greatest problem in this design is the bearing. The original concept was to use a simple teflon bearing similar to the drill stop used for ¼-inch drills. This idea was not satisfactory. A perfect solution was the discovery that a standard (1¼″) thin-wall chrome-plated drop pipe for a bathroom sink drain (about 6″ long) would accept the standard roller bearing used in the bottom of sliding glass patio doors. Two of these bearings with ¼″ center holes fit exactly into the drop pipe. The bottom one stops inside about 3½″ down. The top one is held in place with a stainless pipe

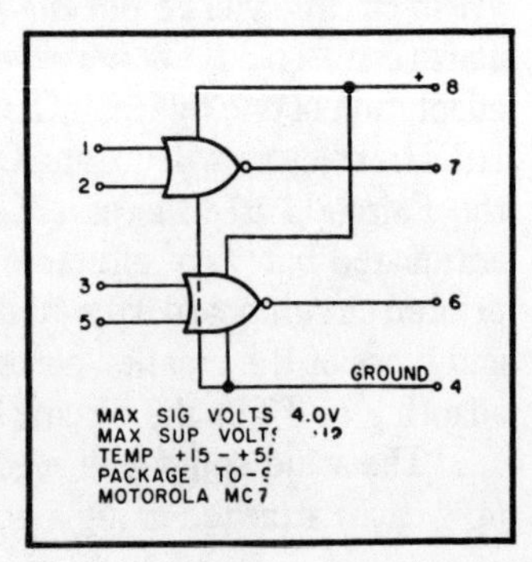

Fig. 9-4. μL914 detail (or Motorola HEP584).

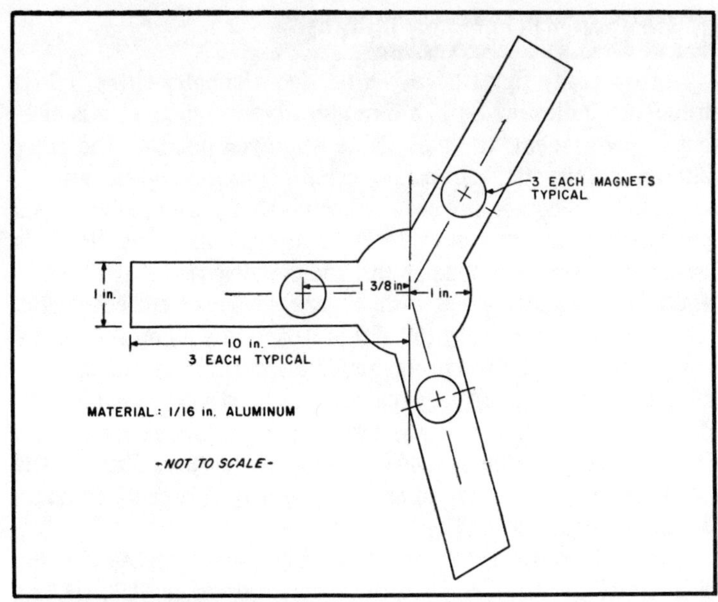

Fig. 9-5. Spinner details layout.

clamp tightened to hold. These bearings come in a brass version and an aluminum-teflon version. I used the teflon at the bottom and the brass one at the top. A ¼″ shaft for the spinner was attached with a sheet metal screw and "Lock Tite" compound. You may want to use a hole and cotter pin in the bottom of the spinner shaft to keep the spinner from crawling out of the bearing holes at higher speeds.

Calibration

I used Hank's technique of a calm Sunday morning, family in auto, father with anemometer protruding over the top of the car from the rear of the station wagon, pencil and paper in hand, and mother driving and calling mph at speeds of 10, 20, 30, 40, 50, and 60. I stopped at 55 and found the relationship linear, and, with the plots for the five points, I extrapolated a curve on some of my daughter's school graph paper. I use the graph, but a scale could be drawn and glued to the meter face.

I'm really pleased with my creation and use it quite often, since the meter sits on top of my antenna coupler. To read it, I just flip the switch, read the meter, and turn off the switch to conserve batteries. It's been in service for more than a year now. See Figs. 9-7 and 9-8.

BUILD A 60-Hz FREQUENCY MONITOR

How many times have you needed a simple, cheap, 60-Hz frequency indicator? Sure, a commercial reed-type meter is nice, but it never seems to be available when you need it. The other alternative, the station frequency counter, is bulky and expensive.

Here is another device—a circuit—which can resolve cycles per minute and costs only a few dollars to build. The circuit is basically a frequency comparator, and the idea can be extended to almost any frequency you wish. Figure 9-9 shows the schematic. The reference frequency (60 Hz) is derived by using a color-TV crystal and an MM5369 programmed divider (integrated circuit). Both crystal and IC are very reasonably priced and useful for many digital clock/timer projects as well. The input frequency is taken from the low-voltage secondary of the power transformer. The power supply is straightforward, using a bridge rectifier and a 5-volt IC regulator, U6.

The reference and the input frequencies are processed by the Schmitt trigger, U2, and fed to the comparator circuit, U3, U4, and U5. U3 provides identical pulse shapes to U4. U4 is a 4-bit counter

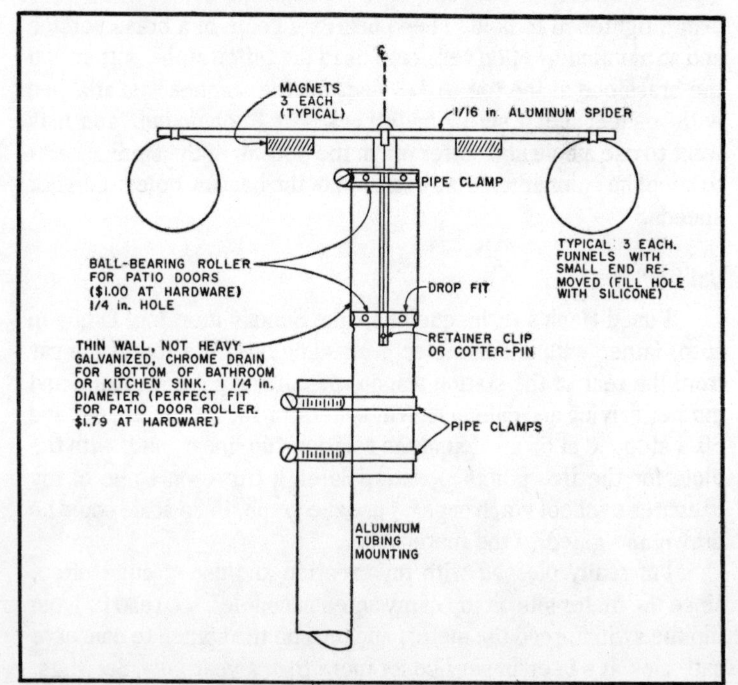

Fig. 9-6. Anemometer assembly detail.

which counts up with one input and down with the other. The counter contents are decoded by U5 and used to light D5 to D8, in sequence. The direction of the sequence will indicate whether the input frequency is fast or slow. For the display, I used a 7-segment readout with a defective segment, but four standard LEDs work as well.

The PC artwork I used is shown in Fig. 9-10, and parts placement in Fig. 9-11. Any method of construction you find convenient to use with ICs is okay. Nothing is particularly critical, but you may need 0.01-μF bypass capacitors for the 7400-series ICs, U6 should have a small heat sink.

It's simple to use. Plug it in and watch the rotation of the LEDs. For the most part, the oscillator trimmer doesn't really have to be adjustable. If you're a purist, the oscillator can be set to 3.579545 MHz with a frequency counter on U1 pin 7, buffered output.

Plug it into your local power company, and you will see a very slow rotation, once every five minutes or so, corresponding to a frequency difference of perhaps four cycles in five minutes. Most power companies rarely hit 60 Hz on a short-term basis because of adjustments needed for demand, etc., but over the long term, all the clocks stay on time. This is why the rotation will be fast (clockwise) at some times while slow (counterclockwise) at others. On an emergency or standby power system, you will see quite wide changes of frequency with load variations.

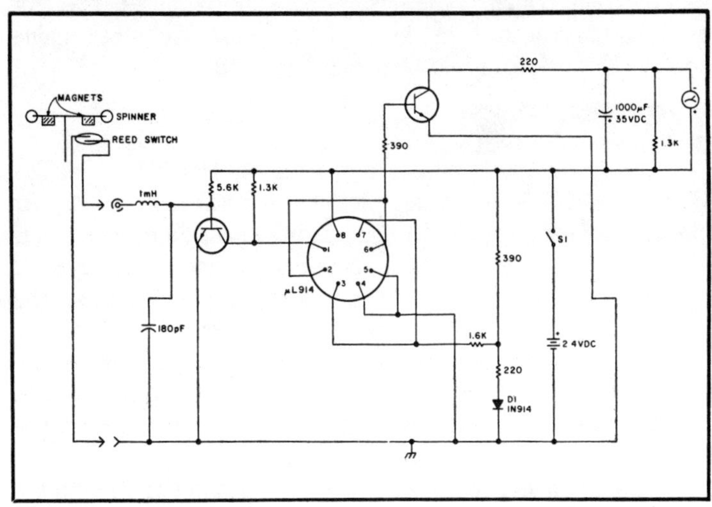

Fig. 9-7. Circuit schematic.

Fig. 9-8. The completed project.

That's all there is to it. One of these devices is in use at a local radio station. (Opening the throttle on a 30-kW diesel generator until the fluorescent lights fire is not the best way to set engine speed!) Try one for Field Day. See Fig. 9-12.

THE SUPER CLOCK

Here is a really unique digital clock. Using the Cal-Tex CT7001 MOS/LSI integrated circuit gives you a 12 or 24 hour clock, four year calendar, 24 hour alarm, 9 hour 59 minute timer, 50/60 Hz operation, failsafe battery operation, and will drive either common cathode or common anode seven segment LEDs. The basic circuit is built on a PC board only 3.3″ by 5″ and will fit in Radio Shack's wood grain utility cabinet. By using a larger enclosure, a back-up battery pack may be added, along with a relay to operate external loads with the timer. Construction cost should be around forty dollars with careful shopping.

The schematic shown in Fig. 9-13 is adapted from both the Cal-Tex and Radio Shack data sheets, plus my own ideas from past

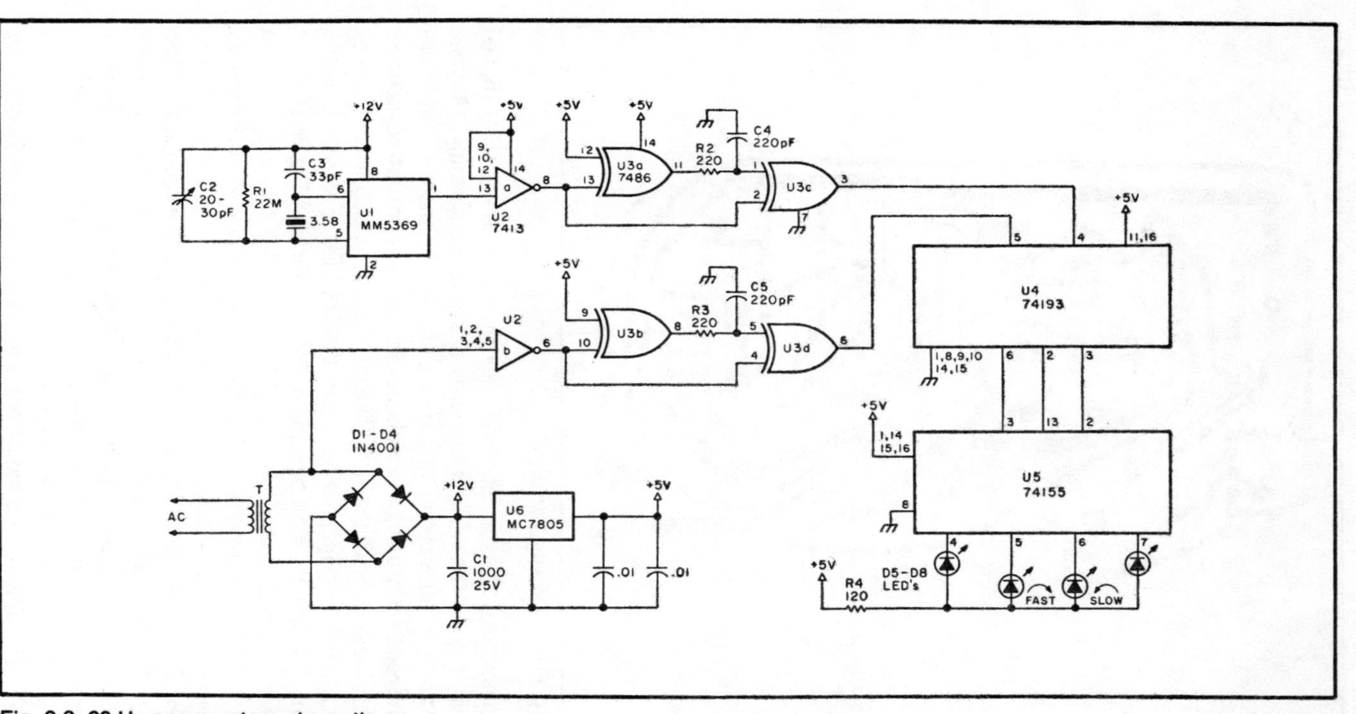

Fig. 9-9. 60-Hz comparator schematic.

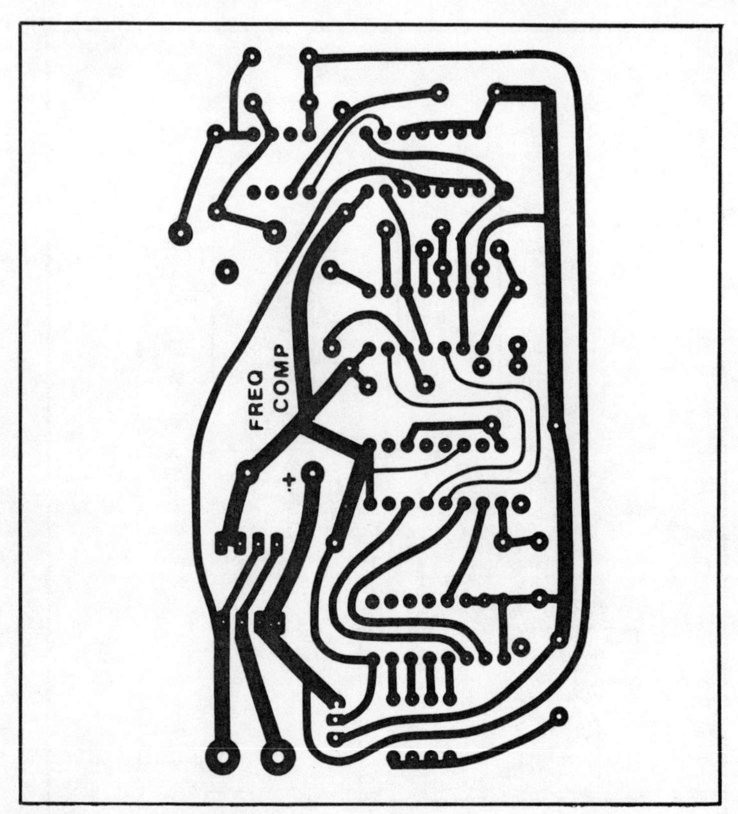

Fig. 9-10. PC board layout.

(unsuccessful) experience. The alarm circuit shown, Q14, 15, and 16, is quite annoying in the early hours of the morning. (Anyone want to beat my record of 5 seconds hitting the snooze from across the room?) I'm using an earphone element from an old telephone for the speaker, and it's loud. The display drive circuit can be programmed with jumpers for either type of LED. Figure 9-14 is the board layout for multiplexed common anode, and Fig. 9-15 is for common cathode.

The power supply is simple and straightforward. However, use a heavy duty nicad battery pack, as the displays are wired to be on all the time. Ac is sampled from one side of the secondary of the transformer through a resistor and diode to drive the chip timers and counters. The power supply also supplies current for the timer relay. Use a low current 12 volt relay with contacts rated at at least 3 amps for driving external loads.

The am and pm indicators will only operate in the 12 hour

mode. They may be omitted if you wire for a 24 hour clock.

Figure 9-16 is the parts layout for the main board. Please note that an insulated washer must be used at the mounting hole next to Q13 if you use common anode displays. It is not needed for common cathode.

Figure 9-17 is a full size negative layout for the main board. Single-sided G-10 is best, but bakelite may be used.

Be very careful with the IC as it can be damaged by static discharges. Once in the circuit it is relatively safe, but can still be destroyed by excessive charges (I found out the hard way).

Figure 9-18 is the front and back panel layout for the clock and should be followed if the same cabinet is used; otherwise things just don't fit. The only one that's not critical is the size of the front panel cutout for the display. It may be slightly smaller or larger.

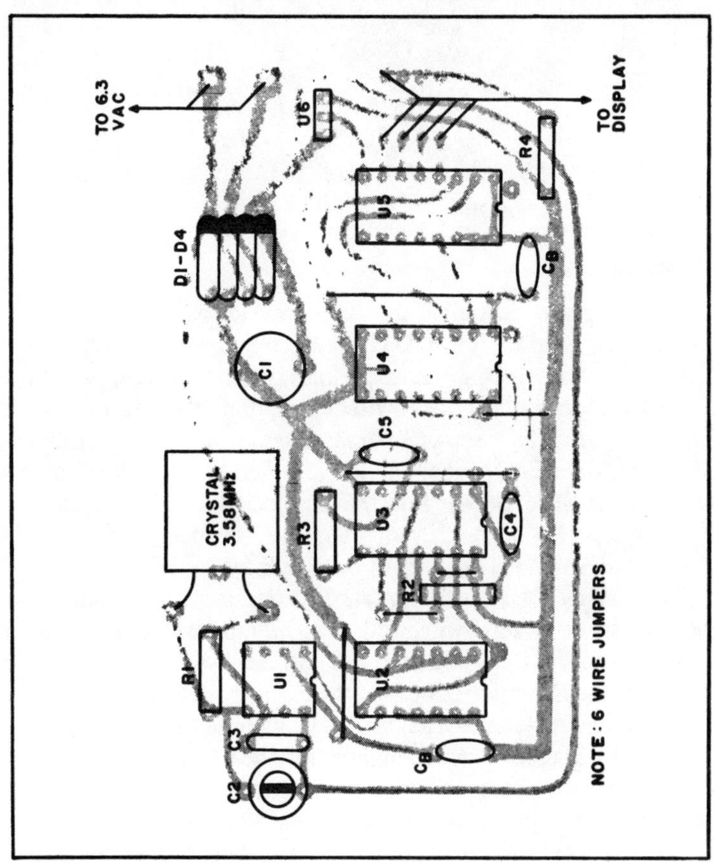

Fig. 9-11. Component placement.

Fig. 9-12. 60-Hz comparator.

Setting the Clock

First, before installing the IC, check for approximately 15 volts at pins 1 and 22. If you don't get 15 volts, look for a bad diode in the power supply. Disconnect the line cord and carefully plug the IC in the socket. Set the switches on the back as follows: Function-Run, Time Set-Off, Timer Enable-Off, Alarm Enable-Off, Time/Date-Center position. Plug the line cord in. The display should show all 8s. Set the Function to Clock Set and Time Set to Minutes/Days. Momentarily depress the Advance push-button. The display should change to all 0s. Don't worry if one or more digits are blank. Depress the Advance switch again and the minutes should start counting. Switch to Hours/Months and repeat to set hours. When the time you have set corresponds with the actual time, turn the Function to Run and the clock should start counting. Pay attention to the am and pm indicators when setting the time, as the calendar changes days at midnight. Set the Alarm, Calendar and Timer in the same manner. The Alarm may sound as you rotate the Function switch, but will stop as the times are set.

The Alarm Time pushbutton is a normally open switch that is wired in parallel with the Function switch so that you can check the

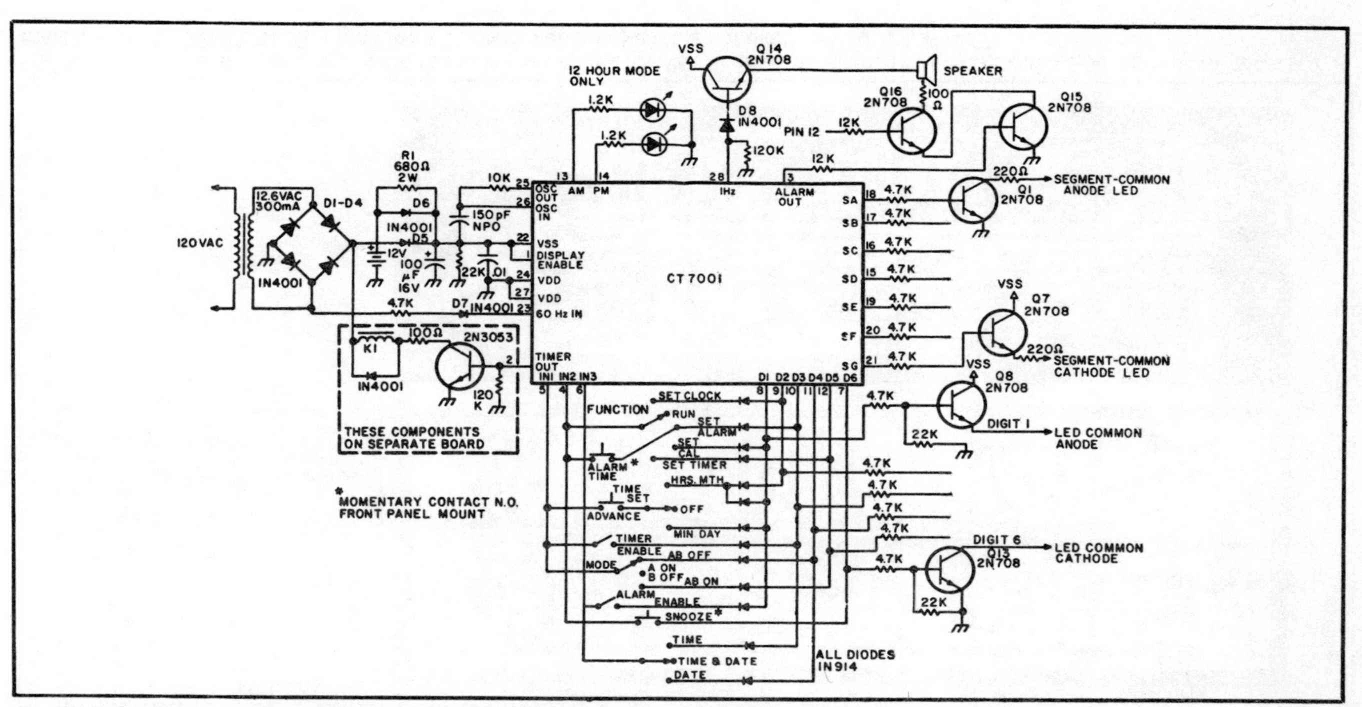

Fig. 9-13. Select appropriate switching for LED display being used. See Fig. 9-16.

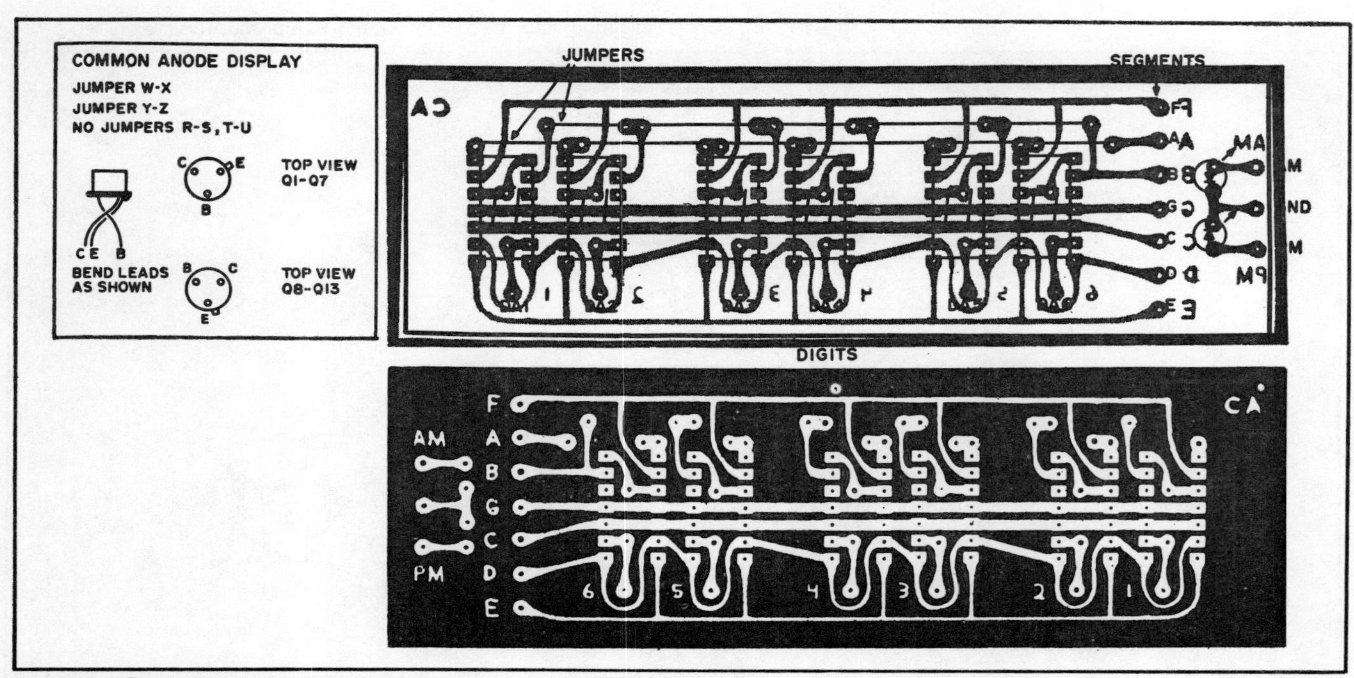

Fig. 9-14. PC board and parts layout for common anode LED display. Board is multiplexed for Monsanto LEDs: MAN 1, 1A, MAN 5, 7, 8, MAN 51, 52, 71, 72, 81, 82. Cut pin 6 on LED (decimal point) as it is not used. Board is 1.5″ × 5″.

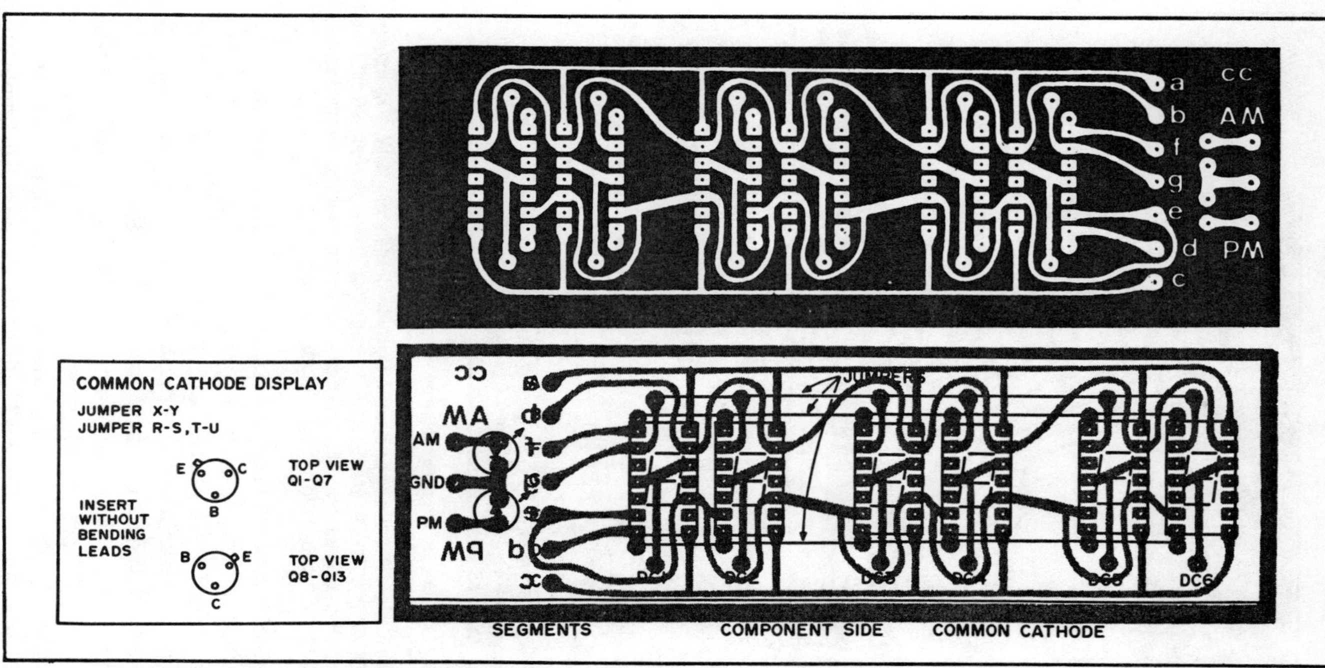

Fig. 9-15. PC board and parts layout for common cathode LED display. Board is multiplexed for Monsanto LEDs: MAN 54, 74, and 84. Board is 1.5″ × 5″.

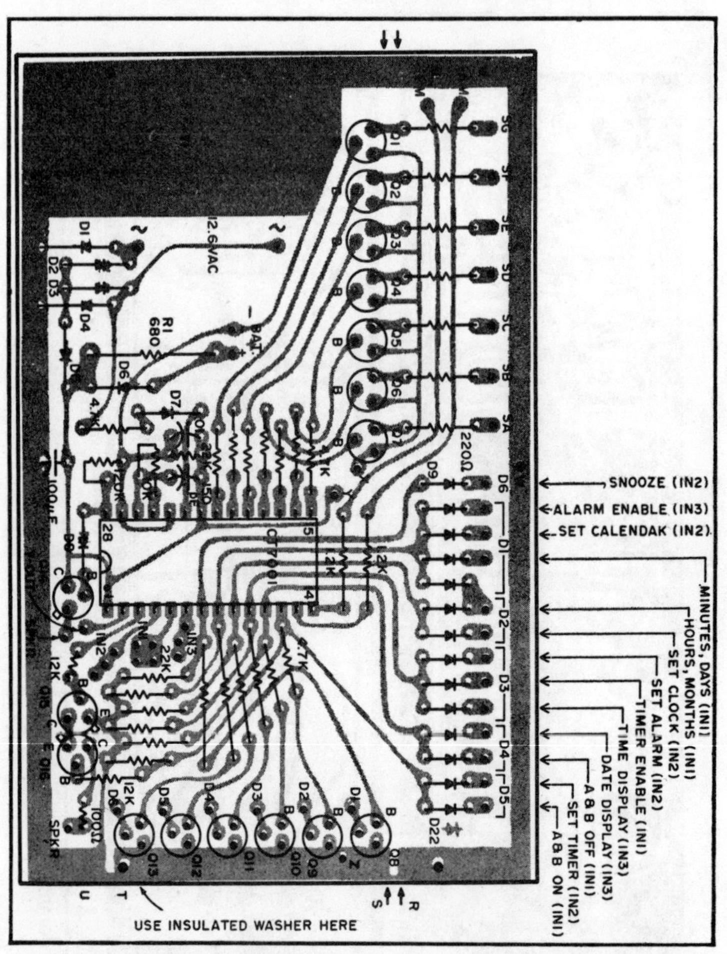

Fig. 9-16. Main PC board and parts layout. All resistors ¼ W, 10% except R1—680Ω 2W. D1-D8 are 50 volt, 1 Amp 1N4001 or similar. D9-D22 are 1N914. Displays are mounted vertically at points marked by arrows, secured to main board with plastic glue, such as Duco cement.

time that the Alarm is set for without fumbling around the back. The Snooze switch is the same, and both are front panel mounted for convenience. The Time/Date switch will force the IC to display one or the other. In the center position the time will be displayed for 8 seconds and the date for 2 seconds. The Mode switch controls how the Timer will function. In the A Off, B Off position, pin 3 will be high for the preset time when Timer Enable is closed.

A On, B Off pin 3 will be high for the preset time and at the Alarm time. A On, B On, pin 3 will be high for the preset time at the

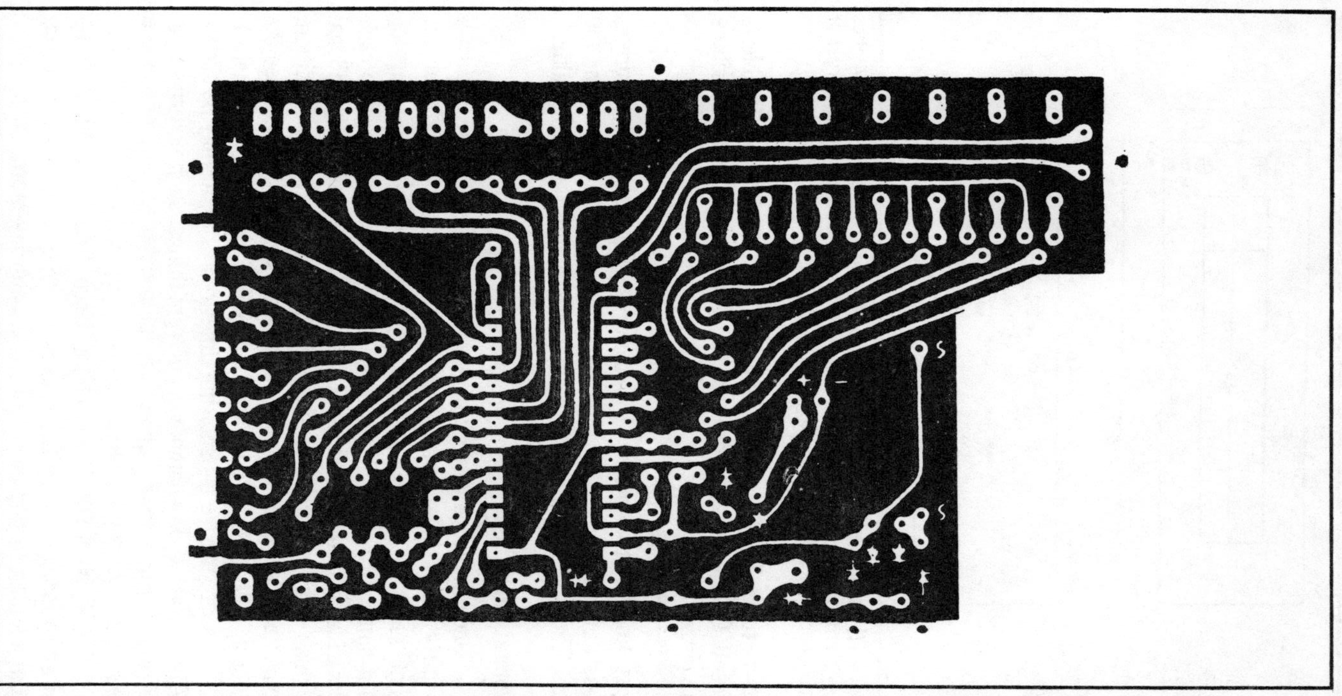

Fig. 9-17. Full size negative of the main board. When mounting T1 to the board, be sure the base of Q1 does not short.

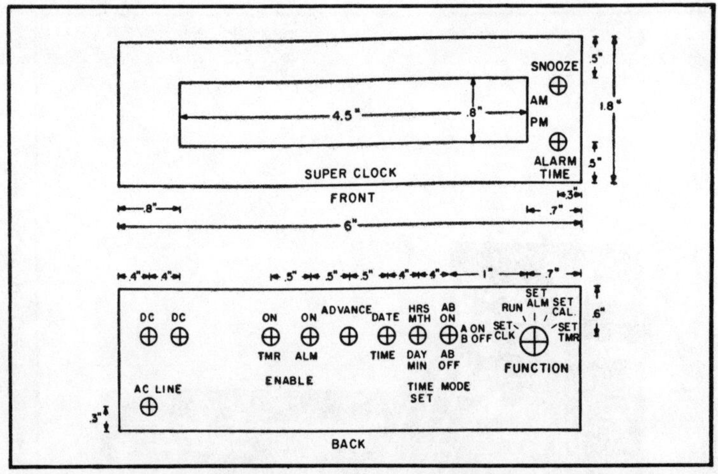

Fig. 9-18. Front and back panel layout. Snooze and alarm time switches may be mounted on either side of display. Display cutout may be larger or smaller, but should be 4.5″ long.

Alarm time. The Timer will only function when Timer Enable is closed. Opening the switch stops the Timer counting, and disables the output.

This IC also has a back-up oscillator to keep time when operating from a battery. If you really want to get it accurate, substitute a 25k pot for the 10k resistor and adjust it as close as you can. Then wire in a fixed precision resistor of the same value.

General Information

Once the clock is working properly and the back-up oscillator is fairly accurate, you may get the idea to use the clock in your car. That is also the reason for two dc inputs on the back: one plugging in a battery to carry it out, and one for operating off the car's electrical system. One word of warning: In some states it is illegal to use red indicators for anything except an emergency condition, so use a different color for the display if you ever plan to use the clock mobile. (All you people with digital tachometers, take note.)

If a larger case is used, the timer relay circuit may be added. With it, you can do a number of different things, such as turn a lamp on and off, turn the rig on at scheduled time, etc., as long as the relay contacts can handle the load.

If you use a DPDT switch for Alarm Enable, an LED can be wired to indicate when the Alarm is active. This could save you from jumping through the ceiling on Saturday morning.

The only time you'll have to manually set the calendar is Feb. 29th. February is programmed into the IC for 28 days. Oh, well, setting a clock once every four years isn't hard. Table 9-2 is the parts list.

TOUCH-TUNE FOR THE VISUALLY HANDICAPPED

A local ham with a visual handicap and limited wrist action needed a means to determine the frequency to which his Drake TR4-CW was tuned. The limitations of his license naturally determine the band limits within which he must operate. And he must do this with a very high degree of precision—to be able to go back to some designated frequency as well as to stay within the limits.

The approach taken is for this amateur to modify his TR4-CW (it can be returned to normal with little or no problem) so that he could operate with relative ease and be sure of his frequency within ± kHz at all times.

The first method tried was to use label tape (Dymo™ or equivalent) with braille markings attached to the face of the transceiver front panel, using the dimple on the tuning-knob skirt as a reference point. This method lacked precision of adjustment and was difficult to use because of the limited wrist action of the ham. The markings can be seen under the new plastic disk in Fig. 9-19.

After studying the design of the tuning assembly on the transceiver in detail, it was apparent that the easiest and most practical form of frequency display would be a disk behind the tuning knob, in place of the graduated skirt. If the braille markings were on the peripheral edge of the skirt and the skirt were transparent, then either a sightless or a sighted amateur could operate the equipment. In addition, the limited wrist action would not be a problem.

Table 9-2. Parts List—"Super Clock."

Qty	Part	Qty	Part
1	CT7001 IC	1	5 position non-shorting rotary
16	2N708 NPN	3	SPDT Center off sub-miniature
1	100 Ohm ¼ W	1	SPST sub-miniature
1	680 Ohm 2 W	1	DPDT sub-miniature
2	1.2k ¼ W	3	SP N.O. momentary contact
14	4.7k ¼ W	1	Ac line cord
1	10k ¼ W	1	Cabinet – Radio Shack #270-260
2	12k ¼ W	1	28 pin DIP socket
7	22k ¼ W	6	14 pin DIP sockets
8	1N4001	1	Main PC board
13	1N914	1	Display PC board
2	LED – Discrete		Hardware, wire, plastic window
6	LED Seven segment displays		
1	12.6 V ac 300 mA Transformer		
1	150 pF NPO		
1	.01 uF Disc		
1	100 uF 16 V Electrolytic		

Most of these parts are available at Radio Shack. If you use the above cabinet, get the sub-mini switches, or they won't fit. The seven segment LEDs are from Poly-Paks (common anode).

Fig. 9-19. Overall view of the TR4-CW with modification in place.

A scrap of ¼″-thick plastic, such as Plexiglas™, was obtained from a local supply house at a very low cost. A local machinist turned a 3½″ disc from this piece and drilled a ¼″ hole in the center to fit the tuning shaft. See Fig. 9-20. A concentric hole had to be drilled partway through the disc at the center to allow for the larger diameter of the concentric shaft on the tuning shaft. This concentric shaft turns the plastic discs on the interior frequency display.

Drake has used a short piece of rubber tubing to transmit the turning motion of the knob to the interior frequency display discs. Since our design for this modification precluded changes which would prevent returning the set to normal for possible later resale, a piece of ruby-red eraser was used to make a new pad that was not as thick as the original piece of tubing (see Fig. 9-21).

When the new parts, rubber pad, and plastic disc were assembled on the tuning shaft, it was discovered that the tuning-knob set screw would not engage the tuning shaft. This minor problem was overcome by machining a 3/16″ × ⅛″ circular flat-bottom groove. This groove allowed the knob to be inset into the new plastic disk far enough for the set screw to engage the shaft and its rough bottom to increase the friction between the knob and the new plastic disc. This machining must be done very slowly, or the plastic will chip along the edges or melt under the cutter. The original aluminum knob skirt was saved, along with the rubber coupling, so that the transceiver could be returned to normal configuration at a later date, if necessary.

Now that the mechanism was working, some method had to be derived so that frequency could easily be determined. Since the tuning system was designed so that one complete turn of the knob was 25 kHz, the perimeter of the new disk was divided at 25 equidistant points (perimeter divided by 25), representing 1 kHz per division. The braille system was used to mark the divisions. The dots were put on label tape, using a braille typewriter that the sightless ham owned. These small pieces of tape were then attached to the periphery of the disk at the appropriate places.

The 1-kHz points were simply marked with a dot while the 5-, 10-, 15-, 20-, and 25-kHz divisions were marked with the appropriate braille symbol and the left-most column of dots aligned as the marker. The starting and ending points were indicated with the 25-kHz braille tape. In order to establish a reference mark on the transceiver face, a simple column of dots was put on a piece of label tape and attached above the window of the face plate (see Fig. 9-19).

Since Drake does not specify the amount of over-travel of the PTO, each transceiver dial will stop at a different frequency. By determining the frequency of the stop (by a sighted ham) and counting the number of turns necessary to get to the edge of the subband, the amateur was able to establish one edge of the subband limit. The set screw hole in the knob can be set so that it is vertical when the edge of the subband is reached, thereby giving the amateur an additional point of reference for counting the number of turns from the PTO stop. The primary reference line is the left-hand edge of dots on the braille 25-kHz mark. With the subband located, the amateur now can stay within the band or operate on a

Fig. 9-20. Detail of plastic disk and rubber spacer.

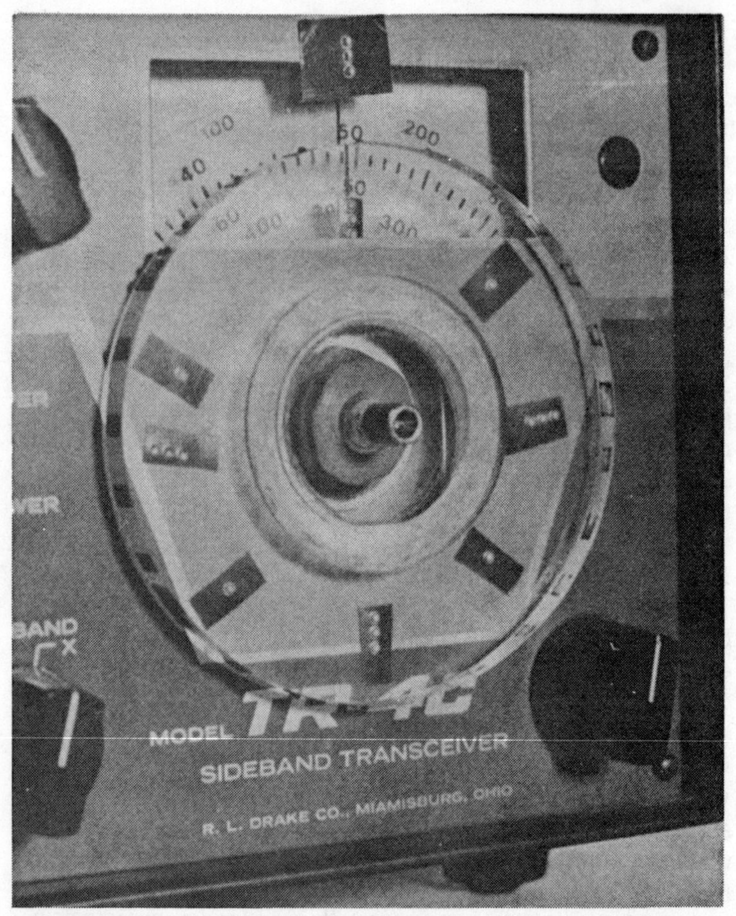

Fig. 9-21. Close-up of plastic disk and rubber spacer mounted on the tuning shaft.

predetermined frequency by aligning the dots used as reference on the front panel with the braille markings on the periphery of the disk. Using this method, the amateur can dial the desired frequency with an error of less than 1 kHz.

The system has now been in use for several months and is working very satisfactorily. The braille dots on the tape attached to the disk edge seem to stay much better than expected. In applying these markers, it is extremely important that the edge of the disk be absolutely clean and that the adhesive on the tapes not be touched with your fingers during application. Small pin heads could be used in the future, but they must be very small so that the ham

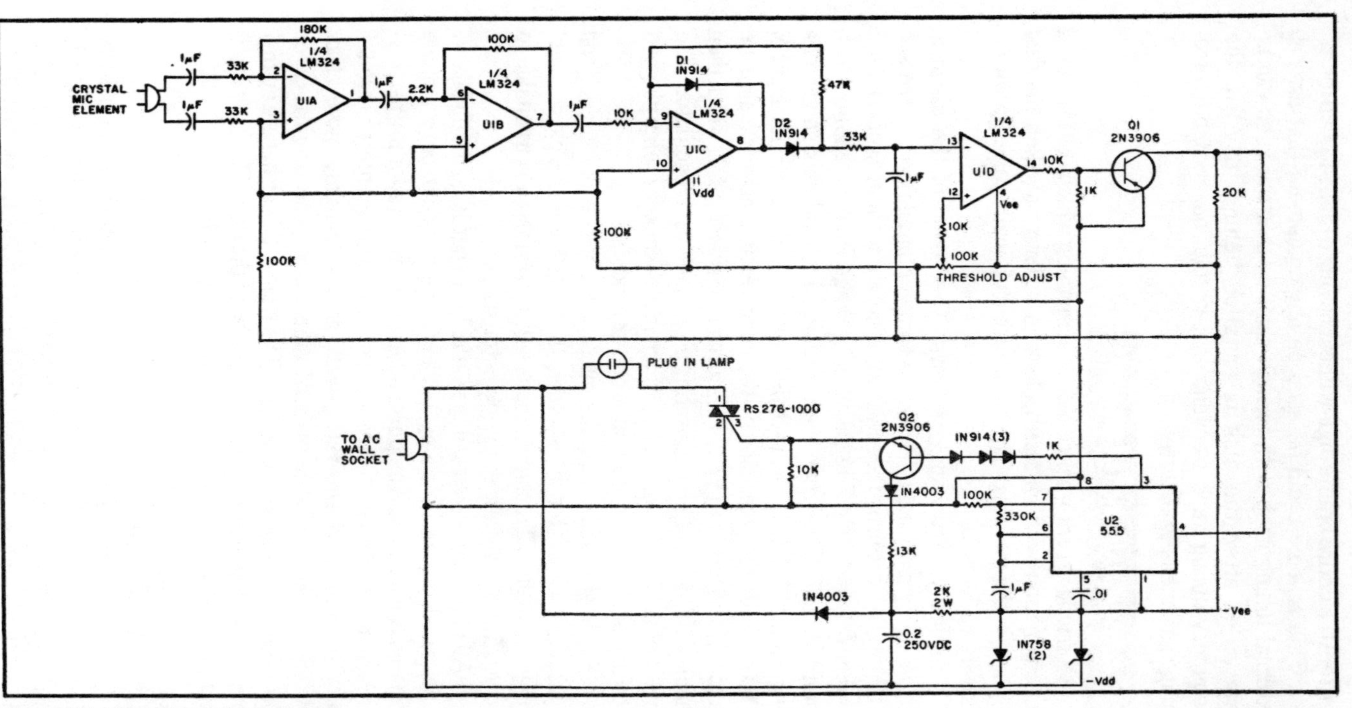

Fig. 9-22. Visual signal schematic.

can feel the entire braille digit without having to hunt around for it.

This scheme could be used on any equipment that uses an external tuning knob with enough shaft so that the knob can be reinstalled after the disk is installed behind it. Also, the frequency-tuning knob frequency-per-revolution must be something that is easily determined.

A VISUAL SIGNAL FOR THE DEAF

A friend recently asked me for something which could be used in conjunction with an alarm clock in order to awaken his deaf nephew in the morning. The circuit which I finally came up with is shown in Fig. 9-22.

Basically, it works as follows: The sound of the alarm is picked up by a crystal microphone element, amplified by U1A and U1B, and detected by the full-wave rectifier consisting of U1C and diodes D1 and D2. The rectified voltage is next applied to a comparator (U1D). When the input signal is loud enough and of a long enough duration, the rectified voltage on the inverting input of U1D exceeds the reference on the noninverting input, causing the output of U1D to go to Vee. This turns on transistor Q1, which then enables U2, a 555 timer operating as a one-second oscillator. The output of the 555 drives a triac, which in turn causes a lamp plugged into the ac socket to flash at a one-second rate.

For best results, tape the microphone to the alarm clock and adjust the comparator threshold so that only the alarm clock sound triggers the unit. Transient noises will not enable the circuit due to the RC time-constant at the detector output.

All of the parts are readily available at your local Radio Shack. I used an LM324 quad op amp, but any quad, two dual, or four single op amps can be used. The pin numbers shown were for my particular layout but, of course, any of the op amps can be interchanged.

This unit performs quite well, and is certainly more of an attention-getter than the original timer used. The original timer was used to simply turn on a light. The triac used can handle up to 6 amps, so several lamps can be plugged into the unit. The lamps will continue to flash as long as the alarm is sounding.

VHF NOTCH FILTER

Having been reintroduced to VHF by modifying an old FM receiver to cover the weather satellite frequencies, I eagerly awaited crystals for a two meter FM rig that I had recently purchased at a hamfest. After plugging the crystals in and erecting a quarter-wave whip in my operating room, I was encouraged to hear the local 19-79 repeater come booming in. Waiting for a pause, I pressed the mike switch to call Normal WB4LJM for an air check. However, the immediate response was noise from the other side of the room.

The weather satellite receiver squelch was being keyed by overload from the two meter rig. It did not take long to determine that the two meter signal was entering the weather satellite receiver through its antenna, all other leads being shielded and filtered. Separating the satellite receiver antenna to reduce coupling is nearly impossible, so I called on the land line to see if we could cook up a filter for the weather satellite receiver. Our first thoughts were for a narrow bandpass cavity, but that idea was rejected for several reasons. First, I am basically lazy and could envision that as a several week project. Second, my operating room, the bedroom, has very limited space and my wife does not think that a copper pipe is as beautiful as I do. Third and finally, the high Q of such a filter would prevent using the weather satellite receiver to monitor police and other services without a coax switch. What was needed was a rejection filter tuned to 146.19 MHz.

A very effective rejection or notch filter can be formed from a wavemeter, i.e., a tuned circuit coupled to the transmission line to couple power out of the line at only one frequency. It was decided that high enough Q could be obtained by using coaxial lines for the resonators. A simple circuit was wired together in a minibox with BNC input and output connectors. Then Norman made a swept frequency response curve. The tap position was changed, and new

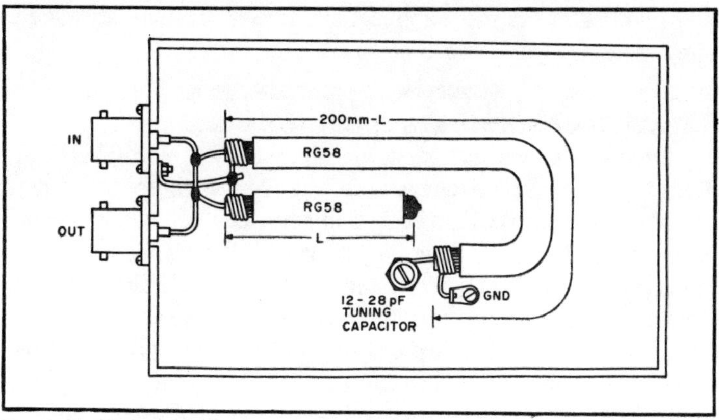

Fig. 9-23. Construction of the rejection filter.

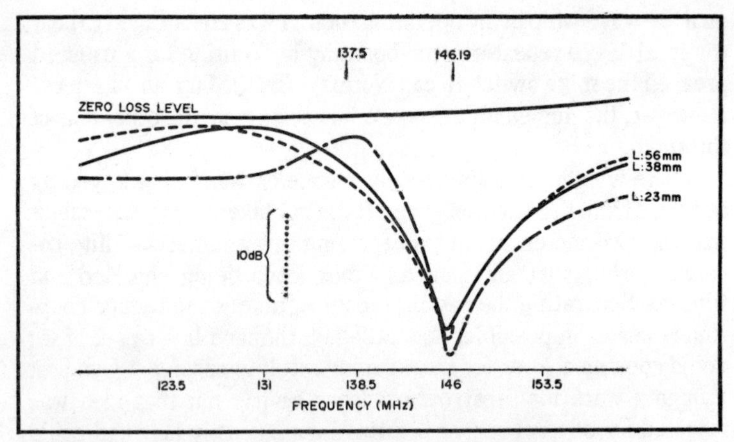

Fig. 9-24. Response of the filter for various tap positions.

curve was measured. The fancy name for this cut-and-try development method is "an empirical investigation."

The actual circuit consists of a shorted one-eight wave coaxial line that is capacitively tuned to resonance with a small trimmer. The input and output taps are made by cutting the cable and splicing it back together, as shown in Fig. 9-23. Figure 9-24 is the response curve of the filter for various tap positions. The rejection bandwidth, or Q, becomes sharper as the tap is moved towards the shorted end of the coax. The final filter has an insertion loss of about 1.6 dB at 137.5 MHz and a rejection of almost 23 dB at 146.19 MHz. The insertion loss might be reduced if rigid coax having a solid outer conductor is used. The rejection frequency is tunable from 127 MHz to 163 MHz. This filter is easily constructed and will probably find more applications as more systems are added to my station.

HOW TO MAKE YOUR OWN CRYSTAL FILTERS

Everyone knows that good, high-frequency (3-9 MHz) crystal filters for use in SSB exciters or accessory CW filters for transceivers are expensive. However, if one has a bit of test equipment and is short on cash but long on patience, it is possible to homebrew very good crystal filters using relatively simple circuitry and without the need for complicated coupling networks.

The crystal filter circuit of Fig. 9-25 has a number of advantages. First of all, no tuned circuits are involved. The crystals are simply paralleled, and from one to six crystals can be used, depending upon whether one wants to construct a simple CW filter with a very sharp response or an SSB filter with a specific bandwidth. The frequency spacing of the individual crystals used is critical; this will be covered later in detail. The crystals are driven from a low-impedance source by the first-stage emitter-follower. At the series-resonant frequency of the crystals, the signal voltage will be developed across the .001-μF capacitor and drive the output amplifier. At frequencies other than those where the crystals exhibit series resonance, the capacitor serves as a bypassing element and helps sharpen the skirts of the filter response. Some signal leak-through will occur because of stray capacitance across the crystals; this is compensated for by coupling some signal from the collector of the first amplifier around the crystals via the 100-pF variable capacitor. A number of general-purpose transistors can be used in the circuit. With those shown, the circuit will have about 10-dB gain.

To make the circuit work properly, the series resonance of the crystals used must be carefully controlled. The only exception might be if one decides to use only a single crystal to form a simple CW filter. However, even for CW reception, the bandwidth provided by one crystal is too sharp and provides uncomfortable reception. Therefore, a controlled bandwidth of 200 to 500 Hz should be used. Such a filter can be constructed using at least two crystals spaced in frequency by the desired bandwidth and centered on the i-f frequency desired. For an SSB filter, at least six crystals should be used. This is because the individual crystal series-resonant frequencies should not differ by more than about 300 Hz. Otherwise, the passband of the filter will not be smooth, as

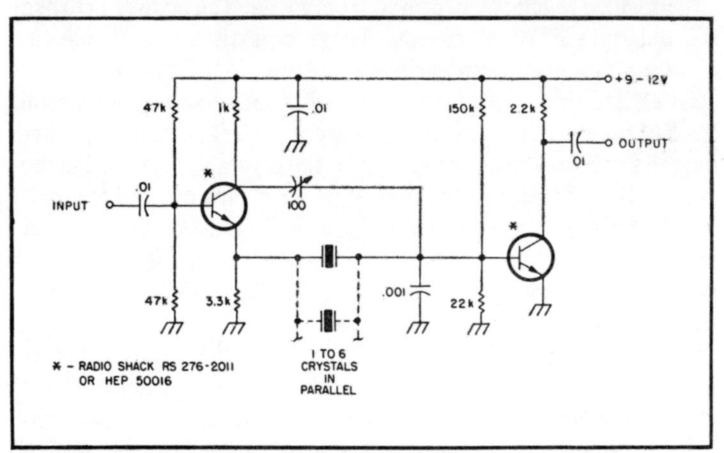

Fig. 9-25. Simple crystal filter circuit does not require any tuned circuits.

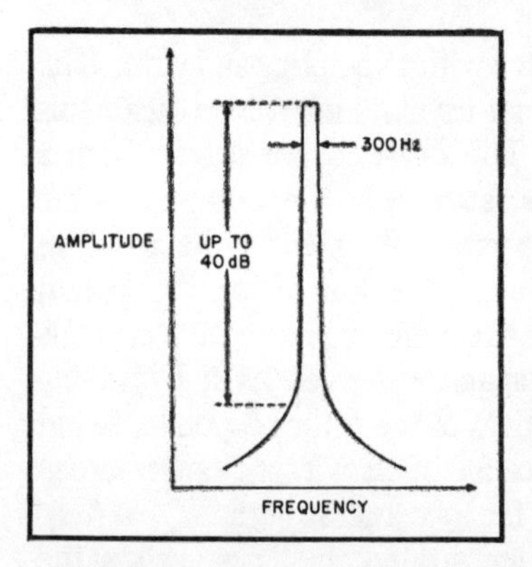

Fig. 9-26. A typical response one could expect from the crystal filter circuit shown in Fig. 9-25. In this case, it would be for a two-crystal circuit with the crystals' frequencies (series-resonant frequencies) separated by about 300 Hz.

it is just a composite of the highly selective passband of each individual crystal.

The type of overall response one might expect from this type of filter is shown in Fig. 9-26. The ultimate rejection that one can achieve with the circuit depends on how carefully the circuit is constructed to prevent stray coupling around the crystals, and on how carefully the 100-pF variable is adjusted. Values of 40 dB can be achieved before the skirts of the filter response start to flare out. Admittedly, this is nothing like the 80-dB out-of-passband rejection of an expensive 8-pole commercial crystal filter. It is sufficient, however, for a simple SSB exciter, and more than adequate for an accessory CW filter in a transceiver when the SSB filter is also left in operation to sharpen the skirts of the overall i-f response.

One may ask how there can be anything inexpensive about a circuit which might require up to six crystals. The answer is to use the old-style FT-243 crystals. These crystals are available for $1.00 or less each from various suppliers (JAN Crystals, for instance), and they allow easy disassembly and access to the crystal itself. The latter is important since there is no easy way of specifying the series-resonant frequency for the crystals to be used in the circuit. So, one has to obtain crystals which are marked approximately for the i-f frequencies of interest, disassemble them, and grind them to the exact frequencies needed. This operation, particularly the "grind" part, is not as terrible as it sounds. In fact, the whole operation is relatively simple.

The circuit of Fig. 9-27 is used to find the series-resonant frequency of a crystal. A signal is applied from a single generator or vfo. The voltage across the crystal is monitored by any high-impedance instrument which will respond at the frequency being used. Usually an oscilloscope is the most suitable instrument. The "bandwidth" of the oscilloscope may be far below that of the test frequency being used, since only an indication of the voltage change across the crystal is necessary. A "5-MHz oscilloscope," for instance, will easily respond to signals up to 10 MHz or more in frequency (although, of course, it cannot be used to analyze the waveform of 10-MHz signals).

As the test frequency is varied, there will be a sharp drop in the voltage across the crystal when its series-resonant frequency is reached, producing a short circuit. The voltage drop is very sudden, and one must vary the test frequency slowly. The series-resonant frequency should be within a few kHz of the frequency stamped on the crystal holder. The FT-243 crystals are easy to disassemble, and the rest of the job now consists of taking the crystal apart, grinding it slowly to raise its frequency, and then testing it back in its holder until one collects the desired number of crystals with the proper series-resonant frequencies.

Although acids can be used to "grind" crystals, it is usually better to use a very slightly abrasive material such as 3M "Trimite" silicon carbide paper (with a fineness of 400 or greater). This paper is obtainable at large hardware stores and comes in 9" × 11" sheets. The paper is taped to any flat surface, and the crystal held flat with a finger and rubbed on the paper with a circular motion. The grinding should be done "wet." That is, the crystal should be kept moistened with water. After grinding, the crystal is carefully cleaned with rubbing alcohol and avoiding any finger marks, put back in its holder.

To go back for a moment, it should be mentioned that although the FT-243 holders are easy to disassemble with just a screwdriver, the operation should be done carefully. The contact plates and spring should be kept in the same order for reassembly. Also,

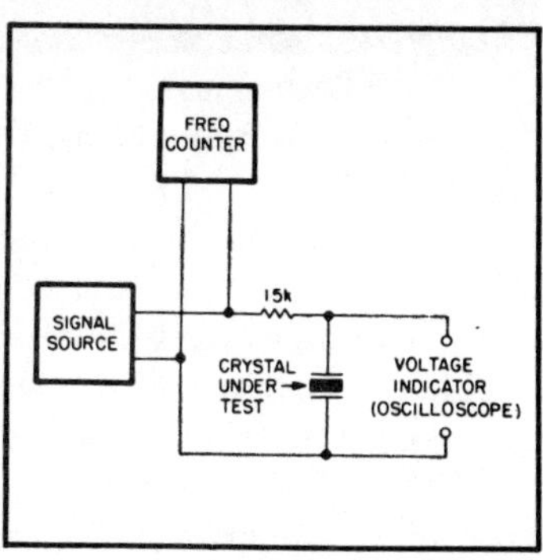

Fig. 9-27. A test setup for measuring the series-resonant frequency of a crystal. Capacitance effects across the crystal must be avoided, so short lads which go directly to the plates of the oscilloscope must be used. A suitable low-capacitance probe may also be used with the oscilloscope or a suitable VTVM.

the crystal should be marked on one corner so that it is set back in the holder in its original orientation.

The amount of grinding a crystal needs to change its frequency depends on the pressure used while grinding, the number of passes while grinding, and so on. However, it takes very little practice to get some "feel" for the changes which take place in the crystal frequency as one grinds it. For instance, with a 8-MHz crystal, one circular pass on the abrasive paper might change the crystal frequency by 150 Hz. One hundred passes might change it as much as 20 kHz. For lower-frequency crystals, more passes will be required for a given frequency change. A 4-MHz crystal might change 60 Hz in one pass and 7 kHz in 100 passes.

Obviously, as one approaches the desired crystal frequencies, the grinding has to be done slowly and patiently. If one goes slightly too far in the grinding process, the crystal frequency can be lowered slightly with a very soft lead pencil. Lightly rub the surface of the crystal with the pencil and then use a soft cloth and a drop of rubbing alcohol to distribute the coating. With a bit of patience between grinding times and possible corrections, one can easily come "right on" with regard to frequency.

Although the acid approach is not really recommended, those amateurs who have access to the chemicals required and who can be careful in their application may want to try it. If so, one should use an ammonium biflouride solution diluted with two parts of water. The crystal is simply placed in the solution (using a small pair of tweezers) for 30 to 60 seconds at a time, removed and rinsed with water, and dried. The crystal frequency is checked and the process repeated as many times as necessary. The whole process is quite easy, but it must be emphasized that the chemical solution is very corrosive and must be used ***and*** disposed of carefully.

OVERVOLTAGE PROTECTION CIRCUIT

The following is in response to the requests for more information about the overvoltage protection circuit which appeared in the March 1978, issue of *73*.

Let me begin by stating that the system was carefully tested *before* it was written up and submitted for publication. Since that time, it has come to my attention that one component could prove troublesome and somewhat improved performance can be obtained by making two rather minor changes.

VR_2 is listed as 5-6 volts and noncritical. However, it is the basis of the abrupt turn-on/turn-off characteristic found in this

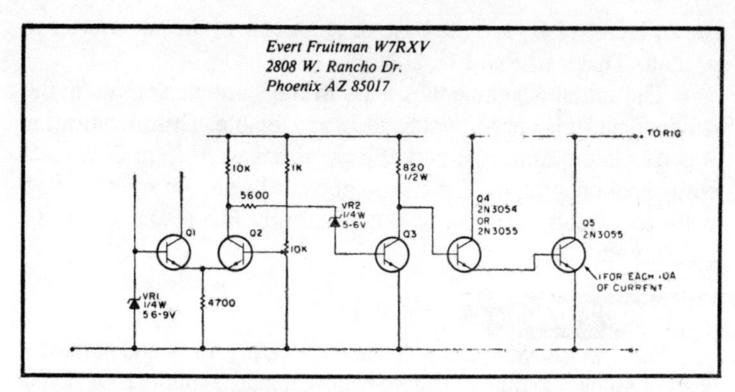

Fig. 9-28. The original overvoltage protection circuit.

system, VR_2 suppresses the 5- to 6-volt offset voltage at the collector of Q_2.

At the low current level found in this circuit, the zener diode may not show sharp enough turn-off characteristic. This property will be helped along in the right direction by connecting 2 or 3 1N4001s (or other cheap silicon diodes) in series with VR_2 as shown in Fig. 2-29. An alternative solution would be to use a 6.8- to 8.2-volt zener for VR_2.

If the voltage is applied to this circuit and *slowly* increased from about 5 volts to whatever level was intended, the system will malfunction. The voltage set/trip control will exhibit some unusual properties. This malfunction will happen only under the above stated conditions. It may be prevented by making the following simple changes:

1. Disconnect the 10k and 5600-Ohm resistors from Q_2 collector.
2. Disconnect Q_1 collector from B+.
3. Connector Q_2 collector to B+.
4. Connect Q_1 collector to the junction of the 10k and 5600-Ohm resistors that used to be tied to Q_2 collector. See Figs. 9-28 and 9-29.

At this point you are almost there. As it stands, after completing the above operation, the fuse will clear if the input voltage is *below* the intended set point. In order to correct that malady and finish the project, it is necessary (and highly desirable) to make the final modification.

Disconnect the zener diode, VR_2 from the base of Q_3, and connect it to the base of Q_6 (another 2N3414, etc.). Connect one end of a 1500-ohm, ¼-watt resistor to B+, and the other end of it to the collector of Q_6 and the base of Q_3. Connect the Q_6 emitter to ground. That's it! See Fig. 9-30.

The initial adjustment is done in the same manner as in the original, and it is very smooth and *very* resetable. This modification was checked out using a power supply that went from 5-25 volts with excellent results. The trip point could be set anywhere from 8 volts to 25 volts and it would consistently fire at the trip point ±0.25 volt.

THE PANIC BUTTON

What would you do if you were operating away and noticed a wisp of smoke coming out of your new linear amplifier? Or if you were working on some gear and felt a whole lot of electrons running between your fingertips and your toes? Or if you were going on vacation and wanted to make sure your station wouldn't be operated in your absence?

You'd want to kill the power (I hope!), and in at least the first two cases, you'd want it off *fast*. And it might be even nicer if this could be done easily by somebody else, such as one of the kids—or even the dog—if you got into trouble and couldn't kill the power yourself. Aside from these situations, there are times when it is merely convenient to know that all your shack power is off.

A properly-engineered kill-switch system thus has a number of advantages, and if you have not yet installed this basic safety/convenience feature, you might consider it. In most cases it is a simple thing to do, it is economical, and whether or not you ever plan on using it, just getting it in place is guaranteed to give you a good feeling. At the very least, your station will have a new touch of class.

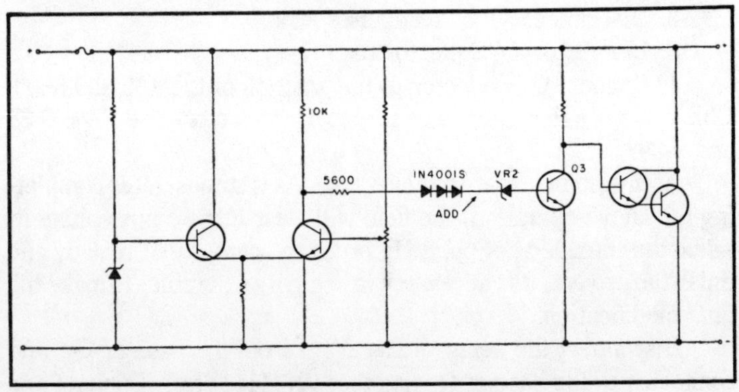

Fig. 9-29. Improved overvoltage protection circuit.

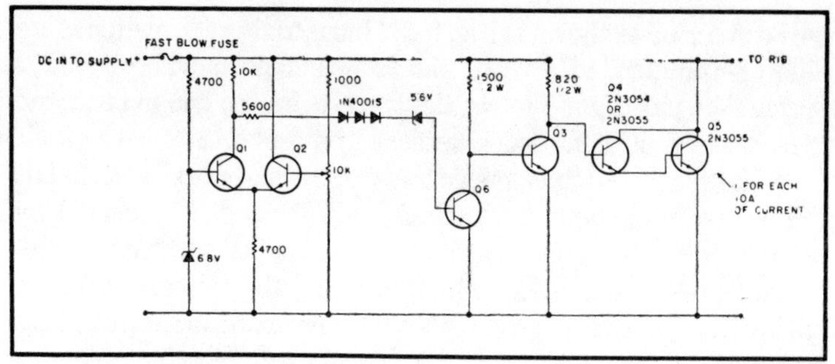

Fig. 9-30. The final modification.

The basic circuit for a kill switch is shown in Fig. 9-31. It consists of three parts: a contactor (relay) in series with the main power line, one or more kill switches, and a reset switch. This last is nice if you ever wish to restore power once it has been killed. The circuit also shows a transformer, the purpose of which will be explained shortly.

The contactor is the heart of the system. It should have high-current capacity contacts at least equal to the trip point of the circuit breaker feeding the shack, and preferably twice that. The reason is that the contacts will carry all the current your shack is using, and if it is necessary to kill power at a time of high-current draw, you will want the contact to open cleanly without arcing. Good reliable contactors (UL listed) are available from your friendly local electrical distributor.

The contactor should have a low-voltage coil; 24 V ac is a readily-available standard. This is so you won't have any unnecessary high voltage running around the shack to the kill switches. The transformer secondary voltage should match the contactor coil voltage; 24-volt bell transformers are readily available at low cost. You also can use 12 V ac, which is a standard and readily available, but the current draw by the contactor coil will be about twice as much and it will tend to get warm. Whatever you use, the contactor should be rated for continuous duty, both in the coil and in the contacts.

The kill switches are normally-closed push-button switches, and they should be located so they are easy to get to and easy to operate. I prefer industrial-grade actuators with big red push-buttons (3″ or so in diameter) so they can be hit easily with an entire hand—literally, "panic buttons." On my own service benches, kill switches are located so they also can be hit easily just by moving a knee. If you use more than one kill switch, they're just wired in series as shown in Fig. 9-31. Industrial-grade switches are highly recommended because what is most important is reliability. It would be mortifying, to say the least, to smash one in panic and then discover it didn't work because you broke it!

The reset switch is a normally-open push-button switch, but can be any other type of switch as well. It needn't be especially reliable or easy to get to, but it should not be located far from the contactor because it will have high voltage (110 V ac) on it. Like the kill switches, the reset switch need have no great current-carrying capacity.

The contactor should be mounted in an enclosure, UL-approved, of course, near where the power line first enters the shack. The transformer and reset switch can be mounted near the enclosure, or on it, using knockouts, depending on the enclosure you get. The whole thing can be placed inside a locked box, if desired, to prevent unauthorized access to the reset button. The 24-volt line (or whatever low-voltage line you have selected) just runs out of the box/enclosure. Very simple.

When wiring up the contactor, a few safety precautions are in order. First, make sure power is off at the circuit breaker, and *verify* that it is off by means of a portable lamp or circuit-tester inside the shack at the point you plan to cut into the power line. Secondly, make sure you insert the contacts into the hot side of the power line, not the cold or "neutral" side. In most localities, the black wire is the hot and the white wire is the neutral. You never, ever, want to put a switch or any other kind of interruption into the neutral.

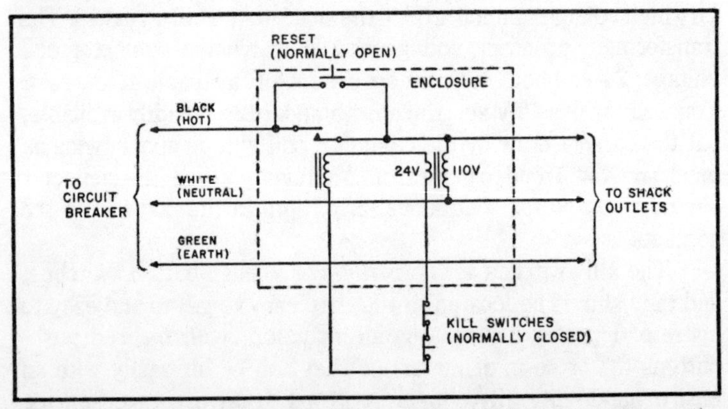

Fig. 9-31. Schematic for the kill-switch system. Only the hot wire may be interrupted—the neutral and earth wires must not be broken. Check with local authorities to verify respective wire colors in your locality.

If you find a green wire also, don't cut it! It is the earth wire, and if it gets broken, you may have all kinds of problems in all sorts of places where you don't want them when you don't want them. (You may even find your trusty soldering iron is frying you as well as the solder, for instance. Not much fun.) If you have any doubts about the color coding, check with your local power company or a local electrical contractor. Note: In some communities it may be required that a licensed electrician actually make the connection for you and/or that the result be inspected by the local building inspector.

Once the contactor is wired up, just run the low-voltage line through the kill switches and you're in business. Now, and only now, you can turn shack power back on at the circuit breaker. The shack will still be dead, however, because the contactor contacts are open and thus there can be no voltage on the coil to energize it. Pressing the reset button applies 110 volts to the transformer, which supplies the low voltage to the contactor coil (through the kill switches), energizing the contactor and closing the contacts. The contactor will remain energized when the reset button is released because the transformer—like all the other equipment in the shack—is now getting its primary power through the contacts.

And there you are! With luck, you've got a nice new convenience for the shack. With somewhat less luck, you've got a simple system that'll keep you from having no luck at all. Just one final point: Teach the family how to kill power if they ever need to. If you ever need assistance, smashing one of the prominent red buttons should be the first thing they do, and such a time is definitely not the time to explain the system to them.

BUILD THIS MINI-COUNTER

Sooner or later, anyone who does much work with communications equipment discovers the need for a accurate frequency counter. Let's face it, crowded communications channels make knowing your frequency more important than ever. This is where this counter project comes in.

Would you like to have a counter that measures frequency up to 135 MHz and beyond—one that is pocket-portable with battery power to make accurate measurements simple anywhere? Of course you would! The Model 304 Communications Counter has these features and much, much more: a low price tag of about $50, easy, one- to two-evening construction, and a superior sensitivity from 4-mV rms at 27 MHz to 12-mV rms at 135 MHz with input protection.

The 304's battery drain is low enough (130 mA) that you get long hours of operation on a set of charged nicads. You get a frequency range of below 1 MHz to over 135 MHz, covering ham, CB, commercial, radio control, and TV frequencies. And, best of all, you can measure the frequency to within 100 hertz of what the transmitter is putting out. If you are in a hurry, you can select a reading to within 1 kHz, and get five readings per second versus one reading per 2 seconds on the 100 Hz range. Accuracy can be up to ± 0.002% ± 1 count if the counter is calibrated on a frequency standard. Now that you know what this counter can do, go out and look at units selling from twice to three times as much as this one. One wonders why no one has thought of this design before.

Construction of this project will be a pleasant surprise because it is very easy. Thanks to a clever PC board layout, this project uses only a single-sided PC board, making duplication from the pages of this magazine a snap. There are only six wire jumpers in the board—for those of you who hate to install jumpers. There are sockets for all ICs and a minimum number of components. Since there are only three ICs and one transistor, your construction time for the board should be about four hours of casual work. Add another evening for the cabinet work and you are all set.

The parts in this counter should be easy to get. The two main CMOS chips have been on the market for nearly three years, and the prescaler chip has been used for well over one year. The rest of the parts are not too critical. All parts are available from the surplus houses. To assist you, however, I am including addresses of typical suppliers of the key parts, plus the addresses of the manufacturers who make the ICs. If necessary, you can write for a local dealer's name and address.

How It Works

This counter project is based on three ICs: One amplifies and divides the signal by 100, another counts it up and displays 7 digits of information, and the third generates precise timing signals to control the second chip. Check Fig. 9-32 for details as you read.

Input signals applied to J1 first encounter a protection network composed of C1 (which blocks dc), R1 (which limits input current) and D1-D2 (which limits input signal). The purpose of this network is to protect the counter from "real world" overloads such as an accidental dose of excessive dc voltage, 120 volts ac, and excessive rf signal. This keeps IC1, the preamp/prescaler chip, safe.

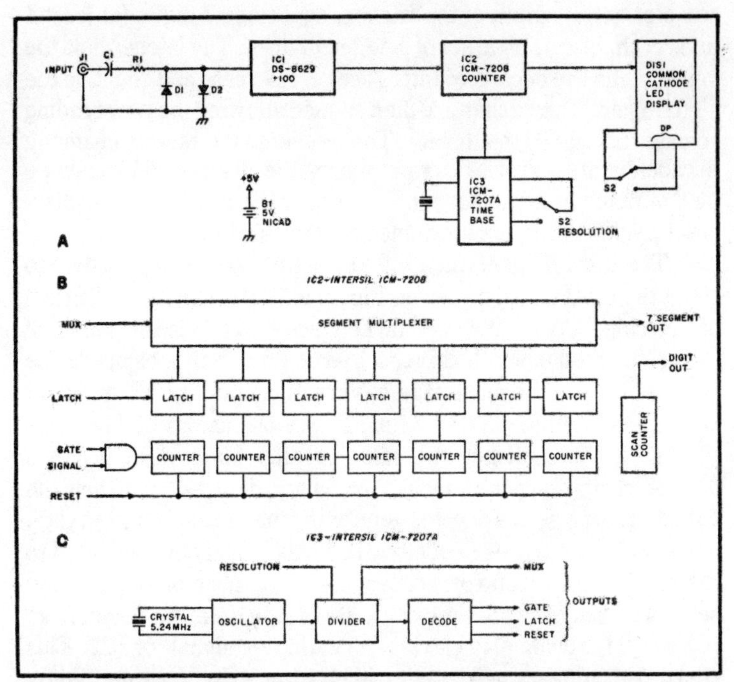

Fig. 9-32. (A) Block diagram of the model 304 counter. It features a 6-digit display, a 135-MHz counting range, and only three ICs. Power is supplied by a 5-volt nicad battery pack. Switch S2 selects either 1-kHz or 100-Hz resolution of the input signal. (B) Inside the ICM-7208 (IC2) counter array. (C) Inside the ICM-7207A timebase controller. An input connected to the divider section controls the gate time, and thus the resolution of the signal being counted by IC2.

Next, the protected signal goes to IC1. Although this IC was designed by National Semiconductor for synthesized FM stereo radios, it works great in a counter. It has a preamp stage for high gain which has sensitivity of up to 4 mV at 27 MHz, then a series of ECL (Emitter Coupled Logic) and low-power Schottky TTL dividers. It divides the input signal by 100, so 100 MHz in equals 1 MHz out. This is necessary because the counter-display chip to be described won't count a TTL signal much beyond 2 MHz.

The divided signal enters IC2, a CMOS/LSI chip that counts it and displays the results on an LED display. Figure 9-32 shows a general overview of this chip. The signal from IC1 passes through seven decade counters. They count up the signal. After it is counted, the data is passed on to the display via the latches (one per decade counter), through the multiplexing circuitry, and on to the display. Actually, the multiplexing circuitry consists of a big data selector at the output of the latches, squeezing 7 digits (at least 4 lines each) into a single set of 7-segment lines. The latches hold the signal while the decade counters are reset to zero and count up the signal again. The latches are then opened allowing the new reading to pass through to the display. This prevents the rapidly changing decade-counter outputs from reaching the display and causing a blur of numbers. The sequence in which the count-latch-display-reset cycles occur is determined by the third IC.

The third IC generates the key signals necessary for IC2 to count the input signal properly. Figure 9-32 shows a block diagram of this chip. A 5.24-MHz crystal connected to IC3 is the source of the proper frequency. It drives a Pierce-type oscillator inside the chip, which has the advantage of very high stability. (The circuit isn't much like the old 6AG7 tube Pierce oscillators of days gone by, but it accomplishes the same result.) The output drives a divider chain composed mostly of binary devices, and then the output drives a series of gates generating the precisely-timed gate, latch, and reset pulses needed by IC2. Also, an input connected to the divider chain controls its length—or number of stages—and selects a 1-second gate time or a 100-ms gate time. An output from IC3's divider chain also carries the multiplex signal for IC2. This signal determines when a digit will be scanned by IC2's multiplexer, and is just as important as the other signals.

Construction Guide

The first step is to round up the parts. If you are lucky, you can probably scrounge most of them from your junk box, but more likely you will have to buy at least the ICs. All of them have been available to the industry for years, and they are starting to pop up in the ads.

The LED display is commonly used in low-cost calculators, and can be almost any such unit; the pinouts have been pretty much standardized by various manufacturers. We have used National NSB-188, Litronix DL-94 displays, and others with success. The HP unit specified is recommended because of its superior appearance, but its price tag may drive you to scrounging a junk calculator instead. Order any parts you need, plus the crystal. If my own experience is any indicator, you should order the crystal first. They are custom made, and take time to get.

The next step is to obtain or make the PC board. Luckily, this one is a single-sided type, and can be made at home using the G-C "Lift It" kit, or the newer printed circuit transfer films such as PCP type A, which is carried by Tri-Tek, Inc. (See Table 9-3 for

Table 9-3. Supplier List.

Item	Factory	Distributor
IC1-National DS-8629	National Semiconductor 2900 Semiconductor Dr. Santa Clara CA 95051	Tri-Tek, Inc. 7808 North 27th Ave. Phoenix AZ 85021 (602)-995-9352
IC2-IC3 Intersil ICM-7208IPI and ICM-7207AIPD. Note: These parts come as a set	Intersil, Inc. 10900 Tantau Ave. Santa Clara CA 95014	Poly Paks, Inc. PO Box 942 S. Lynnfield MA 01940 Stock #92CU4079 (617)-245-3828
Y1—5.24288 MHz crystal	JAN Crystals 2400 Crystal Dr. Ft. Meyers FL 33901 (813)-936-2397	JAN Crystals
DIS1—H-P display	Hewlett-Packard, Inc. Optoelectronics Div. 1501 Page Mill Rd. Palo Alto CA 94304	Poly Paks, Inc.
C7—20 pF trimmer cap	Sprague, Erie, etc.	Tri-Tek, Inc. Stock# CAP9308
Case—LMB-CR-531	Heeger, Inc. 725 Ceres St. Los Angeles CA 90021	

Note on the "Factory" column: These people should be contacted *only* for a local distributor; only JAN Crystals will sell small quantities of parts to individual users.

address.) It might be wise to silver-plate the finished board to improve high frequency performance, but that is up to you. While the improvement will be slight, the appearance will be considerably better. Drill all holes with a number 64 drill, and then drill out the four corner holes with a ⅛″ drill. Now let's start the wiring.

You should install the jumpers and IC sockets before doing anything else. Refer to Fig. 9-33, which shows the component side of the board. Cut up three 1″ pieces of no. 28 solid wire or resistor leads. Then cut two 2″ pieces of wire. Install one 2″ wire between the two holes near the right center of the board, as shown in Fig. 9-33. Install the other 2″ jumper between the two holes near the left center of the board. Install a 1″ jumper near the bottom of the board at the cutout. Move up to the center of the board, and install the remaining two 1″ jumpers. Be sure to get the jumper nearest the left center of the board in the proper holes. An 8-pin IC installs next to it. When done, check to be sure all connections are soldered. Also, trim away the PC board at the bottom to form the trapezoid-shaped cutout, if you haven't already done so.

Next, install the 8-pin low-profile IC socket in the top left corner, next to that small jumper. Move down from it and install a 14-pin low-profile socket near the bottom edge. Finally, install the 28-pin socket next to the right edge of the board. I recommend the use of low profile sockets because they make later installation in the cabinet easier. Molex® pins will suffice, however. Check for solder bridges, fix, and flip the board over. Cut a 1¾″ piece of insulated hookup wire, strip, and solder one end to pin 12 on the 14-pin socket, and the other end to pin 19 on the 28-pin socket. This wire shows up on Fig. 9-34. (Note that there is a "1" to identify pin 1 on each socket.)

The next step is to install the eight segment resistors on the foil side of the board. About ¼″ of each resistor lead appears on the other side of the board for mounting the display, so don't trim the leads off from the row of wires in the center of the board. Start with the A segment resistor, R2. Lay the resistor against the board between the A on the socket (28-pin) and the A pad in the center of the board. Pass the resistor end through the A pad, so that the lead comes through the hole in the center of the board. Remember that this is for the display, so don't trim it. Solder to the pad. Bend the resistor body so that the other lead won't touch any other foil but the A pad on the 28-pin socket.

This process is identical for the other resistors, and is shown in Fig. 9-34. Repeat with installation of the D segment resistor.

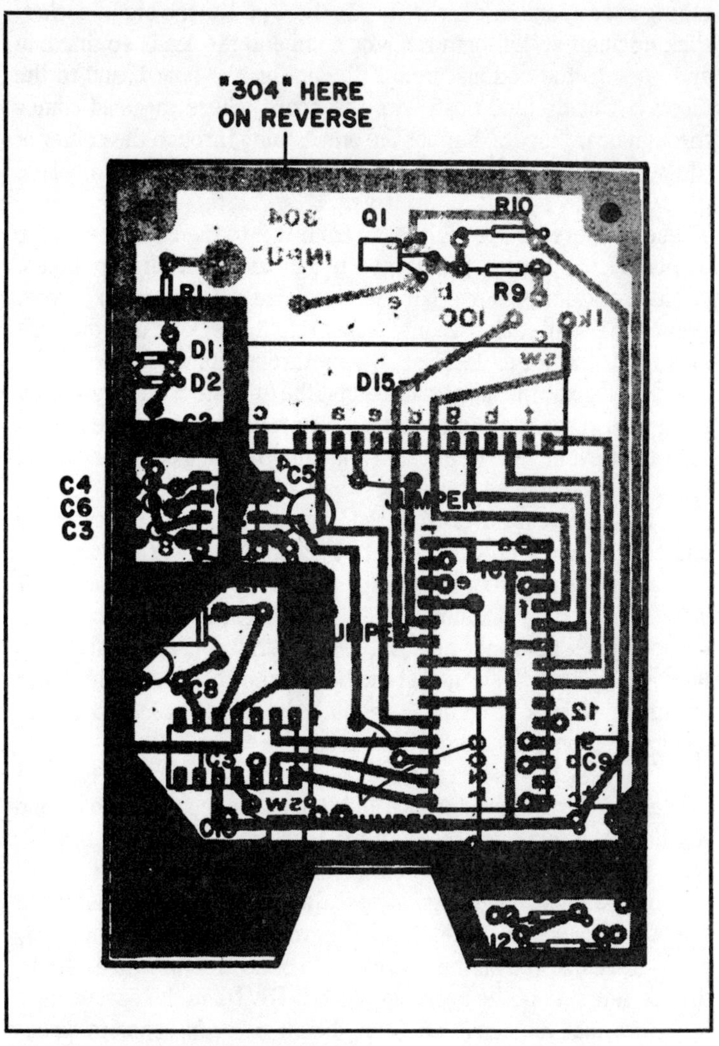

Fig. 9-33. Component side of the board.

Position it over the A resistor you just installed, and bend the leads so they go from the D in the center of the board to pin 2 on the big IC socket. Note that there isn't a D next to pin 2 on the 28-pin socket. (There wasn't room for it.) Continue by installing the E segment resistor in the same manner, leaving at least ¼" of lead sticking through the center of the board to mount the display. Continue with the F, G, B, and one end of the C resistors. Attach a piece of bare wire to the other end of the C resistor, slip a piece of insulating tubing over it, and solder the wire to the C pad on the big IC socket. Pick up another 470-ohm resistor and bend the leads so that one end goes to the pad just below "Input" on the board, and to the display. Cut the end flush with the board where the lead comes through near "Input," but not the one coming through the center of the board. That concludes the resistor installation for the display.

Continue by cutting up seven 1" wires for use as jumpers. These jumpers tie the digit leads from IC2 to the cathodes on the display. Heavy wire, just large enough to fit the display holes, might work better here, but the wire size you use is up to you. Bend one end of each wire into a small right angle, so that the wire looks like the letter L. Pass a wire through the foil side of the board, and place the short arm against the foil pad. Solder carefully. Put a wire in each pad that doesn't have a resistor. Use patience and care to prevent the wires from touching the adjacent resistored pads. This concludes the hardest part of the project.

Turn the board over to the component side and very carefully clip the 15 wires down to about ¼" above the board. Place the display over the wires, being sure to leave the hole on the far *left* side of the display open. All holes on the *right* side must be used. Do not leave a hole open here or the counter won't read right. Press the display down flush against the board, and solder quickly. Use a minimum of heat to avoid loosening the wires soldered to the counter PC board. Clip off the excess wire.

The rest of the wiring needs no comment; refer to Fig. 9-35 for details. Install resistors R9 and R10 at the top of the board and R1 along the left side. Move to the bottom right and install R12. Now you have installed all resistors.

The semiconductors go in next. Install Q1 at the top center of the board, as shown in Fig. 9-35. For your convenience, the E, B, and C were marked on the reverse side of the board. Move over to the top left side of the board and install D1/D2 in the spaces near R1. Note that the bands on these diodes point in opposite directions. Move to the bottom right corner of the board, and install D3. That completes the semiconductor installation. The IC installation will follow later.

Now let's install the capacitors on the front of the board. Refer to Fig. 9-35, then turn to the top left side of the board. Install C2-C3-C4-C6 as close as possible to the board. Jump over the 8-pin socket and install C5 as close as possible to the board. Bend it flush with the board before soldering. Move down the left side next to the 14-pin socket and install C8 near it. Then jump over the socket

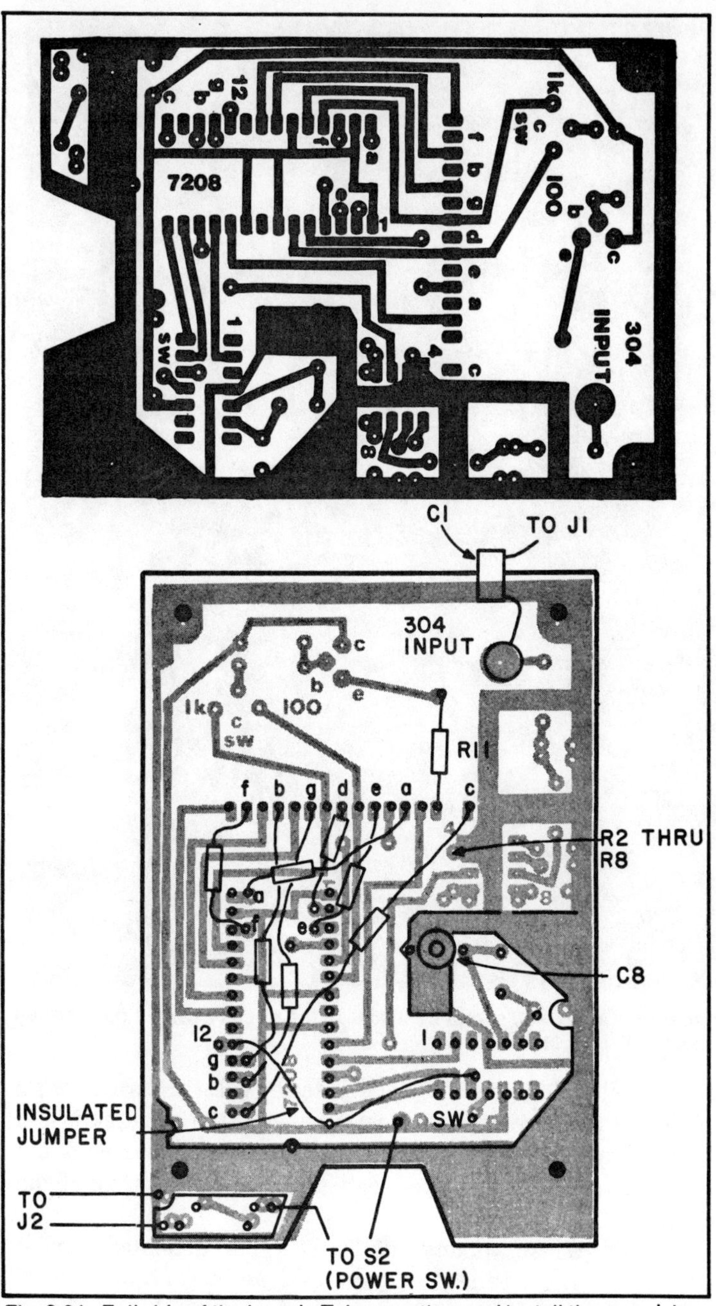

Fig. 9-34. Foil side of the board. Take your time and install those resistors correctly!

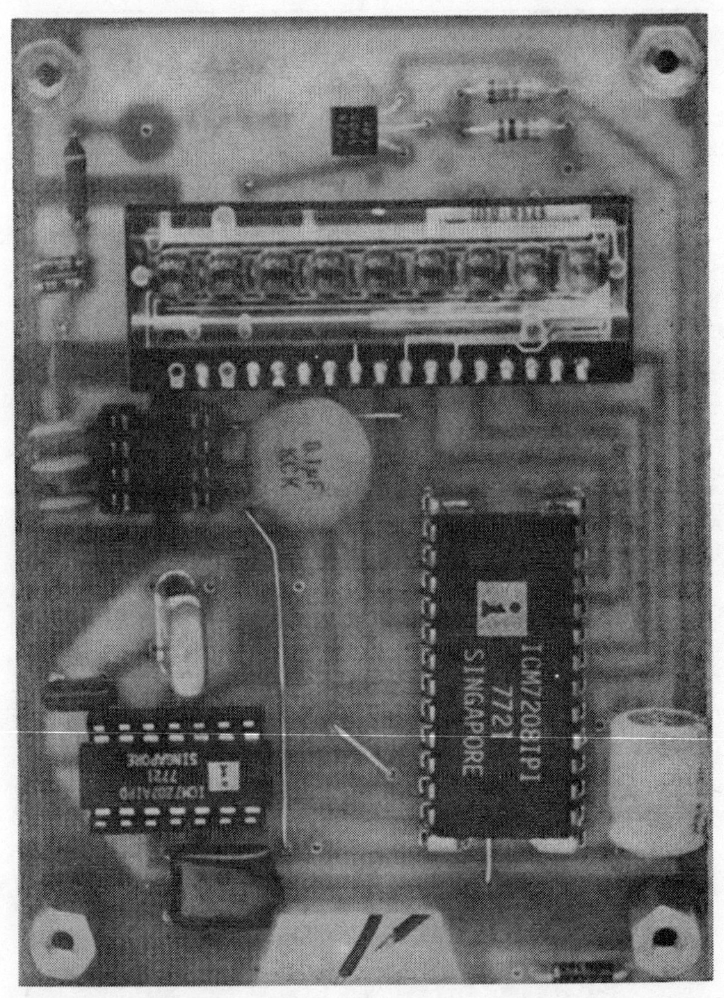

Fig. 9-35. Completed component side of the circuit board. Your finished board will look like this.

and install C10 as shown. Bend it flush with the board before soldering. Skip over to the bottom right side of the board and install C9. Be sure the negative terminal faces the outside edge of the board. That completes capacitor installation on the front of the board.

While you are working with the front, dig out Y1 and install it next to C8. Also, bend the crystal over against the board edge before soldering. Figure 9-35 shows it standing up, but this is not normal practice.

Flip the board over to the foil side so that you can install the remaining capacitors. Install C1, standing on end, at the input pad. Leave the other end free. Install C7 in the spot near the center of the board. Note that on most trimmers there are three leads. The two closest together go to the heavy ground foil on this counter, while the third goes to the crystal and IC3. That completes capacitor installation.

Next, the switch and mounting hardware are installed. Attach six 3″ wires to the terminals of S2. Then wire up this switch as shown in Fig. 9-35 and Fig. 9-36, cutting the wires to size as you go along. Install four spacers in the corners with 4-40 hardware and lockwashers. Note that the spacers are on the front side.

That just about wraps things up on the module. Finish by installing the ICs. Be careful about pin 1 orientation. Remember,

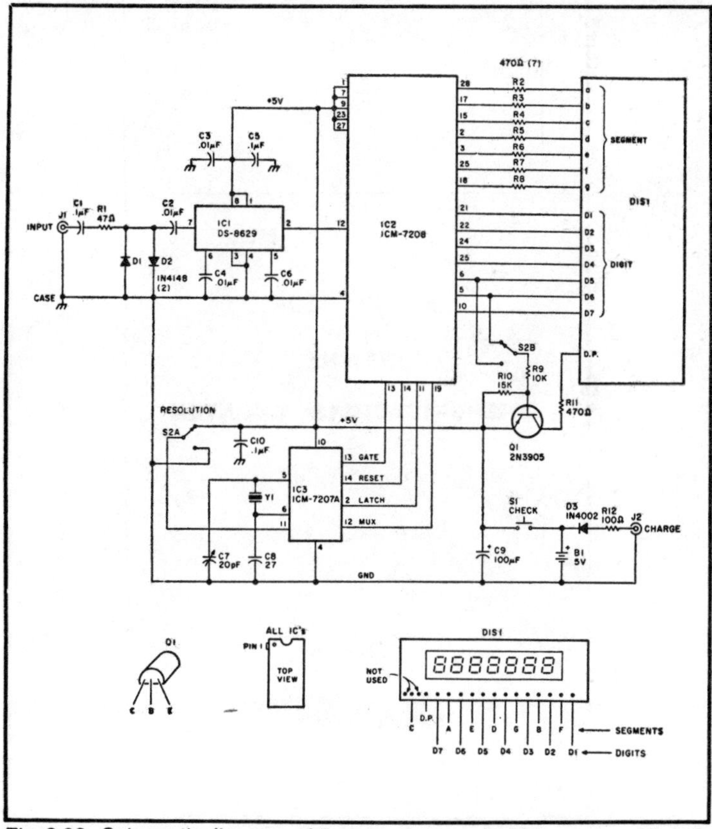

Fig. 9-36. Schematic diagram of the complete model 304 counter. All parts except the batteries and jacks and switch mount directly on a PC board.

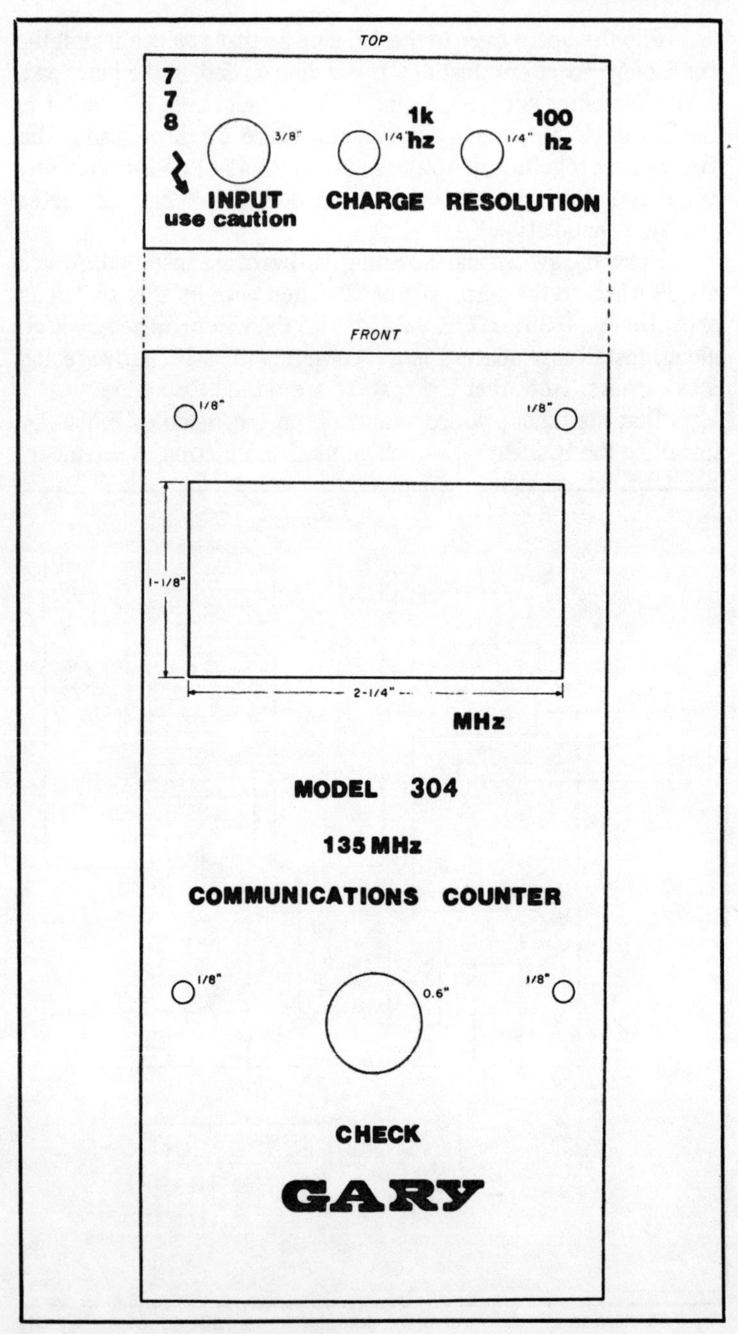

Fig. 9-37. Cabinet drilling template.

pin 1 is identified on the PC board if you get confused. Then cut two 1" pieces of hookup wire, strip both ends, and install at the bottom of the board near the cutout. See Fig. 9-34. Also, cut a 5" piece of the same wire and attach near the end in the bottom left corner of the board. That completes the module assembly. It can be tested by applying 5-volts dc to the wire off the bottom center of the board.

Now you get to work the case. Figure 9-37 shows a drilling template for the front of the box. Drill out the box and nibble the display hole. Clean up the box, label it with press-on transfers, and spray it with clean acrylic spray to preserve the finish.

Complete the assembly by installing the module in the box and wiring it up. First, install J1 and J2 in the top of the box. Then slip

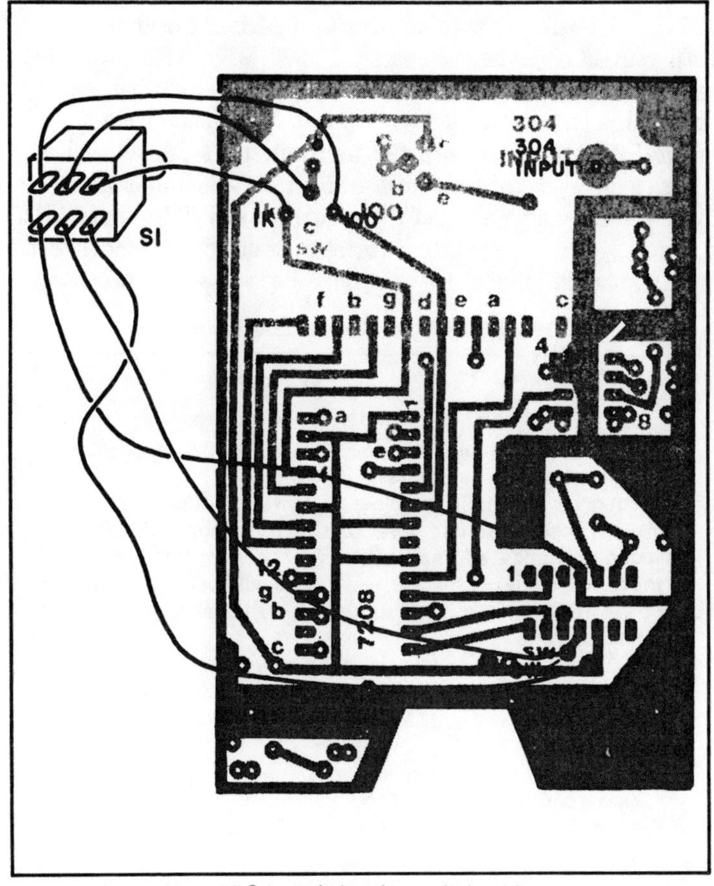

Fig. 9-38. Rear view of PC board showing switch wiring.

the module into the box and secure it to the front with screws and lockwashers. Mount S2 on the top of the box, too. Snap in S1 and wire it up. Install the battery pack and wire it up to the other side of S2 and ground. Finish up the job by wiring up J2 and C1 to J1. Don't close up the box until you finish calibration. Charge up the batteries and you are all set!

Calibration and Operation

There are several ways to calibrate your counter, with the simplest being to measure a known frequency. This is how I calibrated the model 304 counter. Connect a signal of 50 mV or so at 10 MHz from a frequency standard to J1. Your standard should be accurate to at least ±0.001%; ±0.0005% or better would be highly desirable. Press the check button and set the resolution switch to 100 Hz. You will get a reading close to 10 MHz. Adjust trimmer C7 until you get a reading of exactly 10.000 MHz. That completes calibration with a standard. If you don't have access to a frequency standard, a transmitter will do as well. You should have a high-quality counter to measure the transmitter. A CB set will do. Attach a piece of coat-hanger wire about 12" long to J1. Then key the transmitter after you set the resolution switch to the 100-Hz position. Key the transmitter in brief steps and do not modulate it. A walkie-talkie is ideal for this since little power is necessary.

Table 9-4. Model 304 Counter Specifications.

Model 304 Counter Specifications

1. Accuracy: Adjustable to up to ±0.0002% or better, short term. Also ±1 count error on all measurements.
2. Frequency Range: From below 1 MHz to over 135 MHz.
3. Gate Times: Switched 1 second or 0.1 second. Display correspondingly updated 2 seconds or 0.2 seconds respectively.
4. Input: About 500 Ohms at low signal levels (below 0.5 volt).
5. Power requirements: 5 V dc at 130 mA, no signal. Power supplied by nicad batteries.
6. Resolution: 100 Hz and 1 kHz of the measured frequency.
7. Sensitivity: Typical rms sine wave sensitivity for stable count is 55 mV at 1 MHz, 4 mV at 27 MHz, 5 mV at 50 MHz and 12 mV at 135 MHz.
8. Special Features: Hand-held portable unit. Diode-protected input, easy construction, long battery life, etc.
9. Display: 0.105" LED calculator-type with 7 digits.

Table 9-5. Parts List for Counter.

Parts List

B1—4 "AA" Nicad batteries in square holder
C1—0.1-uF, 200-volt Mylar capacitor
C2, C3, C4, C6—0.01-uF, 50-volt disc capacitors
C5, C10—0.1-uF, 25-volt disc capacitors
C7—6-to-20-pF trimmer
C8—27-pF mica capacitor
C9—100-uF, 6.3-volt electrolytic
D1, D2—1N4148 diodes
D3—1N4002 rectifier diode
IC1—National DS-8629N VHF prescaler
IC2—Intersil ICM-7208IPI counter chip*
IC3—Intersil ICM-7207AIPD timebase controller chip*
J1—BNC connector, UG-1094, Amphenol 31-221
J2—RCA jack
Q1—2N3905 PNP silicon transistor

Resistors: All resistors ¼-Watt film except for R12.

R1—47-Ohm resistor
R2 through R8, R11—470-Ohm resistors
R9—10k resistor
R10—15k resistor
R12—100-Ohm, ½-Watt resistor
S1—SPST Push-button switch
S2—DPDT miniature toggle switch
DIS-1—Hewlett-Packard 5082-7441 9-digit LED display, "Industry Standard" calculator LED display, Poly Paks 92CU2954, or similar.
Y1—5.24288 MHz crystal in HC-18/U holder with wire leads. Loading capacitance is 12 pF, accuracy is ±0.005% at room temperature.
Misc.—9-volt ac/dc battery charger with RCA plug, PC board, Bezel, 3/8" spacers (4 each), LMB CR-531 case, 4-40 screws, wire, etc.
***Available from Poly Paks as a set. Stock #92CU4079.**

PC Boards

The PC board (stock #STGM-279) for this project is available for $5.60 (drilled) or $4.00 (undrilled) from O. C. Stafford Electronic Development, 427 South Benbow Rd., Greensboro NC 27401. Add $1.00 for shipping.

Adjust C7 until the counter reads the exact transmitter frequency when the counter is held near the antenna. That completes calibration of the counter. Close up the counter case and secure it with screws.

Operation is a snap with this counter. Simply connect low-level signals to be measured to J1 and press S1. You then read the frequency off the display. The resolution switch, S2, is designed to offer you two time bases, and, thus, two different gate times. Use the 0.1-second position where you want a speedy display of frequency, and use the 1-second position where you want greater accuracy. This will allow you to read up to the last 1 kHz position on the display with S2 set to 0.1 second, and read to the last 100 Hz position with S2 set to 1 second. As you can see, each has advantages.

A few tips on using this counter: You might want to fabricate a 12″ whip antenna out of a male BNC connector and a piece of coathanger wire. Epoxy the wire inside the plug. With this setup, you should be able to measure a 100-mW CB walkie-talkie at least 15 feet away. If your performance-select IC1, this counter will cover the 2 meter ham band. Finally, whenever you measure a transmitter or signal generator, be sure to operate it in the unmodulated mode, or SSB units in the CW mode. This will give you maximum accuracy in your measurements.

Also, see Fig. 9-38 and Tables 9-4 and 9-5.

Chapter 10
Special Projects and Useful Information

Here are some different types of projects. Did you know, for instance, that you can build your own automobile voltage regulator? Well, you can, and here we show you two different ways to do it. There are other projects too, of course, and some very useful information on home-brewing circuit boards. There's even a project to build your own electronic TV games!

A SIMPLE CAR VOLTAGE REGULATOR

A price of $22 for a new voltage regulator in my automobile gave birth to this simple yet very effective solid state replacement. The circuit should be usable in almost any negative ground system using an alternator.

A few words about the automotive battery charging system will help in understanding the operation of this circuit. An alternator's output voltage, and thus its charging current, is controlled by varying the field current in the alternator. Full voltage on the field winding will give full output from the alternator, and reducing the field voltage will result in reduced output. The output capacity of any alternator is limited by its design, primarily the wire size used. In fact, loading an alternator heavily will rarely damage it, as it will put out just so much current, and beyond that its output voltage will drop, limiting the total power available.

In the usual electromechanical type regulator, a resistor is switched in series with the field winding to reduce the alternator output. When the battery voltage drops due to a heavy load, the resistor is shunted by a pair of contacts on the relay, which is voltage sensitive. Under normal conditions, with the engine running, battery voltage should be approximately 13.6 to 13.8 volts, indicating the battery is receiving a charge.

In order to best understand the electronic circuit, it can be looked at as a switch, either supplying no or full voltage to the field winding. Start with a low battery voltage. Diode D1, a 13.0 volt zener, does not conduct as long as the battery is below 13.0 volts. If D1 is not conducting, no bias is applied to Q1 base, keeping Q1 turned off. If Q1 is off, Q2 is on fully, being biased by R2, 120 Ohms. It then applies full battery voltage to the field winding on the alternator. This of course means the alternator will put out full voltage to the battery, causing it to charge.

Now, as soon as the battery voltage increases to 13 volts, a small amount of bias is applied to Q1 base as the diode begins to conduct. If the battery reaches 13.6 volts, the zener diode will conduct fully, dropping 13.0 volts and leaving .6 volts to bias Q1. When this happens, Q1 turns on fully, reducing the voltage on Q2 base to very close to zero. Q2 turns off, removing all voltage from the field. Now the alternator output is reduced to zero, so the battery receives no charging current.

In actual use, this entire process happens very rapidly, and the constantly changing alternator is in effect smoothed out by the battery. Battery voltage will always be the zener diode voltage

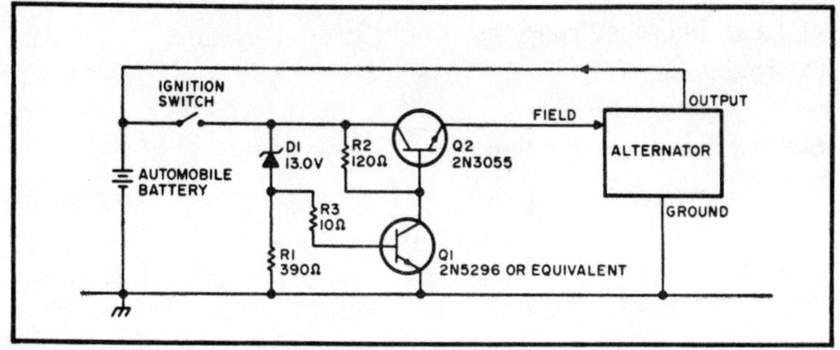

Fig. 10-1. Simple car voltage regular schematic.

plus the drop across the base-emitter junction of Q1, about .6 to .7 volts. Resistor R3 limits the base current to a safe value.

My unit was built on a large heat sink with all parts being supported by the transistor leads. This is not really necessary, as both transistors are operating as switches and consume very little power. The circuit has survived the past winter in my car, and has always kept the battery properly charged, and at a fraction of the cost of a new regulator. It might be a handy circuit to keep in mind the next time you are left with a dead battery due to a faulty regulator, or if you are doing any experimentation with windmill power and surplus auto alternators.

SOLID-STATE CAR REGULATOR

Many car owners, at one time or another, experience electrical system problems usually resulting from a dead battery. In many instances, the battery is blamed for the malfunction when, in actuality, the electromechanical type voltage regulator is the real cause of the problem. This is usually the case, even though the voltage regulator may appear to be functioning properly.

To understand why this happens, consider the fact that a properly charged and maintained lead-acid storage battery should last the life of your automobile. When an early failure occurs, it's usually due to the voltage regulator consistently undercharging or overcharging the battery in the system. In fact, more battery failures result from improper voltage regulation in automotive electrical systems than for any other reason.

Excessive undercharging will cause the battery plates to become covered with lead sulfate, commonly referred to as "sulfating." On the other hand, overcharging a storage battery raises the temperature of the electrolyte, resulting in extreme oxidation of the plates, which eventually crack or buckle. The end result of both of these improper charging conditions is the same . . . *a dead battery.*

Electronic Voltage Regulation

To overcome the above problems, it's necessary to regulate the charging voltage at the proper level. It's up to the voltage regulator to maintain the proper system voltage and, for many years, this task has been accomplished with an electromechanical device. The main disadvantages of these devices are voltage variations due to temperature changes, unadjustable voltage settings, and mechanical type failures.

Many auto manufacturers have recognized these problem areas and as a result are switching over to solid state designs. In fact, if you own a late model car, it may already have an electronic voltage regulator. However, there are still many cars in existence today with the old style electromechanical regulator. If yours happens to be one, you can easily update it with a precision, electronic voltage regulator.

For less than $10 in electronic components, you can build your own solid state voltage regulator that should outperform any electromechanical regulator on the market today.

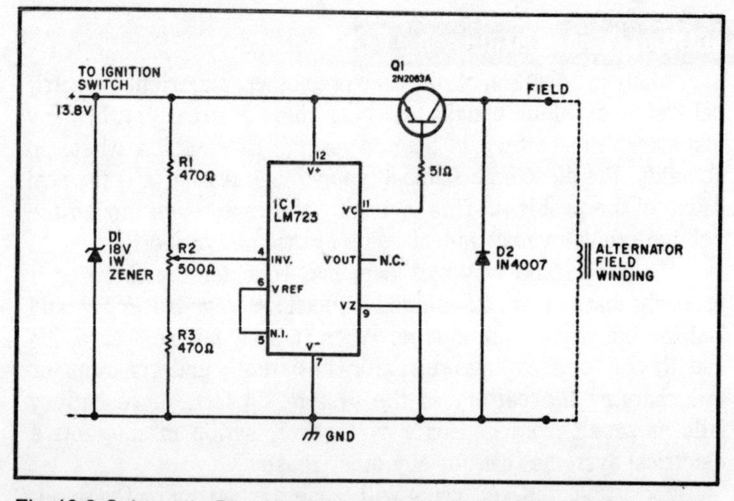

Fig. 10-2. Schematic. D1—18 volt zener diode, 1 watt; D2—1N4007, 100 piv, 1 Amp rectifier; IC1—LM723 voltage regulator (14 pin, DIP); Q1—2N2063A (SK3009) 10 Amp PNP transistor; R1, R3—470 Ohm, ½ watt, 10% resistor; R2—500 Ohm, 10 turn trimpot; R4—51 Ohm, ½ watt, 10% resistor; Miscellaneous—T0-3 transistor socket, 14 pin DIP socket, barrier terminal strip, T0-3 mica washer kit, PC board, minibox, optional relay (see text).

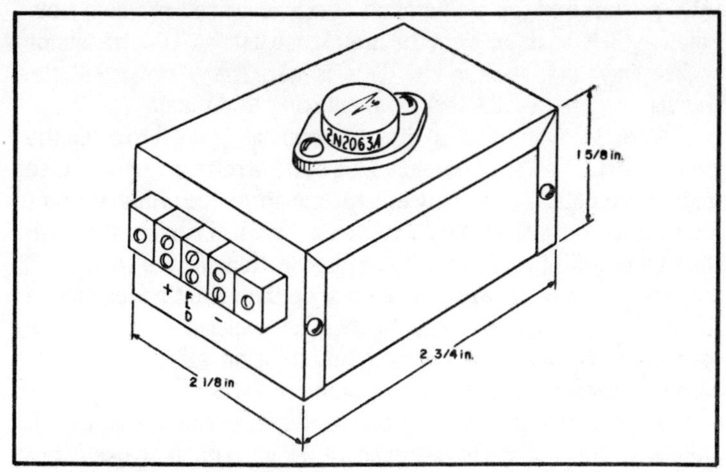

Fig. 10-3. Construction details.

How It Works

As indicated in the schematic diagram (Fig. 10-2), this solid state automotive regulator uses a minimum of components to achieve high performance without sacrificing reliability. The heart of the unit is the LM723 precision voltage regulator IC, which is internally temperature compensated. This integrated circuit is connected as a switching type regulator to control current flow to the field of the alternator. Resistor R2 is adjusted to maintain a system voltage of 13.8 volts, the fully charged voltage of most standard car batteries.

If the alternator tries to produce a voltage above the set level, the LM723 turns off the pass transistor, Q1, thereby cutting off field excitation in the alternator. When this happens, the output voltage from the alternator begins to drop. As soon as the output level drops below 13.8 volts, the regulator turns the field current back on to raise the output voltage. This cycle is repeated hundreds of times a second to maintain the alternator's output voltage precisely at the set level.

The external pass transistor, Q1, is required to handle the large field current of most alternators (approximately 3 amps), since the LM723 has a maximum output current capability of 150 mA.

Construction Details

The solid state voltage regulator may be built in a small minibox (2-¾″ × 2-⅛″ × 1-⅝″) as shown in Fig. 10-3. Transistor

Q1 is mounted on top of the minibox, which is used as a heat sink. Insulate the transistor from the metal case using a T0-3 transistor socket and mica washer kit. This is necessary to prevent the transistor's case (collector) from shorting to ground.

A barrier type terminal strip (3 terminal) is used to bring the BATT, GND and FIELD connections out. If a relay is required (see installation details), you may elect to construct the unit in a larger minibox to house the relay. Also, a six terminal barrier strip will then be required to make external connections to the relay.

In some installations, depending on the mounting location of the regulator, you may want to seal the enclosure for moisture protection. However, if the mounting location under the hood is carefully chosen, this should not be a problem.

The external pass transistor is not critical and almost any 10 amp, PNP transistor will be adequate. However plan to use only a DIP version of the LM723 and not the T0-5 version. The reason for this is that the DIP version has an internal reference zener diode (Vz) and the T0-5 version does not. The T0-5 may be used, but you will have to add an external zener reference diode. Also, the printed circuit board layout (Fig 10-4) has been designed for the DIP version.

How to Install Your Electronic Regulator

First, try to obtain a copy of the schematic diagram for your automotive electrical system. Most local libraries will have automotive manuals containing this type of information. You should become thoroughly familiar with this diagram before proceeding with the installation.

Referring to Figs. 10-5 through 10-8, determine which system best fits your own car. Four basic types of alternator systems are illustrated: Ford/Autolite, Delcotron/GM, Motorola/AMC, and the Chrysler/Plymouth system with an ammeter. With the exception of Chrysler/Plymouth, most systems will require an external relay to maintain the alternator charge indicator light function. However, if you install an external ammeter you can eliminate the requirement of the relay. Simply connect the regulator as shown in Fig. 10-5.

The next step is to find a suitable location under the hood to mount the electronic regulator. Preferably, this location should be near the battery and away from areas subject to moisture or excessive heat.

Disconnect the old regulator and mark each of the connecting wires for future reference, and use crimp-on connectors to connect

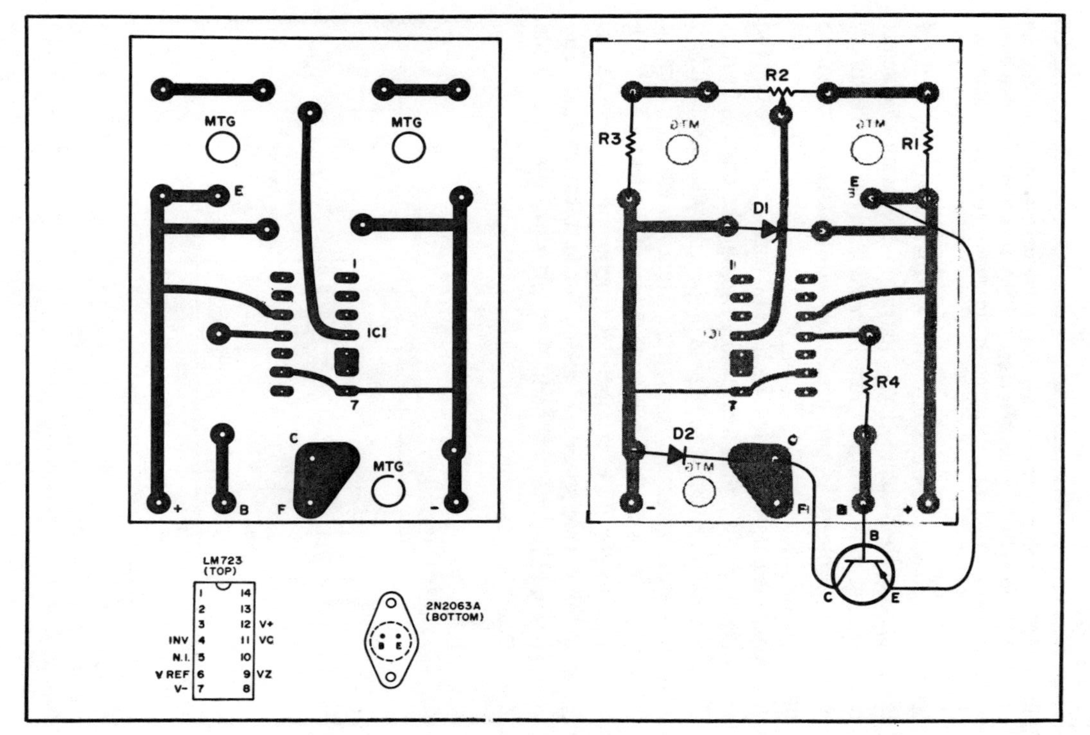

Fig. 10-4. PC board layout.

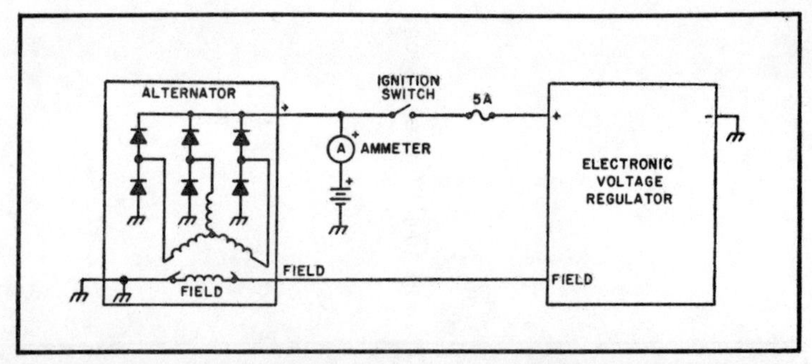

Fig. 10-5. Simplified diagram for a typical electrical system containing an ammeter in lieu of the alternator indicator light. This type of system does not require an external relay to convert to an electronic voltage regulator.

the new regulator to the system. This will maintain the integrity of the original system connections should you ever want to convert back to the original configuration. If an external relay is required, mount it in a protected space, preferably with a dust cover or within the regulator enclosure.

After the unit is installed, recheck all wiring to insure that the system is properly connected. Before starting the engine, turn off all loads until the system voltage is properly adjusted and stable. After the engine is started, adjust the system voltage (with trimpot R2) for 13.8 volts at the positive terminal of the battery.

Check to see if the regulator is functioning properly by increasing the engine speed and adding loads to the system. The voltage should remain constant. Note: At slow idle, with loads

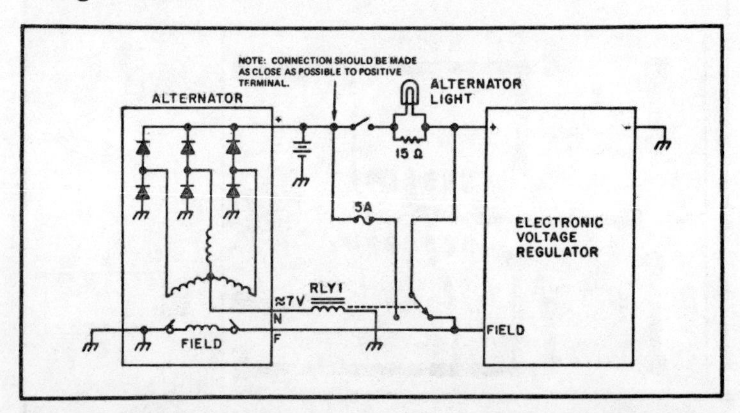

Fig. 10-6. Simplified diagram for a typical Ford electrical system with a charge indicator light. This type of system requires an external relay to maintain the function of the indicator light. RLY 1—any 6 volt relay with 3 Amp SPDT contacts.

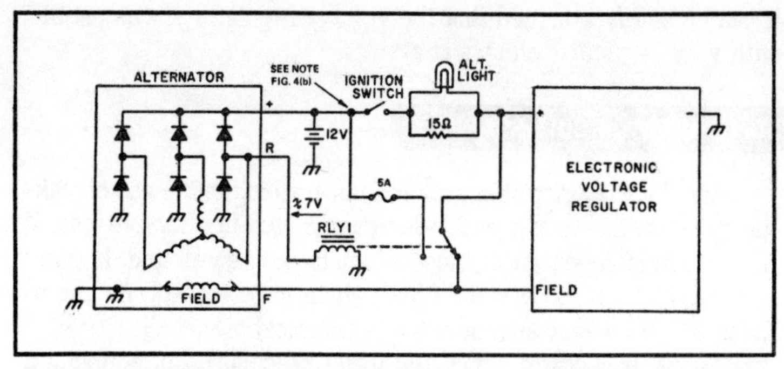

Fig. 10-7. Simplified diagram for a typical Delcotron (GM) electrical system with a charge indicator light. This system also requires an external relay if you want to maintain the function of the indicator light. RLY1—any 6 volt relay with 3 Amp SPDT contacts.

turned on, the voltage may drop slightly, since the alternator is not producing at its rated output. At cruising speed, however, the correct voltage should be maintained if the system is operating properly.

Conclusion

This completes the installation and check-out of your electronic voltage regulator. It should provide many years of trouble-free operation in addition to extending the life of your lead-acid battery.

As a final suggestion, you may want to monitor the system voltage on a continuous basis for the first few weeks after installation. If no problems are experienced during this initial trial period,

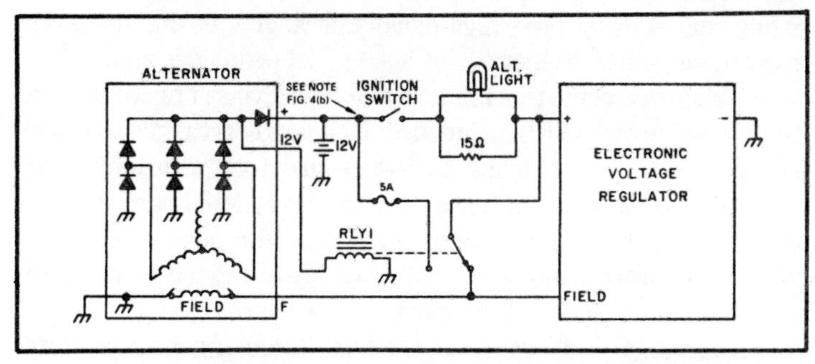

Fig. 10-8. Simplified diagram for a typical Motorola (AMC) electrical system with an internal isolation diode. An external relay will be required to maintain the function of the indicator light. RLY1—any 12 volt relay with 3 Amp SPDT contacts.

it can be safely assumed that the voltage regulator is compatible with your particular electrical system.

HOME-BREW CIRCUIT BOARDS

Hand wiring circuit boards is a not-uncommon means of making quite satisfactory boards which can match in effectiveness, if not always in appearance, those which have been etched. It is my purpose to pass along some ideas which have worked for me to make the job easier and/or more economical.

Some projects in which the usual procedures of applying a resist (mechanically or photographically) and then etching do not pay, include: one-of-a-kind devices with complex patterns, those for which commercially-available boards are not available or are inordinately expensive, those in which the builder wishes to make changes from a published design, and those for which the builder does not have facilities for applying resist, etching, and drilling. Of course, there is always the tinkerer who stubbornly insists on doing a job in his own way, trying something new and different from the established methods of handling a project.

One has to start with a board of some kind. The phenolic, prepunched board sold by Radio Shack is good, given the shortcomings of phenolic. It is punched .100 × .100 inches for IC sockets and other small components. This is a 2-¾″ × 6″ board and two larger sizes are also sold. You will probably need to cut boards to size for specific use; a fine hacksaw or model railroad track saw works well with phenolic or epoxy boards.

Higher-quality board, usually glass epoxy, is available as surplus cards from computers and other devices. The trick here is to buy boards which have high-density packaging so that there are many ICs mounted on them. This leaves a lot of holes from which to select when you come to arranging your own layout. I once found an 8-½″ × 14″ board which included 82 ICs, neatly arranged in rows, plus a 1.8 MHz crystal and associated transistors, capacitors, and resistors, all for $1.75. A better buy was an etched board without any components for a couple of cents. Cut into pieces (after the components were removed from the first board), these boards have provided numerous smaller boards for various projects.

A surplus board must be cleaned off. Removing the capacitors and resistors is an easy trick, even if you want to be careful enough to use them again. Slip a small screwdriver under the component, heat one lead on the other side of the board, pry up gently, and repeat with the other lead. Test all items before using them again. You don't know *why* the board was declared surplus.

ICs are another matter. It is almost impossible without a special soldering iron, to heat seven or eight pins all at once. The solder can be removed with one of the de-soldering "wick" products on the market. I use small size shielding from unwanted mike cable or the like, dip it in soldering paste, lay it on the line of pins, and heat it with an iron. It sucks the solder up so that the IC can be pried off, one side at a time. The heat is likely to ruin the ICs, of course, but these are usually house-numbered and you do not know what they are; thus, they are of no use to you anyway. If you have facilities for determining IC types, you would probably be testing the devices at the same time, so you can discover what they are and whether they are good all at once.

Now you have a board with a lot of empty holes and connections among them, and the next job is to remove the excess foil. The quickest and easiest way to do this is to dump the board (or any portions that you have cut off for use) into PC etchant and let it do its dirty work. Dab spots of resist (fingernail lacquer, paint, candle wax, etc.) on the pads at each IC location. This will permit the pads to remain while everything else is etched away, and will give you something to anchor the IC sockets to when you begin soldering your own circuit. It may be handy to save a ground (common) bus, or pads for external connections to the board, if these will not get in your way. I have tried to compare the pattern on a board with the pattern of the circuit on which I am working, in an effort to save any connections which may be useful, but I do not recommend this. Especially on a doublesided board, or in a complex circuit, it is the road to instant insanity.

Follow the safety notes and instructions on the etchant bottle carefully. The stuff stains hands and clothing, and is definitely injurious to eyes and other sensitive skin areas. It is convenient to have a pail of water handy to dip the board into as a rinse to check things as you go along. Likewise, it is worthwhile to use a pair of plastic tweezers (photo print tongs, for example) to handle the board in the solution. Do not bother to heat the solution according to the instructions unless you are in a hurry. When all of the unwanted foil has etched away, wash the board, clean off the resist with chemical solvent, a "Rescue" pad, or with fine steel wool, and wash the board again.

I am a great believer in IC sockets because I am not a believer in the specs of the bargain ICs I buy. Sockets are cheap—30¢ or less by mail—and they save hours when you have to change the ICs around to find which ones do not work. Molex sockets are a dubious bargain. They do not slip into the holes easily, and, after a few insertions and removals, they are not reliable. At first glance, it would appear that IC sockets with long pins for wire-wrap applications would be easier to solder to, but I find that the extra length gets in the way of precision work and I wind up with two or three pins soldered together.

A word about the holes in the board. You will find many of them plugged with leftover solder, and you will find need for a few where the original manufacturer neglected to foresee your requirements. Stop at your friendly neighborhood hobby shop and buy a couple of fine drills; number 62 or smaller is good. At the same time, if you do not have one, get a cheap pin vise. With this combination, you can ream out plugged holes and drill new ones. You can do this while holding the board in one hand and the pin vise in the other, and thus make holes after you have begun mounting components. You cannot do this safely with a hand drill or drill press.

At the hobby shop, you can also get a small eggbeater-type hand drill made by X-Acto® which is light and, if handled carefully, will not break too many of the little twist drills. However, buy several twist drills at a time as they go fast!

The wire that you use should be as fine and as flexible as you can find. Teflon™ insulation is good for resisting any tendency to shrivel up when the heat of soldering hits it. Again following the surplus route, I have had luck in finding cables made up of many leads of fine stranded wire, number 24 or 26 or so, with various colors of insulation, which I have separated into individual leads. Solid wire is a nuisance.

Wiring this kind of board by hand is admittedly no fun. There are too many repetitious operations, since so many ICs use the same connection schemes. After a while, it becomes a question of "Did I wire pin number five and which pin is number five?" If you have an etching pattern for the device, make a Xerox® copy of it to follow, inking in each connection as you solder it. If not, do the same on a circuit diagram. Yes, this is the same way that beginners did it in the old days of metal chassis and 240-volt transformers, but it works.

Solder two or three corner pins of each IC socket to the pads which you thoughtfully left on the board. This will hold the sockets in place while you get down to serious business. Then, again following the old octal socket tradition, wire in the "high-voltage" leads first, then the ground (common) leads. After that, tackle the rest of the wiring.

As with any other complex wiring job, color coding will help keep matters straight so that you know where you are. I use yellow or red for Vcc, black for common, and one or two additional colors for other leads. Leads coming off the board for interconnection to other boards, switches, etc., must be color-coded or you can get hopelessly lost. If you do not have enough colors of insulation for this, use bits of colored insulation from larger sizes of wire, slipped on the leads and snugged up against the board end of the leads. And make notes as you proceed, reminding yourself of what each color indicates.

You will want a wire stripper, and the small plier-type with a notch to cut the insulation, but not the wire, works well. Set the closure adjustment carefully, since you cannot afford to take any strands of wire off with the insulation. There is not that much wire! Other handy tools include a couple of small screwdrivers, small longnose pliers, a small pair of diagonal cutters, a small soldering iron, and possibly a small file. The key work in the whole operation is "small."

Preform the leads before you solder them in. Strip about 3/8 inch of insulation, tin the end of the wire, cut the lead to approximate size, then strip and tin the other end. Bend a hook in each end small enough to fit snugly on the IC socket pin and solder it on. Where leads are in the clear, with no possibility of shorting to other leads or pins, you can strip and tin a longer piece of wire. Bend a hook in it as before, solder it to a pin, pull it against whatever else it is to be connected with, solder it, and clip off the excess. This speeds things up a bit. If you have an etching pattern, try to follow it fairly closely with your wire leads. The designer may have had some reason for lead placement.

A circuit board wired in this manner will probably win no prizes for neatness, and some unkind friends may compare it with a rat's nest. But the system works and is a suitable substitute for drawing artwork, sensitizing, exposing, developing, etching, and drilling boards, which may be beyond the experimenter for one reason or another. See Figs. 10-9 through 10-14.

TV GAMES

There are TV games sold in every discount store and almost every major electronics periodical has had a construction article on

Fig. 10-9. Jumper wires on the component side of the board are useful at times to avoid too much clutter on the foil side. Push-in terminals (Radio Shack 270-1392) were used here as junctions for anchoring the light-colored 5-volt line, and as soldering points for connection to the IC sockets on the other side of the baord. Terminals are also used along the front edge of the board to bring out leads which will later be soldered to them. Board here is 2-½" × 5-¼".

them. The commercial units cost between $75 and $100 and the construction kits are about the same cost, but there are considerable differences among them. Some may be using older designs which may have as many as 100 chips to perform only one game, or at the other extreme one chip to perform six games. Obviously, Wilbur would have been better off buying a unit which performed

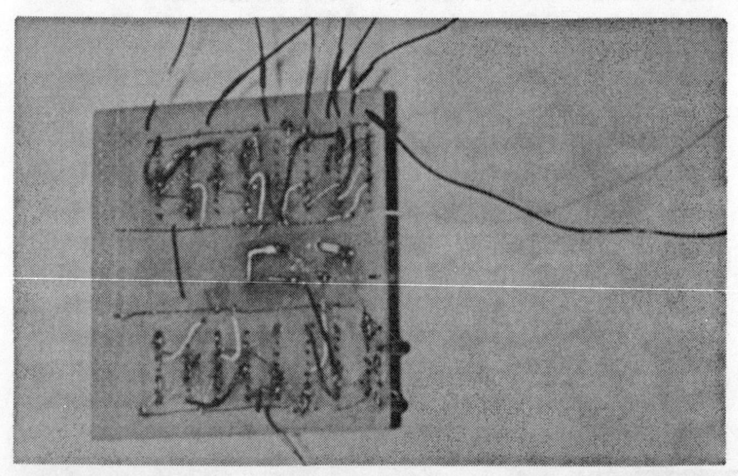

Fig. 10-10. Bare wire can often be used to an advantage, as shown in the thin horizontal lines on this board. The center two lines are the 5-volt dc supply; lines at top and bottom edges are common bus. Leads coming off the board also show small sleeves of colored insulation slipped on for color-coding. Board is three inches square.

Fig. 10-11. Section cut from a larger computer board and then etched clean, leaving pads for anchoring IC sockets, foil strips on edges for ground bus, as well as other leads which seemed to be useful. Board is 2″× 3½″.

more than one game and allowed variations within each. Fewer parts would necessarily mean a lower price and increased reliability.

The AY-3-8500-1 made by General Instruments is a 24 pin MOS integrated circuit TV game chip capable of playing six different TV games. The features are as follows:

1. Six selectable games—tennis, hockey, squash, single player practice, and two rifle shooting games
2. Automatic scoring
3. Score display on TV screen: 0-15
4. Selectable bat size
5. Selectable ball speed

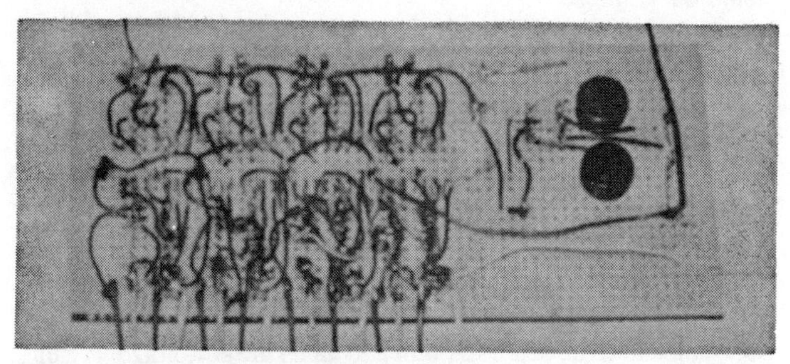

Fig. 10-12. Circuit built on a 3″× 6″ Radio Shack board 276-1395. It looks like a rat's nest of wiring, but color coding and a little care made everything come out right. This kind of sloppy wiring is permissible only when there is no rf or audio circuitry involved.

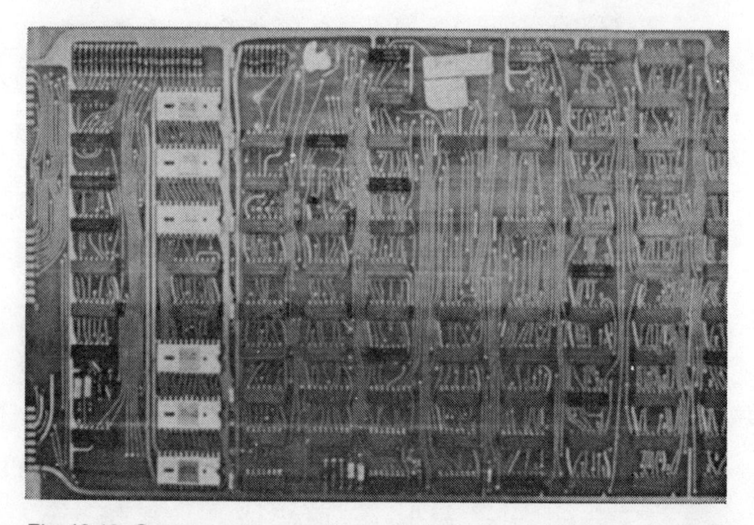

Fig. 10-13. One example of the "raw material"—a surplus computer board measuring 8-½"×14". ICs are not removed until a smaller piece is hacked out for a specific purpose. More than three dozen resistors, diodes, capacitors, and transistors were carefully taken out for future use. A board such as this would make an excellent "mother board" for a project which is made up of several smaller modules.

6. Selectable deflection angles
7. Automatic or manual ball service
8. Realistic sounds
9. Shooting forwards in hockey game
10. Visually defined playing area for the four ball games.

Game Descriptions

Tennis. The tennis game picture on the TV screen will be as shown in Fig. 10-15. There will be one bat or player per side, a playing field boundary and a center net. Scoring position is as illustrated. After reset is applied, the score is 0 to 0 and the ball will serve arbitrarily from one side toward the other. It is the opposing player's objective to intersect the path of this ball and deflect it back toward his opponent. If no intersection occurs, a point will be automatically scored against the erring player and the ball will again be automatically served toward him again. Serve will not change until he scores a point and gains the advantage. A game concludes when one player's score totals 15 points.

The exact details of the game are a function of the optional speed, size, and angle selections. While the game is in progress,

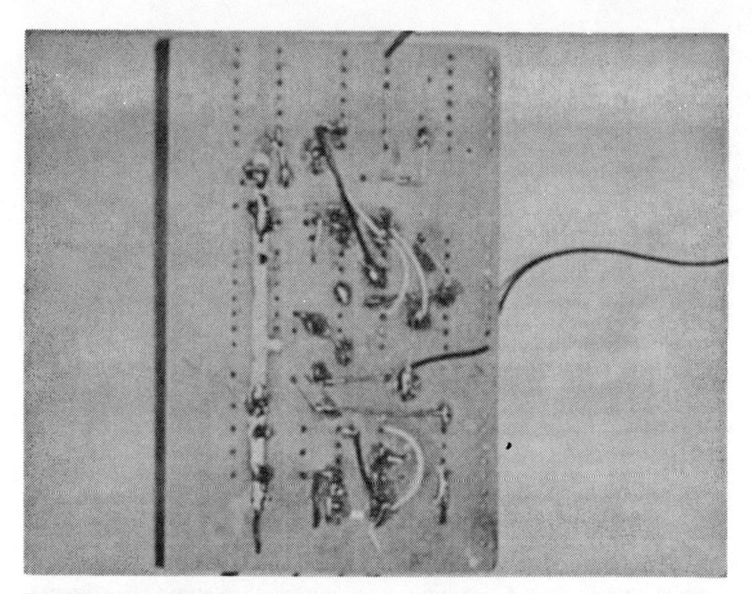

Fig. 10-14. Unused holes which remain after etching a board clean may be used to anchor leads and other components. On this 2-¼″×3-½″ scrap of board, a strip of foil was allowed to remain as a ground bus. Etchant crept under the resist, leaving a ragged edge, but enough foil was left to be useful.

Fig. 10-15. Tennis game with ball in play.

three audio tones are output to indicate boundary reflections, bat hits and scores.

Hockey. The rules of the hockey game are exactly the same as the tennis game except that each human player controls two bats or players on the screen. These players shown in Fig. 10-16 are referred to as the goalie and the forward respectively. The goalie defends the goal, while the forward is located in the opponent's playing area. When the game starts, the ball will be arbitrarily served from one goal toward the other side. If the opponent's forward can intercept the ball, he can shoot it back toward the goal and score a point. If the ball is missed it will travel to the other half of the playing area and the opponent's forward will have the opportunity to deflect the ball toward the goal. If the ball is "saved" by the goalie or it reflects from a boundary, the same forward will have an opportunity to again try to deflect the ball back toward the goal. This method of jamming the ball between the forward and the goalie is a very effective scoring method and makes for an exceptionally exciting game. Scoring and audio are the same as the tennis game.

Squash. This game is illustrated in Fig. 10-17. There are two players who alternately hit the ball against a back court boundary. Scoring and audio are the same as the tennis game.

Practice. This game is illustrated in Fig. 10-18 and is similar to squash except that there is only one player.

Fig. 10-16. Hockey game with ball in play.

Fig. 10-17. Squash game with ball in play.

Rifle. The rifle game is illustrated in Fig. 10-19. Rifle 1 game results in a large target which randomly shoots across the screen while Rifle 2 requires that the target bounce around within the area defined by the TV screen. External circuitry listed in Fig. 10-20 conditions optical input to a photocell located in the barrel of a toy pistol or rifle which is aimed at this random target. When the

Fig. 10-18. Practice game with ball in play.

Fig. 10-19. Rifle games 1 and 2 target.

trigger (PB3) is "pulled" the shot counter is incremented. If the rifle is on target, the hit counter is incremented. After 15 shots the score is displayed.

Circuit Description

The simplest circuit utilizing this game chip is illustrated in Fig. 10-21. A DIP switch (S1-S8) is used for rarely changed func-

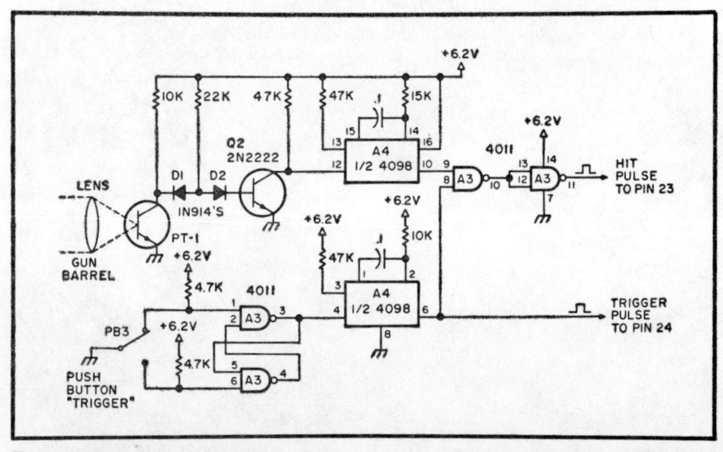

Fig. 10-20. Rifle circuit. PT-1 photo transistor TIL64 or equiv.; 4098—dual monostable; 4011—quad 2 input NAND; all resistors ½ W 5%; all caps min. 25 V dc ceramic.

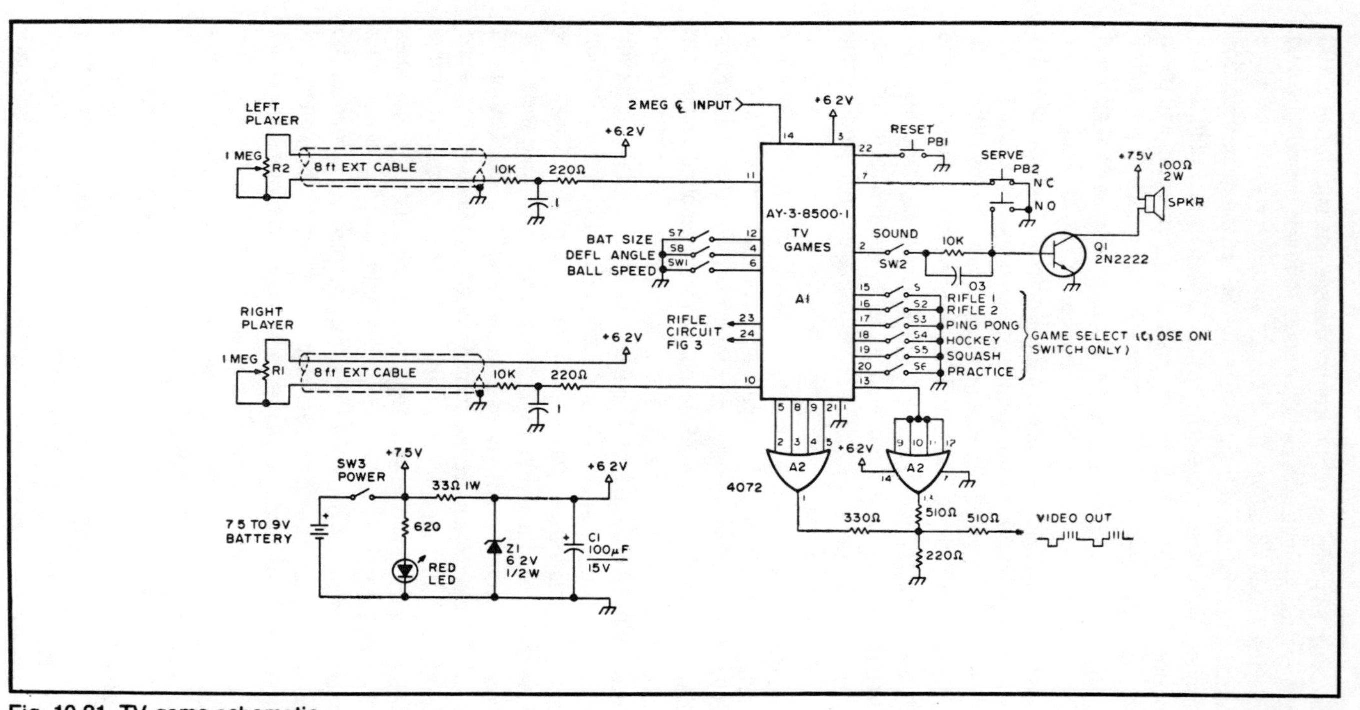

Fig. 10-21. TV game schematic.

tions such as game selection, rebound angle and bat size. A $2.00 eight section switch such as this serves to lower overall costs by replacing about $8.00 worth of toggle and rotary switches while maintaining miniaturization. S1 through S6 are the game selection switches. Only one of the switches is enabled or placed in the ON position. The others must be left open or the game chip will try to play more than one game simultaneously. The correct procedure for selecting a game is to turn the currently programmed game off (all six switches open) and then close the particular switch for the desired game. Switches 1 through 6 will select the following games respectively: Rifle 1, Rifle 2, tennis, hockey, squash, and practice.

Bat size and ball deflection angle are controlled by DIP switch sections S7 and S9 respectively. With S7 open the larger bat size is selected. On a 21" television screen this will appear to be about 2". When this switch is in the closed position, small bats of approximately half the previous size will be displayed. All paddle game photos illustrate the large bat selection.

When first playing a TV game, a player may want to find his bearings and fine tune his eye-hand coordination. For just this reason General Instruments provided for selectable bounce, or deflection angles. When S8 is open, three rebound angles are enabled—plus and minus 20 degrees and straight back at 0 degrees. With S8 closed, five rebound angles are possible—plus and minus 20, plus and minus 40, and 0 degrees. This latter selection requires considerable player skill and dexterity and adds new dimensions to otherwise repetitious games. If that were not enough, selectable ball speed is also available. The ball speed switch SW1 is used more often than the game select switches and therefore should be a more easily used slide switch. When this switch is open, low speed is selected. In this mode the ball takes 1.3 seconds to traverse the screen. When the switch is closed, high speed is chosen and the ball will dart across the screen in .65 seconds. There is a complete understanding of the concept of human fallibility after playing a game which combines small bat size, full rebound angles, and a fast ball speed. With this combination, the cure for boredom becomes electronically induced insanity.

If these features were not sufficient, there are more—realistic sound and automatic scorekeeping. All games consist of 15 points with both players starting with a score of zero after pushing the game reset button (PB1). With pin 7 grounded through the manual serve push-button (PB2), play will resume automatically upon the release of the reset button. Automatic start is signified by the game ball being arbitrarily served into the playing area, and each time a point is scored, the ball will come into play into the court defended by the player having scored the point. If automatic start is not desired, the reset and serve buttons should be pressed simultaneously when resetting a game. The reset button is then released while still depressing the serve button. This will allow complete player readiness and will only put the ball in play when the serve button is finally released. Score is incremented (up to a high of 15) each time a player fails to deflect a ball away from goal.

All of this rebounding and scoring results in some very interesting game sounds. A ball hit upon a paddle results in 32 milliseconds of 976 Hz tone. A boundary reflection is 32 milliseconds (msec) of 488 Hz tone and score is 160 msec of 1.95 kHz tone. This square wave oscillation is amplified by a 2N2222 transistor and applied to a 100 ohm.2 watt speaker. (An 8 ohm speaker may be used with proper current limiting in the collector circuit.) SW2 is provided to switch off the sound without having to shut off the game. Player positioning is remotely controlled through cables attached to pins 10 and 11 of the game chip. Each player control consists of a 1 meg pot and .1 microfarad capacitor which combines to form a variable time constant utilized by internal timing circuitry. Longer or shorter time constants will result in relatively different vertical player positions. To reduce noise, this extension cable should be shielded; otherwise, a display malady referred to as "herringbone effect" will result.

For a TV game to be properly displayed on a raster scan television, the proper video signal, similar to that of any commercial TV station, must be applied to the antenna. Such a video signal results from synchronized dividers inside A1, which divide the 2 MHz master clock (Fig. 10-22) and output the required 60 Hz vertical and 15750 Hz horizontal sync signals. These signals from pin 13 are combined with those of the ball output, right player output, left player output, and score and field output (pins 5, 8, 9, and 21 respectively) in a two bit digital to analog converter formed with a 4072 CMOS dual 4 input OR gate. This type of video output is referred to as composite video output and is suitable only for use on video monitors and not standard televisions. This video output may in turn be used to amplitude modulate an rf carrier suitable for a standard television receiver. Figure 10-23 illustrates a sample circuit of this basic type of modulator. With the components cho-

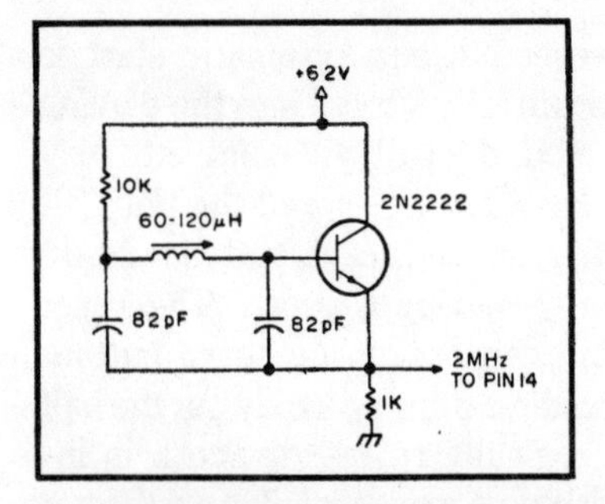

Fig. 10-22. 2 MHz oscillator. Miller 9055 miniature slugtuned coil; all resistors ¼ W 5%; all caps min. 25 V ceramic.

sen, the frequency is approximately that of VHF channel five. (This circuit is intended for illustration only and acceptability by the FCC as a proper class 1 rf modulator is not inferred.) The modulator output is connected directly to the TV antenna terminals, with the antenna disconnected, and adjusted for the best reception.

This game is a marvel of engineering ingenuity through which General Instruments has succeeded in enlightening the average American to the latest advances in electronic technology. It is easy to overlook 16K bit RAMs and microprocessors, but it is hard to ignore such a marvelously exciting TV game when presented on your own home television. Also, see Figs. 10-24, 10-25 and Table 10-1.

NEW LIFE FOR OLD TRANSFORMERS

When amateur radio was very young, it was necessary to build everything from scratch. You couldn't just buy the components and

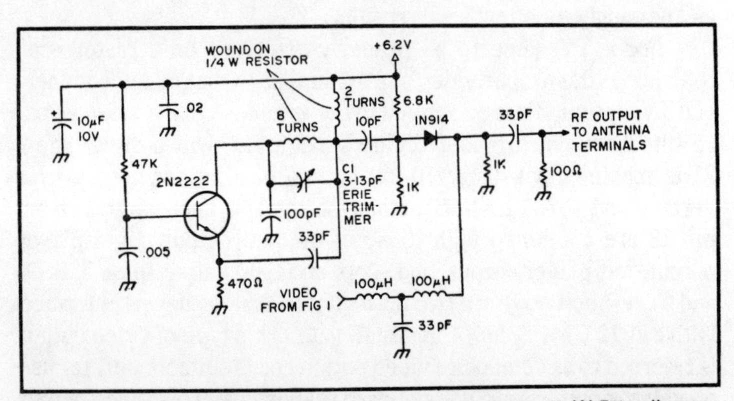

Fig. 10-23. VHF modulator sample circuit. All resistors ¼ W 5%; all caps min. 25 V ceramic unless otherwise noted. NOTE: THIS IS AN ILLUSTRATION OF A SAMPLE VHF MONITOR. THIS CIRCUIT HAS NOT BEEN APPROVED BY THE FCC.

Fig. 10-24. Inside of author's game unit illustrating parts layout.

home brew a receiver or a transmitter; you actually had to make the components themselves. If you needed a coil, you wound one on an oatmeal box. Rectifiers were jars of some foul-smelling electrolyte with the anode and cathode inserted. Capacitors were built up and so on. The reason was, of course, that the necessary parts were not available.

Today, kit-building is popular among hams and computer hobbyists for economic reasons and not because assembled units are hard to get. Building is great fun, and I enjoy it, and the money savings certainly enter into the fun. I'm not suggesting that

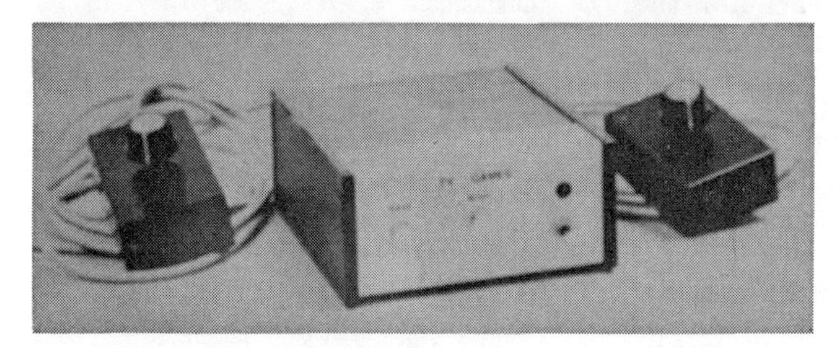

Fig. 10-25. Game circuit built by author.

Table 10-1. Parts List for Figs. 10-20 and 10-21.

A1	AY-3-8500-1 MOS game chip General Instruments
A2	4072 Dual 4 input OR gate CMOS RCA
A3	4011 Quad 2 input NAND CMOS RCA
A4	4098 Dual monostable CMOS RCA
Q1, Q2	2N2222 or equiv.
S1-S8	8 position DIP switch Gray Hill or equiv.
PB1, PB3	SPST momentary push-button C & K Subminiature
PB2	DPST momentary push-button C & K Subminiature
SW1, 2.	SPST slide switch Alco Subminiature
SW3	SPST toggle switch C & K Subminiature 3 A 115 V ac
PT-1	TIL 64 phototransistor or equiv. Texas Instruments
D1, D2	1N914 diode Texas Ins.
C1	100 uF electrolytic 15 V dc
Z1	1N753A or equiv.
R1, R2	1 meg composition potentiometer 2 Watt Allen-Bradley or equiv.
SPK	100Ω .2 watt speaker
LED	NSL5053 LED or equiv.
All resistors are ¼ Watt 10% unless otherwise indicated.	
All capacitors are ceramic type with min. voltage ratings of 25 V dc unless otherwise indicated.	
MISC	extension cable, batteries, box, hook up wire, etc.

everyone start fabricating their own ICs or keyboards, but I think that some might find the following information useful.

When I decided on the Processor Technology SOL System, I ordered only the PC board kit in order to stay within my budget. This is a complete microcomputer on a board with video and keyboard interfaces, cassette controller, serial and parallel I/O ports, RAM and ROM, and much more. All I needed to get it flying, other than the actual construction of the kit, was a keyboard, video monitor, cassette recorder, and a power supply. The first three items I had left over from an amateur radio slow scan to fast scan TV project, and my junk box held all of the components for the power supply, except one. I didn't have a suitable power transformer.

Requirements

In order to allow for some future growth, I had decided on the following capacity: +8 volts at 10 amps, +16 volts at 2.5 amps, and

Fig. 10-26. Typical TV power transformer.

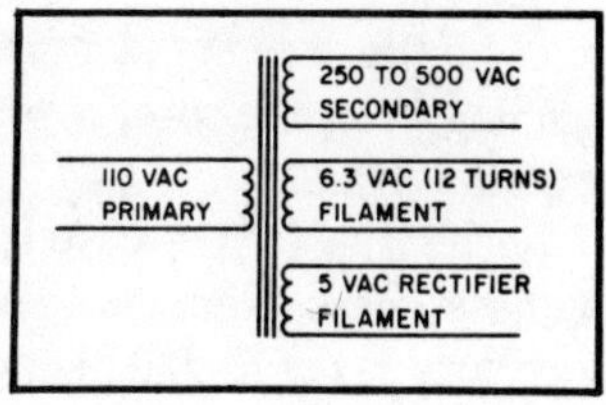

–16 volts at 2.5 amps. A tour through all of my magazines and catalogs didn't disclose exactly what I was looking for, so I decided to make my own. Besides, I wanted it right now! This was not a case of getting carried away in my desire to fire up my machine, since I've built (rebuilt) transformers before.

Only three ingredients are needed: a roll of wire, some tape, and an old power transformer. Retired tube-type television sets are the best source for good transformers. They usually run 300 watts or more and are designed for continuous duty. Discarded TV sets are free for the asking at almost any TV repair shop. Since I never throw anything away, I just happened to have two on hand.

Identify the Windings

The whole operation is comparatively simple; mine took one hour and ten minutes. The first step is to determine what you have to work with. Figure 10-26 is a drawing of a typical TV power transformer. Some have more than one 6.3-volt winding because of the great number of tubes in the set.

Most, but not all, transformer leads use a standard color code (Table 10-2). To determine how yours is laid out, connect a line cord to the leads you suspect are the primary, and clip a voltmeter to the 6.3-volt filament leads. In case you guess wrong, start out with the meter on its highest scale to protect it. Insulate all of the other leads and plug the cord into a 110 V ac outlet. The meter should read between 6 and 7 volts. If not, try some other leads until the right combination is found.

For safety's sake, unplug the line cord while making connection changes. This may sound elementary, but it is easy to forget in the heat of an experiment. I once cut a line cord in half while it was hot and put two big notches in the jaws of my diagonal cutters before they sailed across the room.

Dissection

Once you have found the primary and one of the 6.3-volt filament windings, you are ready to start dissecting. Record lead colors if they are not standard. What you have to do is remove the windings from the core, rewind them, and then put everything back together again.

The transformer case is usually held together with four bolts at the corners. Remove these, and the shells will come off. The core consists of E- and I-shaped metal pieces, glued and laminated together. At first, it would appear impossible to separate and remove the individual pieces, but, once a few are taken out, the rest will practically fall out.

Inserting a knife blade or thin screwdriver between the layers and moving it from one end to the other will separate them, and the I segments can be easily removed. Use the blade or one of the I segments under the top E segment all the way around and through the middle of the winding, and it should pull out with a pair of long-nose pliers. Wiggling it back and forth while pulling should do the job. Once the first segment is out, the rest can be loosened and pulled out without problems.

You want to get rid of all of the windings except for the primary, which you'll reuse. It should be closest to the center of the winding pileup. You can determine this by tracing the leads you identified earlier. Starting from the outside, unwind the layers one at a time until you get down to the level you identified as the 6.3-volt filament winding. As you unwind this layer, count the turns of wire carefully. This is important! This tells how many turns each of the new windings must contain.

Mine had 12 turns, which means that this transformer was designed to have two turns of wire per volt. Using the simple ratio formula P/S = VP/VS, where P is the number of turns in the primary, S is the number of turns in the secondary, VP is the voltage in the primary, and VS is the voltage in the secondary, it is possible to determine how many turns each of the other windings had, but it isn't necessary. All you want to know is how many turns

Table 10-2. Power Transformer Lead Color Code.

Primary	black
High voltage secondary	red
Center tap, if used	red and yellow striped
Rectifier filament	yellow
6.3-volt filament	green
6.3-volt filament #2, if used	brown
6.3-volt filament #3, if used	slate

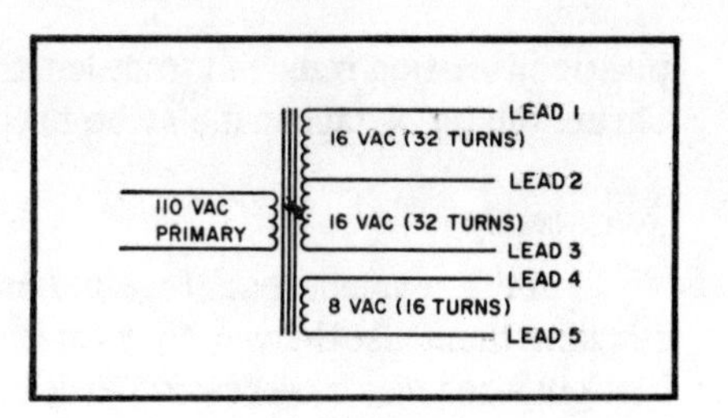

Fig. 10-27. Rewinding guide.

per volt your transformer has. If you were building transformers from scratch, you could use whatever value you desired, but two seem to be standard, as it has been the same on all of the transformers I've rewound.

Remove all the rest of the windings except the primary, and, if it is covered with a copper foil shield, leave it in place. Wrap the primary winding with tape, and you're ready to start rewinding.

Rewinding

The number of turns needed for each winding is the turns per volt times the desired voltage. Figure 10-27 shows how I rewound my transformer. For the 16-volt windings I used # Formvar insulated wire, which will handle more than 3 Amps (Table 10-3). Plastic or enamel insulated wire or smaller sizes can also be used if your current requirements are lower.

Wind the higher voltage winding first, leaving a foot or so for a connecting lead. Wind tightly and neatly until you reach the center tap (32 turns on a 2 turns-per-volt transformer), leave a loop about a foot long for the center tap, and then wind the second half of the winding with another one-foot lead at this end. If all of the turns won't fit on one layer, double back toward the beginning. Wrap this winding with tape.

Wind the 8-volt winding next, using wire large enough to carry the required current (Table 10-3). I didn't have any #10 but did have plenty of #16, so I used 3 strands of #16 loosely twisted together, which should handle more than 10 amps. Slip some insulation over the leads, and tag them for identification. I used

Wire	Capacity
#10	14.8 Amps
#12	9.3 Amps
#14	5.8 Amps
#16	3.7 Amps
#18	2.3 Amps
#20	1.4 Amps

Table 10-3. Current-Carrying Capacity of Wires in Transformers.

plastic insulation removed from lengths of regular house wiring. Shrink tubing or tape can also be used.

Reassembly

Put the segments back together in the windings by alternately stacking them just the way they came out. A hammer and a small block of wood may be necessary to drive in the last segment or two. Do not use the hammer directly on the segments—you'll flare the edges and they will not fit together closely. Make sure that the holes in all of the segments are lined up, and replace the covers and any insulating material they may have contained. Tighten the bolts good and tight. The first transformer I rewound had an audible hum until I really clamped the segments together.

Now you've done it and can check the voltages with a meter. They may differ from our design voltages a little, but, after rectification and filtering, you'll have plenty for your regulators. Also, see Figs. 10-28 through 10-34.

NEW LIFE FOR OLD KLYSTRONS

What? Use a vacuum tube in this modern era of solid-state electronics? Reading about Gunn diodes, Impatts, and other such exotic devices is fine, but with the exception of the Microwave Associates Gunnplexer, there are no complete solid-state oscillators available to the amateur. This became very apparent while I

Fig. 10-28. Retired television transformer with the covers removed.

Fig. 10-29. The winding pileup of the transformer sitting on top of the E and I segments.

was trying to put together a receiver for direct reception of television signals from geosynchronous satellites at 3.7 to 4.2 GHz.

A multi-stage oscillator-multiplier chain was ruled out because of insufficient time. If everything works, the final receiver

Fig. 10-30. The winding form taped and ready for rewinding. All of the windings except the primary have been stripped away.

Fig. 10-31. The end of the center-tapped 32-volt winding. Tape is used to hold all of the loose ends in place. All leads should come off the same end of the form as the primary so that they will match up with the holes in the transformer case.

might have a solid-state local oscillator, but then again, maybe not—why discard a working circuit? Reflex klystron oscillators of the 723 and 2K25 class (see Fig. 10-35) have been available on the

Fig. 10-32. The 8-volt winding—in this case, 16 turns of 3 strands of #16 wire.

Fig. 10-33. The winding form completely rewound, taped, and ready to be mounted back in the core.

surplus market for over thirty years. Output frequencies from below 3 GHz up through 9.6 GHz are currently available with output powers in the neighborhood of 10 to 150 milliwatts. These

Fig. 10-34. The transformer reassembled. It has been said that the segments should be shellacked before reassembly, but that seems like a messy job, and I haven't found it necessary. All that remains is to insulate the leads and replace the covers.

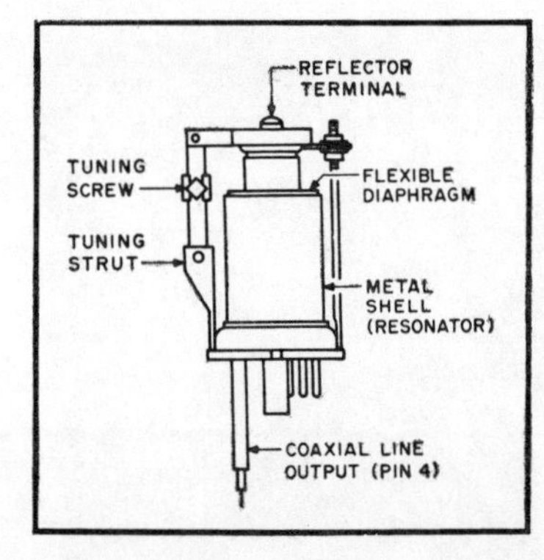

Fig. 10-35. 723-class reflex klystron.

klystrons have found applications in the past in polar-plexers and as pump oscillators for amateur and commercial parametric amplifiers. The tubes listed in Table 10-4 are similar in construction and can be mounted in octal sockets that have had pin 4 removed and bored out with a number 24 drill. The output is via the small diameter, rigid coax line which terminates in a short probe. The probe was used to directly excite a waveguide or was capacitively coupled to a coax cable.

In normal operation, the tube shell is operated at 250 to 300 volts above ground. This puts the output cable, which is connected directly to the shell, at a hazardous potential. However, by operating the cathode at a negative voltage, the shell and output cable can be maintained safely at ground potential.

Table 10-4. Reflex Klystrons From 2.7 to 9.66 GHz.

Tube	Output frequency(GHz)	Output power(mW)
726C	2.7-2.96	100
726B	2.88-3.18	150
726A	3.18-3.41	100
2K22	4.3-4.9	115
6115	5.1-5.9	100
2K26	6.25-7.06	100
2K25*	8.5-9.66	30
723A/B	8.5-9.66	25

*Improved version of 723A/B, both of which can reach 10 GHz by stretching the cavity.[6]

Fig. 10-36. 726A klystron with coax output and terminal strip.

Figure 10-36 shows a 725A klystron mounted on an aluminum plate for use as a local oscillator for the satellite receiver. The tube is clamped by a simple split block to allow conduction cooling and provide a mechanically stable support. This degree of mechanical rigidity is not required but was convenient in this case. Also, free-standing operation with convective cooling is alright. The oscillator covers a measured frequency range of from below 3.2 GHz to above 3.5 GHz.

The output frequency of this particular tube as a function of the rotation of the tuning screw is shown in Fig. 10-37. By changing the

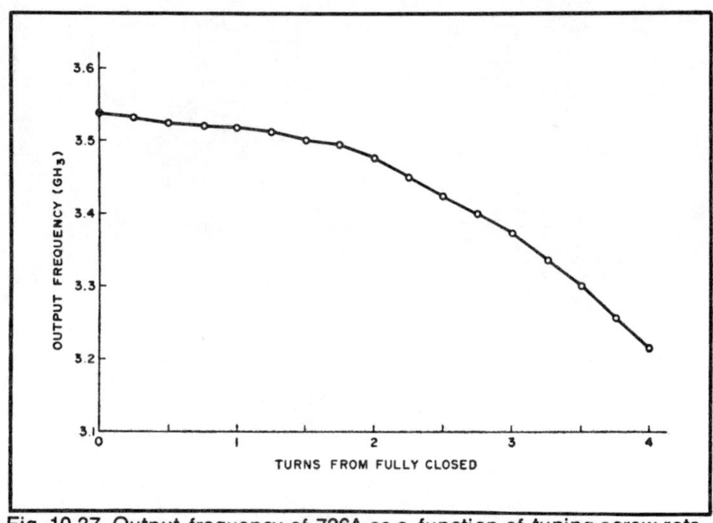

Fig. 10-37. Output frequency of 726A as a function of tuning-screw rotation.

reflector voltage, the output frequency could be shifted electrically plus or minus fifteen megahertz from the mechanically set frequency. This FM characteristic can be used for fine tuning or for a form of afc with simple circuitry.

The output probe is terminated with an SMA connector. A cross section of the connector assembly is shown in Fig. 10-38. A BNC connector could be used with equal success. The output probe center conductor is clipped close to the insulated sleeve, the ferrule is slid back onto the probe. Then the center conductor is soldered to the coax connector. Following this, the ferrule is slid forward, screwed to the connector with two 2-56 screws, and then sweat-soldered to the probe.

Power leads run from a barrier strip to the tube. Connections to the tube are made with a modified octal tube socket—see Fig. 10-39. While normal operation requires −300 to −700 volts at zero current for the reflector, operation at reduced potentials and output powers is not only possible but desirable in that power dissipation and heat are reduced substantially. This adds to the tube life-expectancy.

Operation has been at cathode voltages as low as −150 volts at 10 mA and reflector voltages of −150 to −300 volts. With a

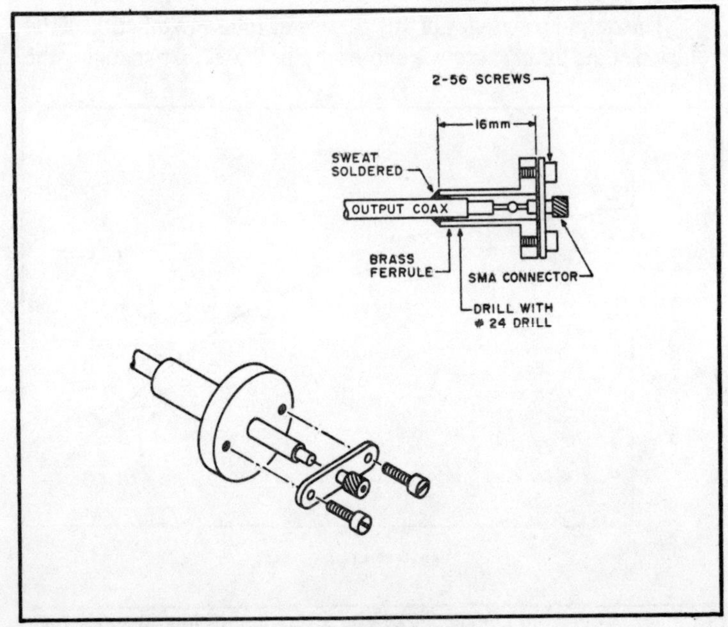

Fig. 10-38. Coax connector detail and assembly.

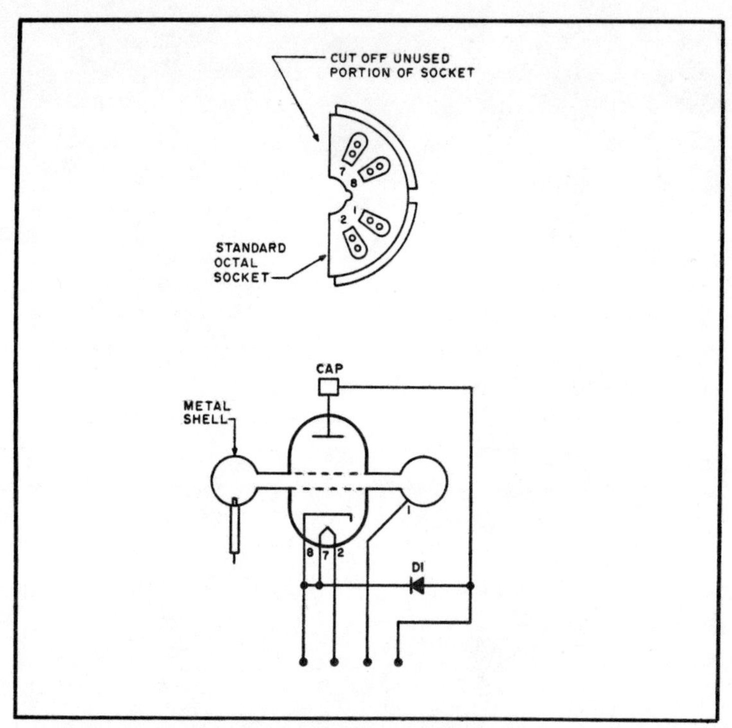

Fig. 10-39. Socket modifications and wiring diagram.

−300-volt cathode supply, output powers in excess of 60 milliwatts were readily obtained. With a −150-volt cathode supply, the output dropped to about 4 mW, which is still sufficient for use as a local oscillator. Even at the high powers with only free convection cooling, life expectancy is high. Tubes pulled out of service after hundreds of hours of service and stored thirty years or more are still operating.

A word of caution: The reflector must *never* become positive with respect to the cathode. If it does, it will draw current, heat up, and outgas, ruining the tube. To prevent this from occurring, merely connect a rectifier diode between the cathode and reflector as shown in Fig. 10-39.

The theory and operation of reflex klystrons is available elsewhere and therefore is not covered here. What I hope I have accomplished here is to remind other amateurs of the availability of relfex klystrons as packaged sources of microwave power that are rugged, cover a wide spectrum, and are economical.

Chapter 11

Easy One-Evening Projects

Okay, if the projects are easy, why are they in the back of the book? Well, if you have read this far and still haven't built one of the projects presented in the first 10 chapters, this chapter will get you started for sure. All of these projects can be easily completed in one evening and some can be put together in only a few minutes.

WORLD'S SMALLEST CONTINUITY TESTER

Using a rubber-type two-pronged plug, the few components that make up this tester are mounted inside the plug. The hole in the end of the plug through which the cord enters is used for the NE-2 neon bulb.

Two small holes opposite each other are made near the base of the plug for the probe wires to extend, and the two 100k, ½ watt resistors within the plug cavity, being in series with the probe cords, prevent shock. A piece of ⅝ inch i.d. aluminum tubing ½ inch long, placed over the bulb end of the plug, with a ⅝ inch o.d. clear plastic lens pressed into the tube, protects the bulb and enhances the appearance of the miniscule tester. Make the probe wires long enough to suit your needs. See Figs. 11-1 through 11-3.

SIMPLE FIELD-STRENGTH METER

For many years I have used two field-strength meters, and they are still in use. I shall give credit to Jo Jennings W6EI (deceased), for he is the person who showed me the simple circuit. This little gadget is non-frequency selective. I have used it from 2 meters through 160 meters. The telescoping antenna may be adjusted to its shortest length when working with 2 meters to keep the needle on scale. I use this field-strength meter to adjust all my 2 meter Js, base-loaded ⅝ wavelengths, beams, etc.

The meter used should be a 100 microamp up to a 500 microamp movement. The diodes may be germanium type, such as

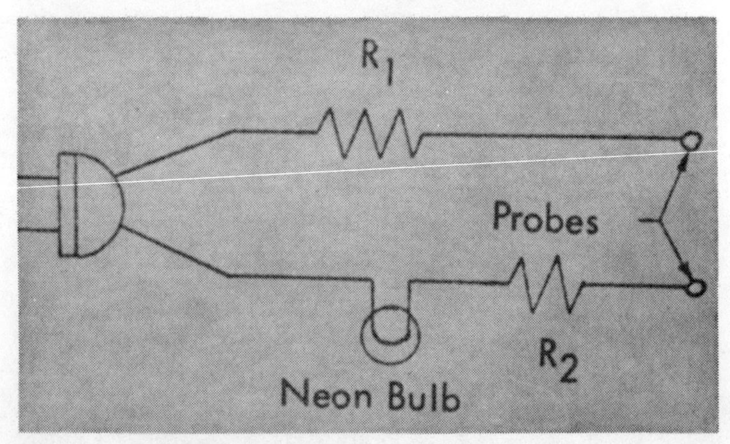

Fig. 11-1. Continuity tester schematic.

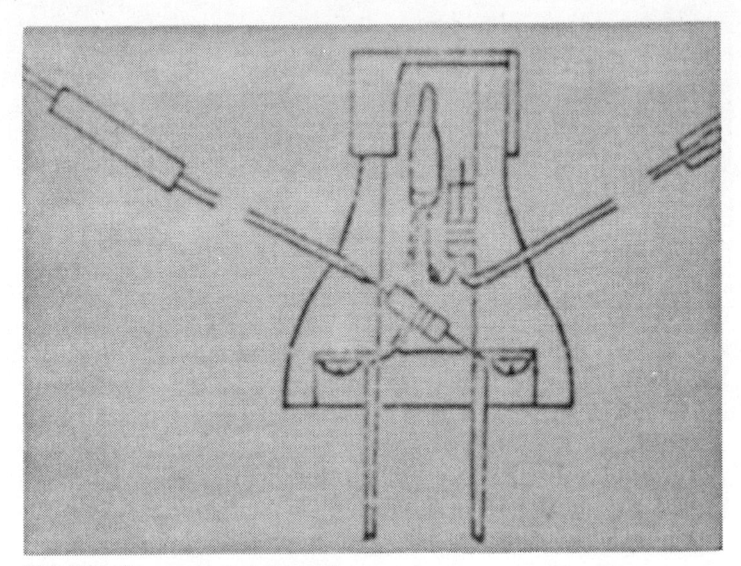

Fig. 11-2. Component placement.

1N34, etc. Silicon diodes will also work, but are a bit less sensitive. The diode leads may be left their normal length. The sloped meter box is ideal. The box does *not* have to be metal. See Fig. 11-4.

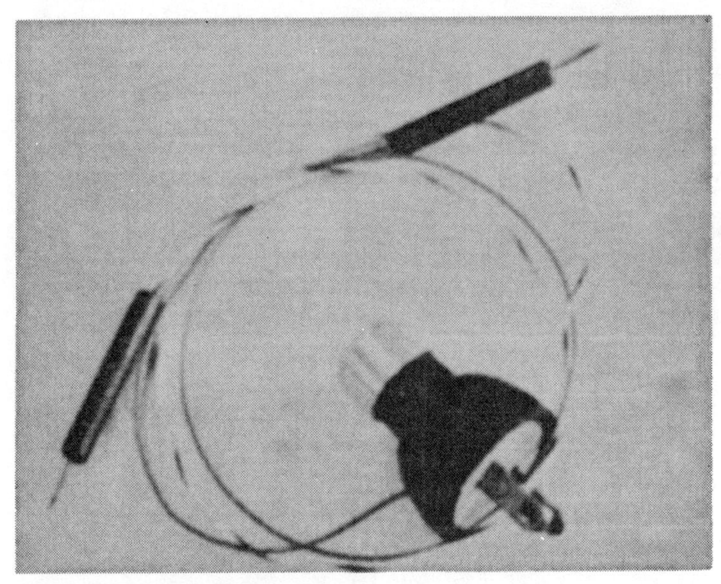

Fig. 11-3. The finished tester.

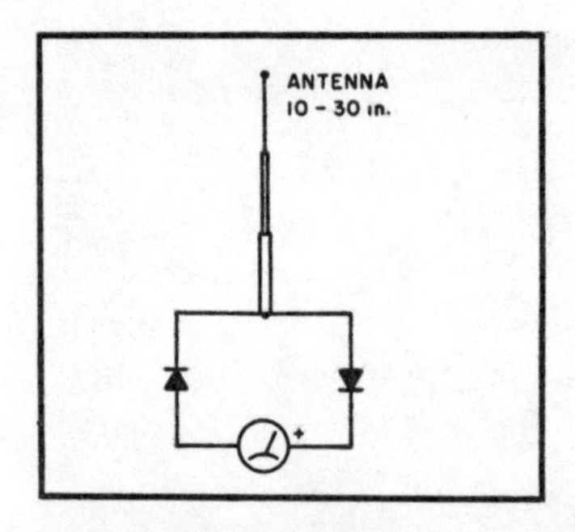

Fig. 11-4. Simple field-strength meter.

A two-way coax switch is a real convenience for switching between two antennas, or for grounding an antenna, or for switching one antenna to another rig. Take a small minibox, Bud #5A30005 or #5A3014, and with a ⅝" drill or a Greenlee punch cut three holes as shown in the drawing in Fig. 11-5. These are cut on the three sides of the U-section.

Mount three SO-239 connectors on the three sides of the U-shaped section. Then punch a ½" hole in the other section and mount a heavy-duty toggle switch—SPDT. I used a DPDT and strapped the two sections together to take heavier current.

Connect as shown in Fig. 11-6. Use a flexible piece of coax shield from RG-58/U for connection to the "common" connector so that the box can be put back together easily.

Use #12 wire or the center conductor from RG-8/U coax for the other connections.

The total actual parts cost was $5.02, saving about $35 over a commercial switch of the same type.

The same idea can be used without the toggle for an emergency T connector, by connecting all three SO-239s together.

SUPER SIMPLE TT GENERATOR

Figure 11-7 shows a schematic detailing my circuit for the generation of touchtone™ frequencies. It works extremely well, is

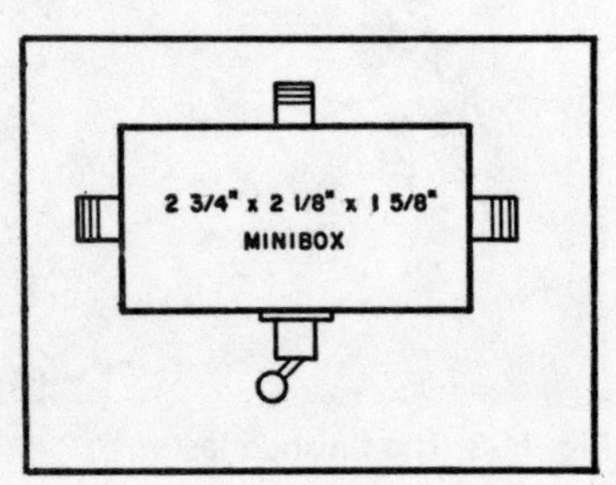

Fig. 11-5. Coax switch.

Fig. 11-6. Coax switch schematic.

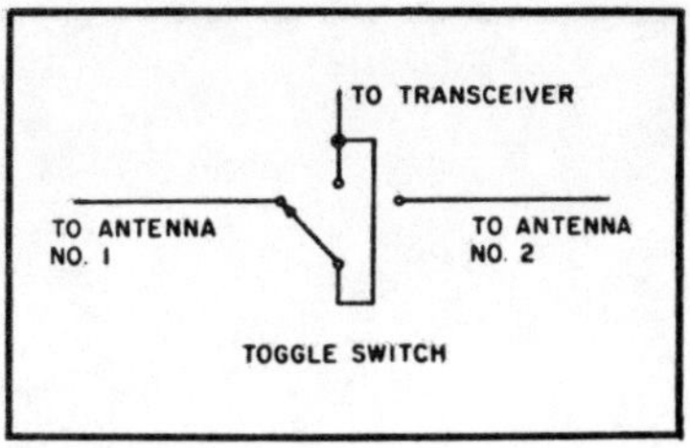

quite easy to construct, and is inexpensive. Total cost for the project was under nineteen dollars.

The oscillator is a Motorola MC144110CP chip using a 1 MHz crystal. The chip generates both the high and low tones, feeding the energy to the amplifier through the 1k resistors and the 1-microfarad capacitor. Values for the output resistors can vary from a few hundred ohms to about 60 kilohms. The value of the resistor shunting the crystal can vary from about three to fifteen megohms.

The amplifier consists of an LM-380N. This is one of the handiest linear chips around. It can output as much as two watts of audio with a minimum of components. In the configuration shown (this configuration may be used for just about any audio amp application, as a look around my shack will be evidence), input from the oscillator is fed to the inverting input. The value of this component may vary from about 470 pF to about .003 microfarads. The 4.7k resistor controls the tone of the outputted signal. If desired, a pot may be substituted for the fixed resistor, and any tone which will both work and please the operator's ears may be selected. Output is taken through a 25-microfarad capacitor and an 8-ohm speaker. The 380 is designed to operate into 8-ohms loads. The value of the capacitor may range from about 18 microfarads to well over 100. For 12-volt operation, change the 4.7k resistor between pins 2 and 6 of the LM380 to a 1k and add a .01 μF capacitor between 6 and ground.

Construction is easy. I built mine on a small piece of vectorboard using telephone wire to make connections. A later model was built using a PC board. It is highly recommended that an IC socket be used for the MC14410CP. I used sockets on both chips. They are inexpensive and can save a lot of grief! An old transistor radio box was used to house the unit and keyboard. I used the Chomerics EF-21360 keyboard. The terminals will only stand very brief periods of heat, as evidenced by the fact that I ruined one while overzealously trying to attain a good connection. On the

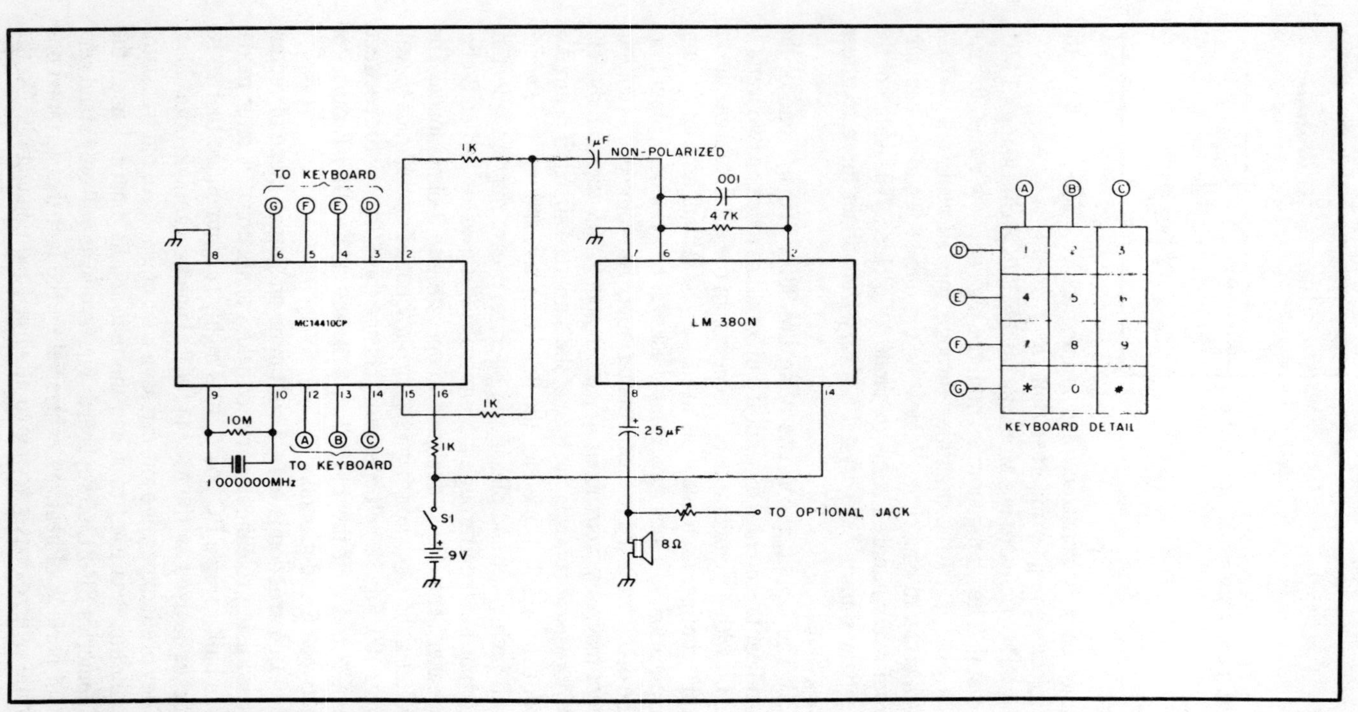

Fig. 11-7. Simple TT generator schematic.

second one, I held the wire against the keyboard terminal with pieces of spaghetti. Heat-shrinkable tubing may also be used. I put a pot into the output line between the speaker and capacitor to reduce the volume, as the 380 can put out quite a lot of audio! I also installed an external jack, as shown in the schematic, to permit direct connection to my rig.

LINE NOISE SUPPRESSOR

One of the most irritating sources of electrical noise, characterized by sharp clicks heard through the speaker of a receiver, is contact noise. The make/break cycles of appliances, aquarium heaters, flashing Christmas tree lights, and other noisy electrical contacts can wreak havoc with radio reception.

Fortunately, there are several options which may be elected to minimize these ear-splitting distractions. Perhaps the simplest is the installation of a 0.1-μF capacitor across the contacts themselves.

Since it is often difficult to find direct access to the offending contacts, an alternate solution is found by bypassing the plug with the capacitor. Probably the simplest way to do this is by rigging a plug-in interference filter as shown in Fig. 11-8.

Fig. 11-8. Line noise suppressor.

For standard 120 lines, select a mylar™ capacitor with a 600-volt rating. Insulate the exposed capacitor leads and connect them directly to the terminal screws of any convenient plug. Insert the plug-in filter into the same outlet as used to power the offending contact device.

The bypass capacitor acts as a smoothing filter for the sharp voltage-spike transients generated by the sparking contact. While it is true that the capacitor might actually resonate an unusually long line cord to enhance the noise at some frequency, in actual practice this is extremely unlikely to happen within the passband of most receiving installations.

SIMPLE DIODE TESTER

For many years I have used my ohmmeter to test diodes. The usual technique was to set the meter on R × 10, fumble with the test leads to get a "low" reading (OK in the forward direction), reset the meter to R × 100 or R × 1000, refumble to reverse the diode, and check for a "high" reading (OK in the reverse direction). Usually, around the second fumble I would lose track as to whether I actually reversed the diode and would have to fumble a few more times to recheck.

Recently, I purchased a bag of 100 untested switching diodes, and anticipating a long evening of lead switching, testing, and retesting, I decided to build a simple diode tester. Construction time was about 15 minutes, a good investment if you occasionally or more than occasionally test diodes.

The circuit shown (Fig. 11-9) was built in a 3″ × 4″ × 1″ plastic box. Three holes are drilled: one for each LED and one for

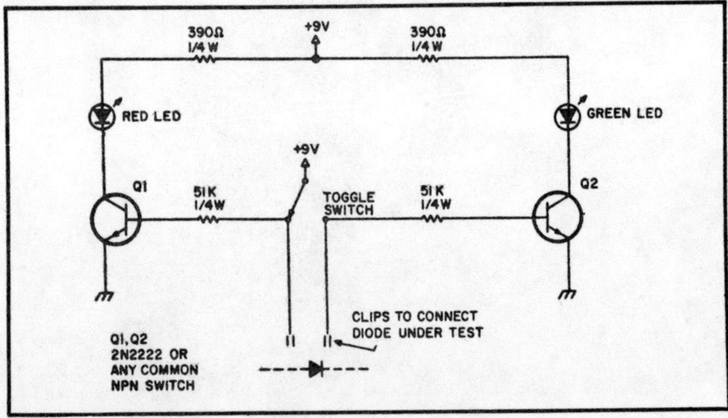

Fig. 11-9. Diode-tester schematic.

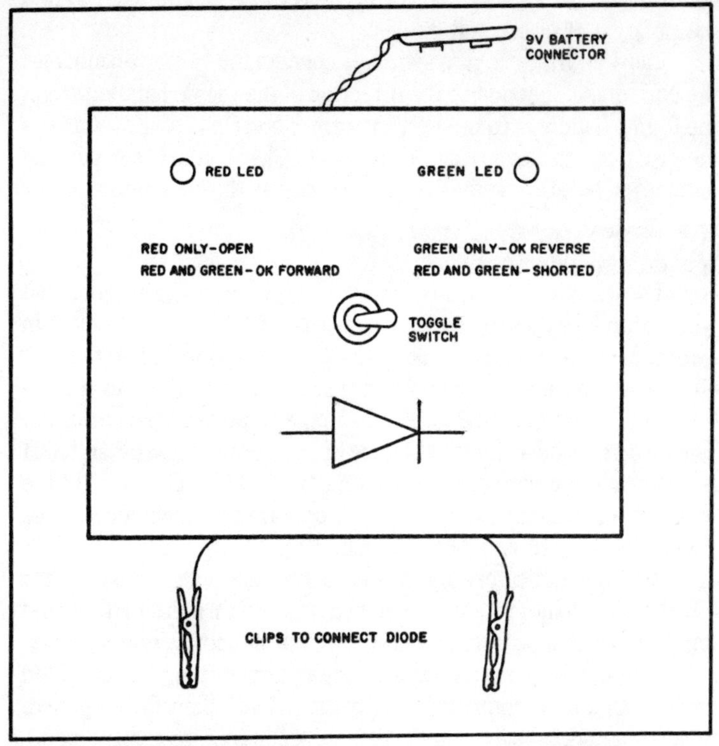

Fig. 11-10. Front-panel layout.

the toggle switch. The "instruction plate" was written on a self-stick label and mounted just above the switch. Two test leads with clips are brought out to connect with the diode as shown by the symbol drawn just below the switch. Figure 11-10 shows the physical arrangement of the front panel.

A standard 9-volt battery connector is brought out the back to power the circuit. I recently have standardized all of my small test instruments on 9 volts and bring out the connectors. When I want to use an instrument, I simply connect a 9-volt battery, thus saving on power supplies and multiple batteries for instruments I only occasionally use.

Operation of the tester follows the "instruction plate." Connect the diode, throw the toggle switch to the left, and check that both the red and green LEDs are on. If only the red LED fires, the diode is open. Now throw the toggle to the right; only the green LED should be on. If both the red and green LEDs fire, the diode is shorted. Note: If the diode fails both tests, you probably have connected it backwards. Thus, the tester also can be used to find the polarity of unmarked diodes.

Caution: Many toggle switches connect the center terminal to the end terminal opposite the direction of the toggle handle. When the toggle handle is to the left, the center and right rear contact (as viewed from the front) are connected. Check yours out with an ohmmeter before wiring it in accordance with the schematic.

THE CAPACITOR COMPARATOR

The simple circuit here is an easy one-evening project that when completed will provide an audio tone comparison of a built-in reference capacitor to an unknown capacitor connected to the test clips. Bearing in mind that the larger the capacitor the lower the tone, it is a simple matter to establish the value of unmarked units. The circuit, as described, will identify caps between .5 pF and .001 μF by providing tones between 8 kHz and 100 Hz. The heart of the tester is a 555 timing IC and may be operated from any dc voltage source between 8 and 14 volts. See Fig. 11-11.

An LED indicator is provided for testing values larger than 001 μ which do not produce a tone but merely turn the LED off and on. A .1 μF unit will trigger the indicator at approximately 5 Hz.

Piston, compression, and rotary trimmers may be identified by first making a comparison fully closed and then fully opened. Small gimmick caps made from twisted leads are also easily sized.

The LED is, of course, optional, as well as the number of reference capacitors. Any NPN switching or audio transistor may be used in place of the MPS6512. If the LED is not needed, the transistor may be eliminated and the speaker with its 1 μF coupling capacitor is connected directly to pin 3 of the IC.

SIMPLE TR SYSTEM

How many of us start our first ham station with manual transmit-receive switching? You know, "real" manual switching. First the other guy turns it over to you, then you throw the receiver into standby, then you throw the transmitter out of standby, and finally you frantically flip that little knife switch that does all the antenna changeover work. Maybe you've even forgotten all about it by now, but chances are that many of the Novices who are reading this are pretty familiar with how it feels to wonder if the other fellow gave up on you during the intermission. It's not an impossible situation to remedy, though. For about two thirds the cost of the popular Dow Key "60" coaxial reply, you can build a 1000 W

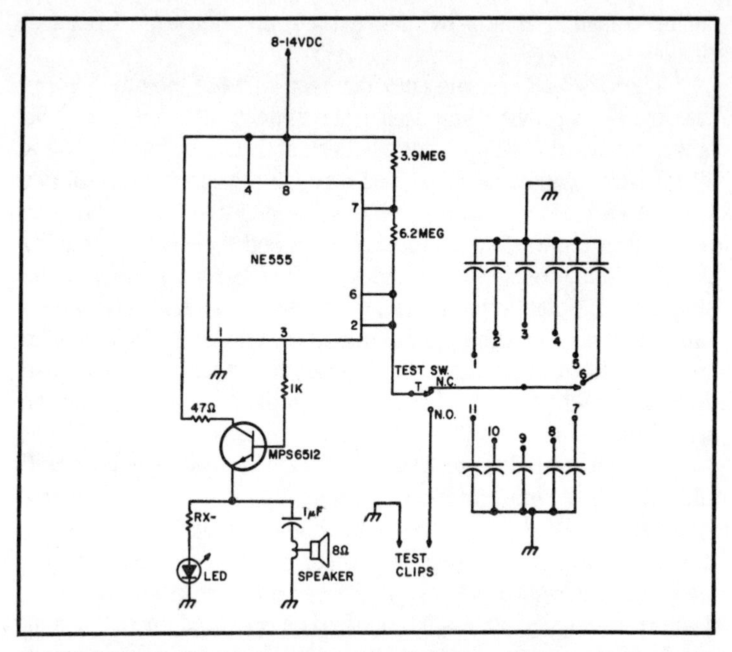

Fig. 11-11. LED indicator and RX—see LED specs. Capacitor bank: 1 = .7 pF; 2=3 pF; 3 = 5 pF; 4 = 10 pF; 5 = 25 pF; 6 = 50 pF; 7 = 100 pF; 8 = 330 pF; 9 = 470 pF; 10 = 680 pF; 11 = 820 pF. Test switch—SPDT push-button.

capacity transceive switch that gives you convenient one switch operation.

The heart of this little gadget is a Kurman 115 V rf power relay. This may be expensive, but compared to virtually any similar rf relay, the cost is a really solid investment. It has quite impressive specs.

The relay coil itself draws about a watt at 115 V, but if you prefer odd coil voltages, models are available with 6, 12, and 110 V dc coils. The DPDT, self-wiping contacts are capable of handling 1.5 kW at frequencies up to 450 MHz. Standard G-7 insulating bridges provide excellent rf isolation between contacts. The relay can be easily mounted on one side of your enclosure by using the four tapped #6-32 mounting holes provided. See Fig. 11-12.

A four pole, three position telephone lever switch is used for the front panel transceive switching. The middle position provides a standby condition for both transmitter and receiver. Two of the DPDT contacts are used to switch the transmitter and receiver standby circuits, while the third energizes the relay. On my model, the fourth contact was used to short the receive antenna terminals during transmit, thus providing extra protection from front end burnout.

Three SO-239 female coax connectors were mounted on the rear panel to provide standard interconnection to any rig. The power cord is also mounted on the back of the Radio Shack 5-¼″ × 3″ × 5-⅞″ cabinet, being passed through a grommet for safety's sake. Additionally, a four pin socket is mounted below the coax connectors for hookup to the receiver and transmitter standby circuits. Since transmitter standby is often just a switching of the primary of the plate transformer, the socket pins must be capable of handling 2 A or so. An optional power switch can be installed, as indicated on the schematic, but it's not really necessary because the relay will be turned off anyway when the lever switch is in the center position.

For most applications, the .001 μF bypass capacitors will adequately prevent rf from riding out on the incoming power leads. However, leads should be kept short on the capacitors bypassing the relay coil since such coils tend to pick up a lot of rf at high frequencies and subsequently put it right into the power line. Also, if the transceive switch is intended to be operated into a constant impedance, then coax is suggested for all rf carrying leads within the box.

This switching system allows considerable versatility at nearly any power level and with virtually any equipment. I experienced only one difficulty. It may be necessary to weight down the entire assembly to prevent it from moving all over the place when the lever is thrown.

Otherwise, you should have a reliable switch that will serve your needs for many years to come, no matter how much your other equipment changes.

BLOWN FUSE INDICATORS FOR LOW VOLTAGE

At first glance the unit to be described would appear to be strictly a luxury, but use it once, and you'll wonder how you ever got along without it.

It is a well known fact that individually fused circuits not only give increased protection to equipment, but simplify troubleshooting by isolating the defective circuits. But it also means more fuses to check when troubleshooting.

In ac equipment it is a simple matter to connect a neon bulb directly across each fuse; thus, when all is well all bulbs are out (being shorted by the fuse). When a fuse blows, the bulb is placed in

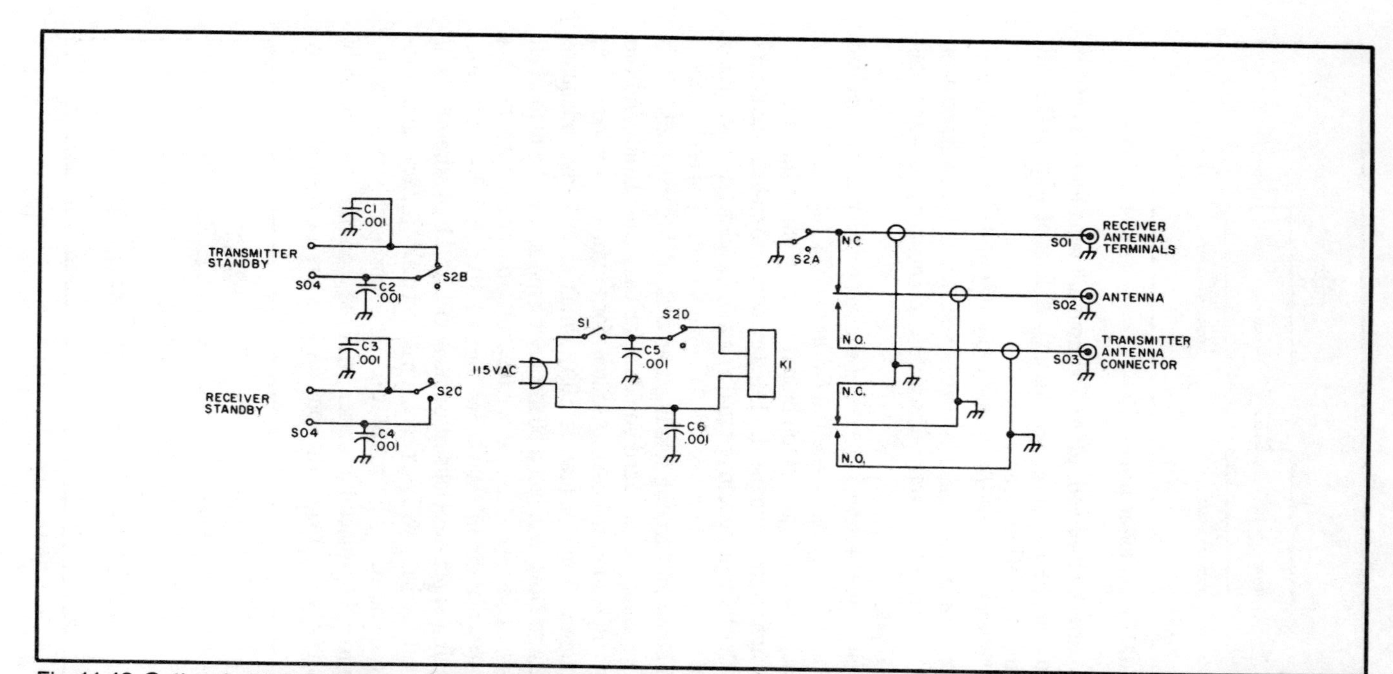

Fig. 11-12. Optional switch S1 (SPDT toggle) is shown in the off position, while S2 (4PDT telephone lever switch) is shown in the transmit position. C1-6: .001 μF, 1 kV, disc ceramic. K1: DPDT rf power relay. S01-3:SO-239 female coax connectors. SO4: 4 pinsocket, all pins well isolated from chassis, capable of 2 A. Enclosure: Radio Shack 5-¼″× 3″ × 5-⅞″ metal cabinet.

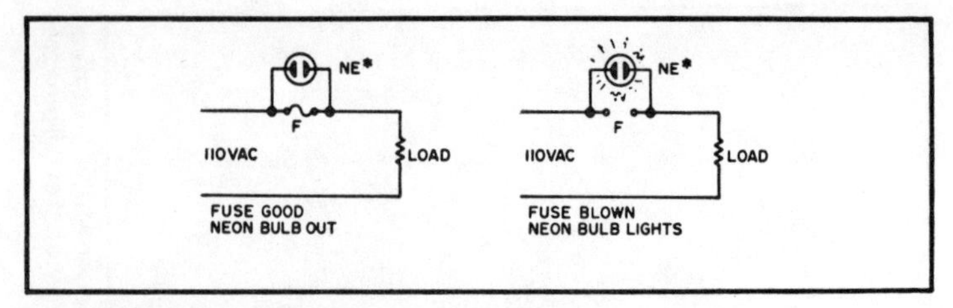

Fig. 11-13. Circuit for ac use.

series with the load and, glows (without allowing enough current flow to damage the defective circuit). A look at Fig. 11-13 will make this operation clear.

I "borrowed" the idea from a piece of military gear several years ago, and have used it ever since. Many hours have been saved by knowing exactly which unit to check when something "went south."

Many times I have wished that such a system could be applied to the mobile rig, but since the 6V in my VW (or the 12 in your Cadillac) will not light a neon bulb it seemed out of the question. Then a light came on in the think department. How about incandescent bulbs? Theory could not find a flaw in the idea (at least not my limited theory) so it was decided to give it a practical test, using a sealed beam unit from a headlight as a "load." It worked!

The values used were selected for easy calculation, but the theory will hold in any case. Assume, for example, a 6A load. The next higher rating of fuse is 10A, which should give adequate protection. Now, connect a 60 mA pilot lamp across the fuse. If all values are considered to be exact (of course they never are, but this is theory) our circuit may now conduct 10.06A before the fuse blows. If a short develops at point A in Fig. 11-14, the fuse will blow, but the lamp will merely have its normal brilliance, announcing to one and all that F1 has blown. If the short is only momentary, the bulb will be placed in series with the load, but application of Ohm's law will show us that our 6A load has a resistance of only

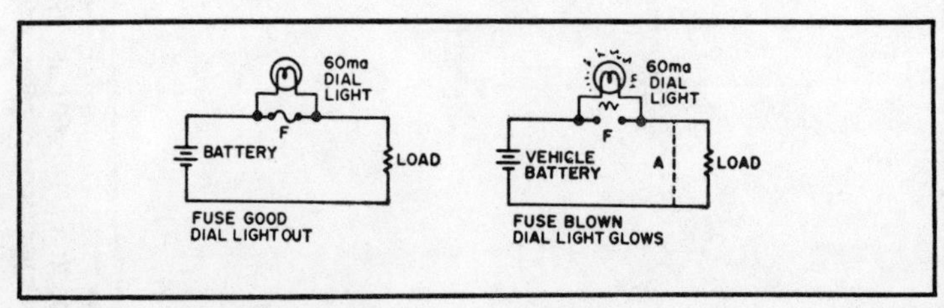

Fig. 11-14. Circuit for dc use.

1.0Ω, while the 60 mA bulb has a resistance of 100Ω, thus most of the applied voltage will be dropped across the bulb, and it will still light.

This method worked out so well on the mobile rig that it was also applied to the entire automobile, with all the bulbs installed on a single panel just below the dash.

Unless you have a large supply of pilot light sockets and jewels on hand, this would at first appear expensive, but there is a sneaky way around this. Bulbs mount nicely in a grommet set into a hole, and coating them with a special paint made for putting orange parking lights on older model cars makes a very attractive installation. In the event the bulbs tend to creep out of the grommets, a drop of Duco cement applied to the bulb and grommet will cure this, and replacement is so seldom that connection may be made by soldering directly to the bulbs.

GRAVITY DETECTOR

The complete circuit is shown in Fig. 11-15. The load resistor, R2, is selected to set the operating point on the negative resistance slope of the thermistor's characteristic curve. Typical characteristics for an 8 or 10,000 ohm bead are shown in Fig. 11-16. R1 is used to prevent applying an abrupt surge of current to the bead, since these take appreciable time to reach operating temperature.

The thermistor should be mounted in a small clear plastic case, using extreme care not to break the fragile leads, or lose the thing by sneezing when you open its container. Flexible leads connect to the rest of the circuit. All connections should be soldered, since if clip leads are used and one happened to come loose, the natural tendency is to clip it right back on. That can be disastrous.

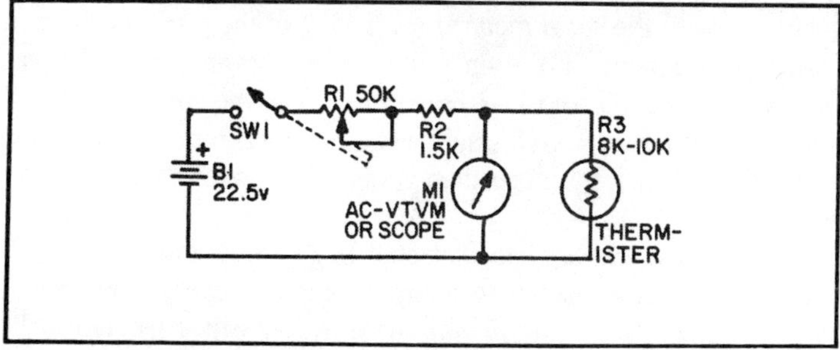

Fig. 11-15. Schematic of gravity detector.

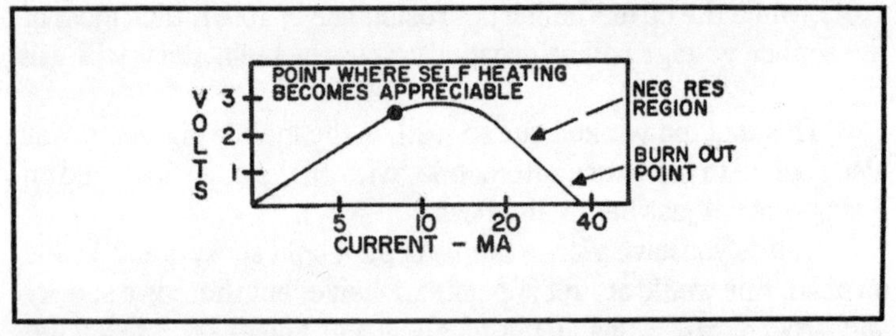

Fig. 11-16. Characteristic curve for a bead thermistor. Values may vary widely.

Rather than tell your friends how it works, let them try to figure it out. The theory is that the thermistor bead has a little chimney of convection currents of heated air rising from it. Moving it sidewise lets the chimney follow the bead, since the air in the plastic box is trapped there. Moving it up or down changes the rate at which the heated air rises from the bead, modifying its temperature. Because all a thermistor does is change its resistance with temperature, a phase sensitive output signal is obtained, which is observed on a scope or voltmeter.

SOLAR-POWERED ALIGNMENT TOOL

Two common methods of calibrating the direction of a beam antenna with respect to true north (or south) are: to align the boom in the direction of the polestar, or to apply the variation correction to the magnetic north (or south) reading of a compass. Unfortunately, there is no accommodating star at the south celestial pole for observers in the Southern Hemisphere. The variation correction depends upon one's latitude and longitude.

The method I shall describe here is simpler; it is based upon the sun's meridian passage at any locality in the world. All one needs to know is one's approximate longitude obtained from a world map and the local mean time (LMT) of the sun's meridian transit. At this moment, the sun is at its maximum altitude and is on a north-south line. Table 11-1 lists the LMT of the sun's meridian passage on the first, tenth, and twentieth of each month. These values do not vary by more than about one minute from year to year.

Since our clocks are based on standard or zone time and not on local time, it is necessary to apply a longitude correction, converted to time units. Table 11-2 allows this, to the nearest standard meridian. The standard meridians theoretically are spaced 15°

Table 11-1. LMTs of Sun's Meridian Passage. These Times Basically Correspond to the Sun's Transit over the Greenwich Meridian, Taken from the American Ephemeris and Nautical Almanac. Because the Sun's Apparent Eastward Daily Motion is of the Order of 1° or Less, the Slight Difference Between the Greenwich and the Local Mean Time of the Sun's Meridian Transit May be Neglected.

Date	Jan	Feb	Mar	Apr	May	June	July	Aug	Sept	Oct	Nov	Dec
1	1202	1213	1211	1203	1156	1157	1203	1205	1159	1149	1143	1148
10	1207	1213	1209	1200	1155	1159	1204	1204	1156	1146	1143	1152
20	1210	1213	1207	1158	1156	1201	1205	1202	1153	1144	1145	1157

apart to the east or west of the Greenwich prime meridian. If the station longitude is east of the standard meridian, subtract the difference in longitude in time units between your station and the nearest standard meridian from the LMT; if the station longitude is west of the standard meridian, add the longitude difference in time units to the LMT. Thus, standard or zone time = LMT plus or minus the difference. Because the time zones have ragged boundaries, it may be necessary to add or subtract one hour and, in some instances, one-half hour, as the custom dictates.

To demonstrate the simplicity of the solar method, two examples are chosen.

(1) What is the standard time of meridian passage of the sun at longitude 114° 20′ W on October 15? From Table 1 we interpolate a

Table 11-2. Difference of Longitude Conversion.

Arc	Time (minutes)	Arc	Time (minutes)
0° 00'	0		
0° 15'	1	4° 00'	16
0° 30'	2	4° 15'	17
0° 45'	3	4° 30'	18
1° 00'	4	4° 45'	19
1° 15'	5	5° 00'	20
1° 30'	6	5° 15'	21
1° 45'	7	5° 30'	22
2° 00'	8	5° 45'	23
2° 15'	9	6° 00'	24
2° 30'	10	6° 15'	25
2° 45'	11	6° 30'	26
3° 00'	12	6° 45'	27
3° 15'	13	7° 00'	28
3° 30'	14	7° 15'	29
3° 45'	15	7° 30'	30

value of 1145 LMT. The nearest standard meridian is 120° W. The difference in longitude between the station and the nearest meridian is 5° 40′. From Table 11-2, this amounts to 23 minutes. Since the station is *east* of the standard meridian, the Pacific standard time of the sun's meridian passage is 1145—0023 = 1122 PST.

(2) What is the standard time of meridian passage of the sun at longitude 25° 40′ E on March 25? From Table 1, LMT = 1205. The difference in longitude between the station and the nearest standard meridian of 30° E is 4° 20′, which from Table 11-2 is equivalent to 17 minutes. Since the station is *west* of the standard meridian, the standard time of the sun's meridian passage is 1205 + 0017 = 1222.

At the standard time the sun is on the meridian, that is, due north or south, depending on your latitude, line up the antenna boom with the sun or parallel to any shadow cast by a vertical structure (pole, tower, etc.). An error of 4 minutes in time amounts to a change in the direction of the sun of only 1°. Set the direction indicator of your rotator to 0°. Make certain that the radiating element of the antenna is on the correct side of the boom—otherwise you could be 180° off. That's all there is to it!

Index

Edited by Roland Phelps